FOREST PRODUCTS

THEIR SOURCES, PRODUCTION, AND UTILIZATION

THE AMERICAN FORESTRY SERIES

HENRY J. VAUX, *Consulting Editor*

Allen and Sharpe · An Introduction to American Forestry

Baker · Principles of Silviculture

Boyce · Forest Pathology

Brockman · Recreational Use of Wild Lands

Brown, Panshin, and Forsaith · Textbook of Wood Technology
   Volume II—The Physical, Mechanical, and Chemical Properties
   of the Commercial Woods of the United States

Bruce and Schumacher · Forest Mensuration

Chapman and Meyer · Forest Mensuration

Chapman and Meyer · Forest Valuation

Dana · Forest and Range Policy

Davis · American Forest Management

Davis · Forest Fire: Control and Use

Duerr · Fundamentals of Forestry Economics

Graham and Knight · Principles of Forest Entomology

Guise · The Management of Farm Woodlands

Harlow and Harrar · Textbook of Dendrology

Hunt and Garratt · Wood Preservation

Panshin, de Zeeuw, and Brown · Textbook of Wood Technology
   Volume I—Structure, Identification, Uses, and Properties of the
      Commercial Woods of the United States

Panshin, Harrar, and Bethel · Forest Products

Preston · Farm Wood Crops

Shirley · Forestry and Its Career Opportunities

Stoddart and Smith · Range Management

Trippensee · Wildlife Management
   Volume I—Upland Game and General Principles
   Volume II—Fur Bearers, Waterfowl, and Fish

Wackerman · Harvesting Timber Crops

WALTER MULFORD WAS CONSULTING EDITOR OF THIS SERIES FROM ITS INCEPTION IN 1931 UNTIL JANUARY 1, 1952.

FOREST PRODUCTS

THEIR SOURCES, PRODUCTION, AND UTILIZATION

A. J. PANSHIN

Chairman, Department of Forest Products
Michigan State University

E. S. HARRAR

Dean, School of Forestry
Duke University

J. S. BETHEL

Head, Special Projects in Science Education Section
National Science Foundation

and the late **W. J. BAKER**

SECOND EDITION

McGRAW-HILL BOOK COMPANY   **1962**

New York   San Francisco   Toronto   London

FOREST PRODUCTS

PREFACE

The American forest-products industries contributed much to the successful prosecution of the Second World War. In the process many of these industries assumed roles that were very strange for them. Furniture factories built gliders; lumber manufacturers produced truck bodies; and plywood plants fabricated boats. As they undertook a variety of, for them, unusual functions, they developed new and exciting technologies that were effective in war and in some instances promised to be equally useful when directed toward peacetime manufacturing objectives.

The years immediately following the Second World War were notable for display of imagination in reorganization of the forest-products industries, for improvisations, and for adoption of new technologies on a trial basis. The first edition of "Forest Products," which appeared in 1950, dealt with "the state of the art" as it existed in this transitional postwar period.

During the decade 1950–1960 the forest-products industries experienced perhaps the most rapid advancement in technology in any decade in United States history. Many of the war technologies took root in peacetime enterprises, resulting in new products, better processes, and more efficient operations. It is appropriate that the text of "Forest Products" be revised to reflect these changes in wood science and technology.

All the chapters have been completely revised, and many new sections and illustrations added. The chapter on Thermal and Sound-insulation Materials has been replaced by a chapter on Wood Composition Board, and the chapter on Cellulose Filaments and Film with that on Cellulose-derived Products. New chapters on The Lumber Industry and Christmas Trees have been added.

Again because of space limitations many significant developments have had to be treated lightly, and others have not been dealt with at all. The authors accept the responsibility for the decisions that had to be made in the interest of space conservation.

Special acknowledgments are made to W. A. Stacey, American Wood-Preservers' Association, for photographs and literature dealing with treated wood; to C. M. Sporley, Valentine Clark Corporation, for photographs and information on pole-incising machines; to W. H. Sardo, Jr., National Wooden Pallet Manufacturers Association, for information and illustrations on wooden pallets; to C. L. Schopmeyer, U.S. Forest Service, for assistance in rewriting the chapter on Naval Stores. A new section on the wood-turning industry has been written by R. M. Carter, School of Forestry, North Carolina State College.

Grateful acknowledgment is due A. J. Stamm, North Carolina State College, for reviewing the chapter on Cellulose-derived Products; R. W. Jenner and

v

A. W. Goos, Cliffs Dow Chemical Company, for reviewing the chapter on Carbonization and Destructive Distillation of Wood; E. G. Locke and J. F. Saeman, U.S. Forest Service, for reviewing the chapter on Wood Hydrolysis; C. E. Libby, North Carolina State College, for reviewing the chapter on Pulp and Paper; W. H. Koepp, Michigan College of Mining and Technology, for information on tannins; A. B. Anderson, University of California, and T. R. Frost, Weyerhaeuser Timber Company, for information on bark utilization; Roy Huffman, Wood Briquettes, Inc., for information on wood briquetting and illustrations of Pres-to-logs stoker fuel machine; Henry I. Baldwin, research forester, State of New Hampshire, for illustrations of the New Hampshire charcoal kiln; and W. R. Byrnes, The Pennsylvania State University, for illustrations of shearing Christmas trees.

Appreciation is also expressed to George G. Marra, Washington State Institute of Technology, for reviewing the chapter on Wood Composition Board; to H. O. Fleischer, John F. Lutz, and Richard F. Blomquist, U.S. Forest Service, for reviewing the chapter on Veneers and Plywood and for supplying illustrative material; to William Groah and C. E. McDonald, Hardwood Plywood Institute, for information and illustrations on hardwood plywood and furniture; to H. A. Peterson, treasurer, Douglas-fir Plywood Association, for information on softwood plywood; to Franklin Bradford, director of the Editor's Division, American Forest Products Industries, Inc., for illustrations on veneer and plywood, lumber, and fiberboard and particle board; to F. J. Rovsek, Hardboard Association, for illustrations on fiberboard; and to Donald G. Coleman, U.S. Forest Service, for a variety of illustrations.

The authors again express their appreciation to firms, associations, and individuals for allowing the use of illustrative and tabular material. Appropriate acknowledgments accompany such material.

The authors hope that this revision of material will make this volume a more useful textbook.

The Authors

CONTENTS

Preface   v

Part I   **Economics of Forest Utilization**   1

1   *Timber Resources and Their Importance in the Economy of the United States*   3

2   *Wood Residues and the Related Problems of Wood-utilization Research*   16

3   *Relation of the Properties of Wood to Wood Utilization*   41

Part II   **Wood Products**   53

4   *Round Timbers*   55

5   *Mine Timbers*   77

6   *Railroad Ties*   94

7   *The Lumber Industry*   106

8   *Veneers and Plywood*   120

9   *Wood Furniture*   167

10   *Shingles and Shakes*   184

11   *Wood Containers*   201

12   *Wood Composition Board*   247

13   *Wood Flour*   265

14   *Sawdust and Shavings*   272

15   *Wood Fuel*   281

16   *Secondary Wood Products*   295

Part III   **Chemically Derived Products from Wood**   319

17   *Pulp and Paper*   321

18   *Cellulose-derived Products*   392

19   *Carbonization and Destructive Distillation of Wood*   404

20   *Wood Hydrolysis*   425

Part IV   **Derived and Miscellaneous Forest Products**   433

21   *Naval Stores*   435

22   *Maple Sirup and Sugar*   466

23   *Christmas Trees*   480

24   *Wood in the Plastics Industry*   488

25   *Minor Forest Products*   500

Index   525

PART I

ECONOMICS OF FOREST UTILIZATION

TIMBER RESOURCES AND THEIR IMPORTANCE IN THE ECONOMY OF THE UNITED STATES

Wood is unique among the world's important raw materials. Virtually everyone uses it in many ways and in many forms, and yet few have a real understanding of its structure and properties. Most of the human beings making up the world's population live in wood houses, sit in wood chairs, eat at wood tables, and are warmed by wood-stoked fires. They use wood constantly in many ways and yet they really have little understanding of it. The average schoolboy is taught much more about iron, aluminum, copper, and the world's great chemical resources than he is about wood. Though man used wood long before he had discovered and used these other materials, its complex molecular structure is still not understood, while the relatively simple structure of the most important companion materials is much more completely known.

As Glesinger [1] has pointed out, wood has three attributes that collectively make it unique among the world's raw materials. It is universal, abundant, and essentially inexhaustible.

Timber has played a major role in the growth and development of the United States. It is estimated that in excess of 3 trillion board feet of lumber was cut between 1776 and 1960. During these same years great quantities of wood were also used for pulp and paper, veneer and plywood, fiberboard, posts, poles, piling, ties, fuel wood, and in the form of many other products. Without wood this country could not have attained its present stature so rapidly, if indeed it could have attained it at all.

I. TIMBER RESOURCES OF THE UNITED STATES AND THEIR UTILIZATION

For approximately 200 years following the erection of the first sawmill in the United States in the early 1600s, timber supplies were considered inexhaustible. As early as 1819, however, there appeared dire warnings of

3

depletion, and as late as 1922 official government pronouncements predicted imminent timber famine. Neither view has proved correct, although depletions of growing stock in most parts of the country have reduced timber resources below desirable levels. Of even greater significance, the quality of timber has been lowered to the point where many primary industries requiring high-quality logs are finding it difficult to obtain adequate supplies. Regionally, drastic steps are required in some instances, but nationally there appears to be sufficient timber to meet immediate needs. Many adjustments are needed, however, to bring all commercial forest lands, both public and private, into the highest possible productivity to grow the type of forests required to furnish a balanced supply of raw materials for all wood-consuming industries, and to provide a reserve capable of meeting all emergencies. Fortunately, the realities of economic forces are pointing to the need for permanency of timber ownership. In times past it has not always been economically feasible for mill owners to retain ownership of forest lands. Now large segments of the industry realize that permanent ownership and the management of forest lands on a sustained-yield basis offer the only sure means of survival. Thus it seems that forces are now in motion that have promise of providing a *realistic* solution to the problem of present and future timber supplies and a balanced national timber budget. But forestry costs money. In most private timber enterprises the costs of sound timber management must come from profits derived from the sale of wood products. The problem is, therefore, to utilize as completely and efficiently as possible all available timber supplies that can be grown on the forest lands of the nation. Attainment of these two broad objectives will make the growing of trees more profitable and encourage better forest practices throughout the country.

To accomplish these goals it is necessary to know the quantity, quality, form, and location of the timber resources available for utilization. Careful planning with respect to all these factors is essential prior to establishing even the simplest wood-using enterprise. Also, the time has come when consideration of the sociological aspects of industrial developments can no longer be ignored. The many ramifications of this topic are beyond the scope of this text. The reader is referred to the selected references at the end of this chapter for more complete presentations of national forestry problems, including complete statistical data on growth and drain, major uses, and future requirements of wood in the United States. An understanding of these factors and of the manner in which they are interrelated is essential to all utilization planning.

A. Major Uses of Wood in the United States

Table 1-1 shows the allocation of the forest harvest in the United States among the important roundwood products, excluding fuel wood, for the period from 1940 to 1957. Adequate annual data on fuel wood production *

---

* "Timber Resources for America's Future," in a production estimate for 1952 that included fuel wood, estimated that 52 per cent of the roundwood went into lumber, 22 per cent into pulpwood, 16 per cent into fuel wood, 4 per cent into veneer and plywood, and 6 per cent into other roundwood products.

Table 1-1. Production of Roundwood Products, Excluding Fuel Wood, in the United States, 1940 to 1957, Millions of Cu Ft [a]

Year	Sawlogs	Veneer logs	Pulpwood	Other products [b]	Total
1940	4,845	235	950	965	6,990
1945	4,365	250	1,150	845	6,615
1950	5,905	345	1,510	770	8,535
1953	5,710	475	1,885	750	8,825
1954	5,650	490	1,890	740	8,770
1955	5,810	575	2,190	780	9,355
1956	5,860	580	2,475	730	9,645
1957	5,220	555	2,370	680	8,825

[a] SOURCE: U.S. Department of Commerce, Bureau of the Census. Statistical Abstract of the United States. 1959.

[b] Other products include cooperage logs, poles and piling, fence posts, hewed ties, round mine timbers, box bolts, and a miscellaneous assortment of similar items.

are not available. Historically lumber has been the wood product that has made the greatest demand upon American forest resources. While the demand for sawlogs has increased somewhat during the period from 1940 to 1957, the demand for veneer logs and pulpwood has more than doubled. The pattern of wood use has changed substantially during the period from 1900 to 1960.

Figure 1-1 illustrates the production of lumber from 1904 to 1960. It will be noted that the annual output of lumber varies over a very wide range. In general the level of lumber production reflects the nation's economic condi-

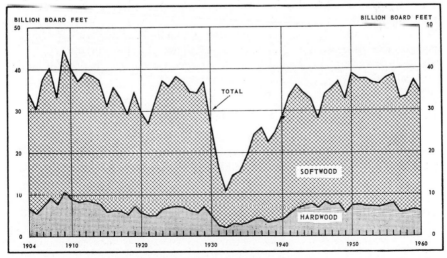

Fig. 1-1. Production of lumber from 1904 to 1960. (*From National Lumber Manufacturers Association.*)

tion. The low point in lumber production occurred during the height of the great economic depression of the 1930s. The lumber industry is more sensitive to economic conditions than many other industries. It is made up of many manufacturing units, the great majority of which are very small. Capital investment in equipment is relatively small, and required operating capital is minimal. Hence they can and do stop production during periods of poor demand and resume operations when demand picks up. The level of activity of the construction industry is a good index of demand for lumber. Wood is still the principal structural component of most homes and many other types of buildings. In addition, the demand for furniture, another large consumer of wood, is likely to be high when the number of housing starts is high.

The demand for various millwork items reflects not only the general level of activity of the construction industry but also current trends in architecture and design. Table 1-2 gives the production of various items of millwork

Table 1-2. Production of Millwork Products, 1950 to 1958 [a]

Item	Unit of measure	1950	1954	1955	1956	1957	1958
Hardwood flooring............	Millions of bd ft	1,077	1,145	1,268	1,166	954	927
Douglas-fir plywood..........	Millions of sq ft ⅜″	2,676	3,871	4,947	5,191	5,379	6,136
Insulating board.............	Thousands of tons	926	1,008	1,093	1,102	994	1,032
Hardboard..................	Thousands of tons	344	465	530	540	569	598
Ponderosa pine doors.........	Thousands of units	3,330	2,285	2,253	2,035	2,015	1,928
Hardwood doors.............	Thousands of units	1,221	5,940	6,786	6,404	5,486	4,500
Open sash..................	Thousands of units	9,414	11,054	12,733	10,551	9,867	9,605
Exterior frames.............	Thousands of units	4,644	5,791	7,259	5,679	5,279	5,762

[a] SOURCE: U.S. Department of Commerce, Bureau of the Census. Statistical Abstract of the United States. 1959.

for the period 1950 to 1958. The increasing use of plywood for sheathing, subflooring, exterior and interior walls, and for many other uses is reflected in the substantial increase in plywood production during this eight-year period. The decreasing use of pine panel doors and the corresponding increase in the use of hardwood flush doors reflect a change in popular architectural styles.

B. Future Requirements for Wood by Wood-using Industries

If industries are to develop in a sound and orderly manner they must attempt to predict future demand for their products. The wood-using industries are no exception. Several studies were completed during the 1950s whose objective was to predict the United States demand for wood products. The Stanford Research Institute was commissioned by the Weyerhaeuser Company to forecast demands for forest products through 1975. The results of this study were published in 1954.[3] The U.S. Forest Service published its "Timber Resources for America's Future" in 1958. This document, too, contains an assessment of future demands for wood in the years 1975 and 2000. Table 1-3 indicates the estimated consumption of timber products in the

Table 1-3. Estimated Consumption of Timber Products in the United States [a]

Product	Standard unit of measure	Volume in standard units, million			Volume in roundwood
		1944	1950	1952	1952
					Million cu ft	*Per cent*
Sawlogs (lumber, sawed ties, etc.) [c]..	Bd-ft lumber tally	34,600	40,850	41,462	6,419	52.3
Veneer logs and bolts..............	Bd-ft log scale	1,533	2,730	2,647	451	3.7
Pulpwood [d]......................	Standard cords	21	34	35	2,697	22.0
Cooperage logs and bolts...........	Bd-ft log scale	737	690	355	73	0.6
Piling............................	Linear feet	45	32	41	28	0.2
Poles............................	Pieces	4	7	6	88	0.7
Posts (round and split)............	do	275	230	306	194	1.6
Hewn ties........................	do	25	12	10	67	0.5
Mine timbers (round)..............	Cu ft	150	100	81	81	0.6
Other industrial wood [e]...........	do	250	250	227	168	1.4
All industrial wood..............	Cu ft roundwood [b]	8,257	10,145	10,266	10,266	83.6
Fuel wood.......................	Standard cords	70	62	59	2,008	16.4
All timber products..............	Cu ft roundwood [b]	11,632	12,272	12,274	12,274	100.0

[a] SOURCE: U.S. Department of Agriculture, Forest Service. Timber Resources for America's Future, Table 207. 1958.

[b] The roundwood (logs and bolts) volume of pulpwood, of "other industrial wood," and of fuel wood includes only that cut directly from trees. Plant residues utilized for such products are part of the roundwood volume principally of sawlogs and veneer logs and bolts.

[c] Estimates of apparent consumption based on estimated production, less exports, plus imports, and changes in lumber stocks.

[d] Includes net imports of pulpwood, also of wood pulp and finished paper expressed in terms of pulpwood.

[e] All other timber products not including fuel wood.

United States for the years 1944, 1950, and 1952. These are the base statistics from which "Timber Resources for America's Future" estimates were projected. This study developed estimates for two levels of demand: lower projected demand and medium projected demand. The medium projected demand indicated that requirements for all timber products would increase over 1952 consumption, 32 per cent by 1975, and 83 per cent by 2000. While this is a large total product increase, when compared with anticipated population, it represents for the medium projected demand a decrease in the per-capita consumption by 1975 and a slight increase in per-capita consumption by 2000. For lumber the estimate is for a decrease in per-capita requirements, whereas for pulpwood a substantial increase in demand is projected.

Table 1-4 illustrates the estimates of consumption of forest products from 1952 until 1975 as projected in the Stanford report.[3] The major conclusions of this report are:

1. Substantial population increases, technological advances, and increases in living standards will expand the size and activity of the United States economy by 1975.

2. This general economic expansion will result in corresponding increases in activity in construction (especially residential), shipping containers, and manufacturing—the major markets for forest products.

3. Despite these expanded markets, substantial increases in the supply of lumber, either from domestic production or from imports, are

Table 1-4. Estimated Consumption of Forest Products in the United States, 1952 to 1975 [a]

Product	Unit of measure	Quantity	Consumption				
			1952	1960	1965	1970	1975
Lumber	Bd ft (lumber tally)	Billion	40.7	40.7	41.4	43.1	44.6
Hardwood	....	....	8.5	8.5	8.7	9.0	9.4
Softwood	....	....	32.2	32.2	32.7	34.1	35.2
Softwood plywood and veneer	Sq ft, ³/₈″	Billion	3.3	4.6	5.5	6.6	7.5
Hardwood plywood and veneer	Bd ft, local log scales	Billion	1.0	1.1	1.2	1.3	1.4
Pulpwood	Cords (roughwood basis)	Million	26.5	34.8	39.4	43.6	48.8
Pulp total	Tons (air-dry)	Million	18.1	24.3	28.2	31.8	36.1
Unbleached sulfate	Tons (air-dry)	Million	6.3	7.4	8.2	9.1	10.0
White pulps [b]	Tons (air-dry)	Million	6.3	8.7	9.2	9.9	10.7
Groundwood	Tons (air-dry)	Million	2.6	3.5	4.1	4.6	5.1
Defibrated and exploded	Tons (air-dry)	Million	1.1	1.7	2.5	2.6	3.1
Semichemical	Tons (air-dry)	Million	0.8	1.6	2.6	3.9	5.2
Dissolving and other nonpaper pulp	Tons (air-dry)	Million	0.8	1.2	1.4	1.5	1.8
Screenings, etc.	Tons (air-dry)	Million	0.2	0.2	0.2	0.2	0.2
Paper and paperboard	Tons (air-dry)	Million	29.0	37.4	42.7	47.9	53.5
Paper	Tons (air-dry)	Million	16.9	20.8	23.3	25.8	28.6
Fine	Tons (air-dry)	Million	1.3	1.6	1.8	2.0	2.3
Coarse	Tons (air-dry)	Million	3.1	3.8	4.2	4.6	5.0

	Unit						
Newsprint	Tons (air-dry)	Million	6.0	6.7	7.3	7.9	8.6
Book and other printing	Tons (air-dry)	Million	3.4	4.2	4.8	5.4	6.0
Sanitary, tissue, and other	Tons (air-dry)	Million	1.9	2.7	3.1	3.5	4.0
Building paper	Tons (air-dry)	Million	1.3	1.8	2.1	2.4	2.7
Paperboard	Tons (air-dry)	Million	12.1	16.6	19.4	22.1	24.9
Containerboard	Tons (air-dry)	Million	5.6	7.4	8.5	9.6	10.7
Boxboard and other paperboard	Tons (air-dry)	Million	5.2	7.1	8.3	9.4	10.6
Building board	Tons (air-dry)	Million	1.3	2.1	2.6	3.1	3.6
Shingles and shakes	Squares	Million	6.2	5.0	4.8	4.6	4.4
Cooperage	Bd ft, lumber equiv.	Million	504	511	490	452	390
Piling	Cu ft	Million	28	21	22	23	23
Poles	Cu ft	Million	103	110	112	121	128
Fuel wood [c]	Cords	Million	50	34	28	26	26
Wood for distillation and charcoal	Cords	Thousand	500	525	550	575	600
Miscellaneous	Cu ft	Million	250	250	250	250	250
Mine timbers (not sawed)	Cu ft	Million	100	100	100	100	100
Railroad ties (hewed)	Pieces	Million	15.8	15.2	14.6	14.2	13.7

[a] SOURCE: Stanford Research Institute. America's Demand for Wood, 1929–1975. Palo Alto, Calif. 1954.
[b] White pulps include bleached and unbleached sulfite, bleached and semibleached sulfate, and soda pulps.
[c] Log and bolt requirements from living commercial timber only.

likely to be forthcoming only at costs that will encourage the substitution of competing materials for lumber. The price of lumber relative to competing materials will increase, although less rapidly than in the past, and lumber will lose part of its markets.

4. The price of plywood will increase (but less rapidly than that of lumber) relative to competing materials; the prices of pulp and paper products will continue in about their present relationship to prices of competing materials; the prices of hardboard and insulating board will decline relative to those of the materials with which they compete.

5. Domestic production of lumber will increase only moderately by 1975, with all the increase in softwood lumber production occurring in the West, accompanied by more moderate increases in hardwood lumber production in the South and East.

6. Despite higher lumber prices, there should be a market for all the lumber produced in the United States, plus a moderately higher level of imports (primarily Canada) and increased reuse of salvaged lumber.

7. Major increases are expected in the domestic production and consumption of pulp, paper and paperboard products, plywood and veneer, and hardboard and insulating board.

8. A major decline is probable in fuel wood consumption. By 1975 this product will be used primarily in fireplaces, with actual disappearance of its uses in residential heating and cooking.

9. Only moderate changes from 1952 levels are foreseen for the domestic consumption and production of other forest products by 1975.

10. Total domestic requirements for sawtimber delivered to mill sites are expected to increase by only 3.4 per cent between 1952 and 1975. This modest increase reflects the relatively mild increase in probable lumber production, more extensive use of small logs below present minimum sawlogs standards, and a substantial reduction in the use of sawtimber for fuel wood.

11. Total domestic requirements for all timber delivered to mill sites, in cubic feet, will increase by about 14 per cent between 1952 and 1975. This relatively greater increase in cubic volume, as compared with board feet of sawtimber, reflects partly the expected increased relative importance of pulp in total timber use and partly the probable general shift in all regions toward the use of smaller logs.

12. Increases in the use of mill residuals to produce pulp and other fiber products will be substantial in the West and South, but less pronounced in the East. This increased use of mill residuals will meet about a third of the increase in domestic pulpwood requirements.

13. Improved pulping methods will result in greater use of hardwoods, particularly in the South and East.

14. The major increases in timber use in the South will be for pulp and for hardwood lumber production; in the West for softwood lumber

and plywood production. The East will show moderate increases for lumber and pulp uses of timber, but a decline in total timber use.

Many authorities disagree with the projections made in these two studies. Nevertheless, these forecasts represent the most definitive attempts at quantitative evaluation of future demands for wood available at this time.

II. THE RELATIVE IMPORTANCE OF THE WOOD-USING INDUSTRIES

It is difficult, if not impossible, to determine accurately the relative importance of one industry in terms of other industries. The U.S. Department of Commerce, Bureau of the Census, issues the only comprehensive statistics with which various industries can be compared. Excellent as this information is, it falls short of providing complete information on some industries. There are numerous reasons for this, among them being (1) the necessity of combining some industries in order to avoid disclosing the identity of certain individual companies, and (2) for the sake of simplifying the census, the necessity of classifying each establishment as a whole in some one industry, according to its product of chief value, although, in many instances, a single establishment may manufacture more than one product. For example, it is impossible to determine how many wage earners are engaged in cutting and transporting pulpwood. Further, many small wood-using industries are classified under miscellaneous industries in a manner that makes it impossible to determine what percentage of the total production in the classification is made from wood; examples of such classification are Pianos and other Musical Instruments, Sporting Goods, Pens and Pencils, Signs, and Patterns.

It seems desirable, however, that the reader have some concept of the relative magnitude of the lumber and wood-using industries. Table 1-5, based upon figures of the U.S. Department of Commerce, Bureau of the Census,[7] shows that wood-using industries ranked first in number of establishments in 1954, the last date for which these data were available, if the Lumber and Wood Products and Pulp and Paper Products production units are combined. The same industrial groups in 1957 ranked fifth in number of wage earners, sixth in total wages paid, seventh in expenditures for new structures and additions, third in expenditures for new machinery and equipment, and eighth in value added by manufacture. If the large fraction of the Furniture and Fixture industries that uses wood primarily and the many other wood-using industries among those classified as Miscellaneous Manufactures were added, the total wood-using industrial group would represent an even more important segment of the nation's industrial economy.

"Timber Resources for America's Future" points out that the proportion which wood makes up of the total mix of physical-structure raw materials is also a criterion of wood's importance to the economic life of the nation. It states:

> During the early 1900's timber products (other than fuelwood) comprised close to one-third of total consumption of physical-structure

Table 1-5. Comparison of the Principal Industries of the United States [a]

Industry group	1954		1957	
	Number of establishments	Rank	Wage earners, average for the year	Rank
Food and kindred products...................	42,374	1	1,134	3
Tobacco manufactures......................	627	20	81	20
Textile-mill products.......................	8,070	11	893	6
Apparel and related products.................	31,372	4	1,123	4
Lumber and wood products...................	41,484	2	579	9
Furniture and fixtures......................	10,273	10	311	16
Pulp and paper products.....................	5,004	15	458	13
Printing and publishing.....................	32,531	3	534	10
Chemicals and products.....................	11,075	9	508	12
Petroleum and coal products.................	1,262	19	135	19
Rubber products...........................	1,406	18	205	18
Leather and leather goods...................	4,845	16	323	15
Stone, clay, and glass products...............	11,162	8	437	14
Primary metal industries....................	5,957	12	1,053	5
Fabricated metal products...................	22,516	6	880	7
Machinery, except electrical.................	25,601	5	1,266	2
Electrical machinery.......................	5,758	13	795	8
Transportation equipment...................	5,348	14	1,401	1
Instruments and related products.............	3,142	17	212	17
Miscellaneous manufactures.................	17,010	7	514	11

[a] SOURCE: U.S. Department of Commerce, Bureau of the Census. Statistical Abstract of the United States. 1959.

materials. The proportion grew steadily less for the next 20 years, from 1910 to 1930. In the 1930's and early 1940's, it diminished still further, but then the trend was reversed. During 1950–52, timber comprised about 20 percent of the total physical-structure raw materials intake, which is about the same as it comprised during the period 1925–40. Thus there appears to be no current trend downward in the importance of timber products in the national economy.

III. THE STABILIZING INFLUENCES OF FORESTS AND WOOD-USING INDUSTRIES

The numerous values obtained from forests before they are harvested are every bit as important in our national economy as the final products derived from wood. Thus the two are interdependent and cannot be separated in a discussion of their importance and the stabilizing influence they exert. This is particularly true when it is recognized that wise and profitable wood utilization can assure the perpetuation of forests at a high level of productivity. The same principle applies to government and private ownership. The products from government-owned forests, as well as from private forests, must eventually be sold to justify the existence of the forests, except in those instances where they are set aside primarily for watershed protection, recreational activities, or other specialized uses.

Since this text is devoted to the subject of wood utilization, space does not permit a complete discussion of the benefits derived from forests alone,

1957		1957		1957		1957	
Wages, dollar amounts in millions	Rank	Expenditures for new machinery and equipment, dollar amounts in millions	Rank	Expenditures for new structures and additions	Rank	Value added by manufacture, dollar amounts in millions	Rank
4,295	4	678	6	266	6	16,022	2
240	20	36	19	6	20	1,233	20
2,632	8	237	12	54	12	5,181	11
2,867	7	93	16	14	18	5,969	9
1,723	14	161	13	51	15	3,296	14
1,049	15	59	18	35	16	2,467	17
2,010	11	704	4	209	8	5,642	10
2,469	9	250	11	95	11	7,723	8
2,337	10	1,103	2	338	4	12,086	5
771	19	275	10	625	2	3,009	15
954	16	125	15	23	17	2,380	18
939	18	25	20	7	19	1,870	19
1,803	13	506	7	227	7	4,810	12
5,440	3	1,483	1	775	1	13,063	4
3,803	5	414	8	140	10	9,329	7
6,061	2	689	5	349	3	15,442	3
3,292	6	353	9	196	9	9,398	6
7,175	1	852	3	322	5	18,235	1
947	17	92	17	54	13	2,726	16
1,826	12	131	14	52	14	4,638	13

An example may serve, however, to demonstrate the need for discussing the relationships that exist between the products obtained from forests and the forests themselves. Coal is an invaluable resource, but until it is mined it contributes nothing to the national economy, other than to serve as a reserve supply of energy and useful products. The difference between this and the case of forests is immediately apparent when the numerous values obtained from forests before they are harvested are considered.

A. The Stabilizing Influences of Forests

Values derived from forests that are not directly related to wood products include (1) controlling erosion and stream flow, (2) providing range for livestock, (3) maintaining wildlife, and (4) providing suitable areas for recreation. It has been stated that approximately one-half the forested areas of the United States exert a major watershed influence and that another one-quarter exert a moderate influence. Forested watersheds retard runoff, protect potable water sources, retain water for irrigation, increase supplies of hydroelectric power, reduce erosion, and help prevent disastrous floods that affect rural and city populations alike. In the great livestock-raising regions, particularly in the West, the industry is dependent upon forested areas for summer range. Without this source of feed, meat, leather, and wool would be more costly than they are at the present time. Wildlife, when properly managed, does not interfere with the livestock industry and fulfills some of man's need for recreational activities. Forested areas, with their mountains, streams and lakes, trees and wild flowers, fish and game, lead all other areas in their spiritual and recreational appeal. And recreation is vital in all walks of life.

Forests are beneficial in another way. They usually occupy land that is unsuited, or poorly suited, for other purposes. Experience has shown that the remaining forest lands are, for the most part, unsuited for agriculture. In fact, replacement by forests is the only efficient utilization of marginal farm lands, particularly those that have been abandoned. Forests, then, provide their many services on what might be called a noncompetitive basis; and they still provide products of inestimable value.

The general benefits derived from forests are shared by all. The multiple-use management of all forests, both government-owned and privately owned, is essential to the national welfare. In the case of private forestry, some or all of the inherent values of forests, such as livestock, range, and recreational activities, can be used to reduce the costs of management. Intelligent multiple-use management is essential to the perpetuation of all the values of forests in their proper proportions.

The above is an extremely limited discussion of the stabilizing influence of forests. The subject merits a good deal more study, and every student of wood utilization should have a thorough grasp of the relationships that exist between forests and the final products that stem from them. The selected references at the end of this chapter will provide a starting point for such a study.

B. The Stabilizing Influence of the Wood-using Industries

The direct benefits of the wood-using industries are derived from the useful goods they produce and the employment they give to large numbers of workers and the industries that serve them. Less direct, and again rather closely affiliated with the forests themselves, are the opportunities that stable wood-using industries afford in balancing the economic and social development of rural communities. Wood-using industries provide farmers with off-season employment in adjacent forests and wood-using plants and direct returns through the sale of forest products from farm woodlots. Farm populations are heavy users of forest products for homes, barns, fence posts, containers for shipping farm produce, and many other uses; and the presence of wood-using industries creates local outlets for farm products. They also create a tax base in rural areas where farm properties would otherwise be called upon to bear the entire tax burden. Stable wood industries are essential to the success of many other major industries. To mention only a few, railroads, mining, and communications depend heavily on forest products for their success.

All the benefits that can accrue from the presence of forests and the industries that obtain their raw materials from them are effective in full measure only under intelligent, long-range planning in sustained-yield forest management. All the elements of these values exist in only a very few of our forested areas at the present time. They have been lost temporarily in those areas where unwise exploitation has left ghost towns in the wake of timber depletion. Abuse of the forests in some areas has resulted in the decline of agriculture, the abandonment of railroads, migration of farm workers to urban centers, and community decadence. Wood-using industries based on sustained-yield management, on the other hand, can form the nuclei of stable communities, with established service industries, permanent social institutions, and normal municipal growth based on permanent values. Such communities stabilize agricultural pursuits by providing local markets for farm products. Other industries, indirectly dependent on forest products, are attracted to permanent communities of this character. Their stability is a bulwark against depressions and a means of reducing migrations to urban areas.

SELECTED REFERENCES

1. Glesinger, E. The Coming Age of Wood. Simon and Schuster, Inc., New York. 1949.
2. Horn, S. F. This Fascinating Lumber Business. The Bobbs-Merrill Company, Inc., Indianapolis. 1943.
3. Stanford Research Institute. America's Demand for Wood, 1929–1975. Palo Alto, Calif. 1954.
4. U.S. Department of Agriculture, Forest Service. Timber Resources for America's Future. Separate No. 1, Summary. 1958.
5. U.S. Department of Agriculture, Forest Service. Timber Resources for America's Future. Separate No. 3, Growth and Utilization. 1958.
6. U.S. Department of Agriculture, Forest Service. Timber Resources for America's Future. Separate No. 11, Basic Statistics. 1958.
7. U.S. Department of Commerce, Bureau of the Census. Statistical Abstract of the United States. 1959.

CHAPTER 2

WOOD RESIDUES AND THE RELATED PROBLEMS OF WOOD-UTILIZATION RESEARCH

The term *wood waste* requires careful definition. It is often used loosely. Conservationists sometimes consider as waste every part of a tree not converted to lumber or some higher use. Under such a classification, tops of trees left in the woods and mill residues used as fuel are called waste. Many lumbermen, on the other hand, give little thought to tops and limbs for which no markets exist under any particular set of economic conditions. They also view sawmill residues as a logical fuel for the generation of steam at the mill, although, in some cases, higher uses might be found for the material that is burned. To arrive at a reasonable definition of wood waste it is necessary to examine some of the factors that play a part in determining the marketability of wood residues generated in the forest and at wood conversion plants. The most important of these factors are time and place.

In colonial times timber was often considered an obstacle to agriculture, and valuable species like oak and walnut were sometimes felled and burned when land was cleared, since there were no markets for oak and walnut logs in those days. The same situation existed in all our frontier settlements. It is conceded that such practices were wasteful in the light of present standards, but much unwise exploitation of timber resources still occurs in spite of conditions that permit more complete utilization. Even now, our best forestry practices are considered wasteful by Europeans. It is a question of time and economic development; the practice of forestry in Europe is several centuries older than it is in this country. Countries that have experienced timber scarcity recognize the wisdom of maintaining forest lands in a high state of productivity and of utilizing every portion of a tree to the highest possible degree.

Time and place often combine to dictate the extent of wood utilization. A sawmill located near a large settlement might, at one stage in the regional development, dispose of all its residual wood and sawdust for fuel at a reasonable return. At the same time, a mill located some distance from the settle-

16

ment might be unable to dispose of its residues in this manner because the cost of transportation would preclude a profit. This situation could, however, be reversed. A pulp mill might be erected adjacent to the isolated mill, providing it with even more profitable outlets than the fuel market of the city mill. If, because of more extensive timber supplies, a group of diversified industries, such as fiberboard plants, plywood plants, and others, should be located in the vicinity, the more isolated mill might find highly profitable uses for all its residues, while the city mill might make no great gains in utilization. The proximity of the pulp mill and the development of processes such as the hydrolysis of wood to produce wood-sugar alcohol and molasses might also make profitable the harvesting of logs and tops formerly left in the woods. Thus, place and time would have a great influence on the extent of utilization of residues in the instances cited.

Fluctuations in economic conditions also have a marked influence on the extent to which residual materials are utilized. In times of unfavorable market conditions, it may not pay to transport sawdust to a fuel market or slabs to a fiberboard plant. When prices are at a low level, it often does not pay to haul marginal materials from the forest, and more material is left unused. An example from the Douglas-fir region may serve to illustrate this point. The demand for lumber was active during the Second World War and prices were high. This resulted in the removal of logs that would have been unprofitable on any previous market. "Wood logs," previously left in the woods, were a common sight in mill ponds. The demand was so active, in fact, that relogging of cutover lands and salvage logging on old burns gained much impetus. With the first break in prices after the war, these activities stopped abruptly; mill operators went into the woods to "reeducate" their bucking crews to prepare for removal only those logs considered profitable under the new price structure.

The foregoing examples are only a few of many combinations of circumstances that can have an effect on the degree of utilization of residues at any time or place or under changing economic conditions. They serve, however, to illustrate the need for careful thinking in expressing the meaning of wood waste.

The U.S. Forest Service has defined wood waste as follows: "Throughout this report the term waste wood means wood material from the forest which does not appear finally in marketable products other than fuel wood, regardless of whether it is economically or technologically feasible to utilize it." [31]

For this chapter, the following definitions will be used:

1. *Waste wood is marketable wood of any form that is not used at the highest level available to a given logging or manufacturing operation, under current economic conditions and at the current stage of technological development.* Marketable wood that is not used for any purpose comes within the scope of this definition.
2. *Wood residues are all forms of wood that cannot be marketed at a profit from a given logging or manufacturing operation, under current economic conditions and at the current stage of technological development,*

The Forest Service definition quoted above views wood residues from the standpoint of the conservationist, labeling as waste all parts of trees that are not converted to salable products, irrespective of whether or not it is economically feasible to convert them. From a long-range standpoint this is tenable, and no one can object to the goal that this definition expresses. All elements of the wood-producing and wood-using industries, including the Forest Service, should be directing every effort toward more complete utilization of materials now destroyed or not utilized to the greatest economic advantage. From the standpoint of the national economy, tops left in the woods and sawdust destroyed by burning because there is no market constitute grave waste, but individual operators cannot be expected to harvest or manufacture such materials at a loss when no profitable markets exist or when no processes have been developed to utilize them to advantage.

Genuine wood waste, however, has been tremendous. Many sawmills have destroyed, and still are destroying, slabs and much other material that would have yielded usable commodities.

The prospects for improvement in yield in the manufacture of basic wood products do not appear to be promising. Advances in technology have, for the most part, been offset by adverse changes in the size and quality of the trees, logs, and bolts that make up the raw material. Provincial Commissioner Sloan of British Columbia has reported [28] that coarse residues resulting from the conversion of Douglas-fir logs to lumber represented essentially the same fraction of log volume during the period 1955 to 1957 as they did in 1928. Guthrie and Armstrong [15] report that in the western forest industries "it appears that, on the average, current sawmilling practices result in less than two-fifths of the merchantable volume in a tree moving into the hands of the consumer in the form of lumber." They also note that comparable recovery data for the West Coast veneer and plywood industry indicate that the yield is essentially the same as that reported for the lumber industry. Efficient utilization of raw material is also relatively low in the pulp and paper industry.

It is apparent then that if losses in raw material in the basic processing industries are not likely to be substantially reduced, increased efficiency in the utilization of raw material must come from greater use of the residues of these industries. It is in this area that the most spectacular advances in wood utilization have occurred since the Second World War. Commissioner Sloan suggests in his report on British Columbia that half of the present woods losses are recoverable and that about two-thirds of solid mill waste can also be recovered.

During the 1950s the use of residues as raw material for the West Coast pulp industry increased to the point where it had changed the whole complexion of the industry. On the Pacific slope residues of the lumber and plywood industries represented 40 per cent of the raw material requirement of the pulp industry by 1960. Other residues are used in the manufacture of particle boards. Progress in the use of residues for fiber and chip products has been less spectacular in other sections of the United States, but it has nonetheless been substantial.

The greatest opportunities for progress in residue utilization by the pulp industry appear to lie with the utilization of lignin. About one-third of the weight of wood is made up of lignin, and in the pulp industry much of this is currently unused residue. As Guthrie and Armstrong [15] point out: "The major loss involved in the pulping process is that occasioned by dissolving the lignin, hemicelluloses and extractive materials from the wood. Approximately one-half of the weight of the wood is lost when it is converted to pulp by a chemical process. This weight loss is approximately the same as that involved in converting logs to lumber and plywood, and as yet no economical use has been discovered for the liquor of the pulp mills."

The key to further advances in residue utilization is research. Until the end of the Second World War the wood industries conducted and supported little research related to the utilization of wood. This picture has changed substantially since 1946, but the level of research in the wood-using industries on a per-sales-dollar basis is still far below that of most competitive industries. The pulp and paper industry has led the wood-using industries in the support of research. The lumber industry has been last among the basic processing industries. There are reasons for the dearth of research in the lumber industry. It is made up, to a large extent, of numerous separate units, many of which are too small to support substantial research on an individual basis. Many of these small firms are banded together into loosely knit associations, whose primary functions have been trade promotion. A few of these associations have accomplished effective research on a small scale, but they have been the exception. The small components of the wood-using industries must discover ways by which they can support research if they are to make the advances in the use of residues required to be competitive.

A few large timber companies, particularly in recent years, have established sizable research laboratories. The experience of such companies points the way to the resolution of the wood-residue problem. Large companies, having integrated operations based on permanent sustained-yield timber holdings, not only can afford research; they can reduce residues materially through the diversity of numerous utilization units. Such an operation may contain separate units for the manufacture of lumber, paper pulp, fiberboard, plywood, fuel briquettes, and bark-conversion products. This permits segregation of raw materials by species, size, and quality and utilizing each at its highest economic level. At the same time, residues can be distributed to specialized plants designed to convert them into useful products. As new processes are added, residues are even further reduced. By the development of entirely new processes, higher uses may be found for residues previously utilized for another purpose; or for the utilization of materials now considered worthless, such as wood containing decay. The process of integration can be carried on indefinitely through the discovery of new processes and the improvement of old methods, until all values inherent in trees are realized.

Integrated units of the foregoing type, because of their large size and efficiency, are the source of a rather high percentage of the total lumber cut in this country. Unfortunately, however, there are thousands of small- and medium-sized mills that are left, for all practical purposes, without benefit of research. Most mills in this category have little prospect of obtaining the capital necessary for the establishment of integrated operations. More often, they even lack the possibility of obtaining a sustained-yield unit. As time passes, the onus of waste is resting more and more heavily upon these small owners. Being financially incapable of applying technological advances to their operations, they continue to generate outright waste or to utilize residues at a very low level, long after higher economic uses have been developed. The situation is not hopeless, even for these small operators, but its correction will require new thinking and aggressive action. Research is one of the answers, not only in the development of new processes, but also in the improvement of existing practices. Among the problems requiring immediate consideration are (1) improvement in sawmill machinery and handling techniques, (2) improvements in the quality of lumber itself, (3) adoption of production-control methods used in other industries, (4) development of building techniques that will utilize lumber more efficiently, and (5) application of modern market-survey techniques. Such research could easily be provided by trade associations and public agencies.

Some industries expend as much as 10 per cent of total sales value for research, but 2 to 3 per cent probably approaches the average. It is doubtful if the lumber industry, as a whole, spent more than 0.2 to 0.5 per cent for research prior to the Second World War, and the average has not increased appreciably since that time. Not all wood-using industries have suffered from lack of research. Pulp companies have been aggressive in sponsoring research, not only in the technology of the chemical processes involved, but in harvesting methods, cut-up mill and barking practices, the utilization of new species, waste reduction, and sustained-yield forest management. Their technical laboratory * is one of the finest research institutions in the country, and the industry has sponsored much academic and public research. The same has been true to a lesser extent in the fiberboard field. The research activities of the Masonite Corporation are an outstanding example. The lumber industry could well follow the lead of these closely allied industries.

Residue- and waste-utilization plants sometimes work against the goals of forestry and the best interests of the nation. Though originally planned for the utilization of residues, some plants of this type become so large that they compete with sawmills for sawlogs in the race for sufficient raw materials to maintain themselves. The hardwood dimension-stock industry offers another example. Numerous mills, originally designed to produce dimension stock from sawmill residues, now find it more profitable to use lumber as their raw material. The same thing has been true of pulp mills in some regions. It is poor national economy to use sawlogs capable of producing structural materials for low-quality fiberboard and for paper pulp that can be made from tops or from inferior species. Here, again, classification of raw materials is essential to the balanced production of all commodities stemming from wood. Without intelligent planning, the goals of forestry and the general welfare of the wood-using industries may suffer. The goals of forestry and wood utilization are identical in basic principle, a fact often forgotten by lumbermen and foresters alike. The aim of foresters should be the sustained-yield management of all forests with a view to producing as much

* Institute of Paper Chemistry, Appleton, Wisconsin.

high-quality timber as possible, together with enough other wood to supply the needs of all the wood-consuming industries that complete the pattern of integrated utilization. The aims of the wood-using industries should be much closer utilization, more intensive remanufacture, greater diversification, and more complete integration, all of which will reduce waste to a minimum. If these aims can be achieved, the future of the lumber industry can be assured and uncertainty can be replaced by stable and permanent production and employment.

As Watts [34] has pointed out: "About the turn of the century the motivating influence (in forest products utilization research) was to conserve the forests; today we study wise and efficient use of wood to conserve manpower and capital and to promote more intensive management of the forest." Utilization of wood residues must be justified on the basis of sound economic practice. Gruen [14] made the economic case for utilizing more wood residues in the following terms:

1. The greater use of wood residues would have the effect of bringing down the prices of primary products—either through direct competition of substitutes made from the by products, or through processors and stumpage owners passing on some of the benefits in order to expand their markets.
2. The greater use of wood residues would have the effect of creating more wealth and useful employment through expansion of the total production of goods and through more intensive management of the forest—without increasing the commodity drain on the forest.

It is hoped that this chapter will impress on all students of wood utilization, if nothing else, the great need that exists in the lumber industry for progressive leadership, imagination (in both management and research), and intelligent nationwide planning. Only in this way will it emerge from its present status of a low-value raw-material industry to become the highly diversified and specialized industry that it could and should be.

I. SOURCES OF RESIDUAL AND WASTE WOOD

The major sources of residual and waste wood are (1) wood left in the forest after logging, and (2) residual and waste wood generated in the manufacture of primary products such as lumber, veneer, and shingles. Other sources include secondary manufacturing operations of all kinds, planing mills, materials unutilized during building construction, and many minor sources. Bark is not usually included in estimates of residues, but perhaps it should be. The bark of some species appears to offer just as great potentialities for utilization as some forms of residual wood. The sum of wood residues and wood actually converted to salable products is called the commodity drain on the nation's forests. An added drain, called the noncommodity drain, results from fire, decay, and insects. In 1944 the noncommodity drain was approximately 12 per cent of the commodity drain.

A. Logging

Residual wood generated in logging consists of trees not cut because of poor form, mechanical defects, and decay; tops and limbs; trees damaged or destroyed in logging; and residual trees destroyed during slash disposal. The amount of wood destroyed or unused after logging depends on species, age and composition of the stand, logging methods used, type of terrain, primary product for which the logs are being harvested, distance of the producing area from its principal markets, size and location of conversion plants, and many other factors.

Hooker,[18] after studying logging residues in the hardwood forests of northern Michigan, stated as a rule of thumb that 75 cu ft of residue might reasonably be expected for each 1,000 bd ft (Scribner Decimal C) of logs harvested. Jenkins,[19] in reporting on studies made by the Ottawa Forest Products Laboratory, stated:

> On selected sawlog operations in Ontario, Quebec, and the Maritime Provinces . . . the four principal sources of logging waste are wood left in the form of (a) high stumps; (b) long butts, including felled trees left on the ground; (c) sound wood left in uncut defective trees of merchantable diameter; and (d) merchantable wood left in large tops.
>
> The results of these studies indicate that more than twice the volume of merchantable wood is left in the form of large tops than from the other three sources combined. Nearly 19 percent (18.2) of the merchantable volume of the stand is left as waste after logging. Approximately 13 percent represents the waste contained in large tops. There remained after logging, from stands averaging about 2300 cubic feet per acre, 430 cubic feet per acre of merchantable wood material.

According to the U.S. Forest Service,[32] in 1952, 52 per cent of the nation's logging residues occurred in the South, where they constituted 14 per cent of the volume of timber cut. In the same year the North accounted for 15 per cent of the total logging residues, with a ratio of residue to timber cut of 11 per cent, and the West supplied 33 per cent of the logging residues which constituted 12 per cent of the timber harvested in that region. California, with a ratio of residue to total timber cut of 18 per cent, had the highest proportion in the country.

In comparing its 1944 and 1952 data the Forest Service observed that there appeared to have been little improvement in the utilization of logging residues during this period in the North. The ratio of residue to harvest in each year was 11 per cent. The South showed a modest improvement in this respect from 16 to 14 per cent. The West, on the other hand, showed a marked improvement from 34 to 12 per cent.*

* According to the Forest Service some portion of the difference between the 1944 and 1952 figures reflects changes in method of computing residues. But even allowing for this, a substantial improvement in utilization is apparent.

B. Primary Manufacture

The most commonly accepted classification of wood residues produced in the manufacture of primary products is as follows: "Primary manufacturing waste includes: (1) Coarse materials like slabs, edgings, and trimmings from sawmills, cores and trimmings from veneer mills, and culled stave bolts from cooperage stock mills; (2) fine materials such as sawdust and shavings from sawmills and other plants; (3) fibers and chemical components lost in making pulp and paper." [31] The last category is not always included in such a classification, although it seems tenable, providing it is recognized that these products are not waste in those cases where no profitable uses exist. The amount of residual wood at primary conversion plants is slightly greater than the amount left in the woods if pulping losses and the sawdust and other products sold for fuel are included, and about one-half as much if they are not included.

According to "Timber Resources for America's Future," [32] plant residues in 1952 totaled 3.4 billion cubic feet. This represented about 38 per cent of raw material entering the plants in roundwood form. Approximately half of these residues were in coarse form such as slabs, edgings, trimmings, miscuts, cull pieces, veneer cores, and other material suitable for remanufacture or chipping. The remaining 50 per cent consisted of fine residues (too small for chipping) such as sawdust, shavings, wood substance lost in barking, chipper rejects at pulp mills, and veneer clippings.

Each primary industry has its own type of wood residues. Coarse materials found at sawmills include slabs, edgings, trimmings, and planer ends. Fine material consists of sawdust and shavings. In a 1957 study of the Canadian redcedar lumber industry,[22] the Forest Products Laboratories of Canada found that 18.3 per cent of the cubic foot volume of wood entering the mill became residue. Rotary veneer plants generate log trim and sawdust, peeler cores, spur trim at the lathe, roundup, green and dry clippings, breakage, and the like. In some hardwood-veneer plants green residues commonly amount to 1 ton per thousand board feet of log scale, Doyle rule. Shingle mills generate residual wood in the form of log trim and waste in producing shingle blocks, "spalts," shingle tow, and so forth. The Canadian Forest Products Laboratory study of the redcedar industry indicated that residues in the manufacture of 18-in. perfection shingles accounted for 59.8 per cent of the cubic foot volume of logs used. Even in producing round commodities, such as poles, there are unused butts and tops that are removed to improve grade, fine sapwood material removed in barking poles on the recently developed pole-shaving machines, and saw kerf and other wood from framing operations. Many small plants produce characteristic residues in the manufacture of blanks, billets, and cut-up stock for a host of small manufacturing plants.

A significant development leading to appreciable reduction in the amount of unused mill residue is the growing practice of barking logs before sawing. In 1958 an equivalent of more than 5 million cords of pulpwood, or about 15 per cent of the total pulpwood used, was from chips developed largely from slabs obtained from barked logs. (For further information see Chap. 17, p. 336.)

C. Secondary Manufacture

The residual wood from primary plants is fairly well concentrated in large manufacturing centers or adjacent to large sawmills, but countless secondary

Fig. 2-1. Use of debarkers at sawmills utilizing small-diameter logs facilitates conversion of slabs and edging residues into pulp chips. (*From American Forest Products Industries.*)

manufacturing plants are widely scattered in every part of the country. These vary in size, nature of product, and nature of raw material over a wide range from large furniture factories to small specialty plants. It is virtually impossible to determine the amount of wood residues they produce. Since the Second World War there has been a spectacular development of plants manufacturing dry-formed particle board associated with large furniture plants; the particle-board plants utilize the residues to produce panels that are ultimately incorporated into the furniture. It is not uncommon in the furniture industry for 40 per cent of the lumber introduced into the rough mill to be discarded as residue. Other residues result from veneer trim. Despite the favorable development of particle-board plants to use residues from secondary manufacture, the amount so used is still a very small fraction of the total produced. There is no great immediate possibility of utilizing the bulk of the remainder, except locally for fuel. It requires a large secondary manufacturing plant to produce enough residue to maintain a particle-board plant. The cost of concentrating residues from widely scattered plants is great.

D. Chemical Industries

Wood losses in the chemical industries take two forms: (1) wood lost in harvesting and preparing the wood for chemical conversion, and (2) losses of lignin and fiber, and dissolved cellulose, hemicelluloses, and extractives. The pulp and paper industry is the principal source of these losses. It is a common misconception that among wood-using industries the pulp industry is unique in that it uses essentially all its raw-material input and develops very small quantities of residue. According to "Timber Resources for America's Future," [32] prechemical residues consisted of (1) from 2.5 to 6 per cent loss of wood substance due to decay in storage, (2) from 1 to 5 per cent rejects of fines in screening chips, and (3) from 1 to 2 per cent of wood substance lost in barking. In addition to these losses, it was estimated that "an additional 40 per cent of the wood used by all processes of pulping in 1952 was dissolved in the various pulping liquors or the water used for washing and conveying the pulp." About 80 per cent of this is recovered, and some high-value products are obtained, but the bulk of it is used for fuel. In discussing the problem Pearl [26] states:

> For many years, the chemical pulping industry has been utilizing essentially only half the tree as pulp, and has been discarding the other half as a waste or as a source of heat in the recovery of soda. Pressures for the abatement of pollutional effluents, the need for conservation to meet expanding requirements for our raw materials, and the urge for more economical production in the face of higher operating costs were responsible for initiating research on the utilization of the other half of the tree.

The total wood residues generated in distillation and extraction are small in comparison with those of the pulp and paper industry. For the most part, these processes utilize less valuable species or residues of other primary industries. Nevertheless, they generate some residues and in some instances utilize wood at a low economic level. Either no provisions are made to recover acetic acid, methanol, and acetone in many charcoal operations, or the recovery methods are primitive.* Spent chips from the steam and solvent process are an excellent source of fiber for insulating boards, but they are often burned as fuel.

* See Chap. 19.

E. Fiberboard

The extent of losses in the manufacture of fiberboard varies with the process. If a chemical cook is used, some wood constituents are lost by dissolution. Fiber is lost in barking, and some fines escape in the white water. Most fiberboard plants utilize slabs and other residues from sawmills. Some, however, use young second-growth trees and sawlogs. In some instances this is true waste, because the trees are either harvested prior to attaining their maximum growth potential or diverted from a higher use.

II. WOOD RESIDUES AND WASTE WOOD IN THE UNITED STATES

The literature on the amount of residual and waste wood developed in the United States is meager. While intensive surveys have been made of individual primary industries in some producing areas, information on a national basis is limited, for the most part, to statistics collected and published by the U.S. Forest Service. These figures, though admittedly based to a large extent on estimates, are of substantial value to all who are interested in developing means of utilizing wood at the highest possible economic level. Complete data on the amounts of wood residues generated in logging and manufacture, by commodities and by regions, are available in a U.S. Forest Service publication on the subject.[31] For reference purposes, selected data are presented in Tables 2-1 through 2-4. The reader is urged to study thoroughly the original work.

Table 2-1. Volume of Unused Logging Residues by Industry Source, 1952, Thousands of Cu Ft [a]

Industry	Residue volume
Lumber	1,019,727
Veneer	99,716
Cooperage	32,790
Pulp	71,812
Other	140,128
Total	1,364,173

[a] SOURCE: U.S. Department of Agriculture, Forest Service. Timber Resources for America's Future. Forest Resource Report No. 14. 1958.

A knowledge of the amount of wood residues generated, their character, and where they occur is the first step toward an understanding of the problem of more complete utilization.

III. BETTER WOOD UTILIZATION

For practical purposes, the means of improving wood utilization should be divided into two categories: long-range possibilities, and methods that

Table 2-2. Residues as a Percentage of Volume of Timber Cut, for Various Rough-mill Products [a]

Product	Residues as per cent of timber cut
Hewed ties	38
Cooperage logs and bolts	31
Veneer logs and bolts	20
Sawlogs	15
Poles	14
Piling	13
Round mine timbers	6
Pulpwood	4
Fuel wood	4
Other	12
Average of all products	13

[a] SOURCE: U.S. Department of Agriculture, Forest Service. Timber Resources for America's Future. Forest Resource Report No. 14. 1958.

Table 2-3. Volume of Plant Residues from Primary Manufacturing Used in the United States and Coastal Alaska, by Industry Source and Type of Use, 1952, Thousands of Cu Ft [a]

Industry	Fuel	Fiber	Other	Total
Lumber	1,397,066	75,777	145,608	1,618,451
Veneer	131,081	33,951	15,147	180,179
Cooperage	24,309	......	2,356	26,665
Pulp	170,437	......	....	170,437
Other	28,706	15	7,369	36,090

[a] SOURCE: U.S. Department of Agriculture, Forest Service. Timber Resources for America's Future, Forest Resource Report No. 14. 1958.

Table 2-4. Volume of Unused Plant Residues from Primary Manufacturing in the United States and Coastal Alaska, by Industry Source and Type, 1952, Thousands of Cu Ft [a]

Industry	Coarse	Fine	Total
Lumber	658,244	673,131	1,331,375
Veneer	6,641	18,186	24,827
Cooperage	5,123	7,944	13,067
Other	5,315	7,968	13,283

[a] SOURCE: U.S. Department of Agriculture, Forest Service. Timber Resources for America's Future. Forest Resource Report No. 14. 1958.

have good prospects of succeeding under conditions existing at present and in the immediate future. Long-range planning should include (1) basic research on entirely new concepts of wood as a raw material, (2) studies aimed at devising the best means of integrating wood-using industries to obtain the highest economic return and the best balance between essential uses, and (3) research to discover entirely new techniques for the more efficient utilization of wood. Such efforts are one of the primary responsibilities of Federal and other research agencies that are not entirely dependent on the returns from sales of forest products to finance their research activities. Obviously, it is impossible to draw a fine line of responsibility for research, but individual wood producers cannot be expected to assume a large share of basic research costs. All research agencies share the responsibility of applying new developments to the reduction or elimination of residues and waste, and each wood user, whether he has research facilities or not, must adopt all technological advancements that will reduce residues and waste to a minimum. The adoption of new methods is usually slow in an industry made up largely of small enterprises, but in view of the ever-increasing inroads of competition, it seems inevitable that only those operators who obtain the maximum return from the raw material will survive. Under a system of free enterprise, it is axiomatic that industry will adopt without delay any new developments that promise a fair return under existing economic conditions. The problem, then, is balanced research to provide practical solutions for the immediate reduction of bulk residues, and fundamental research that will pave the way to the complete and efficient utilization of every cubic foot of wood that can be produced on the forest lands of the nation.

A. Long-range Research Goals in the Utilization of Wood Residues

When the wood-using industries have attempted to develop uses for wood residues, too frequently these efforts have been empirical rather than scientific. They have consisted of trial-and-error programs designed to convert wood residues into substitutes for other products. It is axiomatic that wood residues can be converted into substitutes for hundreds of existing products. The problem is that it usually costs more to manufacture these items from wood by-products than from raw material currently used and furthermore that the new product rarely exhibits favorable properties that would justify the costs. The really great opportunities for increasing the portion of the tree converted into useful and valuable products lie in achieving a much greater knowledge of the fundamental structure and properties of wood. Many of the wood residues are in the form of small particles, such as sawdust and shavings, that can be reassembled into conventional wood products only at great cost.* In the long run the value of these residues lies in recognizing that they constitute great supplies of chemical raw material. They will be converted into useful products only when the fundamental chemical nature of wood is understood much better than it is today. At that time wood residues will be recognized for the real storehouses of useful products that they are.

* For further information on utilization of sawdust and shavings see Chap. 14.

Examples of the long-range problems requiring solution are numerous, and a few of them will be discussed. One of the more far-reaching research objectives of the U.S. Forest Products Laboratory at Madison, Wisconsin, is the conversion of wood into a liquid source of chemical raw materials through hydrogenation or other processes. This would have obvious advantages, particularly if plants using the process could be located adjacent to large concentrations of timber and waste materials. Logs are costly to ship for any distance, but a liquid raw material can be piped or shipped by tank car to any part of the country for conversion. A statement [30] concerning this development is as follows, under the heading Hydrogenation of Wood:

> It has been shown from pioneer research and work now in progress at the Forest Products Laboratory that lignin dissolved in organic solvents or suspended in water can be made to react with hydrogen gas at elevated temperatures and pressures in the presence of various metallic catalysts. Among the products of the reaction are several new cyclic alcohols that had never been previously described in the literature. These show promise as plastic solvents, antiknock agents for motor fuel, and toxic agents. By varying the hydrogenation conditions, phenolic compounds are obtained that may find uses in plastics and complex neutral oils, together with a plastic-like residue. Wood waste or chips can also be hydrogenated in aqueous suspension to produce soluble lignin decomposition compounds and a cellulose pulp residue. This is a possible new pulping process that will be studied further by the Forest Products Laboratory. Under more severe hydrogenation conditions the cellulose can be converted to glycerine and sugars. In this case the entire wood is converted to liquid products.

Lignin is still an enigma to the wood chemist and to most of the chemical industries that use wood as a raw material. Finding profitable uses for this substance is one of the major problems of wood utilization. During the decade 1950–1960 much basic research was accomplished leading to a better understanding of the nature of lignin—a precursor to increasing its utilization. As a result of this research some breakthroughs have been made in the utilization of lignin residue, and many others can be projected. Processes have been developed by Rayonier Incorporated which permit the production of polyester-type fibers from spent sulfite liquor vanillin. The Crown-Zellerbach Corporation has developed manufacturing facilities for the production of dimethyl sulfoxide, dimethyl sulfide, and methyl mercaptan from spent sulfite liquors. Other research has developed uses of lignin and lignin derivatives for soil-improvement materials, resins and adhesives, and components of rubber and cement products.

The destructive distillation of wood yields products such as tars, acetic acid, and methanol.* In general, less than half the dry weight of wood is recovered as charcoal and salable distillate.[31] The cellulose and hemicelluloses are broken down, and only a small part is recovered. If the lignin is separated from the carbohydrate portion, its distillation produces nearly the entire original yield of methanol from wood and approximately half the charcoal.

It is evident that the process is wasteful of the cellulose and hemicelluloses.

* See Chap. 19.

There is a clear-cut need for research to devise less wasteful means of obtaining the products now derived from destructive distillation. The tars also merit attention, particularly when softwood is the raw material. They have little value at the present time, yet it seems likely that careful research might disclose values that are not now apparent.

There is a tremendous need for an inexpensive adhesive having properties comparable to the better present-day synthetic resins that will set rapidly at room temperatures. Such a discovery would extend the benefits of gluing techniques to many small enterprises, because initial investments in gluing equipment would be so greatly reduced. The lamination of rafters, beams, and numerous other structural members would become a relatively simple matter. This would result in great savings of raw material, since it would make possible the low-cost fabrication of laminated products from small pieces of residual wood from a variety of primary industries. Even laminated railroad ties with wear-resistant surfaces are a possibility. Boatbuilding would be greatly simplified and costs reduced. Major adhesives manufacturers are engaged in active research in this field, but all wood-research agencies should contribute to this important project.

Until new adhesives of the type described above are developed, there will be a continued need for information on radio-frequency dielectric heating. This method is particularly useful in gluing certain types of curved veneer products and laminated beams that are too thick to be glued in conventional hot presses. It is also useful for gluing up furniture core stock, box ends, and other built-up products from small pieces of residual wood. Although costly for many applications, this process is gaining increased attention from wood-fabrication industries and deserves a place in the list of projects requiring research attention.

The microbiological utilization of wood waste also deserves long-range research consideration. McElhanney [23] states the problem as follows:

> Investigations have already shown that thermophilic bacteria are capable of attacking cellulose with the formation of valuable organic chemicals such as formic, acetic, lactic, and butyric acids. Microbiological methods have the advantage of simplicity, do not require expensive equipment, and can be carried out on a large scale. This field seems worthy of further investigation. Attempts should be made to isolate organisms from the digestive tracts of wood-eating insects and the rumen of animals capable of digesting cellulose. In addition to bacteria, protozoa might be worthy of investigation. All of these organisms are of interest not only in relation to the formation of valuable chemicals from cellulose, but also because of their high protein content. It is conceivable that sufficient protein material could be produced in the fermentation of wood waste to act as a binder for the conversion of wood waste into structural boards.

Hajny,[16] in discussing the microbiological utilization of wood sugars, points out that "in the 75 million tons of plant and logging waste, there are upwards of 40 million tons of carbohydrate material. Since the annual world production of raw sugar for the years 1950 to 1955 amounted to 36 to 42 million tons, it can be readily appreciated that wood is potentially one of the most important and widespread sources of carbohydrates." He notes that ethyl alcohol and fodder yeast are now commercially produced from spent sulfite liquor or hydrolyzed wood, but that many more chemical compounds can be produced by microbiological means from wood sugars. He argues though that if "further microbiological utilization of wood residues is to be economical, profitable uses must be also found for the two other major components of wood, lignin and hemicellulose."

While having only an indirect influence on reductions of residues, seasoning and the practice of wood preservation have a salutary effect in lessening the drain on forest resources. Continued fundamental research in these fields is, therefore, justified and desirable.

Research is also needed to find uses for wood containing decay, which is present in large quantities in some regions, particularly in the overmature stands of the Pacific Northwest. Exploratory work is needed to determine the constituents of the bark and wood of all species, including those not now utilized for lumber and other products. The cork in Douglas-fir bark, as a possible substitute for imported cork, offers one example; tannins from numerous species, another. The possibility of extracting useful resins from wood chips prior to pulping and fiber manufacture, and many other prospective extraction processes, also merit consideration.

It has long been recognized that wood residues have a potential value as a soil conditioner and mulch. However, usually they must be treated to be effective. As Wilde [35] has pointed out:

> Sawdust includes tannins and phenolic compounds injurious to plants and soil microbes, and it has a high content of cellulose and, consequently, a high carbon-nitrogen ratio, a property that invariably leads to the impoverishment of soil in nitrates and ammonia consumed by carbohydrate decomposing organisms.
>
> The only fraction of sawdust that is valuable in the maintenance of soil fertility is lignin. The carboxyl groups of this high-molecular aromatic compound have an ability to part with hydrogen and retain absorbed ions of ammonia, calcium, magnesium, potassium, and other bases. In this manner lignin acts as a storehouse of soil nutrients that are preserved from leaching and yet are available to plants through exchange reactions.

A number of procedures have been devised to capitalize on the advantages of wood residues and to eliminate or minimize their disadvantages. Where raw sawdust has been used, nitrogenous fertilizers such as ammonium sulfate have been added in the order of about 200 lb per ton of sawdust to compensate for the nitrogen deficiency that develops when the soil microorganisms consume nitrogen while breaking down the carbohydrate fraction of the wood. The use of rotted sawdust reduces this problem, since in normal decomposition the cellulose breaks down first. Composting with organic materials like animal manures as inoculants has proved effective. Farber and Hind [9] report on a successful commercial procedure for condensing the wood carbohydrates into lignin-like materials by heating the wood in the presence of mineral acids under controlled conditions. The Pope and Talbott Company is thus producing a soil conditioner under the trade name Fersolin.

A fruitful area for further research in the use of residues is the study of bark utilization. The pulp and paper industry, since it has traditionally barked its wood prior to use, has been conducting research in this area for a long time. With increasing use of barkers at sawmills, and with bark an important residue of the veneer industry, this type of research is increasing in importance. Bark varies in its properties considerably from species to species, and accordingly the use potential varies with species. Bark has value as a soil conditioner, probably in excess of raw wood, since its lignin content is usually much higher and its carbohydrate fraction lower. Some barks contain valuable extractives. Hemlock bark was, of course, once a primary source of tannin. Dihydroquercetin, a flavonoid, has been extracted [10, 13] from Douglas-fir bark and has potential uses in quantity in the preparation of pharmaceuticals. Lignins from barks have potential as fillers and extenders for plastics. Effective insulation materials have been manufactured from Douglas-fir and redwood bark. These same barks have been used in the formulation of drilling mud for the oil industry.[25] *

It is encouraging that research in long-range potentialities has increased substantially since the Second World War. It is hoped that the lumber industry will avail itself of the results of these efforts to a far greater extent than it has in the past.

B. Immediate Application of Known Methods

Research aimed at the reduction of residues through the application of existing knowledge is an outright challenge to the wood-using industries. Although wood research has been meager in comparison with the research efforts of many other industries, it is still far ahead of the capacity of the lumber industry to apply known improvements. As pointed out previously in this chapter, this condition stems from the lack of industry-owned applied research facilities capable of converting the results of basic findings to product and process development. This is the general situation; there are some examples of progressive enterprises, large and small, that have taken advantage of research to improve utilization, increase profits, and perpetuate their operations by decreasing the drain on their supply of raw material. The rest of the industry must follow these examples. It is not a case of the complete dissolution of the lumber industry if this is not done, but it is one of elevating it from a one-product raw-material industry, earning low returns, to a multiple-product manufacturing industry that obtains profits all the way to the consumer.

It is not possible to generalize on residue-utilization methods, because the amount, character, and location of residues and waste vary too much with species, regions, primary products, types of conversion plants, distances to markets, and many other factors. Each case requires separate analysis in the light of some or all of these variables. The possibilities of improvement on the basis of present knowledge are numerous, and a few examples will be given to illustrate some of the more obvious avenues of approach.

* For further details of bark utilization see Chap. 25.

1. Wood Now Left in the Forests. The greatest opportunities for reducing the amount of wood left in the forests are found in the Pacific Northwest and in the South, where large sawmills are still found. Both of these regions afford the stability of wood supplies required for the establishment of chemical utilization enterprises. Pulp and paper mills are taking much more material from the forests than formerly. Portable chippers, designed to prepare unused forest wood for utilization, are in the experimental stage. Chemical utilization offers the best ultimate possibilities, but other steps are being taken to reduce residues accumulating during logging. In the Pacific Northwest, the ever-increasing replacement of high-lead logging by tractors has resulted in material reductions in losses from broken and damaged trees. Lighter yarding equipment has made it possible to harvest economically small material that was too costly to remove with the heavier equipment formerly used. "Prelogging" and "relogging," encouraged by high log prices in recent years, have aided in reducing the amount of material left in the woods in the Pacific Northwest. Portable mills, designed to utilize trees and logs left by conventional logging methods, have had the same effect. Considerable research is in progress in the Pacific Northwest and at the U.S. Forest Products Laboratory to find profitable uses for wood containing decay and logging residues and to devise cheaper and more efficient harvesting and transportation methods. Much progress is being made in the South along lines of integrated utilization, particularly on large holdings, and many improved harvesting techniques have been introduced in recent years.

Efforts to increase the utilization of wood left in the forests should not be limited to the foregoing regions. Although concentrations of residues are smaller in other regions, they should not be neglected. Many small but permanent wood-using enterprises now leaving wood on the ground in the forests could well give consideration to integrated operations. There are numerous opportunities for small-scale plant expansions that will reduce residues and increase profits. The Northeast Wood Utilization Council, New Haven, Connecticut, is making excellent progress in this field.

2. Residual Wood at Manufacturing Plants. Of all the sources of residual and waste wood, the manufacture of lumber offers the most promising opportunities for immediate utilization at a higher level. Perhaps the greatest waste in the lumber industry is the practice of producing full-length lumber, only to be cut up into small products, when thousands of tons of clear pieces in the form of slabs, cuts from crooked logs, trim, and edgings are used for fuel or burned outright. There are countless articles that can be manufactured from such material, often at a greater profit than the lumber itself. The main thing that is needed is a new outlook, an outlook that envisions products other than lumber as the final product of a sawmill. Some enterprising individuals are making substantial profits from the manufacture of broom handles, garden and surveyor's stakes, chair spindles, oak wedges, adding-machine-tape rolls, toys, ladder rungs, tool handles, mouse traps, molding, door rails, and an endless list of such products from sawmill residues. But the number of companies so engaged is too small, and the amount of such material being burned is too great. Supplying secondary remanufacturing plants with blanks and dimension stock offers another large outlet for wood residues.

The edge-gluing of sawmill residues to produce furniture and door cores, kitchenware, chair bottoms, box ends, ironing boards, and numerous similar items also offers great possibilities. This use is increasing, but there is room for much greater expansion. Much wood suitable for this form of utilization is being burned daily in the waste burners of the local mills. More sawmills could well enter this field or at least furnish the raw material to remanufacturing plants.

Fig. 2-2. Sawmill trim accumulated at a medium-sized Douglas-fir mill. (*From Bull. Series 17, Engin. Expt. Sta., Oregon State College.*)

The box-and-crate industry is one of the largest consumers of wood in this country, and much of the raw material it uses comes from full-length boards and cants. Much wood that is now hogged for fuel or destroyed could be used for this purpose. The Wooden Box Institute, San Francisco, California, has been more alert to this possibility than has the lumber industry and is engaged in constructive research along these lines. High-speed slicers that produce box-shook veneer from short lengths of wood have been introduced recently and have already contributed to the reduction of residues at some western sawmills.

The building industry has accepted lumber as produced for years without contributing significant advances in the use of wood for construction purposes. There is evidence to indicate that many wood structures are overdesigned and that substantial savings of wood could be effected by improvements in present designs. The same applies to lumber grades. Excessively high quality is often demanded where lower grades would serve just as well. Indirectly, this results in leaving logs containing large quantities of lowgrade lumber in the woods, because such lumber is often produced at a loss. Revisions of grade standards would provide increased outlets for this ma-

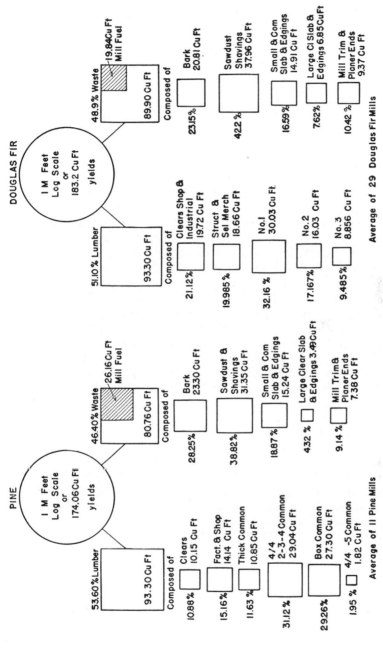

Fig. 2-3. Character of waste and grade of lumber produced in sawing 1,000 ft of logs. (*From Bull. Series 17, Engin. Expt. Sta., Oregon State College.*)

terial. Improved design would no doubt permit the increased use of shorter lengths, as well. The lumber industry cannot depend on architects and construction engineers for improvements in design. Fabricators of competitive building materials have been exceedingly active in this field, and they supply complete data on design factors for their products. The lumber industry, as a whole, has too long neglected this type of selling activity, though it offers potentially lucrative dividends.

The production of fiber products from sawmill residues also offers immediate prospects for reducing wood residues. Much wood, not usable for any other purpose, can be reduced to fiber to produce the filler for asphalt roofing materials and for the manufacture of insulating and hard-pressed fiberboards. The volume of fiberboards from all sources has increased phenomenally during the past 10 years, and there is no reason why residual wood cannot be the raw material for a large part of this output. Barking of sawlogs now practiced at many mills, especially in the South and West, has opened an important outlet for utilization of mill residues for pulp chips (see Chap. 17, p. 336). High concentrations of residual wood at sawmills offer a cheaper source than corn and sugar-cane stalks now used extensively for these products.

The problem of residues in most primary wood-using industries is comparable to that in sawmills, although the character of residues may differ somewhat. Cores constitute a rather large proportion of the residues at veneer plants. Much of the volume of the cores left by standard production lathes can be salvaged by the installation of "pony" lathes to cut the cores to the smallest possible diameter. If no pony lathe is available, the cores from the larger lathes can be cut into 8-ft lumber; yet much of this material is cut into short lengths for fuel wood. Clipper waste can be reduced to fiber, but little of it has been used for fiberboard to date. The waste from veneering operations could easily be used for fiber cores in products not requiring high strength. This would extend supplies of veneer for more exacting uses and lengthen the life of plywood and veneer operations.

The foregoing examples are only illustrations of the many challenging opportunities of utilizing residues at primary manufacturing plants by known methods. It cannot be repeated too often that greatly accelerated research and a conscientious effort on the part of the lumber industry are needed to put the utilization of large quantities of unused wood on a practical and profitable basis.

3. Wood Lost in Chemical Processes. Chemical utilization offers the greatest promise for the future conversion of wood residues into useful products, but prospects for immediate savings are less spectacular. In the pulp and paper field, some improvements can be made in present pulping practices, and steps can be taken to reduce losses in harvesting pulpwood. Research is showing the way to the use of inferior species for pulp. The use of sawmill residues and even wood containing decay is on the increase, particularly in connection with integrated operations. Hydraulic barkers and whole-log chippers are reducing residues, and many pulp mills are remodeling their wood rooms to accommodate shorter logs and smaller material. In a few instances, wood from thinnings and improvement cuttings is being utilized. The practice of segregating logs for various grades of paper is resulting in better utilization. Losses of cellulose can be reduced by milder chemical treatments and by semichemical pulping. All these developments should be adopted as widely as possible throughout the entire industry. The conversion of the wood sugars in sulfite waste liquors to ethyl alcohol or yeast should also be practiced to the full economic limit.

The wood-distillation and steam-and-solvent extraction processes are in need of intensive and extensive research. Some immediate improvements can be made in the former by increasing the recovery of condensable materials and in the latter by finding more profitable uses for spent chips. Increased demands for charcoal may accelerate activities of the distillation industry, but competitive factors indicate a decline unless entirely new principles are evolved.

IV. SUMMARY

The steps required to reduce wood residues will be summarized to focus attention on the important problems that must be solved. The responsibilities for the necessary action will also be discussed.

Of primary importance is the assumption of responsibility by the lumber industry for the applied research necessary to assure that the industry keeps pace with technological developments in all fields of science. Federal, state, and educational agencies should continue and even accelerate their present research activities, in both fundamental and applied fields, but the principal responsibility of applying fundamental results and of developing new products and processes rests squarely upon the lumber industry. The plastics and metal industries do not depend on the results of outside agencies when they develop new products that compete with wood. They do the research themselves. The manufacturers of steel joists and door frames do not wait for architects to suggest new uses for the materials they manufacture. They design them and provide architects and builders with specifications and dealers with advertising information, and they are selling their products at the expense of wood. No one should know the properties and potentialities of wood better than the people who are engaged in the business. It is up to them to launch an applied research program, to support the constructive efforts of all wood-research agencies, and to see that public funds commensurate with the importance of the industry are expended for wood-utilization and forestry research. Such research should include market and economic surveys, customer acceptance tests, and cost analyses, as well as searches for new techniques and improvements in present equipment and manufacturing methods. Some liaison is needed between public and industrial research, and means should be provided to disseminate technical knowledge as widely as possible to all wood-using industries. If the wood-using industries provide sufficient research facilities of their own, there is little doubt that they will be well informed of all research that can be applied to advantage to practical problems.

Another factor of great importance is the more complete integration of all wood-using industries. This is a complex problem requiring action on a national as well as a regional basis. It is closely associated with the sustained-yield management of forest lands, the management policies of Federal agencies having jurisdiction over forest lands, and the interrelationship of the timber requirements of the various types of wood-consuming industries. The pattern of integration is increased efficiency, the reduction of residues through the proper segregation and allotment of raw materials to the highest use, and provision of sufficient raw materials for balanced production to all interdependent units. Large companies having sufficient timber to establish sustained-yield units are already making rapid progress in this direction. Not only do they have the permanence provided by a continuing supply of raw material, but also they have the resources to provide new processes, some of them very costly, that make more complete utilization possible.

The problem is not so simple in the case of small enterprises. Many of them have no prospects whatever of obtaining sustained-yield holdings. Without such backlogs of timber, the large investments required for fiberboard and chemical utilization units cannot be justified. Yet integration is more important to these small enterprises than it is to large companies with assured futures. Without it, many of them will be eliminated by the competition of the more efficient integrated organizations.

Cooperative pooling and segregation of raw materials and distribution on the basis of the most efficient and profitable use would help in some localities. Research, again, is the answer for isolated sawmills and other woodworking industries that lack full opportunities for integration. Many residue-using processes often cost more than the unit producing the residue. Research should, therefore, be directed toward methods of utilizing small accumulations of residues. Some progress is being made along these lines, but more effort is needed.

Increased activity in forest management and related research is just as important as improving the standards of utilization. Economically sound sustained-yield units should be expanded materially, and there is an acute need for bringing all forest lands of the nation into the highest possible state of productivity. Certainly the waste of unproductive forest lands is just as great, if not greater, than the waste now found in the manufacture of forest products. Farm woodlands, on the whole, are badly in need of regeneration, and every effort should be made by all agencies involved to increase the yield of wood from this large source of timber supplies. Quantity, however, is not the only factor. Forest management should aim for balanced production, with sufficient supplies for all uses. Particular emphasis should be given to supplying the high-quality logs required by lumber, veneer, and other important primary industries.

Above all, solving the problem of wood residues demands a realistic approach, combined with the marshaling of all technological resources available to the wood-using industries and to the agencies interested in their welfare. All possible advances must be analyzed in the light of existing economic conditions. It would be folly to put complete reliance on one process to solve the problem. If all the 108.9 million tons of waste and residual wood generated in logging and manufacturing in the United States in 1944 were converted to alcohol at 55 gal per ton, the total would be nearly 6 billion gallons of alcohol; nearly 10 times the peak wartime production year. Obviously, it would be impossible to convert all residues to alcohol, but the example serves to indicate the need for thinking in terms of balanced utilization of residues. The problem, then, is one of concerted action against all sources of residues and the adoption by industry of as many diversified methods and end uses as have chances of success under existing economic conditions.

SELECTED REFERENCES

1. Bethel, J. S., and W. B. Woodrum. Increased Lumber Core Yields with a Patched Strip Core. *Forest Prod. Jour.*, **5**(3):165–167. 1955.
2. Brauns, F. E. Lignin and Its Possible Chemurgic Significance. Chemurgic Papers, No. 483. 1946 Ser.
3. Burkhart, R. L. Interior Trim from Wood Waste. Proceedings, Forest Products Research Society. 1948.
4. Burton, R. E. Bark as a Trickling-filter Media. *Forest Prod. Jour.*, **9**(4):19A–22A. 1959.
5. Colucci, G. Special Machines for Utilization of Waste Slabs for Glued Core Stock. Proceedings, Forest Products Research Society. 1948.
6. Dosker, C. D. Lumber's Future Depends on Research. Address delivered at the annual meeting of National Lumber Manufacturers Association and Timber Engineering Company, New Orleans, Nov. 7, 1946. Timber Engineering Company, Washington, D.C.
7. Doughty, J. B., F. W. Taylor, and W. T. Henerey. Alkali and Thio Lignins from Pine Barks. *Forest Prod. Jour.*, **6**(11):476–478. 1956.
8. Dunn, S., and J. D. Emery. Wood Waste in Composts. *Forest Prod. Jour.*, **9**(8):277–281. 1959.
9. Farber, E., and R. R. Hind. Sawdust into Fertilizer. *Forest Prod. Jour.*, **9**(10):340–344. 1959.
10. Gardner, J. A. F., and G. M. Barton. The Distribution of Dihydroquercetin in Douglas-fir and Western Larch. *Forest Prod. Jour.*, **10**(3):171–173. 1960.
11. Carratt, C. A. Postwar Opportunities in Wood Utilization. Chemurgic Papers, No. 447. 1946 Ser.
12. Grantham, J. B. Salvage Operations in the Douglas-fir Region: Their Present and Future. *Oreg. Forest Prod. Lab., Inform. Cir.* 1. August, 1947.
13. Gregory, A. S., D. L. Brink, L. E. Dowd, and A. S. Ryan. Douglas-fir Bark as a Source of Quercetin. *Forest Prod. Jour.*, **7**(4):135–140. 1957.
14. Gruen, E. D. Economic Aspects of Residue Utilization. *Jour. Forest Prod. Res. Soc.*, **3**(3):46–50. 1953.
15. Guthrie, John A., and G. B. Armstrong. Western Forest Industry. Johns Hopkins Press, Baltimore. 1961.
16. Hajny, G. J. Microbiological Utilization of Wood Sugars. *Forest Prod. Jour.*, **9**(5):153–157. 1959.
17. Hall, J. A. Integrated Utilization-research Task. Report presented at the national meeting of the Forest Products Research Society, Chicago, Ill., Oct. 31–Nov. 1, 1947.
18. Hooker, L. W. Utilization of Hardwood Logging Residue. *Forest Prod. Jour.*, **8**(1):1–5. 1958.
19. Jenkins, J. H. The Utilization of Wood Waste in Eastern Canada. *Jour. Forest Prod. Res. Soc.*, **3**(2):7–9. 1953.

20. Marquis, R. W. Science and Forestry Goals. *Northwest Sci.*, **21**(2). May, 1947.
21. Matson, E. E., and J. B. Grantham. Salvage Logging in the Douglas-fir Region of Oregon and Washington. *Oreg. Forest Prod. Lab., Bull.* 1. August, 1947.
22. McBride, C. F. Western Red Cedar Mills. *Forest Prod. Jour.*, **9**(9):313–316. 1959.
23. McElhanney, T. A. Conservation by Better Utilization. Paper presented to Fifth British Empire Forestry Conference, held in Britain, 1947. Ottawa, Canada. 1947.
24. McElhanney, T. A. Forest Products Research in Canada. Paper presented to Fifth British Empire Forestry Conference, held in Britain, 1947. Ottawa, Canada. 1947.
25. Miller, R. W., and W. G. Van Beckum. Bark and Fiber Products for Oil Well Drilling. *Forest Prod. Jour.*, **10**(4):193–195. 1960.
26. Pearl, I. A. Present Status of Chemical Utilization of Lignin. *Forest Prod. Jour.*, **7**(12):427–432. 1957.
27. Schrader, O. H., Jr. Integrated Forest Utilization in the Pacific Northwest. Contributed by the Wood Industries Division for presentation at the annual meeting, Atlantic City, Dec. 1–5, 1947. Paper No. 47—A-45. The American Society of Mechanical Engineers. 1948.
28. Sloan, G. McG. Report of the Commission Relating to the Forest Resources of British Columbia. Province of British Columbia, Victoria. 1957.
29. U.S. Department of Agriculture, Forest Service. Outlets for Wood Waste. A General Statement of Actual and Potential Uses of Various Kinds of Wood Waste, No. R64. Revised August, 1945.
30. U.S. Department of Agriculture, Forest Service. Proceedings, Conference on State and Federal Forest Products Research. 1946.
31. U.S. Department of Agriculture, Forest Service. Wood Waste in the United States. Report 4 from A Reappraisal of the Forest Situation. Report prepared by Robert K. Winters, Gardner H. Chidester, and J. Alfred Hall. 1947.
32. U.S. Department of Agriculture, Forest Service. Timber Resources for America's Future. Forest Resource Report No. 14. 1958.
33. Voorhies, G. An Inventory of Sawmill Waste in Oregon. *Oreg. Engin. Expt. Sta., Bull. Ser.* 17. July, 1942.
34. Watts, L. F. Utilization: Keystone of Intensive Forest Management. *Jour. Forest Prod. Res. Soc.*, **2**(5). 1952.
35. Wilde, S. A. Marketable Sawdust Composts. *Forest Prod. Jour.*, **8**(11):323–326. 1958.
36. Wilson, S. A. Sawmill "Waste" Developed, Used and Not Used, in Oregon and Washington in 1944. Pacific Northwest Forest and Range Experiment Station, Portland, Oreg., mimeo. January, 1946.

CHAPTER 3

RELATION OF THE PROPERTIES
OF WOOD TO WOOD UTILIZATION

Wood is known to have been used for structural and numerous other purposes since the early recorded history of man. It has served him faithfully
and well down through the centuries, and even in modern times it is one of
the most versatile and universally employed of engineering and industrial
raw materials. It is true, however, that during the past several decades
markets for wood products have been successfully invaded by diverse commodities made at lower cost using substitute materials. In fact, there are
those in authority who aver that further market losses are inevitable unless
management of the threatened industries can maintain a competitive position by matching the technological progress and more exacting mass-production methods achieved by its competitors. An acute awareness of such an
eventuality currently is being reflected in several woodworking plants, where
new and improved technological procedures and product and process developments are replacing older and archaic manufacturing practices. The
implementation of these newer techniques has been made possible by the
application of the results of much fruitful research that has marked the past
10 to 20 years. In the last analysis, however, wood will continue to lose some
of its present markets to competitive materials and products, but the total
demand for wood and wood products will continue to rise, and by the year
2000 there seems little doubt but that the nation's forests, even with improved management practices, will be hard pressed to provide the wood
requirements of an exploding population.

Wood is the principal product of the metabolism of a tree and is indeed
a variable and complex material. In fact, with the several thousand commercial species in the world to choose from, it is no small wonder that the inter-
relationships which exist among the anatomical, chemical, and physical
properties of each species are only too often neither fully understood nor
appreciated. Thus the unsatisfactory performance of a wood or wood product
may be due not to the inherent characteristics of that wood but rather to
factors involving a dearth of information concerning its properties, errors
of judgment, workmanship, or all three. Included here are such common

41

blunders as selection of the wrong wood to perform a specific task; use of timbers in improper dimensions or in inappropriate grades, inefficient designs, poor construction or fabrication; disregard for the effects of changing atmospheric conditions upon dimensional changes and stability; and failure to provide adequate protection against decay, insects, weathering, and other destructive and/or deteriorating agents.

The choice of a wood for a specific use first of all presupposes that a careful appraisal has been made of the requirements for that use. It is only after these have been delineated that a species of timber possessing the properties desired may be selected. Not infrequently more than one species may be found that would fulfill the stipulated specifications, in which instance the choice becomes a matter of availability and cost.

The problems involved in acquiring fundamental information pertaining to the technical properties of wood are admittedly difficult. The measurable values of many specific properties often range between rather broad limits, not only among different species, but also within the same species, and even among boards or pieces of timber from different parts of the same log. Thus, efficient utilization of a given wood requires a thorough knowledge of (1) properties which, within broad limits, are common to all woods regardless of origin; (2) specific properties which characterize a certain species and which not infrequently dictate its use for a particular purpose; and (3) the degree of variability which may be expected to be present within a species, and the extent to which such variations limit use for specific applications.

A. Properties Common to All Woods

Timber, unlike minerals, is a renewable resource. There are also a number of other pertinent characteristics and properties that are common to all commercial woods. Several of these are responsible for wide acceptance of wood in the construction field and in the production of finished goods, but a few of them militate against its general use in other product areas. Summarized below are those properties and characteristics which within broad limits are common to all woods:

1. Wood as it comes from the standing tree is more or less saturated with water. Wet wood has little or no utility; veneer, lumber, dimension stock, and timber must be seasoned adequately after manufacture before further processing. Some moisture must also be removed from poles, posts, crossties, and similar products to assure acceptance of satisfactory amounts of preservative.

2. Wood that has been properly seasoned is light in weight in comparison with most materials of construction and fabrication, is easily handled, and may be transported for long distances at reasonable cost.

3. The influence of temperature extremes upon the expansion or contraction of wood is of little importance. In fact, the effect is usually

obscured by much greater dimensional changes due to changes in moisture content.

4. Wood is a poor transmitter of sound, heat, and electricity, properties highly desirable for many applications.

5. Wood members can be easily and securely fastened together with glues, nails, screws, or bolts to provide good strong joints.

6. The porous-fibrous nature of wood facilitates the holding of paint, lacquer, varnish, and other finishing materials and films.

7. Wood may be worked into intricate shapes using simple tools.

8. Wood absorbs shock loads and vibrations better than most competitive materials.

9. Wood does not rust. Neither does it corrode in the presence of sea water. It is also chemically resistant to most cold dilute alkalies and acids.

10. Wood neither crystallizes nor becomes embrittled, like many metals subjected to repeated stress reversals.

11. Wood, unlike concrete, retains its cohesive characteristics when exposed to exceedingly low temperatures for protracted lengths of time.

12. Defects in wood generally can be detected by visual means. Thus unsatisfactory material can be eliminated in the raw material stage.

13. Wood is highly hygroscopic. During periods of high humidity, seasoned wood will absorb appreciable quantities of moisture, which, when the air becomes drier, are returned to the atmosphere. These gains and losses in moisture are accompanied by unequal dimensional changes. In normal wood there is very little shrinking or swelling in the direction of the grain, but considerable across it, and then 1½ to rarely 3 times more in the tangential direction than in the radial plane.

14. The mechanical and other physical properties cannot be satisfactorily altered or improved by heat treatments of the sort employed in metallurgical work.

15. Wood cannot be extruded or rolled into new shapes. An extension of any dimension of a wood member can be accomplished only by attaching a second piece to the original one using glue or metal fastenings.

16. Wood is limited in its hardness. There are as yet no commercial means by which hardness can be improved without adversely affecting one or more other physical properties.

17. Wood is an organic material and as such is often subject to the ravages of decay and wood-boring insects when used in environments favorable to the activities of these organisms. When such agents pose a threat to normal service life, wood should be impregnated with a suitable toxic.

18. Wood is a combustible material. Once ignited it gives off inflammable gases which further abet combustion even after the original heat source has been removed. It will also ignite spontaneously at 275°C. On the other hand, when the surfaces of large timbers such

as beams, columns, and girders are exposed to fire, they become heavily charred, and the char, acting as an insulation to the inner recesses of the member, effectively reduces the continuing rate of destruction. It is a well-established fact that property losses from fire in buildings erected with large wooden structural members are often less than in those where the structural components were fabricated with steel. This is due almost entirely to the fact that in an intense fire steel quickly loses its strength, and total collapse of the

Fig. 3-1. Comparative behavior of steel and wooden beams subjected to fire. Note that the steel structural members buckled in the intense heat generated by the fire, while the wooden beam, although partially destroyed by fire, was still capable of supporting heavy loads. (*Courtesy U.S. Forest Products Laboratory.*)

imposed loads is inevitable, while wooden members during the same interval are only partially destroyed and still capable of supporting a part of the original loads after the fire has been suppressed (Fig. 3-1).

19. Wood possesses directional strength properties and is usually from 4 to 15 times stronger along the grain than across it.

20. Wood is variable in strength. Not only do different species vary from one another, but great variations in strength also exist within the same species, and even for pieces cut from different parts of the same log.

It should also be noted here that wood is an important raw material for many chemical industries. Because of its fibrous nature and chemical composition it can be reduced by relatively simple means into pulp, which may then be formed into paper or converted into plastics, rayon, transparent films, and similar products.

B. Characteristic Properties of Different Species

The basic properties of wood discussed in the preceding paragraphs are not possessed in an equal degree by all species; in fact, there is a tremendous variation among timbers in both their appearance and their properties. Color-wise, they range from white through shades of yellow, orange, red, brown, purple, and green, to black. In some the pigmentation is uniformly distributed; in others it is largely concentrated in bands or stripes, or in irregular splotches to give a surface a variegated appearance. Some woods are essentially figureless and of little value for decorative purposes; others exhibit many pleasing patterns traceable to pigmentation, distribution and arrangement of the various cell types, alignment of the cells along the grain, or combinations thereof.

There is also great variation among woods with respect to weight; hardness; strength; degree of shrinkage; penetrability; ease of working with hand and power tools; ability to hold nails, screws, and other fastenings; compatibility with paints, lacquers, and other finishing materials and glues; and reaction to pulping agents. Equally notable are chemical and anatomical differences which not infrequently dictate specific uses. As previously indicated, in the utilization of wood the choice of a species must be based upon a combination of the properties desired.

This can best be illustrated by citing a few examples. As a case in point, both red and white oak timbers of essentially the same density possess strength properties that recommend them for use as bridge stringers or crossties. If the wood is not to be given a preservative treatment, white oak is preferred, in fact recommended, because it is characterized by a natural durability not possessed by red oak. If, on the other hand, a preservative treatment is contemplated, red oak is the more desirable, as it is much more pervious to liquids, owing to the few tyloses present in the large springwood pores. The heartwood of both these timbers is also used in the manufacture of tight cooperage. In this instance, however, the impervious nature of white oak is a desirable quality of the wood, since white oak staves, unlike those made of red oak, do not have to be treated with a sealer to prevent leakage.

Ash and hickory combine strength with the good bending qualities which make these woods outstanding for bent work; at the same time, they possess the required toughness and good finishing characteristics which account for their extensive use in the manufacture of tool handles. The natural figure, rich color, and uniform texture of mahogany, combined with strength, durability, dimensional stability, and good finishing qualities, have made it a favorite furniture wood, either in solid or in veneer construction. Uniform texture and hardness, responsible for the ability of persimmon and dogwood to remain smooth under constant wear, are reasons for selecting these woods for shuttles and bobbins. Softness, low shrinkage, and uniform texture, combined with adequate strength and ease of working with tools, account for the extensive use of white pine for pattern stock.

Douglas-fir is widely used for structural purposes because it combines strength with durability and is available in large sizes. Good gluing properties and fairly low shrinkage are important considerations in the choice of yellow-poplar, cottonwood, and red alder for core stock, while the acid-resistant quality of Port-Orford white-cedar was one of the reasons why it was the principal species used in the manufacture of battery separators, when sufficient supplies of this wood were available.

Good screw- and nail-holding ability is a prerequisite of wood used for boxes and crates, but really satisfactory wood for shipping containers should also be strong, light-colored, light in weight, and free from objectionable odors, tastes, and stains. Ease of working, combined with satisfactory finishing characteristics and the ability to take paint or other finishing materials, are important considerations in selecting wood for interior trim. Appearance and finishing qualities of woods dominate the choice of species for exposed parts in furniture; while ability to stay in place, and gluing and nail- and screw-holding characteristics are the deciding factors in selection of woods for concealed parts. The long, strong fibers from such conifers as the southern yellow pines, spruce, hemlock, and balsam firs are preferred in pulp manufacture; differences in the resin content of the woods, however, determine the method of pulping and the appearance and quality of the resulting pulp.

C. Variability of Properties within a Species

Not infrequently too much emphasis is placed on relatively small differences among species, while the variability in properties of wood of the same species is largely ignored.

Although lack of uniformity is a property common to all materials, wood exhibits a greater range of variability than other structural materials. Thus it is essential that not only the extent and the significance of variability in wood properties of the same species be known but also the possible limits of variation within the bole of a single tree. It is well recognized that variations within a species are largely traceable to environmental influences affecting the growth of trees, such as climate, soil, moisture, and growing space. Undoubtedly heredity also plays a dominant role.

The most readily recognizable instances of variability within a species are differences of appearance and weight, while differences in strength properties and workability, although more difficult to recognize, are frequently of much greater practical significance. Fortunately, however, strength properties of normal wood can quite often be closely correlated with specific gravity and with such easily observable factors as the rate of growth and the relative proportion of springwood and summerwood within a growth increment. The correlation of strength properties with the above-noted factors (particularly with specific gravity of normal wood) allows a fairly reliable basis for selecting the material best suited for a given use when strength is a factor of major consideration.

The variability of wood properties within a species is also greatly affected by differences in the moisture content and by the kind, number, extent, and location of defects. The effect of moisture on the strength properties of wood, as well as on shrinkage and swelling, general workability, and the nailing, gluing, and finishing qualities of wood, is so great that the satisfactory use of wood depends on a precise understanding of the altered behavior of wood accompanying changes in its moisture content. Although detailed discussion of this topic is outside the scope of this text, it may be noted in passing that as a general rule wood should be dried to a moisture content commensurate with the expected service requirements; i.e., wood should be dried to approximately the moisture content which it will attain in actual service. If the moisture content of wood is so great that a considerable amount of moisture is lost after the wood has been placed in service, corresponding shrinkage, warping, and other drying phenomena must be expected; wood dried to a much lower moisture content than necessary will pick up additional moisture from the atmosphere, with accompanying swelling and some loss of strength.

The extent to which various defects, such as knots, decay, cross grain, shakes, checks, splits, and sap stain, affect utility is considered in grading lumber. Grading is largely designed to cope with the effects of variability in the properties of wood of the same species. The underlying principles on which these grading rules are based, as well as the proper use of the grades, presupposes a thorough understanding of the nature of defects and their specific effect on the properties of a wood for given purposes.

It is important to recognize also that a considerable equalization of the properties of wood can be achieved by the design of an article and treatment of the wood. Not infrequently better service can be obtained from a well-designed article made of an inferior wood than from a poorly constructed item made of a species with superior characteristics. Differences in decay resistance can be compensated for by preservative treatments or by protecting wood against moisture. Adjustments in the speed of machinery or a change in the cutting angle of knives may assist materially in the production of acceptable stock from species with less desirable working characteristics. Substitution of glue for nails or screws may result in a stronger joint with a species of poor nail- or screw-holding ability. Good design and corrective treatments, therefore, may extend considerably the number of species that can be employed for all but the most exacting uses.

D. Chemical Properties of Wood

Utilization of wood frequently involves treatments which affect some or all of its chemical constituents. The chemical utilization of wood may be grouped under several headings: (1) pulping; (2) extraction; (3) hydrolysis, i.e., conversion of carbohydrates to sugars; (4) destructive distillation; and (5) various chemical treatments designed to improve the properties of wood. A general understanding of wood chemistry is essential to full appreciation of the basic principles underlying any type of chemical utilization. Therefore, the discussion which follows is offered merely for the purpose of acquainting the reader with a few of the basic principles of wood chemistry.

Each wood cell consists of a cavity surrounded by a rigid wall. This wall contains two morphologically distinct layers: the *primary wall* on the outside and the inner, *secondary wall.* All cells are cemented together by a common layer, the *intercellular substance,* also called the *true middle lamella.** The entire cell wall, including the intercellular layer, consists principally of cellulose, lignin, and hemicelluloses. The bulk of cellulose is found in the secondary wall, while lignin occurs largely in the intercellular layer and primary wall. Hemicelluloses are thought to be intimately associated with microscopic cellulose units (called *fibrils*), of which the bulk of the secondary wall is built. Some hemicelluloses are also found in spaces between the fibrils.

In addition to these three major chemical components, a number of extractable substances may also be present. *Extractives* reside largely in the cell cavities and may be removed with suitable solvents without affecting in any way the composition of the residual wall. Extractable materials include, among others, tannins, dyes, resins, essential oils, fats, nitrogenous organic compounds, and traces of organic acids and their salts. When burned, wood yields a small amount of mineral residue, called *ash.*

No exact figures indicating the composition of wood in terms of its different chemical constituents can be given. Such data as are available vary within wide limits. This is owing in no small measure to the methods of analysis employed and the interpretation of analytical data by individual investigators. The composition of wood, however, based on its ovendry weight, is approximately as follows: cellulose, 40 to 55 per cent; lignin, 17 to 35 per cent; hemicelluloses, 15 to 30 per cent; water extractives, from less than 1 per cent in aspen to more than 10 per cent in redwood; ash, about 1 per cent.

Much of the chemical utilization of wood involves *delignification,* i.e., the disintegration of the middle lamella through removal of a large part of the lignin and other noncellulosic materials. Wood pulp thus prepared forms an important raw material for the manufacture of paper, paperboards, and other wood-pulp products, while specially purified forms of pulp are used extensively in the manufacture of rayon and transparent films, and to a lesser extent in paints, lacquers, and similar products.

The carbohydrates in wood may be converted into sugars when hydrolyzed with dilute solutions of several mineral acids, and most of the resulting sugars can be fermented to ethyl alcohol. When decomposed by heat in the absence of air, a treatment called *destructive distillation,* wood will give rise to numerous chemical products, chief of which are methyl alcohol, acetic acid, tar products, and residual charcoal. Oils, wood turpentine, pine tar, and pitch are products of resinous-softwood distillation.

The industrial uses of wood noted above are based primarily on the utilization of the chemical components of the cell wall or their derivatives. An entirely different type of chemical utilization of wood involves removal of extractives such as tannins, essential oils, and dyes. Examples of such utilization are the extraction of tannin from wood of chestnut or the recovery of wood rosin and turpentine from the stump wood of southern pines

---

* Less technically, the term *middle lamella* is applied to the *intercellular substance,* or the *middle lamella* proper, or the *cambial,* or *primary,* walls combined.

by steam distillation. In both instances the extractive-free wood is frequently converted into wood pulp. In the majority of cases, however, no attempt is made to utilize the residual wood for anything but fuel.

1. Cellulose. Pure cellulose is a complex carbohydrate whose empirical formula is expressed as $(C_6H_{10}O_5)_n$. Cellulose is insoluble in water and in the ordinary solvents such as alcohol or ether. It is very resistant to the action of alkalies but can be completely dissolved by strong acids.* The purest form of cellulose known in nature is obtained from the seed hairs of the cotton plant, *Gossypium* spp., more than 90 per cent of which by weight is pure cellulose, or *alpha-cellulose*. In wood, cellulose, as already explained, is associated with other components, chief of which are lignin and hemicelluloses. Moreover, wood cellulose is not all alpha-cellulose, but has considerable admixtures of other cellulose-like materials, termed *beta-* and *gamma-celluloses*. These differ from alpha-cellulose mainly in being soluble in alkaline reagents. When unpurified wood cellulose is treated with a reagent such as 17.5 per cent sodium hydroxide, the cellulose swells and a part of it dissolves. The dissolved portion contains the beta- and gamma-celluloses.† Wood cellulose prepared by commercial methods of pulping may lose as much as 10 to 20 per cent by weight when treated with an alkali, owing to removal of the beta and gamma fractions. The purity of wood cellulose is judged by its alpha-cellulose content, and for many uses, e.g., in the manufacture of rayon, wood cellulose is carefully refined with alkalies to remove its nonresistant portions.

2. Hemicelluloses. This is a term applied to a group of cell-wall carbohydrates found in more or less intimate association with cellulose and lignin. Technically, the hemicelluloses may be loosely defined as those less resistant substances which, though insoluble in hot water, can be removed from the cell wall with either hot or cold dilute alkalies or readily hydrolyzed to constitute sugars or sugar acids by means of hot dilute acids. Theoretically, when hemicelluloses are thus removed from delignified cell walls, only cellulose is left behind. Actually, however, some of the hemicelluloses are water-soluble, ‡ and not all of them are extracted from the cell wall by alkali treatments. In chemical pulping, hemicelluloses are largely removed from the pulp. If acid cooking liquors are employed, the hemicelluloses are converted into sugars (such as galactose, mannose, and others), which remain in solution in the cooking liquor. Approximately 60 to 75 per cent of these sugars could be fermented into alcohol. § If alkalies are used for pulping, the hemi-

---

* Another well-known cellulose solvent is Schweizer's cuprammonium reagent.

† Beta- and gamma-celluloses can be separated by treating the alkaline solution with acetic acid; the beta-cellulose will then form a precipitate, while gamma-cellulose will remain in solution. Some chemists believe that beta- and gamma-celluloses are hemicellulosic materials.

‡ A polysaccharide, galactan, found in the wood of western larch (to the extent of 8 to 25 per cent of the dry weight of the wood) may be considered one of these. The galactan, extracted with water from the wood, can be hydrolyzed into a sugar—galactose —and this, in turn, can be oxidized to mucic acid, which is used in the manufacture of baking powder.

§ Another sugar, xylose, is converted in the course of acid cooking into *furfural*, which ordinarily escapes with the steam. However, under suitable conditions it could be

celluloses are converted into acids (largely glucuronic and galacturonic). The function of hemicelluloses in nature is not yet understood, although a theory has been advanced suggesting that they may be the precursors of lignin, i.e., substances from which lignin is formed.

3. Lignin. This is the third important constituent of the woody cell wall. Its exact chemical composition is still unknown. It can be obtained from wood either by removing all carbohydrates with a strong acid or by removing the lignin and some of the more soluble carbohydrates and leaving the resistant cellulose behind and then separating lignin from the resulting solution. Still another method calls for dissolving all constituents of wood in a mixture of formaldehyde and sulfuric acid and then precipitating lignin by diluting the acid solution with water. In any case, it requires very drastic treatment to extract lignin from wood, and there is no certainty that the lignin in its extracted form is at all identical with that occurring naturally in the cell wall. Furthermore, no two lignin preparations have the same physical properties or the same chemical reactions; likewise, there is considerable evidence to show that the lignins from different species of wood are not identical. This has led to the conjecture that "either the substance known as lignin is a complex mixture of compounds with similar properties but of unrelated chemical structure, or else a mixture of compounds which are similar in structure or have the same basal form but which vary in minor ways." [6]

Whatever its exact chemical nature, lignin will remain a term of convenience to describe the cementing material which is largely removed during chemical pulping of wood. Lignin extracted from wood in this manner constitutes one of the greatest industrial residues of the present time.*

Among industrial uses of wood in which lignin is utilized along with other wood constituents are destructive distillation derivatives and lignin plastics. In the former, the lignin fraction of wood gives rise to methyl alcohol, acetic acid, and a portion of the complex mixture of tarry and oily substances; in the latter lignin serves as a binding agent. A number of aromatic substances are also associated with lignin, among them vanillin and eugenol (the flavoring principle of cloves). Manufacturers of storage batteries incorporate lignin as a part of negative battery plates, with a resulting tenfold increase of current output in zero weather. Purified lignin, hydrogenated in the presence of a catalyst, has been converted by Drs. S. C. Sherrard and E. E. Harris, of the U.S. Forest Products Laboratory, into a thick, viscous, colorless fluid from which, by means of fractional distillation, they have produced methyl alcohol and four new substances. Two of these can be made into plastics; another can be used as a solvent for organic gums, resins, and oils, and has a preservative value; the fourth has adhesive properties. The utilization of extracted lignin is discussed in more detail in later pages under the utilization of sulfite liquor (see Chap. 17). Any real progress toward utilization of the vast amount of lignin available cannot, however, be expected until its chemistry is better comprehended.

---

recovered as one of the products of wood hydrolysis. Furfural is an important solvent and intermediate reagent in the chemical industry.

* Pulp-mill lignin losses amount to several million tons annually.

4. Extractives. Wood extractives, also called the *extraneous components* of wood, are substances which can be extracted from wood by means of suitable solvents or by steam distillation without affecting the composition of the cell wall. Wood extractives, therefore, are not considered to be an integral part of the cell wall, but rather are thought to reside largely in the cell cavities. Most important of the wood extractives are essential oils, resins, fats, tannins, and natural dyestuffs or their precursors. Small amounts of organic acids or their salts are also present in many woods. The most widely distributed acid in plants is oxalic acid, which in wood is usually found in the form of crystals of calcium oxalate. A few woods are characterized by the presence of alkaloids. The wood of *Cinchona* spp., for example, contains small amounts of quinine and related alkaloids, while traces of strychnine have been isolated from several woods of the genus *Strychnos*. Finally, small quantities, about 0.3 per cent, of other nitrogenous compounds, thought to be traceable to proteins in the dried protoplasm, are also found in some woody cells. Nearly all woods have small deposits of starch; some also contain sugars.

Some of these substances, notably tannins, resins, essential oils, and dyestuffs, have considerable economic importance and in a few cases may even constitute the chief value of a given wood. A few examples are described in later chapters. By and large, however, no attempt is made to recover these materials. The largest waste of various extractives occurs in the preparation of wood pulp, where they are removed and usually discarded in the spent cooking liquor, together with lignin and the hemicelluloses.

5. Inorganic Components. The residue which remains after complete combustion of wood is called *ash*. In the North American woods analyzed by the U.S. Forest Products Laboratory, the ash content was found to range between 0.2 and 0.9 per cent. The highest ash content reported elsewhere was for the European olive tree, *Olea europea* L., in which sapwood yielded about 5 per cent. The principal components of wood ash are salts of calcium, potassium, and magnesium; small amounts of sodium, aluminum, iron, and manganese sulfates, chlorides, and silicates are also invariably present. Perhaps the most characteristic component of wood ashes is potash (K_2CO_3). The presence of this salt accounts for the once widespread use of wood ashes for fertilizer and as a raw material in making soap.

In concluding this discussion on the relationships between wood properties and wood utilization, it should be apparent to the reader that many of the present uses of wood are the direct result of research which has led to better understanding of some of its many unique properties. Nevertheless, other potential uses of wood have either gone unrecognized or are exploited in only a very limited manner because several of its fundamental properties are not even yet fully understood or appreciated. For example, drying, preserving, and even finishing procedures and techniques might be greatly revolutionized if we had a complete understanding of the forces and mechanics involved in the movement of moisture in living trees. The chemical utilization of wood offers a promising future, but there is yet much basic information to be developed before its impact can be fully appreciated. The lignin complex, for example, is still somewhat of a chemical enigma. A determination of the precise nature of the chemical bond within the cellulose fibril will provide a basis for the development of good wood-stabilization techniques. A knowledge of the nature of the bonds between cellulose and lignin and cellulose and hemicellulosic materials is also a field for fruitful research. These are but a few of the many basic problems confronting the researcher. Wood will continue to play an important role in the American economy, but its ability to compete with new and improved materials for new markets lies in the hands and skills of wood technologists, chemists, and engineers seeking basic information that will provide a firm foundation in new and improved wood products and process developments.

SELECTED REFERENCES

1. Anon. Wood Handbook. U.S. Department of Agriculture, Forest Service, Forest Products Laboratory, Handbook No. 72. 1955.
2. Anon. Timber Design and Construction Handbook. F. W. Dodge Corporation, New York. 1956.
3. Brown, H. P., A. J. Panshin, and C. C. Forsaith. Textbook of Wood Technology. McGraw-Hill Book Company, Inc., New York. Vol. 1, 1949; Vol. 2. 1952.
4. Harrar, E. S. Hough's Encyclopaedia of American Woods, Vol. 1, 2. Robert Speller & Sons, Publishers, New York. 1957.
5. Koehler, A. The Properties and Uses of Wood. McGraw-Hill Book Company, Inc., New York. 1924.
6. Norman, A. G. The Bio-chemistry of Cellulose, Polyuronides, Lignin, etc. Oxford University Press, New York. 1937.
7. Stamm, A. J., and E. E. Harris. Chemical Processing of Wood. Chemical Publishing Company, Inc., New York. 1953.
8. Tiemann, H. D. Wood Technology, 2d ed. Pitman Publishing Corporation, New York. 1944.
9. Wise, L. E., and E. C. Jahn. Wood Chemistry, 2d ed., Vol. 1, 2. American Chemical Society Monograph 97. Reinhold Publishing Corporation, New York. 1952.

PART II

WOOD PRODUCTS

CHAPTER 4

ROUND TIMBERS

I. POLES

The first commercial telegraph system in the world, erected in 1844, was a line of some 40 miles in length extending between Washington, D.C., and Baltimore, Maryland. The wires connecting these two cities were carried on poles cut from tree-length logs, the harvesting of which marked the beginning of the wooden pole industry in the United States. A few years later when the first central telephone system was put into operation in New Haven, Connecticut, pole lines were found to be the most economical means of connecting the exchange with its 21 original subscribers. The first use of poles in electric power transmission lines occurred in 1882 when Thomas A. Edison constructed a utility station in New York City and sold electric power to nearly 100 consumers who had installed lighting circuits in their homes and/or business establishments.

By the turn of the twentieth century the nation's annual pole requirements were being counted in the millions. Today the more than 80 million poles in service support a vast labyrinth of power transmission and communication wires and cables which not only link the nation's thriving nerve centers but also extend into the most remote of rural areas.

There are currently some 5 to 5½ million poles harvested annually. Many of them are used to replace those no longer capable of rendering satisfactory service, while many more are used in the erection of new lines. The growing popularity of the economical and efficient pole-type construction, of which more will be said later, has developed a new and expanding market for treated poles.

A. Factors Governing the Selection of Poles

Several of our commercial timber-producing species are, for one reason or another, unsuited for the manufacture of poles. Bole form is especially important, and trees which do not develop characteristically straight, full-rounded shafts with a minimum of taper are generally avoided. Moderately lightweight timbers possessing high strength-weight ratios are preferred.

Poles produced from timbers of high density usually possess excellent mechanical properties, but their excessive weight adds to their harvesting, transportation, and erection costs, and in consequence they are seldom used in any appreciable numbers. Timbers of low density are unsuited for poles, as they are wholly incapable of supporting the heavy loads required in many lines, but especially in regions where severe icing along the wires is a common occurrence during the winter months.

The vast majority of poles now placed in service are partially or wholly impregnated with some kind of toxic prior to their installation. Thus, it is highly desirable that they be produced not only from timber of required strength but also from those woods which are also known for their ability to accept and retain adequate amounts of a prescribed preservative material. When specifications call for the use of untreated material, a practice which is still followed in some localized areas, only those pole woods well known for their excellent natural durability should be employed. While the initial costs of untreated poles may appear to be very attractive, such items often prove to be the most expensive because the frequency with which they must be replaced increases maintenance costs beyond all reasonable limits.

Finally, pole timbers must be readily accessible and available in quantities sufficient to harvest, impregnate, and market at reasonable prices.

B. Polewood Species

One privileged to examine pole lines erected prior to the First World War would find that a large variety of timber species were then in use. Chestnut, usually untreated, was one of the prime favorites throughout the whole of the Eastern United States. Baldcypress, Atlantic white-cedar, and southern yellow pine poles * were dominant in the South. Lines in the Pacific Northwest and the Rocky Mountain region featured western redcedar and Douglas-fir, while both round and saw-tapered poles of redwood were prevalent in California. Installations in the Lake states were predominantly of northern white-cedar and tamarack; in the southern Plains states poles of osage-orange were not infrequently used. Other species to be found in appreciable numbers included catalpa, oak, ash, elm, butternut, red mulberry, and in some localized areas even cottonwood, sassafras, and spruce.

During the past quarter of a century, however, hardwood poles for use in communication and power transmission lines have been virtually eliminated. This has been occasioned by a series of events the most significant of which have been:

1. The harvesting of all residual blight-killed chestnut trees in pole sizes
2. Rising stumpage prices paid for hardwoods by other industries whose products bring higher net returns
3. The facility with which most coniferous woods can be made to accept preservative compounds, and the now almost universal insistance that poles receive some form of treatment prior to use
4. And, perhaps foremost, industry and consumer acceptance of care-

* Loblolly, shortleaf, longleaf, slash, and pond pines are all used in pole production.

fully devised standards and specifications governing materials, sizes, and the manufacture of all poles destined to receive a preservative treatment (see page 63)

Poles are currently manufactured primarily from the southern yellow, lodgepole, ponderosa, red, and jack pines; Douglas-fir; western redcedar; several of the western balsam (true) firs; western larch; western hemlock; and northern white-cedar. A few other species still enjoy a very limited local use, but those enumerated above supply at least 97 per cent of the nation's pole requirements.

1. Southern Yellow Pines. The use of southern yellow pine poles has increased steadily since 1915, and this group of pines currently accounts for about 75 per cent of the poles produced annually. These poles are procured throughout the southern pine belt, usually either from scattered tracts of second-growth timber or from small farm woodlots. Loblolly and shortleaf pines are preferred since all southern pine poles are given full-length preservative treatments, and pole-sized trees of these two species usually contain large volumes of sapwood, which is readily amenable to treatment. Few producers own standing timber. For the most part they deal directly with forest owners for those trees which, in their judgment, are suitable for pole manufacture in the lengths and in the classes for which they have markets.

The harvesting of poles in the southern pine belt is essentially a year-round undertaking except in those areas where the incidence of bark-beetle outbreaks is inordinately high following logging operations. In such instances cutting is usually restricted to the winter months.

2. Douglas-fir. The use of Douglas-fir poles has increased steadily since 1906. For many years, however, an ample supply of the more durable, lighter-weight western redcedar in the same producing areas limited the production of Douglas-fir poles. At the present time this is the second most important polewood species and supplies about 7 per cent of the nation's annual pole requirements. Prior to the Second World War consumption of Douglas-fir poles was largely restricted to the Pacific Northwest. Demands for poles in greater lengths than those capable of being produced in the southern pine belt, however, have resulted in markets for these commodities throughout the United States.

Douglas-fir poles are produced in both the Pacific Northwest and in the Inland Empire. As a rule they are harvested as a separate enterprise in advance of heavy logging operations.

3. Western Redcedar. Western redcedar is a tree of gigantic proportions endemic to the Pacific Northwest and montane regions of northern Idaho and western Montana. Most of the western redcedar poles are produced in the Idaho panhandle, however, where the trees are of a size and shape especially well suited for pole production. Several thousand cedar poles are also harvested annually along the western slopes of the Cascade Mountains in Washington and Oregon, where the pole producers remove the younger and smaller trees from uneven-aged stands of mixed coniferous growth as a separate operation, usually in advance of a concentrated lumbering enterprise. Poles 130 ft or more in length are obtainable in this region when desired. This cedar is currently the third most important polewood species and annually supplies about 6.5 per cent of the nation's needs.

4. *Lodgepole Pine*. The lodgepole pine, a tree of wide distribution in western North America, was initially used for poles in 1923 when the Mountain States Telephone and Telegraph Company began experimenting with them in its lines. The Rocky Mountain form of this tree is especially well suited for the production of poles, and while the wood is not particularly resistant to decay, it has been found that a simple open-tank butt-treatment will assure a service life of at least 20 years for a pole placed in the more arid areas of the Rocky Mountain region. Lodgepole pine is presently contributing only 4 per cent to our pole supply; however, it is certain to preempt a position of greater importance in the near future.

5. *Northern White-cedar*. The production of northern white-cedar poles has declined rapidly in the last 10 years, and less than 100,000 are now being produced annually. The heartwood of this species, like that of its western congener, is exceedingly durable in contact with soil, and records reveal that untreated poles have given many years of satisfactory service. The bulk of the poles currently harvested, however, are butt-treated.

6. *Other Species*. Several other species in addition to those enumerated above are used to produce a limited number of poles annually. Western larch, because of its size, strength, and desirable form, has appeared on the market in ever-increasing numbers since the Second World War. Ponderosa pine, the western balsam (true) firs, and western hemlock are others which, because of their form, strength, and ease of treatment, have invaded the pole markets in recent years.

C. Seasoning of Poles

In a normal woods operation the trees selected for pole production are felled, trimmed of all side branches, cut to specified lengths, and barked. Bark removal facilitates rapid drying and thus minimizes the threats of insect attack and blue-stain infection. In most areas the peeled poles are gathered, bunched, and then trucked to concentration yards for seasoning or hauled to railway sidings for delivery to seasoning yards at treating plants. In some Idaho operations western redcedar poles are floated to the yards in flumes.

Air-seasoning is practiced at most facilities. Upon their delivery at the yards, the poles are segregated by lengths and classes and placed in well-spaced, stickered tiers, much in the manner of air-seasoning green lumber, and permitted to dry for 2 to 9 or more months depending upon species and locality. Open sheds or temporary roof decks are sometimes used to protect the piles from rain and the direct rays of the sun. Especial care is taken to provide adequate movement of air through the piles, as improper ventilation not only prolongs the period of seasoning but also abets conditions favorable for decay.

The introduction of the vapor-drying process by Dr. M. S. Hudson in 1942 has provided a rapid and inexpensive means for seasoning timbers in large sizes. It is especially well suited for drying poles that are to be given a full-length preservative treatment, since both procedures can be accomplished using the same chamber. Briefly, vapor-drying consists of placing green wood in an enclosed vessel and exposing it to hot vapors of such organic solvents as xylol or Stoddard fluid. These gases heat the wood and cause it to give up some of its included water in the form of vapor. This vaporized moisture together with the inert gas are drawn off continuously and led into a condenser, where they are liquefied, then drained off into a separatory tank. The water is bled off through a metering device on its way to a sewer, thus permitting an operator to determine the drying rate of the charge. The solvent, on the other hand, is revaporized and returned to the drying chamber. Southern yellow pine poles may be satisfactorily dried in a matter of 10 to 12 hours, and with the use of this process the necessity for tying up substantial sums of working capital for several months in large inventories of slowly seasoning poles is virtually eliminated.

D. Finishing Poles

1. Shaving. The general appearance of poles may be materially improved by subjecting them to a mechanical shaving operation. Shaving not only removes the last vestiges of inner bark left on the surface but also smooths off knots, burls, and other irregularities. The surfaces of shaved poles also permit better and more uniform distribution of preservative while poles are in the treating cylinder.

Two basic types of pole-shaving machines are in current use. With one of them a pole is set into rotation as it is slowly advanced under a series of cutterheads mounted on long actuating arms (Fig. 4-1). With the other type a pole is fed without rotation through a battery of cutterheads which revolve about the pole. The depth of cut for either type of machine is easily regulated, and no more wood is removed than is ordinarily necessary to produce clean, smooth, cylindrical shafts.

2. Framing. Pole framing consists of preboring all bolt holes and cutting the gains and roofs. Gains are the flat recesses let in near the top of a pole and used for seating the crossarm. In practice these are always placed on the face * of a pole. A pole is roofed by making a slanting cut across its top, or by cutting a pair of opposing angled surfaces to form an obtuse, double-beveled peak.

The American Standards Specification and Dimensions for Wood Poles, Designation 05.1—1948, decrees that all poles to be given a full-length preservative treatment shall be framed within the terms of the purchaser's contract prior to treatment. This practice gives assurance that there will be no exposure of untreated wood below the depth of impregnation, a condition which is often unavoidable when pole framing is done after treatment.

3. Incising. Poles which are given only an open-tank butt-treatment before emplacement are usually incised from 12 in. above their predetermined ground line to at least 24 in. below it. Incising is a mechanical operation consisting of punching slit-like holes, ¾ in. long, ⅛ in. wide, and ½ in.

* The concave side of a pole, or that side of greatest curvature having reverse or double sweep between the top and ground line, is designated the face of the pole.

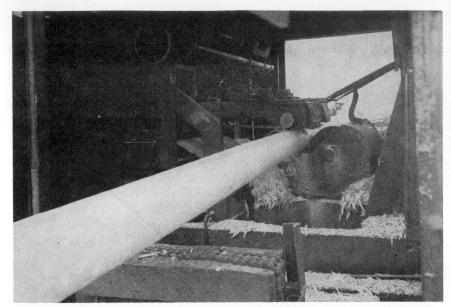

Fig. 4-1. Typical pole-shaving operation. (*Courtesy American Wood-Preservers' Institute, Chicago, Illinois.*)

Fig. 4-2. Incising the ground-line region of a western redcedar pole. (*Courtesy Valentine Clark Corporation, St. Paul, Minnesota.*)

deep, at regular and frequent intervals about the pole's surface within the linear limits indicated above (Fig. 4-2).

E. Preservative Treatment for Poles

Materials used in the preservation of wood fall into one of two broad classes, viz.: (1) oil and oil-borne chemicals and (2) water-borne salts. Standard preservative oils include creosote, creosote-coal tar formulations, creosote-petroleum blends, and pentachlorophenol. Creosote is an inexpensive and highly effective toxic and the one most widely used in impregnating poles. Water-borne salts may be used when poles will not be subjected to leaching and where cleanliness, paintability, and lack of odor are essential requirements. Pentachlorophenol is an excellent preservative for wood poles and posts. The surfaces of poles treated with this toxic are dry, clean, and nonoily; may be painted with facility if so desired; and are especially well suited for buildings and pole-type construction.

Ordinarily all pine poles, as well as those produced from Douglas-fir, western larch, and other pole timbers known for their relatively perishable nature, are subjected to full-length, standardized pressure treatments as prescribed by the American Wood-Preservers' Association.* However, it has been found that lodgepole pine and western larch poles placed in the drier areas of the Rocky Mountain region give many years of satisfactory service when merely butt-treated in accordance with recommended procedures.

Unless otherwise specified, one of the empty-cell methods of treatment such as the Rueping or the Lowry process is used when poles are given a full-length impregnation. When customer specifications call for a full-cell treatment, either the Bethell or Burnett process is recommended.   Table 4-1

Table 4-1. A.W.P.A. Minimum Acceptable Retentions of Preservatives for Poles

Species	Retentions of wood, lb/cu ft		
	Creosote	Creosote-petroleum	Pentachloro-phenol
Southern yellow pines.........	8	..	0.4
Douglas-fir:			
Coastal form..............	8	8	0.4
Intermountain form........	8	..	0.4
Lodgepole pine..............	6	..	0.3
Western larch..............	6	..	0.3
Jack pine...................	6	..	0.3
Red pine...................	6	..	0.3
Ponderosa pine.............	8	..	0.4

* The reader is referred to A.W.P.A. Standards C-1 and C-4, respectively, for details of acceptable practices in pressure treating poles.

gives the A.W.P.A. minimum acceptable retentions of preservatives for the principal polewoods when subjected to empty-cell treatments.

Untreated poles of the species enumerated in Table 4-1, depending on their geographic location and site of erection, cannot be depended upon to last for more than 2 to 10 years. A life expectancy of at least 35 to 40 years may be reasonably anticipated, however, for poles treated in accordance with recommended practices.

Poles produced from timber exhibiting natural resistance to the depredations of fungi and insects do not require full-length pressure treatments to

Fig. 4-3. Open-tank treating plant for western redcedar poles. Note shaved ground-line region of poles in foreground. (*Courtesy J. P. Wentling.*)

assure long service life. Western redcedar is such a timber. In fact, untreated poles of this species placed in service more than 30 years ago are still in a satisfactory state of repair. The sapwood zone of pole-sized trees of this species is seldom more than ⅝ in. wide, and while this zone may be destroyed at and near the ground line in a very few years, the heartwood, which constitutes more than 90 per cent of the volume of a pole, is seldom molested. Nevertheless it is now common practice to give western redcedar poles * at least butt-treatment in creosote oil in accordance with A.W.P.A. Standard C-7.† Accordingly, poles that have been properly seasoned, shaved, and incised through the ground-line zone, are stacked upright in an open tank (Fig. 4-3), and their butt ends permitted to steep for no less than 6 hours in a bath of hot creosote oil held at temperatures ranging from 190 to 235°F. The level of the oil is so maintained that the incised ground-line area is under immersion at all times. Following this cycle, the hot oil is either (1) quickly replaced with cold oil, and the poles held in this bath for 2 additional

---

* Also northern white-cedar.

† Standard for the Preservative Treatment of Incised Pole Butts by the Thermal Process, Western Redcedar and Northern White-cedar.

hours at temperatures ranging from 150°F to that at which solids begin to form in the bath, or (2) allowed to cool for not less than 2 hours during the final 10 minutes of which the temperature shall range from 150°F to that at which solids begin to form. Finally, at the end of cold or cooling cycle, hot oil not exceeding a temperature of 230°F is pumped into the tank and the poles steeped for 1 additional hour. This treatment is usually adequate to obtain complete penetration of the toxic in the incised sapwood zone with oil retentions often amounting to as much as 35 lb per cu ft. When customers demand full-length treatment for cedar poles, provision is made at the treating plant for total immersion during treatment (see A.W.P.A. C8—56, Standard for the Full-length Thermal Process Treatment of Western Redcedar Poles).

F. Pole Specifications

In 1941 the American Standards Association released a series of standard specifications and dimensions for each of the major polewoods then in common use. These standards served as a guide in the production of poles and were often incorporated into a purchaser's contract as a specification governing acceptance of all pole shipments. Several years later (1948) these standards were superseded by a single, all-inclusive standard designated as 05.1—1948 and titled American Standard Specifications and Dimensions for Wood Poles, in which several new polewood species were included.* In the acceptance of this revision recognition was made of the fact that there were a number of inconsistencies in the methods employed for rating various species of wood poles and that there was also a dearth of reliable mechanical data upon which to establish working stresses for some of the new species included. This situation prompted the American Society for Testing Materials Committee D-7 on Wood to organize a Subcommittee on Wood Poles and Cross Arms that would have as its responsibility the planning and execution of an exhaustive research program in which standard methods for testing wood poles would be developed. Two task-force groups were appointed within the committee, one to plan and direct the technical features of the program, the other to provide adequate financing for the project. After nearly 6 years of careful planning and programming the research got under way in February, 1954, with the testing of the first 30-ft Class 6 western larch pole. During the second period of 6 years tests were conducted on more than 600 full-sized poles and on over 14,000 standardized clear-wood test specimens. The results of this research program, probably the most comprehensive of its kind ever undertaken, are embodied in the final report of A.S.T.M.'s wood pole research program.† While a new commercial standard for wood poles has not yet been released, this report will without doubt be the basis for its revision when it appears. A few of the most important conclusions derived from this study are summarized as follows:

* American Standards Association Standard 05.1—1948. Available through the A.S.A., New York, N.Y.

† L. W. Wood, E. C. O. Erickson, and A. W. Dohr. Strength and Related Properties of Wood Poles. American Society for Testing Materials, Philadelphia, 1960.

1. The strength of a pole is related to the specific gravity of the wood.
2. The stiffness of a pole may be represented by the compression modulus of elasticity of the wood in it.
3. The modulus of rupture in small poles is less than that in larger ones.
4. Southern pine poles of rapid growth and low specific gravity for the species are below average strength for the species.
5. There is a significant strength correlation between untreated poles and small clear-wood test specimens, and between treated poles and small clear-wood test specimens.
6. Natural defects, such as knots and spiral grain, as presently limited in American Standards Association specifications are not an important factor in controlling the strength of poles as they are normally used.
7. Preservative treatment has no consistent effect on the surface hardness of a pole.
8. The bending strength of southern pine poles is reduced 20 to 40 per cent by steam conditioning and treatment within the limits of 1960 American Wood-Preservers' Association standards.
9. Reduction of strength resulting from conditioning and treatment of southern pine was greater in poles than in small clear-wood test specimens taken from poles.
10. A supplementary series of strength tests on treated longleaf pine revealed that lowering the temperature of steam and shortening the steam-conditioning period gave strength values closer to those of untreated poles.
11. Treating-plant experiments indicate that reduced steaming temperatures of 240 to 245°F may be used successfully on southern pine poles.
12. The bending strength of Douglas-fir and western larch test poles was reduced about 10 per cent by Boultonizing and treatment within the limits permitted by A.W.P.A. 1960 standards.
13. Lodgepole pine and western redcedar showed no significant change in strength following air-seasoning and treatment in accordance with A.W.P.A. 1960 standards.
14. Reintroduction of moisture into poles that were treated after air-seasoning proved to be difficult.

G. Pole-type Buildings

During the past decade or so, a new concept in design has been employed in the framing of a wide variety of farm and industrial buildings fabricated wholly or in part of wood. Known as pole-type buildings, this new method of construction has been found to be especially well suited for the erection of such structures on the farm as barns, granaries, cribs, livestock shelters, and poultry houses, as well as for a variety of industrial sales and services buildings, canneries, boat shelters, and large warehouses, often of 300 ft or more in length. Unlike their conventional counterparts, which rise from foundations of brick or masonry, pole-type buildings consist of numerous full-length, pressure-treated * poles of predetermined classes and lengths firmly anchored in the ground at specified distances from one another. These poles, when in position, are the main structural components of a building, as the siding (of whatever form), bracing, and rafters are firmly fastened directly to them using ring-shanked nails. No foundations are required, and flooring, when it is used, is independently supported by the ground. Because they are firmly anchored in soil, pole-type buildings are much less apt to be destroyed by winds of hurricane force or by flash floods than those rising from conventional foundations.

Fig. 4-4. Pole-type construction of a 32 by 32-ft hay shed. Built with sixteen 25-ft southern yellow pine poles pressure treated with creosote to final retention of 8 lb per cu ft. (*Courtesy W. A. Stacey, American Wood-Preservers' Institute, Lawrence, Kansas.*)

Pole-type buildings are appearing rapidly and in large numbers in many parts of the country, as this type of construction can be done rapidly and at very substantial savings in both materials and money. Eliminated are expensive foundations, systems of bracing, and the high costs of skilled labor. Many farm buildings can be erected at less than $1 per square foot of enclosed area, while industrial structures requiring heavy slab floors often cost less than $2 per square foot. Figs. 4-4 and 4-5 depict two pole-type buildings in common use.

H. Painted Poles

Treated poles ordinarily present an acceptable surface, but when desired they can be painted as readily as untreated materials. Painting is seldom resorted to for merely decorative effects, but rather when a high degree of visibility is mandatory. Salt-treated poles can be painted with any paint of

---

* Poles treated with either creosote or pentachlorophenol are used, the choice being usually a matter of customer preference.

Fig. 4-5. Erecting a 71 by 300-ft pole-type construction storage building for shelled grain. Poles and basal wallboards pressure treated to final retention of 0.4 lb of pentachlorophenol salt per cu ft. (*Courtesy W. A. Stacey, American Wood-Preservers' Institute, Lawrence, Kansas.*)

quality once the surfaces have been thoroughly dried. The dried surfaces of poles containing creosote or an oil-borne preservative should be primed with aluminum powder in an approved vehicle before painting. A good grade of paint in any color may then be applied.

I. Fish-net Poles

A small but locally important market exists in some areas in the northeastern United States for fish-net poles. These are employed by commercial fishermen in offshore waters, tidal basins, and a few large fresh-water lakes to support pound and gill nets. Such poles seldom exceed a length of 30 ft. They are usually produced from any one of several available species, but production seldom exceeds 10,000 annually.

J. Furnace Poles

About 50,000 green hardwood poles are consumed annually in the refinement of copper along the Atlantic seaboard from New York to Maryland. When the poles are placed into the molten metal, a chemical reaction occurs that removes oxygen and other impurities from the smelt.

Furnace poles vary from 25 to 45 ft in length. No restrictions are placed upon them with respect to the number and kind of defects, but it is mandatory that they be in a green condition at the time of their delivery. Poles of this sort are sold on a weight basis and bring from $7 to $10 laid down at the refinery.

K. Future of the Pole Industry

In many of our metropolitan areas, particularly in congested business districts, poles have been replaced in favor of underground conduits. Ornate reinforced concrete and metal poles support the lighting systems in some of our larger cities, while high-power transmission lines thread their way across the breadth and length of our land on poles and towers of latticed steel. There is little likelihood, however, that these fabricated structures will eventually replace the wood pole, for in the last analysis properly treated wood poles are still the least expensive means for supporting wire and cable.

The annual rate of pole replacements in existing power and communication lines, and the erection of new ones to meet the insatiable demands of a rapidly growing urban and rural population, together with an expanding pole market in the construction field, are ample evidence that pole production will be maintained at current or possibly at slightly higher levels in the foreseeable future.

II. ROUND-TIMBER PILES

Round-timber piles are essentially poles used as columns for the support of heavy structural loads. *Foundation* piles are those driven into the ground in excavations and capped with masonry to provide adequate foundations

Fig. 4-6. Pilings for a bridge support. Note those at left to be driven and those at right in final position. (*Courtesy American Forest Products Industries, Inc.*)

for heavy structures resting on soils that of themselves are wholly incapable of such support. The term *timber* piles, by contrast, is usually applied to the round, vertical components of docks, wharves, trestles, and similar structures, where the greater length of each pile rises above the ground line, or where the piles merely rest on stone or masonry. *Piling* is a collective term connoting an indefinite number of piles or a framed structure fabricated principally with pile timbers.

The use of piles actually antedates the recorded history of man. Many an aborigine, in an effort to secure maximum security for himself and his family, erected his crude abode on piles over the waters of shallow lakes and deep swamps, evidences of which are still to be observed in some of the montane lakes of Switzerland. Piles have been widely used in Europe since early medieval times, and particularly in the support of buildings in the cities of Holland and in Venice. And in modern times the towering structures of brick and steel rising into the skies over New York, Chicago, and other large metropolitan areas rest upon a dense forest of round-timber piles.

Production of piling in the United States during the past few years has exceeded 20 million units annually. Pile timbers are harvested, peeled, and seasoned much in the manner of poles. Prior to use they are given a full-length preservative pressure treatment in accordance with provisions set forth in A.W.P.A. Standards C-1 and C-3, respectively, thus giving assurance of a long and satisfactory service life.

A. Factors Governing the Selection of Pile Timbers

The same general considerations as are employed in determining the suitability of a species for use as poles may be applied in piling. In addition, pile timbers must also possess good shock resistance and the ability to absorb heavy hammer loads imposed upon them during emplacement. End-brooming resulting from the pounding of a hammer is often excessive for some of the softer woods, but this type of failure can be largely overcome by placing steel collars about the head of a pile before it is driven. Piles with excessive sweep are objectionable because they are difficult to drive and only too often incapable of supporting loads expected of them. A rule-of-thumb inspection procedure is to reject those with sweep when a line projected from the top to the bottom centers of a pile falls outside anywhere along its length.

B. Species Used

The southern yellow pines currently account for about 61 per cent of the pile timbers produced in the United States. Douglas-fir is the second most important species and contributes about 26 per cent of the supply annually. The residual 13 per cent is composed of several species including oak and a few other hardwoods; red, ponderosa, and jack pines; western redcedar; and larch.

Mention should also be made of two exotic timbers which enjoy use as timber piles in marine construction. The first, Demerara greenheart (*Nectandra rodioei* Schomb.), a heavy yellowish-green wood often mottled with brown or black pigmentation, is indigenous to British Guiana and several other South American countries. It is well known for its resistance to attack of marine borers and has been used in both European and American tidewaters with equally satisfactory results. Fender systems, dolphins, ferry slips, and wharves are but a few of the many sorts of marine installations fabricated with greenheart timber piles. No preservative treatment is needed for the wood of this species.

Angelique (*Dicorynia paraensis* Benth.) is another South American tree productive of timber notable for its natural resistance to marine worms. Heavy, and russet brown in color, this wood possesses an exceptionally high silica content, which, it is believed, is a deterrent to the destructive activities of marine organisms. Angelique piles are largely produced in French Guiana, where there is an abundance of pile-sized trees proximate to salt water. Marine installations of this timber in the warm, teredo-infested waters of the Canal Zone evinced little or no deterioration after 15 years of exposure, and since the Second World War, angelique timber piles are being used in ever-increasing numbers along the water fronts of both the American and European continents. Like greenheart, this timber requires no preservative treatment.

C. Specifications for Round-timber Piles

In 1927 the American Society for Testing Materials issued Standard Specification for Round Timber Piles, under the designation D25—27. In the ensuing years it has been subject to a number of revisions the latest one of which appeared in 1958 as D25—58. This same specification has also been approved by the American Standards Association under the A.S.A. designation 06.1—1959. Both documents are available for purchase at nominal cost.

III. POSTS

It has been reliably estimated that more than 600 million posts are required annually to meet the national needs. More than 450 million of these are used in the continual erection and maintenance of fence lines, as fencing has been found to be the only practical means for area control on farms and ranches and along rights of way. Nearly another 100 million are utilized for guard posts and as supports for the array of directional and regulatory traffic markers that characterize the nation's highways.

In the days before preservatives were used to prolong the life of posts, it was customary in so far as possible to use only those species of proved durability. Chestnut, black locust, osage-orange, and most cedar posts could be relied upon to last for 20 to 30 or more years provided they were constituted principally of heartwood.* Such posts even after their thin peripheral layers of sapwood had been destroyed in the region of the ground line gave adequate service for many years. Redwood, baldcypress, white oak, and catalpa were usually good for 10 to 20 years, but such woods as ash, beech,

---

* The sapwood of essentially all species offers little resistance to the ravages of termites and wood-destroying fungi.

birch, maple, hemlock, and spruce, even when the posts contained large volumes of heartwood, were usually in a bad state of repair after only 2 to 5 years in the ground. It was not until the availability of durable post-wood timbers began to diminish, however, that consideration was given to the use of preservatives to extend the useful life of the more perishable species.

A. Sources of Posts

Posts are produced in prodigious numbers in all timbered areas of the nation. Several million of them are procured from farm woodlots where they are cut during the winter months when most agricultural pursuits are at a virtual standstill. Many others are by-products of logging operations where treetops and large limbs are fashioned into post-sized pieces. In several sections of the country, and notably in the southern yellow pine belt, which is dominated by thrifty second-growth stands, thinnings in sizes too small for poles provide a multitude of posts, the sale of which more than pays for the improvement cutting. In recent years post producers and distributors have made their appearance in many areas. Unlike farmers who, as previously indicated, usually restrict harvesting operations to a few weeks during the winter months, commercial producers work from 200 to 300 days yearly. Their operations both in the woods and at the concentration yards are mechanized wherever possible, with portable power saws, mechanical debarkers, and treating facilities a part of their standard equipment.

Most round posts are cut in lengths of 7 to 8 ft, with their diameters ranging from 3 to 6 in. Post timbers in large diameters are not infrequently halved longitudinally, and some of the largest are even quartered. The supply of round posts is also augmented by materials from other sources.

Hand-rived timbers of post dimensions are produced in large numbers from large western redcedar and redwood logs too severely damaged in felling or in logging operations to be of use for sawtimber.

Posts with square or rectangular cross sections, products of sawmills, are commonly produced from large cants.

B. Bark Removal

The facility with which bark may be removed from round-timber posts, just as from poles and piles, varies with the season in which they are produced. The bark on winter-cut timber is "tight" and always difficult to dislodge. That on trees felled during the early growing season, however, is seldom troublesome, as the cells in the active cambial region are easily ruptured.

For many years the removal of bark was strictly a hand operation. Depending upon his skill and the tightness of the bark, an experienced operator could peel 100 to 200 posts in a single working day. Hand peeling operations seldom removed all the inner bark, which, when the posts were subjected to preservative treatment, inhibited the absorption of toxic. This, coupled with rising labor costs, caused pole producers to develop mechanical barkers that not only increased production rates but also produced cleaner posts.

Three types of mechanical barkers are in current use. In some producing centers they are mounted on skids and moved from one cutting site to the next as required. Others are set on permanent foundations at post concentration yards.

1. Drum Barkers. Machines of this type operate on the same principle as rossers used in the wood yards of pulpwood mills. These consist of motor-driven drums approximately 5 ft in diameter and 9 ft in length, the shells of which are made of either perforated boiler plate or closely spaced channel irons. In its operation, the drum is loaded with 80 to 100 posts and then rotated at 25 to 35 rpm. The tumbling and rubbing action dislodges the bark which falls out through the perforations, or slots, in the wall provided for this purpose. In the design of one such machine the axles are eccentrically mounted at either end to provide greater tumbling and sliding action.

Southern yellow pine posts as well as those of spring-cut and summer-cut hardwoods can be debarked in 12 to 25 minutes. The use of drum barkers is not recommended, however, for late-fall-cut and winter-cut hardwood posts. Machines of this sort have a daily capacity of from 900 to 1,000 posts depending upon the source of the wood and the season it was cut. The cost of drum debarking varies from $1\frac{1}{4}$ to 2 cents per post.

2. Hammer Barking Machines. Two versions of hammer barking machines are in current use. One of them is actually a machine specifically designed for rossing pulpwood and is composed of a power-driven shaft to which has been attached a number of small, free-swinging hammers. In debarking, posts are fed into and rotated against the flying hammers which literally beat off the bark. This mechanism has a daily capacity of about 3,000 posts, but because of its high initial and operating costs, its use is prohibitive in all except the largest of post-producing enterprises.

The second version of these hammer-type machines is the *chain barker*. Its principal component is a large, steel, power-driven, flanged spool, with several short lengths of heavy logging chain bolted in transverse festoons between the spool's flanges. This is mounted in a heavy metal frame to which are also attached feed rolls to carry the posts over and past the spool. In chain barking the spool is run at a speed of 700 rpm, and the chains, which are slack when at rest, fly outward to strike and dislodge the bark on a post as it is moved over them. Feeding and rotating operations are manual. This machine has a daily capacity of from 300 to 400 pieces.

3. Peeling Machines. Two types of post-peeling machines are in current use. They differ in their work from the other debarking machines previously described in that the bark is removed by peeling rather than by a pounding action. One machine is essentially a diminutive pole shaver with floating cutterheads. The other is a modified lathe. The lathe-type barker peels off not only the bark but also some wood, so that the finished post is a smooth cylinder of wood much like the residual core of a veneer bolt. Four men operating this machine, either in the woods or at a concentration center, can turn out 120 to 200 finished posts per hour.

Lathe-peeled posts have smaller volumes than do tapered poles of comparable length and top diameter. Hence they are lighter in weight and require lesser amounts of preservative than their tapered counterparts. Savings in shipping costs alone often amount to as much as $100 per car. It is questionable, however, if posts with small volumes of sapwood should be lathe-peeled, since it is difficult to obtain satisfactory penetration of preservative in the heartwood of most species.

C. Seasoning

There is little or nothing to be gained by seasoning posts that are not destined for treatment with oil-borne preservatives, since once in the ground they pick up moisture until they reach equilibrium with that in the soil. Moreover, when Osmose salts are used, it is imperative that green material be used for satisfactory results. But to obtain adequate absorption and penetration of creosote and other oil-borne toxics, posts must first be thoroughly seasoned.

In seasoning, the posts should be open-piled or ricked in pens to facilitate the free circulation of air about them. In such species as the oaks, particular care should be taken to avoid excessive checking. A better quality of posts is obtained from trees of these species when they are cut in the fall and seasoned during the fall and winter months when drying ordinarily progresses slowly. Most posts can be adequately seasoned in from 60 days to 6 months depending upon their diameters, season of the year when cut, and prevailing atmospheric conditions during drying.

D. Framing

Little or no supplementary framing is required for the vast majority of posts annually put into service. Those used in fence lines on farms and ranches are squared at either end at the time of their manufacture. In the event, however, that they are to be driven into the ground rather than inserted into prepared holes, the smaller ends are customarily taper-pointed to expedite driving. Right-of-way fence posts are handled in virtually the same manner except that gains are cut into those where bracing timbers are used. Highway guard posts are roofed in one of several ways and then either (1) bored to provide lead channels through which strands of restraining cable can be threaded or (2) gained to provide adequate seating for guard rails. Bolt holes are prebored in posts used to support the larger and heavier highway signs or markers. Posts, like poles, are framed prior to preservative treatment as a safeguard against the possible exposure of toxic-free wood.

E. Methods of Treating

Several procedures are used in impregnating posts with a preservative, but the one most widely employed, particularly for farm and ranch fence lines, is the open-tank immersion method with alternating baths of hot and cool creosote. This process is similar to the one previously described for the preservation of western redcedar poles (see page 62), although the equipment involved is seldom as elaborate. Treating facilities erected at concentration yards and on farms often consist of nothing more than oil drums or iron tanks or vats resting on stone or brick fireboxes.

Properly seasoned posts subjected to complete immersion in such baths may be expected to last for 20 or more years under normal exposure. When merely butt-treatments are desired, they are stood upright in treating tanks, and the toxic is maintained at a level that will assure penetration several inches above a predetermined ground line.

Guard and other highway posts, as well as most of those used in major right-of-way fencing, are subjected to pressure treatments with creosote or oil-borne preservatives in accordance with A.W.P.A. standards,* thus giving every reasonable assurance of maximum protection and a long and satisfactory service life.

During the past decade or so Osmose salts have been used in ever-increasing quantities for the preservation of fence posts. This preservative, which in its formulation includes sodium fluoride, potassium bichromate, dinitrophenol, and bisodium hydrogen arsenate, is supplied in the form of a cream or a paste. No special equipment is needed in its use, for the matrix, when mixed with a small amount of water, can be applied with a brush or, if desired, further diluted in a tank and used as a dipping bath. But unlike other preservative processes where wood must be well seasoned before impregnation, Osmose salts are applied to green material. Immediately after brushing or dipping, the posts are piled in solid stacks and covered with waterproof paper and permitted to stand for about 30 days. During this interval the salts slowly diffuse into the wood. This method of treatment is especially well suited for southern yellow pine posts, which usually include large volumes of sapwood. With such material salt penetration is deep, and the posts are well protected. Retentions of $\frac{1}{4}$ to $\frac{1}{2}$ lb of toxic per cu ft of wood are generally regarded as adequate.

During the past decade or so unscrupulous dealers have marketed millions of fence posts purported to have received an adequate preservative treatment. Yet in reality these materials were merely bathed in cheap, nontoxic mineral oil, often nothing more than crankcase drainings to which just enough creosote was added to provide an acceptable color and aroma. With the view to putting an end to these deceitful practices, the American Wood-Preservers' Institute developed an industry standard which was submitted to the U.S. Department of Commerce for consideration and adoption. Released on February 6, 1961, by the Commodity Standards Division of the U.S. Department of Commerce under the designation CS 235—61, Pressure Treated Fence Posts (with oil-type preservatives), this new industry standard, when properly used, assures consumers of pressure-treated fence posts that may be expected to give a service life of at least 30 years.

The use of this standard, as is the case with all commercial standards, is strictly voluntary. It may be employed as a specification in contractual negotiations or incorporated as a part of a general specification. When it is made a part of a contract, its various stipulations may be enforced by

* See A.W.P.A. C-5, Standard for the Preservative Treatment of Posts by Pressure Processes.

either or both parties concerned. Particularly desirable features of this document are the standardization of dimensions for round, half-round, and square-sawed fence posts, and the provisions for uniform treatment and branding and tagging practices.

F. Cost Considerations

Rising farm-labor costs have made it expedient to use more efficient methods in erecting fence lines, with the result that mechanical post-hole diggers

Fig. 4-7. Driving a 6-in. creosoted southern yellow pine post in fence line along the Kansas Turnpike using a Danuser driver. (*Courtesy W. A. Stacey, American Wood-Preservers' Institute, Lawrence, Kansas.*)

and mobile, powered post drivers have made their appearance in many areas.

Setting posts by hand, which involves digging holes and then tamping the soil around a post once it has been positioned, is both laborious and time consuming. A two-man crew can rarely set more than eight posts an hour when hand labor alone is employed. Mechanical post-hole diggers, usually drills mounted on and powered by a tractor, make it possible to set about 12 posts an hour, but hand tamping is still necessary and is the principal deterrent to efficient operation. Power driving of pointed posts is the most economical method for setting posts yet devised. A two-man crew using a powered hammer (Fig. 4-7) mounted on the rear of a tractor can set about 30 pointed posts an hour. As the posts are driven, the displaced soil is firmly packed against them, and thus supplementary tamping becomes unnecessary.

Untreated fence posts, depending upon their size, species, availability, and geographic source, cost from $0.20 to $1 each. Treated posts, on the other hand, bring from $0.35 to $1.50 or more, contingent not only upon those factors noted above but also upon the nature and amount of toxic retention and the method of impregnation.

It is well for the user of fence posts to ascertain what materials are available in his area and their life expectancy on his particular lands. He should also determine the comparative costs of both treated and untreated posts, and if he decides to use the latter, what his replacement costs may amount to over 5, 10, or even 20 years. With such data in hand, he is in a position to determine the most economical installation consistent with his needs. If the purchase price for treated materials can be reasonably expected to exceed that of untreated stock plus anticipated replacements for those damaged in service, say for a 20-year period, there is little justification for the greater initial expenditure. Under most circumstances, however, the use of treated posts will effect appreciable savings in cash to the consumer.

G. Steel Posts

Prefabricated steel posts used for supporting wire barriers have made their appearance in many rural areas during the past several years. Their initial costs are appreciably higher than those of properly treated wooden posts, and there is no evidence to support the claim that they will outlast the wooden article. While it is expected that their use will continue, it is doubtful that they will be serious competitors with wood posts on the vast majority of American farms.

SELECTED REFERENCES

1. Anderson, I. V. Trends in the Utilization of Pole Species and Their Effect on Forest Management. Proceedings, Society of American Foresters. Washington, D.C. 1947.
2. Anon. A Report on Pole No. 92159. Applied Research Laboratories, United States Steel Corporation, Pittsburgh, Pa. 1955.
3. Anon. Pressure Treated Timber Poles. American Wood-Preservers' Institute, Chicago, Ill. 1957.
4. Anon. Fencing. The Farm Fencing Association, St. Louis, Mo. Undated.
5. Anon. Low-cost Pole Buildings for Industry. The Dow Chemical Company, Midland, Mich. Undated.
6. Bonnicksen, Leroy. Multi Combination Pole-type Construction. *Oreg. Agr. Expt. Sta., Bull.* 557. 1956.
7. Drow, John T. Strength of Western Larch and Its Suitability for Poles. U.S. Department of Agriculture, Forest Service, Forest Products Laboratory Report No. R1758 (revised). 1952.
8. Helphenstine, R. K., Jr. Quantity of Wood Treated and Preservatives Used in the United States. Joint Report of the U.S. Forest Service and the American Wood-Preservers' Association. Issued annually.
9. Hunt, G. M., and G. A. Garratt. Wood Preservation, 2d ed. McGraw-Hill Book Company, Inc., New York. 1953.
10. Keeney, W. D., R. H. Mann, and C. M. Burpee. Pressure Treated Timber Foundation Piles. American Wood-Preservers' Institute, Chicago, Ill. 1955.
11. Lehrbas, M. M. Fence Post Barking Machines in the South. *La. Forestry Comn., Bull.* 3. 1947.

12. Mann, R. H. 54-year-old Creosoted Pine Poles Free from Decay and in Excellent Condition. Reprint from *Wood Preserv. News*. 1952.
13. Neetzel, J. R. Cost of Setting Fence Posts in Minnesota. *Lake States Forest Expt. Sta., Tech. Note No.* 350. 1951.
14. Neetzel, J. R. Power-driving of Wood Fence Posts. *Minn. Forest Notes, No.* 4. 1952.
15. Neetzel, J. R., and S. A. Eugene. What Do Fence Posts Cost on an Annual Basis? *Lake States Forest Expt. Sta., Tech. Note No.* 341. 1950.
16. Ostrander, M. D. Production and Marketing Wood Piling and Poles in the Northeast. *Northeast. Forest Expt. Sta., Paper No.* 57. 1953.
17. Patterson, D. How to Design Pole Type Buildings. American Wood-Preservers' Institute, Chicago, Ill. 1957.
18. Smith, W. R., and R. A. Hertzler. The Preservation of Fence Posts. *N.C. Dept. Conserv. and Devlpmt., Resources—Indus. Ser.* 1. 1946.
19. Stacey, W. A. How to Build Good Highway Fences. *Wood Preserv. News,* **35**(9):6–8, 18. 1957.
20. Stacey, W. A. Pole Construction Is Just Beginning. *Wood Preserv. News,* **36**(1):6, 18–19. 1958.
21. Stanford Research Institute. America's Demand for Wood, 1929–1975. Palo Alto, Calif. 1954.
22. U.S. Department of Agriculture, Forest Service. Potential Requirements for Timber Production in the United States. Report 2 from A Reappraisal of the Forest Situation. 1946.
23. Wentling, J. P. Western Red Cedar: The Ideal Pole. Western Red and Northern White Cedar Association, Minneapolis, Minn. 1938.
24. Wilson, T. R. C., and J. T. Drow. Fiber Stresses for Wood Poles. U.S. Department of Agriculture, Forest Service, Forest Products Laboratory Report No. D1619 (revised). 1953.

CHAPTER 5

MINE TIMBERS

I. USES OF MINE TIMBERS

The term *mine timbers* is a collective designation denoting a variety of wooden supports used in the construction of mine tunnels, shafts, and other openings and chambers. Fay [2] defines *timber* as "any of the wooden props, posts, bars, collars, lagging, etc., used to support mine workings," and *timbering* as "the timber structure employed for supporting the face of an excavation during the progress of construction." The word timbering is sometimes applied to concrete, masonry, and steel supports, as well as to those made of wood, but the general term *ground support* is considered preferable to refer to all types of supporting structures. This chapter will be restricted, in the main, to ground supports made of wood, as well as to other types of wood products used in connection with mining.

The subject of mine timbers is one of considerable importance in the forest economy of the United States, the consumption of wood for this purpose being exceeded only by lumber, fuel wood, pulpwood, railroad ties, and veneer logs.

A. Purpose of Mine Timbers

When an underground excavation is made, whether horizontal or vertical, the removal of soil and rock creates new and unsupported surfaces. Those composed of loose material soon start to slough away. Hard-rock faces exposed in this manner break away from the back along planes of fracture. Material of this latter sort will eventually arch and support itself, but if the back is not supported immediately, caving is apt to become cumulative until a large chamber is formed. In solid, unfractured rock formations, caving is seldom a problem, and the roofs of even large chambers may stand indefinitely without the need for internal supports.

It is quite obvious that the amount of timbering that can be placed in a mine opening many hundreds of feet below the surface of the earth would be totally inadequate to support the tremendous weight of the overlying strata. The primary purpose of timbering is, therefore, to prevent loose material, blocks, or slabs from falling and damaging equipment, injuring workmen, and plugging passageways. Timbering also provides a firm foundation for machinery in soft ground, a base for car rails, linings for elevator shafts, and numerous other uses.

The amount and character of timbering vary widely in different types of ground.* Weaknesses caused by faults, fissures, bedding and cleavage planes in rock strata, blasting fractures, seepage, etc., may even dictate the use of several different methods and intensities of timbering in the same mine. Even after the ground around an excavation has "set," pressures may be great. Such pressures may be either vertical or lateral, or both. Substantial timbering is required to keep some tunnels open, since even large timbers are sometimes broken in a matter of days. In some cases, tunnels must be abandoned and alternate passages cut in areas where pressures are less severe. Occasionally, some mines also require floor timbering to prevent heaving.

B. Mining Terminology

Colloquialisms are common in mining terminology, and many of the terms used are not to be found in a dictionary. To complicate matters further, the same term may have different meanings in various mining regions. Some of the more common mining terms are defined below:

Adit. A nearly horizontal passage from the surface by which a mine is entered and watered. In the United States an adit is usually called a *tunnel*, though the latter, strictly speaking, passes entirely through a hill and is open at both ends. Frequently also called drift, or adit, level.

Cage. A frame with one or more platforms for cars, used in hoisting in a vertical shaft. It is steadied by guides on the sides of the shaft.

Crossbar. A horizontal timber held against the roof to support it, usually over a roadway; a collar.

Crosscut. 1. A small passageway driven at right angles to the main entry to connect it with a parallel entry or air course. 2. A level driven across the course of a vein or in general across the direction of the main workings or across the "grain of coal."

Drift. 1. A horizontal passage underground. A drift follows the vein, as distinguished from a crosscut, which intersects it, or a level or a gallery, which may do either. 2. In coal mining, a gangway or entry above water level, driven from the surface in the seam.

Gob. Any pile of loose waste in a mine.

Lagging. Planks, slabs, or small timbers placed over the caps or behind the posts of timbering, not to carry the main weight, but to form a ceiling or a wall, preventing fragments of rock from falling through.

Level. A horizontal passage or drift into or in a mine. It is customary to work mines by levels at regular intervals in depth, numbered in their

---

* The term *ground,* as used in the mining industry, denotes rock formations in which a mining operation is being conducted.

order below adit or drainage level, if there be one. Rarely applied to coal mining.

Pillar. 1. A solid block of coal, etc., varying in area from a few square yards to several acres. 2. A piece of ground or mass of ore left to support the roof or hanging wall in a mine.

Raise. A mine shaft driven from below upward; called also upraise, rise, and riser (Webster). An opening, like a shaft, made in the back of a level to reach a level above (Standard). The term is in general usage at mines in Western states.

Shaft. An excavation of limited area compared with its depth, made for finding or mining ore or coal; raising water, ore, rock, or coal; hoisting and lowering men and materials; or ventilating underground workings. The term is often specifically applied to approximately vertical shafts, as distinguished from inclines or inclined shafts.

Stope. An excavation from which the ore has been extracted, either above or below a level, in a series of steps. A variation of step (Standard). Usually applied to highly inclined or vertical veins. Frequently used incorrectly as a synonym of room, which is a wide working space in a flat mine.

Stull. 1. The top piece of a set of mine timbers. 2. A timber prop supporting the roof of a mine opening.

C. Mining Methods and Types of Timbering

A mine may consist of (1) a simple horizontal tunnel driven into the side of a hill; (2) a complicated interconnecting system of underground shafts, levels, crosscuts, drifts, pumping chambers, stopes, and rooms located hundreds of feet below ground; or (3) any degree of development between these two extremes. An open pit of the type used in strip or surface mining is also called a mine.

Little timber is needed in open-pit mines other than that used for crossties. The timbering requirements of some few mines cut into solid rock are also small, but in the majority of cases, depending upon the character of the ground, method of mining employed, and length of time the mine is to be kept operative, vast quantities of timber are required to maintain safe operating conditions.

1. Timbering Open Rooms. The simplest form of timbering consists of single posts, or props, mounted with or without sills and topped either with caps or headboards. The cap may take the form of two wedge-shaped pieces of wood that can be driven together to assure a tight fit between the prop and the area of roof that it supports. Such props are most common in flat mines, such as coal mines, where they constitute the largest single item of wood consumption.

2. Timbering Tunnels, Drifts, and Crosscuts. The timbering in tunnels, drifts, and crosscuts may vary from simple, one-piece tunnel sets to complicated, eight-piece arched sets used in ground that swells when exposed to air and moisture. Probably the most common form of timbering is the three-piece tunnel set used in drift and crosscut construction. This type of tunnel set consists of three principal members: two posts and one cap (Fig. 5-1).

The sets form the major framework of tunnel timbering. Additional protection is required to prevent the caving of loose material between the sets. This is accomplished, where necessary, by placing poles, slabs, or sawed lumber on top of the caps and on the outsides of the posts, between the posts and the tunnel walls. This is called *lagging*; that above the caps is known as *top*, or

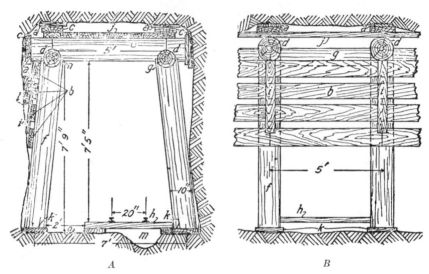

Fig. 5-1. Three-piece set for timbering a drift: *A*, front view of set; *B*, side view of set; *a*, wedge; *b*, lagging; *c*, headblock; *d*, butt cap; *e*, cap, round; *f*, post, round; *g*, girt; *h*, tie; *i*, bridge piece; *j*, top lagging; *k*, footboard; *l*, block; *m*, ditch; *n*, bottom of cap; *o*, bottom of drift. (*Courtesy U.S. Bureau of Mines.*)

working-floor, lagging, that at the sides is called *side*, or *gob*, lagging. The construction of chutes, used for the loading of ore or waste into mine cars, consumes rather large quantities of heavy stull timbers and plank.

3. The Square-set Method of Timbering Stopes. The square-set method is widely used in timbering stopes and rooms. Fay[2] defines a *square set* as follows: "A set of timbers composed of a cap, girt, and post. These members meet so as to form a solid 90-degree angle. They are so framed at the intersection as to form a compression joint and join with three other similar sets." A diagrammatic sketch of square-set timbering is shown in Fig. 5-2*A*; a vertical transverse section through a typical square-set stope is shown in Fig. 5-2*B*.

a. Joints Used in Constructing Square Sets. The most common method of joining the component members of a square set is shown in Fig. 5-3. This is the *step-down*, or *Rocker 4-8-12*, method used extensively in the copper mines in Butte, Montana. This joint has many advantages over the former butted posts and square and beveled framing.

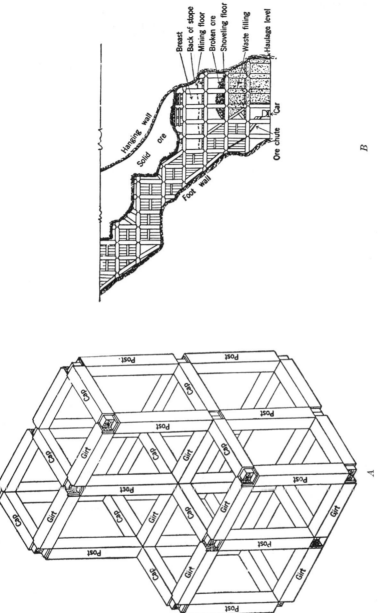

Fig. 5-2. Example of square-set stoping: *A*, square-set timbering; *B*, vertical transverse section through typical square-set stope. (*Courtesy Illinois State Bureau of Mines.*)

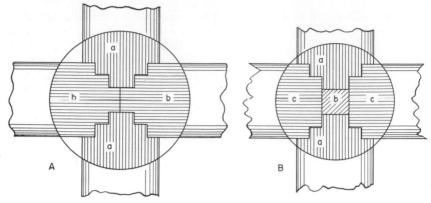

Fig. 5-3. Vertical sections of joint in Rocker method of framing square-set timbers: *a*, post; *b*, cap; *c*, girt. (*Courtesy U.S. Bureau of Mines.*)

D. Avoiding Timber Waste in Mines

It has been estimated that as much as 85 per cent of all timber used underground is employed in temporary working spaces. It has also been reported that at least two-thirds of this timber is abandoned in a relatively short time, often while the wood is still sound. This estimate gives due allowance for timber quantities known to be salvaged, including mine ties. Much of this waste is unavoidable, yet more emphasis on salvage procedures would undoubtedly result in large savings. At least that has been the case in Europe, where the matter has been given a good deal more thought. The development of methods that require less timber and of more efficient salvage operations has resulted in substantial savings of timber in Europe. The fact that much of the timber cannot be salvaged, however, dictates certain requirements that mine timbers must meet, the most obvious being the lowest possible cost consistent with strength properties that will provide maximum safety.

II. CLASSIFICATION AND CHARACTERISTICS OF MINE TIMBERS

A. Classes of Mine Timbers

Cost is the principal factor governing the selection of mine timbers. For convenience, such materials are classified into two general groups, round and sawed, although a limited number of split and hewed timbers are also placed in service. Round timbers constitute somewhat more than half the total amount of timber used. Ordinarily sawed timbers are employed in the more or less permanent installations, such as headframes, entry and shaft timbers, haulageways, certain types of lagging, tipples, trestles, ore bins and shoots, pumping stations, and mine timbers. Square timbers have certain advantages, notably the ease with which they may be framed and the more efficient and economical construction that can be accomplished. When the square-set method of timbering was first originated, it was considered impossible to obtain desired joint strength with round timbers. The development of better joints, however, such as the Rocker joint (Fig. 5-3) has dispelled this belief. Now round timbers, usually untreated, are used almost exclusively where the duration of a working mine is less than the life expectancy of treated wood, and for nearly all temporary timbering where it is impossible to salvage the timbers as the workings advance.

It has long since been demonstrated that round timbers and sawed timbers of the same net cross-sectional area have virtually the same strength, although the former have two inherent mechanical advantages. The axis of a round timber is almost always parallel to that of the component fibers, whereas sawed material is often cross-grained as a result of improper sawing. Knots can be expected to have less effect in reducing strength in round timbers than in sawed material, since in the former they are usually smaller. The above is based on two assumptions: (1) that the round members can be fastened together as securely as sawed timbers; and (2) that the volume of sapwood will not be reduced by decay during the service life of the timber.

B. Qualities Desired in Mine Timbers

Wood for use as mine timbers should possess the following qualities:

1. Strength
    a. High stiffness and compressive strength
    b. Lack of brashness
    c. High strength-weight ratio
2. Durability
3. Availability
4. Low initial and low maintenance costs

Strength is, of course, of vital importance, since the safety of workers is at stake. Stiffness and compressive strength, both parallel and at right angles to the grain, are essential. Most mine props are used as columns, although they may serve as beams under various types of loading. Bearing strength is also important, particularly since the joints most commonly used depend on side bearings to sustain superimposed loads as well as to maintain tight joints. Brashness, which results in timbers breaking at reduced loads and without warning, cannot be tolerated. Cross or spiral grain or combinations of the two, if present to the extent of weakening timbers beyond safe limits, is also dangerous.

The total output of a mine is often governed by the efficiency and capacity of the hoisting system, and the transportation of supplies into a mine is a large part of the total cost of operation. The weight of wood used for timbering is, therefore, of the utmost importance, since wood for timbering represents a large part of the material that must be moved in many mines. Direct labor costs are also involved, since most timbering must be placed by hand under unfavorable working conditions. The preference for Douglas-fir over western larch in the Rocky Mountain mining regions is an example of the effect of weight on the selection of a species. The properties of the two species are essentially equal for use as mine stulls, yet western larch usually enjoys only second choice because of its greater weight.

The importance of durability varies with the length of service required of timbering. In temporary locations it may not be a factor at all. In semipermanent installations the wood used should have sufficient durability to ensure protection from decay for the expected life of the operation. Where long service is required, preservative treatments are essential, and ease of treatment may be of greater importance than natural durability.

Availability and cost may be interdependent factors. If timber supplies are available at or near the mine site, the initial cost of wood is usually far less than that of substitute materials. This is the case in most mining areas in the United States, but there are exceptions. Sometimes mines must compete for supplies of costly timbers that are more valuable for other uses. Low-quality supplies may be expensive because of scarcity. In some arid regions even strength properties may be secondary to availability, making it necessary to utilize local species in order to obtain any timbers at all. In most mining regions, however, adequate supplies of relatively low-cost species are available.

The high cost of maintenance is one of the disadvantages of wood for use as mine timbers. This objection is valid when untreated wood is used in permanent installations and must be replaced frequently because of decay or excessive wear. Often this difficulty can be overcome by preservative treatment. The objection is not applicable in temporary workings or in stoping and other locations where no attempt is made to salvage the timbering.

C. Advantages of Wood and Some Comparisons with Substitute Materials

The advantages of wood over substitute materials are numerous, and they presage the use of wood timbering at rather permanent levels of consumption per ton of ore or coal mined. Steel and concrete have displaced wood in a few instances. These substitutes have found logical uses in specialized installations, but particularly where permanence and fire resistance are factors. Even so, it is expected that wood will continue to be used extensively, even in the one-third of those mines which are regarded as permanent in nature.

The advantages of wood over other materials are (1) relatively low initial cost; (2) availability; (3) favorable strength-weight ratio; (4) gradualness of failure, giving warning of impending danger; (5) adaptability because of ready workability; and (6) sufficient durability for most short-term requirements. Cost, availability, and safety are probably the most important advantages of wood over substitute materials.

The initial cost of wood is far less than that of substitute materials. This is important, since at least two-thirds of all mine timbers are either not salvageable or are used for only short periods. Availability is often a factor in cost, but stulls can be shipped for considerable distances and still maintain favorable price margins. The fact that treated timbers cost no more to ship than untreated timbers is also a factor, since the former have sufficient service life to compete with steel under all but the most unfavorable conditions. Considering the average life of permanent installations in a mine, the cost of maintaining treated materials may not be significantly higher than the upkeep of steel, which must be painted at regular intervals to prevent deterioration. Opinion differs on this point, however; figures are available to support the contention of lower over-all costs for both materials. The fact remains that wood is suitable, and often preferred, for most mining operations. It costs less for most uses and is available at or adjacent to most of the important mining regions of the United States.

The strength-weight ratio of wood compares favorably with that of steel and concrete. The stiffness of wood is also a factor in withstanding the high pressures produced by earth movements.

Adaptability, attributable to a large extent to workability, is an outstanding advantage of wood. It has a marked effect on timbering costs, particularly in temporary workings. Steel tunnel sets designed for temporary workings lack the flexibility of wood in situations that require ready adjustments to variations in the size of tunnels. Also, the joints are usually more difficult to assemble underground. Because of the higher initial cost of steel, losses are greater when timbering is buried as a result of cave-ins. The waters found in many metal mines and some coal mines are highly corrosive. In such places steel suffers a fate similar to that of wood timbers subjected to decay in warm, humid mines, and rapid corrosion may make the use of steel costly. Steel timbering may last indefinitely, however, in comparatively cool, dry mines, with only a minimum of maintenance in the form of occasional coats of protective paints. Durability, because it is so closely related to wood preservation, will be discussed under that heading.

Steel timbering was once hailed as the solution to the rising costs of wood stulls, but its extensive use has never materialized. Both steel and concrete have, however, become standard materials for certain locations in mines, particularly in the coal mines of the eastern United States. Tipples, steel or cast-iron supports for main haulageways, shaft bottoms, and pumping stations, and, to some extent, shaft sets and drift sets, are among the common uses of steel at the present time.

Concrete is most commonly used for machinery foundations, headframes, shaft collars, shafts and linings for shaft and hoisting stations, as a lining for shafts to exclude water, for sumps and the walls of pumping stations, for fire walls, and, rarely, for precast shaft sets and stringers. It is sometimes used for the construction of pillars in large stopes and to reinforce natural pillars to prevent them from falling or crumbling. The most important uses of steel and concrete are for the construction of headframes, shaft and adit openings, and other surface structures where the prevention of fires is vital.

From the foregoing it can be seen that wood, steel, and concrete all serve useful purposes in mine construction, each being adapted to special applications on the basis of inherent properties. There seems little doubt, however, that wood will continue to dominate the mine-timber field. Its favorable properties, low cost, and availability, combined with its adaptability to the short-term requirements of most mining operations, provide advantages that will be difficult to overcome unless prices of substitute materials can be reduced considerably below present levels.

D. Preservation of Mine Timbers

The principal disadvantage of wood for use as mine timbers is its lack of durability, i.e., its inability to resist the attacks of insects and wood-destroying fungi. Insects often attack underground timbers, particularly unpeeled stulls, but the total damage produced does not approach that caused by decay. The decay problem is more critical in mines than it is above ground because the warm, damp atmosphere in most mines provides optimum growth conditions for many wood-destroying fungi. Under such conditions, nondurable woods may have a service life of only 1 year, and even naturally durable woods, such as white oak and the heartwood of Douglas-fir and the southern yellow pines, may last only 3 or 4 years.

As mentioned previously, durability is not a point of issue in the case of timbers that cannot be salvaged. The situation is quite different, on the other hand, when wood is used in permanent and semipermanent structures such as entry shafts and headframes, haulageway timbers, raises, ore bins and chutes, pumping stations, and similar installations. Even though the use of steel and concrete for such structures is increasing, wood is still used extensively. This poses the problem of frequent replacement costs versus wood preservation. For semipermanent construction, the selection of naturally durable species may be sufficient to assure adequate life. In most cases, however, preservative treatment is the only means of reducing replacement costs and providing the service life required in permanent locations. Tests over a long period of time have demonstrated repeatedly the economy of treated mine ties and timbers. The average life of untreated timber, where not subject to crushing or destruction by mechanical wear, is approximately 3 years, or even less in some cases. The implications are obvious: untreated material would have to be replaced at least three times as often, if not more. Properly treated timbers will last 10 to 20 years. Increases in costs over the use of treated timber are significant, since the cost of replacement is often double or more that of the original installation. The increased drain on available supplies of suitable stulls must also be considered. Wood preservation cannot be justified from the standpoint of economy in the many temporary structures used in mining. There can be no question, however, of the far-reaching benefits that accrue to the mining industry from the intelligent application of wood-preservation practices.

Any of the standard preservatives are satisfactory for the treatment of mine timbers. The choice of a particular treatment will depend on the length and character of service the treatment is required to perform, the species to be treated, the location of the mine with respect to treating facilities, the cost of treatment, the extent of fire hazard in a given mine, and other factors. It is sufficient to say, for the purposes of this chapter, that the judicious use of any proved wood preservative will result in material savings to the mine owner.

1. Other Causes of Deterioration. Decay probably causes more than half the timber failures which occur prior to the normal abandonment of mine openings, but breakage and crushing and, to a lesser extent, mechanical wear and fire also cause the destruction of mine timbers. Decay, of course, contributes to losses from breakage and crushing. Crushing and mechanical wear of sound timber can be reduced by proper engineering. Losses caused by fire, although smaller than those caused by other destructive agents, are of vital importance from the standpoint of human safety and, because of this, a matter of concern to anyone interested in the use of wood in mines. Direct losses of timbers as a result of fires are usually small as compared with lost operating time and the cost of fighting the fire and of putting the workings back in operating condition. Most serious of all is the ever-present danger of loss of life, which can be staggering in a serious mine fire. Wood, being inflammable, is suspect from this standpoint.

Probably the best method of eliminating the danger of fire from wood timbering is to resort to fire-retardant treatments. Unfortunately, such treatments are still too costly for general use, even though they can be combined with a wood-preserving process. Other methods consist of spraying timbers with concrete (*Guniting*) and other noninflammable materials. Guniting has given good service in some mines. The use of concrete and steel headframes and other units at the mouths of shafts, where fires are particularly dangerous, and at all other critical points is effective in reducing the fire hazards of mines. The use of large quantities of wood is inevitable, as has already been pointed out, and special precautions must be taken to reduce the dangers of fire. Among these are the use of efficient sprinkler systems; the exercise of great care in the installation of electric wiring; adequate ventilation to prevent the accumulation of explosive dusts; avoiding carelessness with miner's lamps, blasting fuses, combustible fuels, and the like; and providing ample fire-fighting equipment. Decayed wood is more combustible than sound wood. Thus a good preservative has a secondary advantage of aiding in the reduction of fire hazard. Some preservatives, however, or the solvents in which they are applied, increase the inflammability of wood, at least temporarily. This must be considered in selecting a preservative.

Wood no doubt contributes to the spread of fires in a mine, particularly the smaller pieces used for lagging, but once a fire has started, failure of heavy wooden members progresses rather slowly. Steel and concrete will fail rapidly at the temperatures encountered in a major fire. Heavy wood timbers will char on the outside but retain a large part of their strength for a much longer time. The timbers may fail eventually, but the time gained to fight the fire may be instrumental in saving endangered lives.

E. Production of Mine Timbers

The production of mine stulls is almost invariably carried on apart from standard logging operations. Sometimes the stull cutters follow other logging operations, taking the smaller trees, but more often they operate in young stands, cutting the trees that will meet mine specifications. This is true both in the hardwood regions of the East and the softwood areas of the West. In Idaho some stands cut for lumber during the depression have been re-logged for stulls. There are a few instances of stulls having been supplied from thinnings and improvement cuttings, but such cases are rare. For the most part, stull production is carried on without consideration of the aims of sound forest management and often in direct conflict with such aims.

Supplies of round timbers, except in unfavored areas, are most commonly obtained from nearby forests, and most of the sawed material comes from local mills. The anthracite mines of Pennsylvania obtain about 90 per cent of the round timbers and 75 per cent of the sawed material they use from the 15 counties of the anthracite region of the state. This situation is the general rule in the mining areas of the United States. In certain parts of Montana, the Southwestern states, and a few other regions not favored with suitable timber stands, appreciable quantities of timber must be shipped from out-side sources. Douglas-fir and southern yellow pine timbers are being shipped in increasing quantities to Eastern coal mines. The shipment of treated timbers for considerable distances is becoming more common; e.g., mines in Idaho and Montana are purchasing treated material from treating plants in Seattle and Portland.

There are no standard specifications for mine stulls, and few mines have published specifications. Round timbers are usually cut in accordance with an advance purchase order. In the coal-mining region of Pennsylvania, props may be purchased by the linear foot according to top diameter, at random by linear foot and top diameter, or by the ton. The diameters and lengths speci-fied vary with the individual preferences of the mine operators. They are, of course, influenced by the type and size of the ore bodies being mined, the method of mining used, the pressures encountered, and other factors. The re-quirements of one group of mines in Montana offer an example of the range of stull sizes. Stulls are cut in the woods to log lengths of 11, 14, and 16 ft; about 2 per cent are 14 ft long, 5 per cent 11 ft long, and 93 per cent 16 ft long. The 16-ft length is the most economical, since it is a standard log length that can be handled with standard logging and loading equipment. It is also the most versatile, because it will produce the following pieces with a mini-mum of waste: three 5 ft 5 in. caps, one 7-ft post, and one 8-ft 2-in. post; or two 5-ft 4-in. caps and one 5-ft girt; or two of these girts and one cap. An 11-ft stull will cut two caps without waste but it is not adapted to the pro-duction of posts. The 14-ft length is even less economical. The majority of the stulls obtained for this group of mines from immature Douglas-fir and western larch stands come from trees that give top diameters of 8, 9, 10, and 7 in. in the approximate order given. No single top diameter can be con-sidered ideal, but a tree with a diameter of 12 to 13 in. at breast height approaches this. The common practice in the bituminous coal regions is to cut stulls 5 to 15 ft in length and with a minimum top diameter of 3.5 in. inside the bark. Round props are preferred, but split props up to 8 ft long will be accepted. Lagging as small as 2 in. top diameter inside the bark is purchased by some mines.

The production of round stulls is much the same in all parts of the country, although the type of producers involved may vary somewhat. In many parts of the Eastern and Middle Western states, large quantities of stulls are supplied by farmers who work in the woods during the winter months or who harvest trees from their own woodlots. Professional stull choppers are also common, some cutting purchased stumpage, others harvesting props on land owned by mining interests. Jobbers are common; many mines prefer to deal with reliable jobbers rather than with a large number of small producers. In parts of the West, stulls often are cut by loggers in off periods, or the mines may contract for logging with persons who are engaged in stull production as a regular business. Local shortages, the fear of depleted supplies, or competition from other users of logs have prompted some mining companies to purchase timber holdings and have put some mine owners in the sawmill business.

Framing of timbers is done at the surface in modern framing plants at most large mines. Preframing of square-set members requires a precision which would be difficult to accomplish in the crowded quarters below ground. Modern sawing equipment, with automatic log-turning devices and mechanical conveyors, is used extensively in the framing plants. Sawed material finds many uses in a mine. In addition to the sawed material used for permanent structures both within the mine and at the ground level, large quantities of wood are used for wedges, lumber and plank for lagging, ladder stock, mine-car material, shaft guides, and many other items. These, in so far as possible, are purchased from local sawmills or produced in the mine company's own mill. Mine ties may be purchased locally or imported from outlying regions, depending on the suitability of species growing in the vicinity of the mine. Items such as shaft guides, which require special grades of structural material, are usually obtained from outside sources.

1. Species Used. Oak is the most widely used species for all types of mine construction in the Eastern, Southeastern, and Central states. White oak is preferred because of its greater durability. Depending on available supplies, red oak, maple, beech, birch, ash, some hickory, and the dwindling supplies of chestnut are other hardwoods commonly used. Some yellow-poplar is used for sawed stock. Among the softwoods, hemlock, spruce, and the local pines are used to some extent. The use of Douglas-fir and the southern yellow pines is increasing east of the Mississippi River, particularly for structural timbers of all kinds. Available sources of suitable timbers in the Illinois mining area have largely been consumed, remaining sources being of rather low-quality, second-growth hardwoods. Much of the timbering consumed in this region comes from Missouri, but some supplies of tamarack and hardwood are obtained from Wisconsin and Michigan. Here, too, the southern pines and Douglas-fir are being used more extensively. The situation is similar in the Lake Superior region, the local supply of high-quality, durable species being very small.

In the West, the Rocky Mountain form of Douglas-fir is the most common source of mine timbers. In Colorado this species is supplemented by Engelmann spruce, lodgepole pine, and, less commonly, some of the high-elevation five-needled pines. In some of the more arid locations western yellow pine is the only species that is available in any quantity. In the northern Rocky Mountain area of Montana and the Inland Empire lodgepole pine was once used extensively, but Douglas-fir and western larch are now the principal species. Lodgepole pine, ponderosa pine, and some of the spruces are used in areas where local supplies of the preferred species are unavailable. The Southwestern states depend to a large extent upon locally grown western yellow pine, but supplies of this species must be augmented by importations from other states. Local supplies of durable species of suitable quality are decreasing in some regions. Under proper forest management practices, such as have already been demonstrated as practicable, this situation could be corrected in most cases.

2. Selection of Mine Stulls. The selection and handling of stulls and sawed material prior to placement in a mine may have marked effects on the service that can be expected. Timber is usually selected on the basis of species and the reputation it may have for strength and durability. This is not enough in many cases. Low-quality material of a desirable species may be less suited for mine timbers than good-quality material from an inferior species. A knowledge of structural grades and the relationship of grade to strength values is essential. A further factor in the selection of stulls is the assurance that they are sound and not already infected by decay-producing fungi. Ordinarily this dictates immediate delivery to the mine, since logs left on the ground in the woods usually decay rather rapidly, particularly in the summer months.

3. Peeling. Peeling is one of the best means of protecting stulls from deterioration during storage. Unless immediate use is assured and the expected service life is extremely short, all round timbers should be peeled as soon as cut or shortly thereafter. Peeling expedites seasoning and reduces materially the possibilities of insect damage both prior to and after installation. The elimination of insects has double benefits because the spores of fungi find ready access to the interior of wood through insect borings. Numerous experiments have shown that peeled stulls last longer than those that are installed with the bark on. In spite of such experiments, much timber that could be peeled to advantage is put in place without the bark removed.

4. Seasoning. Proper seasoning of mine timber has several advantages. Seasoned timbers are slightly stronger than green timbers, and it has been shown that properly seasoned wood gives longer service than unseasoned timbers. The difference is not great in extremely wet locations but may be appreciable in well-ventilated, comparatively dry mines. Seasoning is, of course, desirable prior to preservative treatment. The saving in weight brought about by seasoning represents distinct savings in freight and handling and in hoisting costs at the mine.

5. Storage of Mine Timbers. In spite of the high cost of mine timbers, outright losses of large volumes of wood have been observed at many mines as a direct result of improper storage. Too often costly timbers are dumped or stacked in large piles on damp, poorly drained soil without any thought being given to the prevention of decay. Piling methods that prevent deterioration also promote seasoning. Proper storage and seasoning methods are the same as for other forest products. The important factors to consider are a properly drained storage yard, elevation of the piles above the ground, and provision of sufficient space to assure adequate circulation of air around each piece of timber in the pile. The increasing costs of stulls are focusing the attention of mine operators on better storage practices, but much improvement is still to be desired.

6. Costs of Mine Timbers. A detailed discussion of stull, prop, and mine-timber prices is purposely eliminated from this chapter as there are no standard specifications for mine timbers. Furthermore, there have been no comprehensive analyses of costs for many years, and the current inflated prices, if quoted, would be obsolete in a very short time.

F. Annual Requirements and Trends

Comparatively speaking, the use of wood as mine timbers is a rather unproductive form of wood utilization because so little value is added to the raw material prior to its ultimate use.

The amount of wood used per ton of ore or coal mined varies with the type of material being mined. For example, in 1935 the total amount of wood used, in cubic feet per ton of ore mined, was as follows: iron ore, 0.158; bituminous coal, 0.382; "other metals," 0.603; anthracite coal, 0.740. The consumption of "all mines" was 0.426 cu ft per ton of ore. The relationships between the consumption of wood in the various types of mines are maintained in the 1946 estimates, but they show a decrease in the proportion of sawed material, an increase in the amount of round material, and an over-all increase in the total amount of wood used per ton of ore.

Coal mines use approximately 80 per cent of the wood used in mining; Pennsylvania, West Virginia, and Kentucky, in the order named, lead all other states by wide margins. As mentioned previously oak is the predominant species used. Of the wood used in coal mines, the distribution is approximately as follows: 80 per cent for single props, 15 per cent for caps or crossbars, and 5 per cent for shaft and repair timbering. In most metal mines the percentages of timbers used for shafts, tunnels, and square sets are much greater, single props being used much less frequently.

The consumption of wood in mines varies directly with ore production, which, in turn, is sensitive to general economic conditions. Prior to the First World War, mining expanded rapidly in keeping with the tremendous industrial development of the times. It was further stimulated during the war years, suffered a heavy decline after the armistice, recovered in 1923, and rose again to a fairly steady level until the depression years, when the total output of all mines fell considerably below the 1905 level. By 1935 production had recovered to the level of approximately one-half the 1928 total. The heavy demands of the Second World War again stimulated production, and again there was a sharp drop at the end of hostilities as a result of labor unrest, scarcities of essential materials, and other factors. As this is written, production is again on the increase, and it may again reach the peaks attained in 1918 and 1928. With prospects of full employment at home and a comprehensive aid program abroad, it seems likely that ore production may continue at relatively high levels for several years, at least. If so, there is every prospect that wood will enjoy a position of heavy demand. The competition for supplies from badly needed house construction and other industries using wood has raised the cost of mine timbers to high levels in some regions. This may encourage the use of substitutes at higher levels than previously realized, but steel, too, is quite costly. On the other hand, should petroleum shortages develop, eventually this may accelerate the production of coal for its conversion to liquid fuels by the Fischer-Tropsch and similar processes. Eventually, of course, there is the possibility that the development of atomic energy may have far-reaching effects on the amount of coal that is mined. According to the already mentioned Stanford report [9] an average of 100 million cubic feet of logs and bolts will be required annually by the mining interests at least through 1975. Not included in this projected figure, however, are tie requirements, which have been in decline for the past several years.

G. Research in Mine-timber Utilization

There are virtually no records of research designed to reduce the amount of wood used in mines or to improve mine structures constructed of wood. It seems unlikely that applications of wood to mine needs have reached a zenith of perfection that makes further improvements unnecessary. Certainly this is not true in other industries in which wood is one of the basic raw materials used. Even wood railroad ties, which have withstood the competition of substitutes since the earliest days, are now under close scrutiny by numerous research agencies. Progress has been made in wood preservation and in treatments for fire resistance, but these improvements have resulted from interest generated in other fields. Some wartime research on mine guides, sponsored by the Office of Production Research of the War Production Board, is a notable exception to the general lack of research mentioned above. This study demonstrated that laminated mine guides, made to replace high-density, straight-grained material that could not be obtained during the Second World War, were superior to the clear material previously used. It was believed that their superiority would probably offset the increase in cost involved. Improvements in fastenings used to connect mining timbers, perhaps by the adaptation of modern timber connectors, would also seem to offer an excellent opportunity for successful research.

SELECTED REFERENCES

1. Anon. Mining Industry Is Large User of Wood. *Timberman,* **67**(8):212. 1940.
2. Fay, A. H. A Glossary of the Mining and Mineral Industry. *U.S. Bur. Mines, Bull.* 95. 1920.
3. Forbes, J. J., and C. W. Owings. Some Information on Timbering Bituminous-coal Mines. *U.S. Bur. Mines, Miner's Cir.* 40. 1939.
4. Helphenstine, R. K., Jr. Quantity of Wood Treated and Preservatives Used in the United States. Joint Report of the U.S. Forest Service and the American Wood-Preservers' Association. Issued annually.

5. Hough, A. F. Silviculture of Mine-prop Cutting in Western Maryland. *Jour. Forestry*, **9**(43):642–645. 1945.

6. Hunt, G. M., and G. A. Garratt. Wood Preservation, 2d ed. McGraw-Hill Book Company, Inc., New York. 1953.

7. Joyce, A. R. Steel and Timber Roof Supports for Mine Timbers. *Wood Preserv. News*, **19**(9):110–117. 1941.

8. Schrader, O. H., Jr. Laminated Mine Elevator Guides. *Timberman*, **48**(3):126. 1947.

9. Stanford Research Institute. America's Demand for Wood, 1929–1975. Palo Alto, Calif. 1954.

10. U.S. Department of Agriculture, Forest Service. Potential Requirements for Timber Production in the United States. Report 2 from A Reappraisal of the Forest Situation. 1946.

CHAPTER 6

RAILROAD TIES

Railroad ties, depending upon the manner in which they are used, fall into one of three general classes, viz.: (1) *crossties*, or *sleepers* as they are more commonly known in many other parts of the world; (2) *switch ties;* and (3) *bridge ties*. Crossties are transverse beams set in a roadbed to position and support railroad rails. Switch ties, which are greater in length, but otherwise similar to crossties, are used under the rails at frogs and switches. Bridge ties are closely spaced and transversely oriented beams attached to the stringers of bridges and trestles and upon which the rails are anchored. Switch and bridge ties are usually sawed on four sides, but crossties may be either sawed or hand-hewed and furnished with two opposite faces, or faced on all four sides. Years ago most crossties were hand-hewed, but somewhat more than 60 per cent of all crossties produced annually are now manufactured at sawmills. In fact, the old tie maker, or "tie hack," with his once familiar broadax is rapidly being replaced by the sawyer.

A. Advantages of Wood as a Tie Material

Over the years more than 2,500 patented ties have been introduced, yet none has ever seriously threatened to replace the wood tie, which in modern times differs but little from those used in the early days of railroading. In 1888 the Commissioner of the Agriculture Division of Forestry released a bulletin which reviewed the successes European, Asian, and African railroads were experiencing with metal ties. The author decried the American railroads' seeming apathy to experimenting with similar structures and vouchsafed that "as the cost of wooden ties increases, and the cost of metal ties is reduced," their attitudes will change. The statistics for 1959 reveal how wrong he was, for of the more than 900 million crossties in maintained tracks of Class I line-haul railways * less than 0.03 per cent were fabricated of steel, concrete, or other structural material.

* Class I line-haul railways include those whose annual revenues total one million dollars or more. These roads make up approximately 95 per cent of the track mileage and employ about 98 per cent of all railway wage earners in the United States.

It is of interest to record here that tests on prestressed concrete ties are currently being followed with considerable interest in quarter-mile lengths of track of both the Atlantic Coast Line and Seaboard Air Line railway systems. These ties, with a concave 12-in. base, cost about $9 each, or nearly three times the cost of a treated wood tie. On the other hand, it is claimed that only 2,000 are needed per mile of track compared with the nearly 3,000 used in conventional installations. Furthermore, because of their greater weight (550 lb), they are expected to give greater stability to the track and prevent sun kinks and buckling of continuous welded rails. A normal service life of 50 years compared with a well-established 30 years for treated wood ties is also prognosticated. It is still far too early to validate these claims, and there are in truth a number of obvious disadvantages associated with their use. Their weight involves added handling and shipping costs, and because of their brittle nature, unloading at trackside becomes a problem involving additional expense. It is also questionable how long such structures can stand up under the repeating impact loads generated by heavy, high-speed rolling stock. Their performance on curves under an entirely different set of stresses and their lasting qualities in regions subjected alternately to periods of high humidity and subzero temperatures are also open to question. Wood, on the other hand, because of its resiliency, strength, economy, availability, electrical resistance, capacity for absorbing severe shock loads, ability to dampen sound, screw- and spike-holding properties, and its non-electrolytic nature, has no peer as a crosstie material. There is little likelihood that the wood tie will be replaced by any other engineering material, at least in the foreseeable future.

Table 6-1 reveals the quantities of tie replacements in Class I railroads in previously laid tracks for the years 1948 through 1958. While there has been

Table 6-1. Ties Replaced in Previously Laid Tracks by Class I Railways, 1948 through 1958 [a]

Year	Untreated wood crossties, thousands of pieces	Treated wood crossties, thousands of pieces	Other than wood crossties, thousands of pieces	Wood switch and bridge ties, thousands of board feet
1948	1,818	35,024	[b]	119,932
1949	1,410	28,875	[b]	107,793
1950	1,154	29,340	[b]	98,400
1951	1,124	27,938	[b]	92,798
1952	815	29,517	[b]	96,917
1953	899	28,910	[b]	99,792
1954	791	22,383	[b]	85,346
1955	419	23,730	[b]	79,098
1956	285	23,361	[b]	74,100
1957	163	21,920	[b]	71,582
1958	178	15,852	[b]	54,985

[a] SOURCE: U.S. Interstate Commerce Commission. Seventy-second Annual Report on Transport Statistics in the United States for the Year Ending December 31, 1958.
[b] Fewer than 1,000 pieces.

a steady decrease in the amount of tie material during this period, it is significant to note that substitute materials have not been able to compete for the wood-tie market. Crosstie renewals in Class I railroads for the year 1960 were estimated at 19,700,000.

B. Disadvantages of Wood as a Tie Material

While wood has several properties that collectively make it the favored tie material, it also possesses others that work to its disadvantage. Principal among these are:

1. It is subject to attack by wood-destroying fungi.
2. It checks and splits in seasoning, particularly in "boxed-pith" pieces. Checking in boxed-pith ties is unavoidable because of the difference in tangential and radial shrinkage of wood. Sawed and hewed ties from small trees and ties sawed from the centers of large logs exhibit this fault. Pith-free, flat-grained ties are usually not subject to severe checking, and they are likely to have greater spike-holding power because, in normal tree growth, the density of annual rings of the wood of most species becomes greater as the distance from the pith increases. This usually means that the wood directly under the rails is denser than that bearing on the ground, for the top of the tie is always considered to be the wide face that is farthest removed from the pith, whether the pith is present or not. For these reasons, the flat-grained, or "half-moon," tie is preferable to any other form.
3. It burns. Engines using coal as a fuel must dump cinders and ashes at designated places because of the fire hazard. Ties at these points are often burned.
4. It is subject to mechanical wear. Crossties that have been treated with excellent preservatives often wear out and must be replaced before they decay. This mechanical wear is thought to be largely due to the actual abrasion of the wood by the movement of the steel tie plates or rails over the surface of the tie, particularly when sand gets between the wood and the metal. To a lesser extent, crushing of the wood fibers contributes to mechanical failure.

C. Factors That Prolong Tie Life

Tie-replacement costs, which aggregate more than 100 million dollars annually, are still one of the largest single railroad maintenance expenditures. There is little doubt but these must be drastically reduced if the railroads are to continue their position of leadership in the transportation industry. All the companies are endeavoring to prolong the service life of ties by adopting new practices and equipment that will inhibit decay and reduce mechanical wear, breakage, spike-pulling, lateral thrust of spikes, and the development of such seasoning defects as splitting and checking.

It is significant that approximately 90 per cent of the ties in maintained tracks of Class I railways are now of treated wood, and that several railroads have no untreated wood ties in their tracks. The Chicago, Burlington, and

Quincy Railroad, in 1909 and 1910, laid ties of 20 species in test tracks. Some of the ties were untreated, some were treated with zinc chloride, some were treated with a mixture of zinc chloride and creosote, and others were treated with creosote only. The average life of the untreated ties was found to be 5.6 years; that of those treated with zinc chloride, 15.5 years; that of those treated with zinc chloride and creosote, 18.8 years; and the estimated life of those treated only with coal-tar creosote, 29.1 years. These data are combined averages for all species. Individual species varied widely from the average. Table 6-2 gives the average life of 20 species of untreated ties and

Table 6-2. Average Life of Untreated Ties and Approximate Number of Times That Two Preservative Treatments Will Increase Average Life of Certain Species [a]

Species	Average life for untreated ties, in years	Approximate number of times that two preservative treatments will increase average life of untreated ties	
		Coal-tar creosote, full-cell process, 10 to 12 lb per cu ft absorption	Zinc chloride, Burnett process, ½ lb per cu ft absorption
White oak................	11.1	2.2	1.5
Chestnut.................	9.7	b	b
Cypress..................	8.3	2.9	2.2
Pin oak..................	6.3	4.7	3.1
Hickory..................	5.5	4.5	3.6
American elm............	5.3	5.8	3.3
Red oak.................	5.3	5.3	3.1
Yellow-poplar...........	5.3	4.1	2.6
Loblolly pine............	5.3	5.1	2.3
Tamarack...............	5.1	4.9	3.4
Ash.....................	5.1	3.8	3.5
Beech...................	5.0	6.5	3.2
Hemlock.................	4.9	4.1	3.4
Sugar maple.............	4.7	5.8	3.9
Redgum.................	4.1	7.4	2.9
Soft maple..............	3.6	7.0	3.8
Yellow birch............	3.6	7.6	3.8
Black tupelo............	3.5	8.4	4.2
Sycamore...............	3.3	10.5	4.1
Cottonwood.............	2.9	11.2	b

[a] Adapted from Report of Thirty-fourth Annual Inspection of Chicago, Burlington, and Quincy Railroad Experimental Ties. *Wood Preserv. News.* January, 1944.
[b] No data.

the approximate number of times that two methods of preservative treatment will increase the average life of these species.

Mechanical wear is being reduced by the use of larger steel tie plates under the rails and by adzing the tops of ties so that the tie plate may have a uniform bearing surface. It has been demonstrated that the cutting action of sifting particles of grit can be practically eliminated under the tie plate by inserting a felt pad composed of asphalt and jute fiber. Breakage is being reduced by the use of larger ties and better ballast and by maintaining good roadbed conditions.

Spike-pulling is being minimized by boring holes prior to preservative treatment or by using power-driven screw spikes. A spike driven into a bored hole of the proper size offers greater resistance to withdrawal than one driven into solid wood. Some railroads, in order to obtain maximum spike-holding capacity of the ties, specify that the top one-fourth layer of the tie shall be of *compact wood*. This provision is applied to ties made from coniferous species and calls for an average of six or more annual rings per inch along any radius in this section of the tie. In addition to the ring count, some specifications decree that each annual ring in the top quarter of the tie shall contain at least one-third summerwood; or, if the section has four to five rings per inch, the tie shall have one-half or more summerwood in each annual ring.

Resistance to lateral thrust of spikes is secured by specifying ties with compact wood in the top quarter of the thickness, by the use of a tie plate which is perforated for the spikes and which has a shoulder that meets the outside edge of the base of the rail, or by the use of special tie plates that have flanges that extend down the sides of the tie. These special plates are bolted to the tie through the flanges.

Antichecking irons, though not entirely satisfactory, are being used to retard checking and splitting during the seasoning process. Common forms of these devices are the S, C, Beegle, and crinkle irons. One of the newest innovations of this sort is the Koppers Company's *Krinkle-Loc* antichecking iron which is available in both S and C configurations (Fig. 6-1). It works like a fishhook and is said to possess five times more holding power than the conventional irons, since it is made with a special rib which locks the iron and wood together.

Severely split ties are being salvaged by the use of bolts, by squeezing the parts together and applying metal straps around the end of the tie, and by the use of spiral metal dowels made of twisted, square, iron bars that are driven into a bored hole of a size that will ensure a tight fit.

Deterioration of ties because of fungus attack during the seasoning period prior to preservative treatment is being reduced by better methods of piling and better yard sanitation. Many railroads have adopted 1 by 7, 1 by 8 (Fig. 6-2), or 1 by 9 piling in tie-seasoning yards. In this method, 7, 8, or 9 ties are laid edge to edge on two parallel stringers that are set on permanent foundations. The tops of the stringers are on a level plane. Two ties are placed on top of this first layer of 7, 8, or 9 ties at right angles to the layer and at the ends of the layer. Next, another layer of ties is placed above these two ties; then only one tie is placed at one end and at right angles to this second layer. The third layer is placed on top of this one tie so that it is in a sloping position. After the third layer is in place, a single tie is placed across it, at its lower end, and the fourth layer is then added. The next "sticker" tie will be placed on top of the fourth layer at the end opposite the sticker

Fig. 6-1. *A*, Koppers' Crinkle S irons. *B*, Driving an iron into the end of a tie. *C*, Koppers' C irons. (*Courtesy Wood Preservers Division, Koppers Company, Inc.*)

99

tie upon which the fourth layer rests. This alternation of sticker ties continues until the pile is completed.

There are two principal advantages to this method of piling. The first is that there is less wood-to-wood contact between the ties in the pile, and the second is that a pile of this type will contain more ties for the same height than one that uses two sticker ties between the layers. That is, if a 1 by 9

Fig. 6-2. Crossties piled in 8 by 1 stacks for drying. (*Courtesy U.S. Forest Products Laboratory.*)

pile contains 100 ties, a 2 by 9 pile of the same height will contain only 86 ties.

Until recently the decision to replace a tie in service was the result of visual inspection, which in many instances was restricted to only the upper face and ends. A tie tester recently developed by the New York Central Railroad uses gamma rays to probe for hidden defects within a tie, with the result that inspection is not only faster but vastly more accurate.

D. Principal Species Used for Ties

Species that are suitable for crossties include the ashes, balsam firs, beech, birches, catalpas, cedars, cherry, chestnut, cypress, Douglas-fir, elms, gums, hackberry, hemlocks, hickories, larches, locusts, maples, mulberry, oaks, pines, poplars, redwood, sassafras, spruces, sycamore, and walnut. Formerly, when untreated ties were used and railroad traffic was not so heavy and fast as it is today, natural durability, cost, and availability were the principal considerations in selecting a species for crossties. The more durable and softer woods, such as cedars, cypress, redwood, and chestnut, gave good service when not subjected to severe mechanical wear. Now, however, factors such as resistance to mechanical wear, crushing, spike-pulling, and lateral displacement of spikes; bending strength; and ease of treatment with preservatives are making the harder and denser woods preferable. The economy of using treated ties has been demonstrated so thoroughly that now few untreated ties are being laid. Thus the item of natural ability to resist fungus attack is secondary, and the ability of the wood to take preservatives is of primary importance.

Of the more than 21,326,000 crossties given preservative treatment in 1958, more than half were of oak. Ranked below are species in percentage of the total number of pieces indicated above.

Kinds of wood	Per cent
Oak	53
Mixed hardwoods	19
Redgum	10
Douglas-fir	10
Lodgepole pine	2
Spruce	1
Ponderosa pine	1
Larch	1
Mixed softwood	1
Not specified	1

Red oak is now considered by many to produce the most satisfactory tie to meet present conditions because of the ease with which it may be treated and its superior combination of desirable physical and mechanical properties. Of the oaks given preservative treatment, more than half the total is red oak.

E. General Specifications for Crossties

Although some railroads publish separate specifications for ties, the following American Standards Association specifications [1] are followed rather closely. With certain exceptions, ties must be free from defects that would impair the strength or durability of the wood, such as decay, large splits, large shakes, large or numerous holes and knots, and cross grain having a slope of more than 1 in. in 15 in. In cedar and cypress ties, a limited amount of pipe or stump rot and peckiness is permitted.

The rail-seat areas on a tie are important sections. These areas are located between 20 and 40 in. from the mid-length of the tie on each end of standard-gauge ties and between 15 and 25 in. from the mid-length of the tie on each end of narrow-gauge ties. Defects such as large knots (greater in average diameter than one-fourth the width of the tie) are accepted when they are located between the ends of the tie and the rail-seat areas. Ties that are to be used untreated may contain sapwood if it is not more than one-fourth the width of the tie at the rail-seat areas. Ties having sapwood in excess of this amount should be given preservative treatment and are designated "sap" ties; the others are designated "heart" ties.

Ties are divided into two general classes on the basis of necessity for preservative treatment for satisfactory performance. Class U ties may be used untreated, and Class T ties should be treated. Each of these general classes is subdivided into four groups in order to permit tie manufacturers to pile several species together when the purchasing railroad will permit such grouping. The grouping in each class of ties is as follows:

Class U—"Heart" Ties That May Be Used Untreated

Group Ua
Black locust
White oaks
Black walnut

Group Ub
Douglas-fir
Pines
Larches

Group Uc
Cedars
Cypresses
Redwood

Group Ud
Catalpas
Chestnut
Sassafras
Red mulberry

Class T—Ties That Should Be Treated

Group Ta
Ashes
Hickories
"Sap" black locust
Honey locust
Red oaks
"Sap" white oaks
"Sap" black walnut

Group Tb
"Sap" cedars
"Sap" cypresses
Balsam firs
"Sap" Douglas-fir
Hemlocks
"Sap" larches
"Sap" pines
"Sap" redwood
Spruces

Group Tc
Beech
Birches
Cherry
Hard maple
Redgum
Tupelos

Group Td
Butternut
"Sap" catalpas
"Sap" chestnut
Elms
Hackberries
"Sap" mulberry
Poplars
"Sap" sassafras
Sycamore

All ties must be straight, cut square at the ends, free of bark, well sawed or hewed, and must have tops and bottoms parallel, within certain tolerances. A tie will be accepted when (1) a straight line along the top from the mid-point of the width of one end to the mid-point of the width of the other end lies entirely within the tie, and (2) when a straight line along a side from the mid-point of the thickness of one end to the mid-point of the thickness at the other end lies everywhere more than 2 in. from the top and bottom of the tie. The top and bottom of the tie are considered to be parallel if any difference in the thickness at the sides or ends does not exceed ½ in. A tie is not well hewed or sawed when its surfaces contain score marks that are more than ½ in. deep or when its surfaces are uneven.

Lengths, widths, and thicknesses specified are considered to be met when they are within the following limits: lengths not more than 1 in. shorter, and not more than 2 in. longer, than the specified length; widths not more than ¼ in. narrower, and not more than 3 in. wider, than that specified; and thicknesses not more than ¼ in. thinner, and not more than 1 in. thicker, than the specified thickness. Ties more than 1 in., but not more than 2 in., thicker than the maximum ordered are accepted as one size lower than the largest tie ordered. Thicknesses and widths are determined at the rail-seat areas, but the dimensions of the tie are not averaged. Width determinations are made on the top of the tie, and for this purpose the top of the tie is considered to be the narrower of the two horizontal faces or the face with the narrower or no heartwood when both horizontal faces are of the same width.

Standard lengths of narrow-gauge ties are 5, 5½, 6, 6½, and 7 ft; standard-gauge ties are 8, 8½, and 9 ft long.* In cross-sectional dimension, seven standard sizes of ties are recognized.

Size	Thickness, in	Width on top, in.
0	5	5
1	6	6
2	6	7
3	6	8
4	7	8
5	7	9
6	7	10

These dimensions apply to ties that are sawed or hewed on top, bottom, and sides, as well as to ties that are hewed or sawed on top and bottom only. In standard-gauge ties, sizes 0 and 1 are not acceptable when sawed or hewed on top, bottom, and sides. In the case of ties that are hewed or sawed on top and bottom only, a 7-in.-thick by 7-in.-wide tie may be accepted as a

* Some railroads now do not use ties less than 9 ft long, and others are considering adoption of the 9-ft tie.

size 3 tie. If the railroad wishes to have 7 by 7 ties kept separate from 6 by 8 ties in size 3, it should designate the 7 by 7 ties as size 3A.

F. Advantages of Sawed and Hewed Ties

Sawed ties have several distinct advantages not possessed by hewed ties. Their principal advantages are:

1. There is less waste of raw material in the manufacture of sawed ties. Much material that would be wasted in the form of chips in hewing can be converted into lumber by sawing. It has been estimated that at least 35 per cent of a tree converted into hewed ties is wasted.
2. Sawed ties are more uniform in size, particularly when sawed on four sides. This means a saving in transportation costs per tie because a greater number can be loaded into a given car. At the preservation plant, a greater number of ties per charge can be put into the treating cylinder, and it is possible to determine more accurately the amount of preservative needed to attain a given absorption per cubic foot of wood.
3. Ties sawed on four sides are less expensive to replace in tracks because there is less disturbance of the ballast in the process of removing the old tie and placing the new one.
4. Ties sawed on four faces usually have less sapwood than hewed ties.
5. Sawed ties are more easily adzed and bored prior to preservative treatment because of the parallel faces.
6. In the absence of adzing, sawed ties provide a better bearing surface for the tie plates and rails.

Hewed ties have several advantages, the chief of which are:

1. Hewed ties are likely to be more straight-grained than sawed ties coming from a mill that does not saw its logs parallel to the bark.
2. For a given size classification, a tie hewed on two faces has a greater volume of wood than a tie with four sawed faces. This advantage is lost in comparisons of hewed and sawed ties that have the same number (whether it be two or four) of hewed or sawed faces.
3. Hewed ties often have more sapwood than sawed ties, and because sapwood absorbs preservatives more readily than heartwood, the hewed ties are easier to treat.

G. Crossties Required per Mile of Track

One mile of track requires from 2,600 to 3,328 crossties, depending on the size of tie and the service the road must render. Class I railways use an average of 2,994 ties per mile.

SELECTED REFERENCES

1. American Standards Association. American Standard Specifications for Cross Ties and Switch Ties. 1926.

2. Anon. What Price Cross Ties. *Cross Tie Bull.*, **28**(9):9–19. 1947.

3. Anon. 21,326,034 Cross Ties Treated in 1958. *Cross Tie Bull.*, **40**(6):22. 1959.

4. Dick, M. W. What's Ahead in Railroad Maintenance of Way in 1960. *Cross Tie Bull.*, **41**(2):9–10. 1960.

5. Fivaz, A. E. Trends in Foreign Markets for Cross Ties. *Cross Tie Bull.*, **41**(11):11–12. 1960.

6. Hoskins, R. N. Railroads Future Demands for Timber Products. *Cross Tie Bull.*, **40**(2):28–32. 1959.

7. Hunt, G. M., and G. A. Garratt. Wood Preservation, 2d ed. McGraw-Hill Book Company, Inc., New York. 1953.

8. Jeffords, L. S. The Nine Foot Cross Tie. *Cross Tie Bull.*, **27**(8):20–22. 1946.

9. Magee, G. M. Advantages and Disadvantages of Prestressed Concrete Relative to Treated Wood Ties. *Cross Tie Bull.*, **41**(11):30–34. 1960.

10. Putnam, J. A. Management of Southern Hardwood Forests Relative to the Supply of Railroad Stock. *Cross Tie Bull.*, **40**(2):18–27. 1959.

11. U.S. Interstate Commerce Commission. Seventy-second Annual Report on Transport Statistics in the United States for the Year Ending December 31, 1958. 1959.

12. Woolford, F. R. Changes in Cross Tie Replacements. *Cross Tie Bull.*, **40**(2):26–32, 36. 1959.

CHAPTER 7

THE LUMBER INDUSTRY

The production of lumber represents one of the first manufacturing industries established in North America. The early American colonists needed to clear the forest cover from the land to prepare it for the growing of foodstuffs. At the same time they required housing for themselves, their businesses, their churches, and their governments. What was more natural than that they should convert the timber cleared from the land into useful construction material? A further incentive to the production of lumber in colonial days was the fact that the Old World quickly recognized the value of American lumber, and lumber became one of the most important of the early export commodities.

The date of the establishment of the first sawmill in America is not well documented, but it seems possible that lumber manufacturing was begun in Virginia as early as 1608. Horn [5] says that the wealth of "pynes" impressed the British backers of the Virginia Company causing them to send sawmill workers to the New World in 1608 for the purpose of erecting sawmills. Mills operating in Jamestown, Virginia, in 1625 and Berwick, Maine, in 1631 are more commonly reported as the first in North America.

As the lumber industry developed in the white pine region of New England, it had its center in Maine. With the depletion of the original New England forests, the industry moved to new areas of virgin timber. The center of production shifted from Maine to New York, thence to Pennsylvania, the Lake states, the South, and finally to the remaining virgin forest lands of the Pacific Coast. Lumber production in the United States increased rapidly with the opening of new lands until it reached its peak production in 1909, when 45 billion board feet was produced. Since then lumber production has varied in the range of 30 to 40 billion board feet per annum.

While the lumber industry in the United States originally developed around the use of large-sized virgin timber, it converted to the use of smaller, second-growth material when the virgin timber became scarce. This conversion has been associated with increased stumpage and manufacturing costs. Accordingly, the price of lumber has increased much more rapidly than the prices of competitive materials. The President's Materials Policy Commis-

106

sion [6] found that "while the wholesale prices of all commodities increased 109 percent from 1940 to 1950, the price of lumber advanced 218 percent. During this same period, the prices of aluminum, iron ore, nickel, and copper showed an increase less than the average of all commodities." The Stanford report [7] noted an increase of 286 per cent in the price of lumber during the period from 1929 to 1952. According to this study the price of other competing construction materials during this period increased about 90 per cent.

Despite this unfavorable price situation, all available evidence suggests that the demand for lumber will continue to increase. "Timber Resources for America's Future" [8] estimates that the demand for lumber in 1975 will

Fig. 7-1. Large sawmill in state of Washington. (*Courtesy American Forest Products Industries, Inc.*)

be in the range 47.6 to 55.5 billion board feet. The President's Materials Policy Commission [6] found that: "Despite increasing use of many competing materials, lumber remains indispensable in building construction, shipbuilding, furniture manufacture, and dozens of other enterprises."

A. Classification of Lumber

Among the important raw materials of commerce and industry, lumber has one of the most complex classification systems in common usage. There are a number of explanations of this phenomenon. Lumber is produced from a living organism—the tree. Lumber-producing trees themselves represent a large number of species, forms, and sizes, and this variety is reflected in the characteristics of the lumber manufactured from them. Lumber is used in many different ways and for many different purposes, and it is produced to best accommodate these uses. Finally, lumber specifications developed empirically, and customs and habit have militated against any radical simplification of specifications.

Basically lumber is classified according to broad botanical origin, size, and surface appearance. The major subdivision is into hardwood (angiosperms) and softwood (gymnosperms) types. Within these two broad classes size, standards, and grading rules further subdivide the product.

Grading rules are systems of classification used to separate lumber into classes on the basis of the number, size, and character of such physical features as knots, splits, checks, wormholes, wane, pitch streaks, and pockets. *Yard lumber* is lumber that is graded and marketed on the basis that it will, or at least may, be used in the form in which it is sold. Accordingly all yard-lumber grades are based upon the establishment of minimum permissible defects. *Factory* or *shop lumber* on the other hand is graded and marketed on the assumption that it is to be cut up before it is used. Factory-lumber grades are therefore based upon the specification of a minimum fraction of the surface area that will yield cuttings of specified sizes and qualities.

1. Softwood Lumber Grades. Most softwood lumber is manufactured according to yard-lumber grades. This is the type of lumber that is commonly used in house construction, and the sort usually stocked by retail building-supply dealers. *The structural grades* are special groups of lumber grades also based upon the assumption that the piece will be used as purchased. Structural lumber is used where strength is of major importance; in fact structural-lumber grades are not uncommonly referred to as *stress grades*. Softwood factory lumber is used where individual boards are to be cut up to recover small clear pieces. Here the location of defects is much more important than their size. This type of lumber is commonly used in the manufacture of such millwork items as sash, doors, moldings, cabinets, etc., where small and relatively defect-free cuttings are required. The specific requirements of the various grades of softwood lumber are stated in the grading rules established by the trade association concerned with the species in question. In general all softwood grading rules conform to the broad criteria established in "Recommended American Lumber Standards." [3] Table 7-1 illustrates the general basis for softwood lumber grading common to most grading rules for softwood lumber.

2. Hardwood Lumber Grades. Basically, all hardwood lumber is graded on the assumption that it will be used in the form of cuttings of specified quality and size. When the hardwood lumber is sold in the form of boards, the defects remain in the piece, and factory-lumber grades are applicable. These grades are prepared and issued by the National Hardwood Lumber Association. They provide for the classification of boards on the basis of the fraction of the surface area that will yield clear, or sound, cuttings, meeting or exceeding certain minimum specified sizes. Sometimes a lumber manufacturer processes boards using ripping and crosscutting operations to eliminate unacceptable defects and converts the lumber into cuttings before shipment. Such cuttings are distributed as *hardwood dimension stock* and are sold subject to the grade requirements of Commercial Standard CS 60—48,* Hardwood Dimension Lumber, a grade classification developed and

---

* The second number in the designation of a commercial standard refers to the date that the revision was promulgated. Users should always be certain that they have the most recent revision.

Table 7-1. General Classification of Softwood Lumber ᵃ

Softwood lumber is divided into three main classes—yard lumber, structural lumber (often referred to under the general term "timber"), and factory and shop lumber. The following classification of softwood lumber gives the grade names in general use by lumber manufacturers' associations for the various classes of lumber

Class	Type	Subtype	Detail	Grades
SOFTWOOD LUMBER (This classification applies to rough or dressed lumber: sizes given are nominal)	YARD LUMBER (lumber less than 5 in. thick intended for general building purposes; grading based on use of the entire piece)	Finish (4 in. and under thick and 16 in. and under wide)		A, B, C, D
		Common boards (less than 2 in. thick and 1 in. or more wide)		No. 1, No. 2, No. 3, No. 4, No. 5
		Common dimension (2 in. and under 5 in. thick, and 2 in. or more wide)	Planks (2 in. and under 4 in. thick and 8 in. or more wide)	No. 1, No. 2, No. 3
			Scantling (2 in. and under 5 in. thick and less than 8 in. wide)	No. 1, No. 2, No. 3
			Heavy joists (4 in. thick and 8 in. or more wide)	No. 1, No. 2, No. 3
	STRUCTURAL LUMBER (lumber 5 in. or more thick and wide, except joists and planks; grading based on strength and on use of entire piece)	Joists and planks (2 to 4 in. thick and 4 in. or more wide)		
		Beams and stringers (5 in. or more thick and 8 in. or more wide)		
		Posts and timbers (5 by 5 in. and larger)		
	FACTORY AND SHOP LUMBER (grading based on area of piece suitable for cuttings of certain size and quality)	Factory plank graded for door, sash, and other cuttings (1¼ in. or more thick and 5 in. or more wide)		Association grading rules should be referred to for standard grades and sizes
		Shop lumber graded for general cut-up purposes		

ᵃ SOURCE: U.S. Department of Agriculture, Forest Service. Timber Resources for America's Future. 1955.

accepted by a majority of the trade concerned and issued by the U.S. Department of Commerce. Most hardwood lumber is sold in the form of boards according to factory-lumber grades, with only a small fraction marketed as dimension stock. Table 7-2 gives the basic requirements of the factory-lumber grading rules for hardwood. Individual species have special requirements, and the user should refer to the published grading rules for these exceptions and special requirements.

3. Measurement. The standard unit of measurement for lumber 1 in. or greater in thickness is the *board foot*. Lumber less than 1 in. in thickness is commonly sold on the basis of surface measure. Lumber is cut to nominal inch thicknesses. Shrinkage during drying and surfacing reduces the thickness of individual boards, but for marketing purposes the volume is still computed on the basis of the nominal dimension. During manufacture lumber may be sawed to even-inch widths or it may be ripped to random widths. Softwood lumber is customarily produced in even widths and hardwood lumber in random widths. Softwood yard lumber is usually manufactured to standard lengths in 2-ft multiples. Softwood factory and shop lumber and

Table 7-2. Standard Hardwood Grades [a]

Grade and lengths allowed, ft	Widths allowed, in.	Surface measure of pieces, sq ft	Amount of each piece that must work into clear-face cuttings, %	Maximum number of cuttings allowed	Minimum size of cuttings required
Firsts: [b]	6+	4 to 9	91⅔	1	4 in. by 5 ft, or
8 to 16 (will admit 30% of 8 to		10 to 14	91⅔	2	3 in. by 7 ft
11-ft, ½ of which may be 8		15+	91⅔	3	
and 9-ft)					
Seconds: [c]	6+	4 and 5	83⅓	1	4 in. by 5 ft, or
8 to 16 (will admit 30% of 8 to		6 and 7	83⅓	1	3 in. by 7 ft
11-ft, ½ of which may be 8		6 and 7	91⅔	2	
and 9-ft)		8 to 11	83⅓	2	
		8 to 11	91⅔	3	
		12 to 15	83⅓	3	
		12 to 15	91⅔	4	
		16+	83⅓	4	
Selects:	4+	2 and 3	91⅔	1	4 in. by 5 ft, or
6 to 16 (will admit 30% of 6 to		4+	[d]		3 in. by 7 ft
11-ft, ½ of which may be 6					
and 7-ft)					
No. 1 Common:	3+	1	100	0	4 in. by 2 ft, or
4 to 16 (will admit 10% of 4 to		2	75	1	3 in. by 3 ft
7-ft, ½ of which may be 4 and		3 and 4	66⅔	1	
5-ft)		3 and 4	75	2	
		5 to 7	66⅔	2	
		5 to 7	75	3	
		8 to 10	66⅔	3	
		11 to 13	66⅔	4	
		14+	66⅔	5	
No. 2 Common:	3+	1	66⅔	1	3 in. by 2 ft
4 to 16 (will admit 30% of 4 to		2 and 3	50	1	
7-ft, ½ of which may be 4 and		2 and 3	66⅔	2	
5-ft)		4 and 5	50	2	
		4 and 5	66⅔	3	
		6 and 7	50	3	
		6 and 7	66⅔	4	
		8 and 9	50	4	
		10 and 11	50	5	
		12 and 13	50	6	
		14+	50	7	
No. 3A Common:	3+	1+	33⅓ [e]	[f]	3 in. by 2 ft
4 to 16 (will admit 50% of 4 to					
7-ft, ½ of which may be 4 and					
5-ft)					
No. 3B Common:	3+	1+	25 [g]	[f]	1½ in. by 2 ft
4 to 16 (will admit 50% of 4 to					
7-ft, ½ of which may be 4 and					
5-ft)					

[a] SOURCE: U.S. Department of Agriculture, Forest Service. Timber Resources for America's Future. 1955.
[b] Inspection to be made on the poorer side of the piece, except in selects.
[c] Firsts and seconds are combined as one grade (FAS). The percentage of firsts required in the combined grade varies from 20 to 40 per cent, depending on the species.
[d] Same as seconds.
[e] This grade also admits pieces that grade not below No. 2 Common on the good face and have the reverse face sound.
[f] Not specified.
[g] The cuttings must be sound; clear face not required.

hardwood factory lumber are manufactured to standard lengths in multiples of 1 ft. Such millwork items as trim and molding are commonly sold by the linear foot.

4. Flooring. Flooring is manufactured from both hardwoods and softwoods. Softwood flooring is usually produced from the denser conifers such as Douglas-fir, western larch, and the southern pines. Among hardwoods, oak is the species used most commonly in the manufacture of flooring, but maple, birch, and beech are also widely used.

The major part of the flooring manufactured is produced to a standard match pattern. Vertical-grained and flat-grained flooring are available in both hardwood and softwood types.

Softwood flooring grades are *B and Better, C Select,* and *D Select.* Hardwood flooring grades vary with the species used. Oak flooring is available in the standard grades *Clear, Select, No. 1 Common,* and *No. 2 Common.* Hard maple flooring is available in four grades: *First Grade White, First Grade, Second Grade,* and *Third Grade.*

Hardwood flooring is most frequently sold as strip flooring, tongued and grooved and end-matched. Strip flooring is produced in random lengths. The most common thickness is $^{25}\!/_{32}$ in. with face widths of $1\frac{1}{2}$, 2, and $2\frac{1}{4}$ in. Hardwood flooring in the form of parquetry and plank is used in smaller quantities.

B. Manufacture of Lumber

The manufacture of lumber involves five primary processing functions: breakdown, ripping, crosscutting, drying, and surfacing. In any given circumstance all the basic functions may not be involved. For example, lumber sold rough and green has not been dried or surfaced. It is also not uncommon for the basic operations to be conducted in two or more separate manufacturing installations. A small ground mill may perform only the breakdown and ripping functions. The product of this mill may go to a concentration yard where crosscutting and drying are accomplished. It may go still later to a planing mill for surfacing. In contrast, at a large lumber-manufacturing installation all these operations will be performed in the same factory. Similarly the degree of sophistication of the operations may vary from one situation to another. When lumber is processed in small plants, the breakdown may be completed in a single operation. In a larger factory two or three separate operations may be required to perform the same function.

1. Breakdown. To convert round logs into lumber it is necessary first to reduce the log to boards, planks, or timbers. This is the breakdown. There are three basic types of log breakdown equipment: the circular headsaw, the band headsaw, and the gang headsaw.

a. Circular Headsaw. The circular headsaw consists of a circular saw of large diameter mounted on a mandrel. The log is supported on a carriage that runs on a double track situated approximately parallel to the plane of the saw blade (Fig. 7-2). Most circular headsaws are of the inserted-tooth variety. The carriage is driven either by a cable or by a hydraulic, steam, or air piston. The log is passed through the saw resulting in the removal of a slab or a board, plank, or timber. As successive cuts are made the sawyer examines the exposed face to determine how the turning and cutting operation should proceed in order to obtain the most advantageous breakdown in terms of yield and lumber grade. The circular headsaw is used in the largest number of sawmills in the United States. This type of headsaw is most appropriate to a small sawmill. Its principal disadvantage is that it removes a very wide kerf.

b. Band Headsaw. Large sawmills in America are most frequently equipped with band headsaws. The band saw consists of a continuous wide steel band

Fig. 7-2. Small sawmill in Minnesota. (*Courtesy American Forest Products Industries, Inc.*)

with teeth on one or both edges. Band saws have solid teeth as contrasted with the inserted teeth ordinarily found on circular headsaws. The band saw is mounted on two wheels. The lower wheel is powered, and the upper one free-running. The band-saw carriage is essentially the same as the carriage used with a circular headsaw. Since band saws are ordinarily used on much larger saw mills than circular saws, the carriages of band mills are usually much more elaborate and automatic (Fig. 7-3). This is, however, a function of mill size rather than headsaw type. During recent years fully automatic small circular mills have been built and effectively used.

Most band headsaws are single cutting; i.e., they have teeth on only one edge. Some band headsaws are constructed with teeth on both edges and are accordingly double cutting. With a double-cutting band saw a cut is made with the movement in each direction of the carriage. In contrast the single-cutting band saw cuts only on the forward stroke.

c. Gang Headsaw. Gang headsaws differ from circular and band headsaws in that the cuts made to achieve log breakdown are accomplished simultaneously instead of consecutively. Some gang saws consist of a battery of flat steel reciprocating saw blades mounted in a sash. With a round-log gang mill a feed carriage grips the log and forces it through the sash frame, where it is cut into boards or planks. A support carriage at the rear of the sash frame carries the boards away from the saw.

Gang headsaws represent the principal type of sawmill throughout most of the world. They are used more extensively outside the United States than within. Gang saws are, however, increasing in popularity in this country. The principal disadvantage of the gang saw is that it does not permit turning the log on the carriage to take maximum advantage of the opportunities to obtain higher grades of lumber. This disadvantage is most important when the logs being sawed are from old-growth timber. As the supply of virgin timber in the United States is depleted, the advantage of the circular and band headsaw over the gang headsaw is diminished. Gang headsaws are noted for the accuracy with which they cut lumber.

Fig. 7-3. Automatic air-operated carriage for large sawmill. (*Courtesy Heaps, Waterous Limited.*)

d. Resawing. In large sawmills the log breakdown may be accomplished in several operations. Large timbers or cants may be cut on the headsaw and then further reduced in a resawing operation. Resaws may be of either the single saw or gang saw variety. Single resaws sometimes use circular saws, but most frequently they utilize band saws. Mills are designated as horizontal or vertical depending upon the orientation of the saw blade. Small band mills, called pony rigs, are similar in construction to a band headsaw and are used as resaws in very large sawmills. Pony rigs are very fast and not uncommonly of the double-cutting variety.

2. Ripping. The ripping operation has as its function the removal of any wane or bark edges and the dimensioning of the board or plank to width. On very small ground mills ripping is sometimes done with the headsaw. The boards are stacked on the carriage, wany edges out, and then passed through the headsaw. In most sawmills the ripping operation is accomplished with an edger (Fig. 7-4). The typical edger has two or more saw blades mounted on a mandrel. The left-hand blade is usually fixed on the mandrel, while the remaining blade or blades are movable and may be spaced in each case to make the edging cut where required.

Fig. 7-4. Gang edger. (*Courtesy American Forest Products Industries, Inc.*)

Fig. 7-5. Trimmer. (*Courtesy American Forest Products Industries, Inc.*)

Careful ripping operations permit the manufacturer to take full advantage of the opportunities to obtain higher grades of lumber.

3. Crosscutting. The crosscutting operation has three principal functions: squaring the ends of the board, cutting to desired length, and removing defects. Machines used for crosscutting are called trimmers. They vary in size and complexity from the simple-swing cutoff saw used with a portable ground mill to the automatic multiple-saw trimmer used in a large sawmill (Fig. 7-5). Perhaps the most common form of trimmer in use is the two-saw type. This machine has two saws mounted on a long steel mandrel. The board is passed sideways through the saws and trimmed to length. As in the case of the edger, the left-hand saw is usually fixed and the right-hand saw adjustable on the mandrel to permit trimming to the desired length. Careful attention to the crosscutting operation is also essential if the maximum yield of upper grades is to be obtained.

4. Drying. Wood in the living tree contains a considerable quantity of water, some of which it will lose after manufacture into lumber unless it is

Fig. 7-6. Lumber packages piled prior to shipment. (*Courtesy American Forest Products Industries, Inc.*)

kept constantly wet. Some lumber is used green and permitted to dry in use. In the more discriminating uses of lumber it is necessary to dry the wood before utilizing it. Drying is normally accomplished in one of two ways: air-drying or kiln-drying. In some cases both air-drying and kiln-drying are used.

a. Air-drying. Lumber is air-dried by stacking it in such a way as to expose the individual boards to the atmosphere. The most common procedure in air-drying is to build up stacks of lumber flat piled in such a way that the individual boards in a layer are separated by about an inch, and the layers are separated from each other through the use of cross stickers again by about an inch. Piles of lumber stacked for air-drying are ordinarily sloped to permit rain to run off, thus speeding up the drying. The stacks are raised from the ground on foundations to permit free flow of air below the stack and to minimize decay and stain occurrence. Stacks are usually separated from each other by several feet to allow air movement along the sides and ends of the lumber pile. Temporary roofs are placed on the stacks to protect them from rain.

Variations from the flat-piling technique of air-drying are end-racking and crib-piling. End-racking permits faster drying, but it also results in greater warping and in the occurrence of more degrade from stain and decay. Crib-piling is usually used in the drying of short-length specialty products such as staves and dimension stock.

b. Kiln-drying. Kiln-drying of lumber is accomplished by placing it in a chamber where the atmospheric conditions are subject to more precise control than is possible in air-drying. The factors normally subject to control in kiln-drying operations are dry-bulb temperature, relative humidity, air velocity, and time of exposure. The nature and extent of control possible depend upon the kind of kiln that is used. Generally lumber dry kilns are characterized in terms of the method of charging and the nature of the air circulation provided.

Based upon charging method used, kilns are classified either as *progressive kilns* or as *compartment kilns.*

Progressive kilns are usually several hundred feet long. They are charged at one end and discharged at the other. At any given time all the conditions of wet- and dry-bulb temperature required by the drying schedule are present in some part of the progressive kiln. Normally the green, or charge, end of the kiln is maintained at a relatively low dry-bulb temperature and high relative humidity. The dry, or discharge, end of the kiln is maintained at a higher temperature and lower relative humidity. As the lumber moves from the green to the dry end of the kiln, it is subjected to increasingly severe drying conditions. Because lumber in all stages of drying is moving through the kiln at the same rate, the degree of control that can be exercised over the drying operation is somewhat limited.

In *compartment kilns* the drying chamber is filled with green or partially dry lumber and may then be subjected to a sequential series of temperature and relative humidity conditions most conducive to rapid drying with a minimum of drying-induced defects. Because all lumber in the kiln at a given time can be at approximately the same condition of dryness, it is possible to use much more severe drying schedules than can be applied in a progressive kiln. If rapid drying results in development of stress or non-uniform moisture content within the charge, these deficiencies can be corrected through the application of conditioning and equalizing treatments at the end of the drying schedule.

On the basis of air circulation lumber-drying kilns are characterized as natural-draft or forced-draft kilns. Natural-draft kilns depend for air circulation upon maintaining air density differentials between one part of the kiln and another. The heating units are ordinarily under the lumber load. Air passing over the heat source increases in temperature and decreases in density. As it becomes less dense, it rises in the kiln, while the heavier cool air flows downward. As the hot air rises through the load, it becomes cooler and denser, as it gives up its heat to the lumber. This process continues, providing for a natural draft or gravity circulation. Usually the lumber is stacked and baffled in such a way as to allow the warm air to move up through the load and the cool air to move down alongside the load and escape the kiln through vents or chimneys. The vented air is replaced by fresh air introduced into the kiln below the heat source.

In forced-draft kilns the air is forced past the heat source and through the load by means of power-driven fans. The load is stacked and baffled in such a way as to ensure an air circulation most conducive to rapid and uniform drying.

Relative humidity is controlled in a dry kiln in two ways. Through adjustment of the vent openings, the moist air leaving the load can either be vented and replaced by dry air from outside or be recirculated in the kiln. If more moisture is required than is supplied from the evaporation of water from the wood, this can be introduced into the air stream by means of water or steam sprays.

The most modern dry kilns are forced-draft, compartment kilns. These are subject to much more precise control of drying conditions than are the natural-draft and progressive kilns. Virtually all new installations are of the forced-draft, compartment type.

5. Surfacing and Shaping. Following the drying operation, lumber may be surfaced and/or shaped. These are machining operations that are used to manufacture the lumber to more accurate dimensions, improve its surface characteristics, or impart to it a desired shape. Modern surfacing and shaping machines utilize high-speed cutterheads to perform the desired machining operations.

The most simple production surfacing machine is the *single-headed planer* or *jointer*. In this machine the rough board is moved across a cutterhead by means of live feed rolls, and a measured cut is removed from one surface of the board. In this machine only one face of the board is surfaced at a time.

The *double-headed planer* differs from the single-headed planer in that the rough board is fed between two parallel cutterheads. The board thickness obtained is determined by the spacing of the two cutterheads.

The *planer* and *matcher* is a machine equipped with four cutterheads. It can surface all four faces of the rough lumber in a single pass. The cutter-heads that surface the faces are installed ahead of those that surface the edges. This is essentially a machine for squaring up a rough board. It is possible to use profile attachments on a planer and matcher to achieve various shapes of finished product.

The *molder*, or *sticker*, is a machine similar to a planer and matcher but specifically designed to permit the manufacture of shaped products. The molder has at least four and frequently more cutting heads. They provide for a wide variety of cuts, and it is possible on this type of equipment to work four or more sides of the stock to shape in a single pass. Such items as flooring, molding, dowels, furniture stock, and sash and door items are commonly produced on the molder.

C. Status of the Industry

The lumber industry is characterized by a continuing decrease in the size and quality of its raw material, coincident with an increase in stumpage prices and labor costs associated with processing. To combat these unfavorable factors the industry is in the midst of an evolution that may permit the maintenance of a favorable competitive position in the market.

According to "Timber Resources for America's Future," [8] in 1952 about 1 out of every 4 ft of timber cut, or available from other sources, was not utilized. During recent years important strides have been made in the direction of improving the utilization of sawtimber. Many of the most progressive lumber-manufacturing enterprises are now utilizing slabs, edgings, and trimmings for the production of chips for the pulp and paper industry and the particle-board industry. Development of better methods of quality control and improved machinery are reducing waste. Improved technology has made it possible to convert into lumber logs that would have been rejected for lumber production a decade ago.

Labor costs are being reduced substantially through the development of automatic equipment, through increased use of conveyers in lumber plants, and through better plant layout.

While the major improvements in technology are in evidence in only the most modern factories, the number of plants that are taking advantage of technological advances is rapidly increasing. It is reasonable to assume that modernization of manufacturing facilities will accelerate and that further technological improvements will be forthcoming.

SELECTED REFERENCES

1. Bethel, J. S., and A. C. Barefoot. Can Lumber Compete? *Forest Prod. Jour.*, **8**(7):9A–14A. 1958.
2. Brown, N. C., and J. S. Bethel. Lumber, 2d ed. John Wiley & Sons, Inc., New York. 1958.
3. Central Committee on Lumber Standards. Recommended American Lumber Standards. U.S. Department of Commerce Report. 1923.
4. Glesinger, E. The Coming Age of Wood. Simon and Schuster, Inc., New York. 1949.

5. Horn, S. F. This Fascinating Lumber Business. The Bobbs-Merrill Company, Inc., Indianapolis. 1943.
6. President's Materials Policy Commission. Resources for Freedom. 1952.
7. Stanford Research Institute. America's Demand for Wood, 1929–1975. Palo Alto, Calif. 1954.
8. U.S. Department of Agriculture, Forest Service. Timber Resources for America's Future. Forest Resource Report No. 14. 1958.

CHAPTER 8

VENEERS AND PLYWOOD

The art of veneering is nearly as old as civilization itself and actually ante-dates the birth of Christ by more than 1,500 years. Exquisitely designed and skillfully fabricated pieces of plywood furniture found in the tombs of the Pharaohs give silent testimony to the dexterity and artistry of the ancient Egyptian cabinetmaker. A number of the essential features of a modern plywood panel were found in the headboard of a bedstead reputed to have belonged to the grandparents of Tutankhamen's wife. Made of laburnum wood and lavishly embellished with gold and precious stones, it was as sound as though it were made but yesterday. How the wood was cut into thin, delicate sheets of veneers or what was the nature of the adhesive that stood the tests of 35 centuries, however, is still a matter of speculation.

In the years that followed, Assyrian, Babylonian, and Roman craftsmen, influenced by the Egyptian stylists, made several significant advances in the use of veneers. The Romans developed the art of matching figured veneers to produce unusual decorative effects as well as beautiful patterns of warmth and charm. In his "Natural History," Book XVI, Pliny revealed that the root-wood of certain species was often used, thus indicating that the orna-mental value of stump wood and root-wood has long been recognized. He also recorded that among the many personal treasures of Julius Caesar, the one held in highest esteem was a beautifully veneered table.

During the Middle Ages, oppressive ecclesiastical and political orthodoxy effectively stifled creative thinking, cultural interests, and practice of the arts, but toward the end of the seventeenth century a revival of learning became manifest, and in the years that followed some of the world's finest masterpieces in sculpture and painting were created. Interest in woodworking and marquetry was reborn, and in 1769 Riesener completed for Louis XV his "Bureau du Roi," said by many in authority to be the finest specimen of veneered furniture of all times. The eighteenth century witnessed the peak in period stylings and with it the creations of such renowned craftsmen as Duncan Phyfe, Hepplewhite, Chippendale, Sheraton, and the Adams brothers.

Until the middle of the nineteenth century the finest furniture was hand-made, and into each piece went all the skill, ingenuity, artistry, and pride

120

of the master. Quality, not quantity, was his watchword. The work was slow, painstaking, and laborious; the furniture costly and available only to the wealthy.

About 1840, furniture manufacturers began to mechanize their shops in order to produce goods on a mass-production basis. In some of these plants quality was readily sacrificed for quantity. Ornate inlays and a variety of highly figured veneers were applied over a poorly fabricated article merely for the purpose of concealing shoddy workmanship. Eventually this practice affected adversely the sale of all veneered furniture, because buyers began to regard such merchandise as an inferior substitute for comparable articles constructed of solid wood. This concept is still held by many uninformed even today. Our leading dictionaries do little to dispel this view, defining veneer as "a layer of more valuable or beautiful material over an inferior one."

Veneering as practiced today, however, needs no defense, but rather an intelligent understanding of the principles, purposes, and advantages of veneered construction. Wood technologists, engineers, chemists, and designers working together harmoniously have pooled their knowledge and skills to produce a superior product. This is plywood. How it is made, its advantages over solid wood, and its many and diverse uses will now be reviewed.

I. VENEERS

Veneer is a thin sheet of wood of uniform thickness produced by peeling, slicing, or sawing logs, bolts, or flitches. Veneer produced by peeling a bolt on a lathe is commonly referred to as rotary-cut veneer. It exhibits a flat-grained figure. Veneer manufactured with a slicer, a stay-log, or a veneer saw may be flat-grained, vertical-grained (quartered), or on some grain bias to give a desired figure. Veneer cut on a slicer is referred to as sliced veneer and that cut on a stay-log as half-round veneer. While some veneer is used as such in the production of containers such as baskets, hampers, crates, boxes, drums, and a few specialty products, the principal reason for cutting logs into veneer is for the purpose of producing plywood panels.

A. Manufacture of Veneers

There are at present five methods employed in the manufacture of veneers. Prior to the twentieth century all veneers were produced by sawing. In this process logs were first cut into flitches, and the flitches in turn were reduced to thin sheets of wood by driving them against a circular saw. This was both an expensive and a wasteful procedure, since much of the best and clearest wood in a log was removed as slabs and more than half the potential veneer in the flitches was reduced to sawdust in the cutting operation. Veneer saws long since have been largely replaced by modern rotary lathes, and less than 5 per cent of all veneers now manufactured in the United States are sawed.

1. Rotary-cut Veneer. Today substantially more than 90 per cent of all veneers are rotary-cut. In this method of cutting, a bolt of wood is centered between two chucks on a lathe. The bolt is turned against a knife extending across the length of the lathe. Fig. 8-1 is a diagrammatic sketch of the modern veneer lathe. The nose bar, or pressure bar, presses against the log and compresses the wood just ahead of the knife. As the log turns and the compressed wood contacts the knife-edge, a thin sheet of veneer is peeled off.

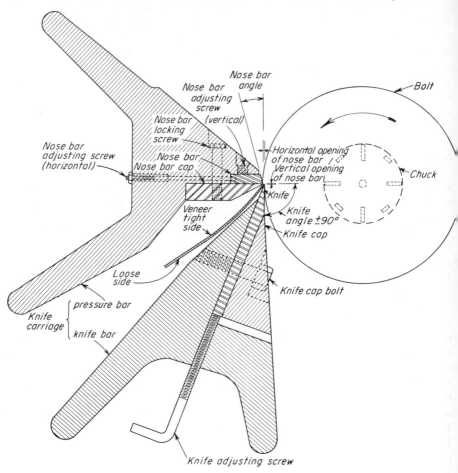

Fig. 8-1. Diagram showing relationship of lathe carriage and veneer bolt. (*Courtesy U.S. Forest Products Laboratory.*)

In rotary cutting, a bolt is literally unwound (much as one might unroll a bolt of wrapping paper), and the veneer is produced in a long, continuous ribbon of wood. Giant lathes capable of peeling logs 12 to 16 ft in length are used in a few plants. More commonly, however, veneer is cut in lengths ranging from 2 to 8 ft. The bolts that are turned into veneer are usually cut from 4 to 6 in. longer than the length of veneer to be processed from them. The veneer is cut to width by means of spur knives mounted on the pressure-bar carriage. Because Fig. 8-1 is a cutaway diagram, the spur knives are not shown.

The quality of the veneer that is cut is determined in large part by the setup of the lathe. The principal adjustments made in setting up a lathe are knife height, knife angle, vertical-bar opening, and horizontal-bar opening. When the knife is placed in the lathe, it is raised through the use of the knife-adjusting screw until the knife-edge is parallel with the chuck centers and at the same elevation with them. The knife angle, i.e., the angle subtended by the face of the knife and a line from the chuck center to the knife-edge, must be correctly set. The proper knife-angle setting depends upon the species being cut, the condition of the bolt, the concave of the knife, and the diameter of the bolt. When the knife angle is excessive, the knife chatters in cutting and produces a short ripple in the veneer. When the knife angle is too small, the heel of the knife rides on the bolt resulting in a long, undulating wave in the veneer. Both errors result in the production of veneer that is nonuniform in thickness. Since the optimum knife angle is not the same for all bolt diameters, most modern lathes automatically adjust the knife angle as the diameter of the bolt decreases.

The best quality veneer must be uniform in thickness, smooth, and free of excessive checking. Rotary-cut veneer is characterized by the presence of small checks, called lathe checks, on that side of the veneer sheet that was originally nearest the center of the block. The lathe checks are formed when, in the cutting process, the veneer is bent sharply as it passes between the knife and the nose bar (Fig. 8-2A). Since the sheet of veneer is essentially a circumference of a circle, when it is flattened out for use the checks on the knife side of the veneer are opened up. In the veneer industry the knife side of the veneer is called *the loose side*, and the bar side is called *the tight side*. Veneer that has many deep lathe checks is called *loose-cut* veneer (Fig. 8-2B). Veneer that has shallow lathe checks is called *tight-cut* veneer. Proper setting of the nose bar results in the production of tight-cut veneer (Fig. 8-2C). To produce tight veneer, the nose bar must be set so that the distance between the edge of the bar and the edge of the knife is less than the thickness of the veneer being cut. If the nose-bar opening is too great, the veneer will not be compressed sufficiently and will be uneven in thickness and loose. If the nose-bar opening is too small, the veneer will be compressed beyond the proportional limit of the wood and will be too thin (Fig. 8-2D). In the extreme case of too close a nose-bar setting, the force required to push the veneer between the bar and the knife may exceed the power of the lathe, and it will choke down and refuse to cut. Under these conditions the chucks are likely to be twisted out of the block.

Where it is desired to cut smooth, tight veneer, logs of some species are commonly heated in hot water or steam before cutting. The hot wood is more plastic, can be cut smoother, and will bend over the knife with minimum checking. High-density woods such as oak, maple, birch, gum, etc., require heating to produce smooth, tight veneers. Even low-density woods such as poplar are heated if smooth veneer is required in substantial thicknesses (usually above $\frac{1}{8}$ in.). Care must be exercised to avoid getting low-density woods too hot since this causes the veneer to be fuzzy.

The practice of heating logs prior to cutting has been observed almost universally by the hardwood-plywood industry. At one time this was also

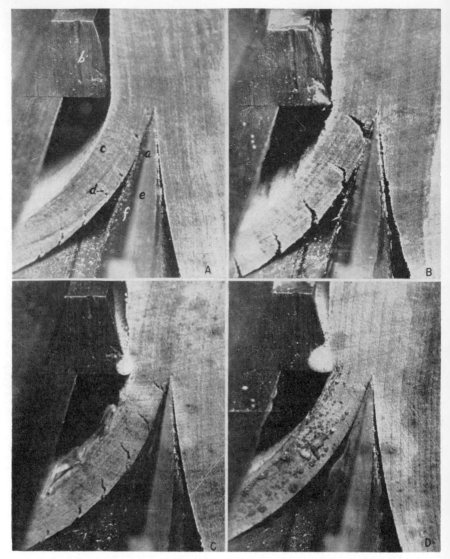

Fig. 8-2. *A.* Cutting action and relationship of knife, nose bar, veneer, and block during cutting: *a*, knife; *b*, nose bar; *c*, veneer; *d*, lathe checks; *e*, face of knife; *f*, back of knife. *B.* No nose-bar pressure, veneer very loose. *C.* Nose bar set at 93 per cent of veneer thickness, veneer moderately loose. *D.* Nose bar set at 86 per cent of veneer thickness, veneer very tight, overcompressed. (*Courtesy U.S. Forest Products Laboratory.*)

a standard practice in the softwood-plywood industry. Heating of veneer logs serves not only to improve the cutting properties of the wood but also to facilitate bark removal. With the introduction of the roller nose bar and the mechanical barker, most softwood-plywood plants ceased the practice of heating logs before cutting. In recent years there has been a revival of interest in the practice of heating logs on the part of softwood-plywood manufacturers, and several West Coast mills have reinstituted this manufacturing practice.

In the manufacture of plywood it is common practice to have the tight side of face veneers on the outside unless it is necessary to reverse it to match grain or figure. To make certain that the tight side is exposed in plywood panels, it is customary plywood-plant management that in all stacking, handling, and processing operations, the tight side be up. To facilitate this orientation, a scribing device on the lathe, a pencil or crayon, marks the

Fig. 8-3. Stay-log cutting of a walnut stump. (*Courtesy Wood Mosaic Company, Louisville, Kentucky.*)

sheet on the nose-bar side. This mark may be sanded out or trimmed off later in the finishing of the plywood panel.

2. *Stay-log Cutting.* This is merely a modification of rotary cutting and was developed primarily for producing fancy face veneers from quarterflitches and irregularly shaped materials such as burls, crotches, and stumps (Fig. 8-3).

A stay-log is a long, flanged steel casting mounted in eccentric chucks of a conventional lathe. To produce veneer, a flitch or section from a stump, crotch, or burl is securely fastened to the flange with several heavy lag screws, and then turned against the knife. Large flitches and even half-bolts mounted and cut in the manner illustrated (Fig. 8-4*A*) are called *half-round cut*. The manner in which the bolt in Fig. 8-4*B* is mounted produces *back-cut* veneer. Back-cut veneers obtained from stump wood and crotches are usually handsomely figured because fiber alignment in materials of these kinds is very irregular. In fact much fiber is often cut across rather than along the grain with the result that the sheets of veneer are very brittle and special care must be exercised in handling them.

Stay-log cutting permits the manufacture of wider sheets of veneer than it is ordinarily possible to produce in conventional slicing operations described in section 4.

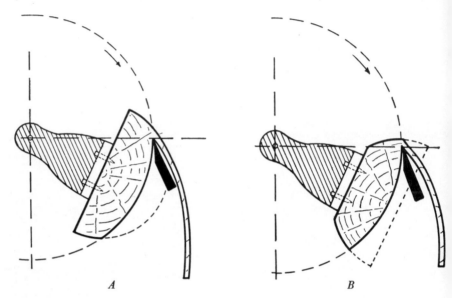

Fig. 8-4. Diagrams showing stay-log cutting: *A*, half-round; *B*, back-cut.

3. Cone-cutting. Cone-cutting produces circular sheets of veneer by taper-peeling a bolt in a manner similar to that of sharpening a wooden pencil. The contact angle of the knife determines the degree of taper, the approximate width of the sheet for a bolt of a given diameter, and the number of revolutions the bolt must make to produce a completely circular sheet. Only a very small amount of veneer is produced in this manner, as such veneers are ordinarily "short-grained" and hence quite brittle. Beautiful "wheel" or stellate figures result, and the veneers are used in the fabrication of panels for fancy, circular table tops.

4. Sliced Veneers. Veneer slicers were developed in an effort to eliminate the wasteful practice of sawing. With numerous improvements of the original slicing device, most of the figured woods that were formerly sawed into face veneers are now cut on slicers. While slicers can be employed to produce veneers of all types, they are used largely for cutting only figured woods for face stocks.

Sliced veneers are usually cut in lengths of from 12 to 16 ft, although cutting is not restricted to these sizes. Logs to be used for the production of sliced veneers are first halved lengthwise on a circular or band saw and then not infrequently cut into smaller flitches at the option of the producer.

An experienced and skillful operator is usually in charge of preparing such flitches. Upon inspection of an opened log he must be able to visualize the type of figure it possesses and to indicate how it is to be flitched in order that this figure may be best brought out in the veneers subsequently cut from it. If in his judgment radial slicing is desired, the half log is further reduced to six or eight flitches for the production of quartered veneers, obtained when the slicing is parallel to the wood rays and at right angles to the growth rings. If, on the other hand, "flat-cut" veneers * are to be sliced, no further flitching is ordinarily required. Before the actual slicing operation begins,

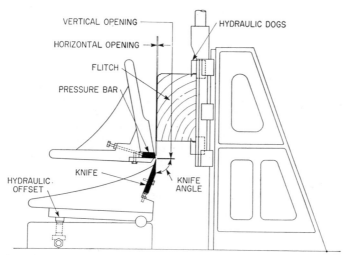

Fig. 8-5. Diagram showing relationship of knife, pressure bar, and flitch on vertical veneer slicer. (*Courtesy U.S. Forest Products Laboratory.*)

it is necessary to soften the stock in steam or hot-water vats in a manner similar to the treatment of bolts for rotary cutting.

The most commonly used slicer in the veneer plants of the United States is the vertical type (Figs. 8-5 and 8-6). In practice a flitch is firmly "dogged" to a heavy metal bedplate which is fastened to two angling slides set in a heavy metal frame. The flitch is driven down against a pressure bar and knife, and the resulting sheet passes out through a slot in the knife carriage in a manner similar to that of a rotary lathe. On the upstroke of the flitch, the knife carriage moves forward and thus is in position to slice off another sheet with the next cutting stroke. The angling slides give the flitch a slight lateral movement as it advances toward the knife. This makes for smoother veneers, since the cutting action is somewhat along the grain of the wood rather than directly across the fibers. Such a shearing action is particularly helpful in producing smooth surfaces when woods with curly, wavy, or interlocked grain are sliced.

In most European veneer factories sliced veneer is produced on horizontal

---

* Flat-cut veneers are made by slicing half-sections of logs. In such a case both quarter-grained and flat-grained figures are commonly present in the same sheet.

slicers. These are massive machines with a production rate that is much slower than that normally achieved by the American high-speed vertical slicers. It is claimed that these horizontal slicers produce veneer that is more accurately dimensioned than that manufactured on vertical slicers.

There are also several special types of slicers designed largely for production of battery separators and shooks for fruit, vegetable, and cheese boxes and crates and similar containers.

In the slicing of face veneers the sheets from a given flitch are turned over

Fig. 8-6. Slicing a mahogany flitch. (*Courtesy The Mahogany Association, Chicago, Illinois.*)

as they come from the knife and stacked consecutively, exactly in the order cut. Such a bundle is also called a *flitch;* in subsequent handlings it is regarded as a unit, and special care is exercised to preserve sheet sequence at all times. The reasons for this will be explained later.

As in rotary veneer, the side of a sliced veneer next to the knife contains numerous fine checks. Therefore, whenever possible, care must be taken to spread glue on the loose side; in matched veneers, however, in order to produce a balanced figure, some pieces must be glued loose side up.

5. Sawed Veneers. As previously indicated, little veneer is produced by sawing at the present time. Ordinarily, sawing is restricted to cabinet woods that are either highly refractory and unsuitable for slicing (ebony, oaks, satinwood, knotty eastern redcedar) or that cannot be softened in cooking vats without impairing their color.

The modern veneer saw is known as a *segment saw.* It consists of a heavy metal tapering flange (mounted on an arbor) to which are bolted several thin, steel saw segments along its periphery. In this way the kerf is considerably less than that of a conventional circular saw blade of similar size. Despite this, about half of a flitch is reduced to sawdust in the production of $\frac{1}{20}$-in. veneer (Fig. 8-7).

6. Veneer Thicknesses. Veneers are cut in a wide variety of thicknesses ranging from $\frac{1}{110}$ to $\frac{3}{8}$ in. Most of the rotary veneers are either $\frac{1}{7}$, $\frac{1}{8}$, $\frac{1}{10}$, $\frac{1}{16}$, or $\frac{1}{20}$ in. thick. During the Second World War, rotary-cut

Fig. 8-7. Veneer saw in operation. (*Courtesy Hardwood Plywood Institute, Arlington, Virginia.*)

mahogany veneers $\frac{1}{64}$ in. in thickness were used in the manufacture of wing skins for aircraft. Sliced veneers usually range from $\frac{1}{20}$ to $\frac{1}{40}$ in., but the bulk of face veneers so cut are either $\frac{1}{20}$, $\frac{1}{28}$, or $\frac{1}{32}$ in. in thickness. The very thin sheets of Spanish-cedar used to wrap fine cigars individually and some other specialty veneers are often only $\frac{1}{100}$ to $\frac{1}{110}$ in. thick.

Sawed veneers vary from $\frac{1}{4}$ to $\frac{1}{32}$ in. in thickness, with $\frac{1}{20}$-in. stock being the most common.

7. Green-clipping. The ribbon of veneer coming from the lathe is cut to rough green size, and defects are removed at the green clipper. Two general types of clippers are employed for this operation: *the shear clipper* and *the anvil clipper*. The shear clipper (Fig. 8-8) consists of a large knife that moves past a clipper bar with a shearing action. The knife is activated electrically, pneumatically, or hydraulically. The anvil clipper has a knife that strikes a hardwood anvil when activated. The veneer is clipped by placing it between the knife and the anvil. Green clippers may be activated by an operator who manipulates a switch when a cut is appropriate or they may be equipped to cut automatically at a preset interval of distance.

The veneer-cutting operation is an intermittent one. It continues while the block is in the lathe and stops while a new block is being placed in the lathe and chucked. Since the clipping operation is relatively continuous, it is necessary to devise special veneer-handling operations between the lathe and the green clipper. Three methods are commonly used to accomplish this purpose: stack-and-bulk-clip, storage-deck, and reel.

Fig. 8-8. Shear-type veneer clipper. (*Courtesy Merritt Engineering and Sales Company, Inc., Lockport, New York.*)

a. Stack-and-bulk-clip Method. This is the oldest method of handling rotary-cut veneer behind the lathe. The veneer is carried by workmen from the lathe to a table equipped with a chain conveyer. When the end of the table is reached, the sheet of veneer is broken, and the next sheet is carried back on top of the first. When a stack is built up, it is moved up to the clipper by means of the chain conveyer, and a new stack is built up. The completed stacks are clipped in bulk to rough green widths without regard to defect, since only the surface of the top sheet is visible to the clipper operator. When this method is used defects are removed in the dry-clipping operation behind the drier.

b. Storage-deck Method. Since the stack-and-bulk-clip method does not permit removal of defects in green-clipping, it is considered less efficient and more wasteful of veneer and of drier time than methods that permit single-sheet clipping. The storage-deck method was developed in the softwood-plywood industry to facilitate rapid single-sheet clipping. It has been adapted by some of the largest hardwood-plywood plants. In this method a series of decks equipped with belt conveyers is placed behind the lathe. A *tipple,* also equipped with conveyer belts, is installed at each end of the storage decks. The tipple is a section of conveyer that bridges the space between the lathe and the storage decks. At the lathe end the tipple is hinged at the rear of the lathe below and behind the carriage. It is raised or lowered from one storage deck to another so that veneer coming from the lathe can be stored on an empty deck as a single sheet awaiting clipping. A similar tipple at the clipper end of the storage decks permits the veneer to be removed from a filled deck and conveyed to the green clipper, where defects are cut out and the veneer is clipped to an appropriate size. The storage-deck is more efficient than the stack-and-bulk-clip method since it permits single-sheet clipping. Its principal disadvantage is that it requires a considerable amount of factory space. The storage decks in the largest plants are not uncommonly hundreds of feet in length. This method is feasible only in the largest factories.

c. Reeling. A method of handling green veneer used in some hardwood plants and where space is at a premium is the reeling method. Here the veneer sheet is carried from the lathe and attached to a reel mounted on a shaft. As the veneer comes from the lathe, the reel is turned, and the green veneer is wrapped up on the reel. The loaded reel can be removed and stored ahead of the clipper until it is unwound and the veneer clipped as single sheets to remove defects and to dimension to the proper size. In the more primitive operations the reel may be turned by hand. In mechanized operations the reel is equipped with some type of variable-speed drive and is turned mechanically. The reel system was developed in Europe but has become increasingly popular in American mills since the Second World War.

B. Veneer Drying

Freshly cut veneers are ordinarily very wet and in that condition are wholly unsuited for gluing. Moreover, while green they are readily susceptible to the attack of molds and of blue-stain and wood-destroying fungi. It is necessary, therefore, that excessive moisture be removed from them as rapidly as possible, consistent, of course, with good drying practices. Exception to this general rule may be found in the basket industry. Since wood is much more pliable when it is wet, it may be bent to much sharper curvature without danger of rupture. Hence the basket manufacturer commonly works with wet veneers and dries them only after the product is fabricated.

Several methods for drying veneers are in common use. The selection of one of them for any given plant is usually dependent upon the ultimate use of the stock and the facilities available to the manufacturer.

1. Air-drying. The cheap grades of rotary (and occasionally sliced) veneers used for fruit and vegetable crates, egg cases, and packing-case materials are sometimes dried in the open. In such instances they are dried merely enough to prohibit the growth of fungi and to reduce shipping weight. When air-drying is practiced, the sheets are laid up between tiers of stickers or loosely stacked on end in finger racks in order to permit free circulation of air over their surfaces.

2. Loft-drying. A loft is a well-ventilated room with or without humidity control. Wet sheets of veneer are hung on clips from rafters or edge-stacked in finger racks. Here they remain until the moisture content stabilizes at from 12 to 15 per cent. In drying thin sheets of the order of $\frac{1}{20}$ in. in thickness, only a day or two is required to bring the moisture content down to within these limits.

3. Kiln-drying. The conventional progressive-type and compartment-type lumber kilns are also used in drying veneers; usually several sheets are bulked between each tier of stickers. One plant in North Carolina producing $\frac{1}{8}$-in. rotary redgum veneer bulks 10 sheets between stickers and dries them to a 6 per cent moisture content in 48 hours.

4. Veneer Driers. At the present time most veneers are dried in carefully designed and engineered driers that permit rapid and uniform seasoning without adversely affecting the sheets. Two types of such equipment, viz., conveyer driers and hot-plate driers, are in common use.

Conveyer driers are chambers 50 to 300 ft or more in length, fabricated of sheet metal and suitably provided with heating equipment and two to eight banks of power-driven rolls or belts to move the veneers longitudinally through them.

Hot-plate driers consist of a bank or battery of heated platens. Sheets of veneer are placed on the upper faces of the platens, and then the platens are brought together to keep the sheets flat while drying. During the drying interval the platens are intermittently opened and closed to allow moisture to escape as it is driven from the wood. Driers of this type are commonly used as redriers. In this capacity they serve to bring veneer that has been previously dried to a final satisfactory moisture content just prior to gluing. They are commonly referred to as *plate driers* or *plate redriers*.

Still another type of modern drier combines the use of platens and conveyers. In installations of this kind, paired live rolls alternate with heated platens. The platens open and close like the platens of a plate drier. On the open phase of the cycle the veneer is advanced through the drier via the paired live rolls. On the closed phase the rolls are stopped, and the veneer is stationary. Good flat veneer is produced in this manner, but these driers lack the high capacity of continuous driers.

The majority of high-temperature (above 212°F) veneer driers depend upon steam as a heat source. They are usually equipped with radiators below each deck. More recently, direct-fired oil and gas driers have been introduced.

Most of the veneer produced in the United States is dried in *continuous-type, high-temperature driers* (Fig. 8-9). These driers are similar in basic principle to the progressive kilns sometimes used in drying lumber and are subject to many of the same limitations. They provide less control over the conditions of drying than is possible in compartment-type driers, and accordingly more attention must be paid to preliminary segregation of veneer into uniform groups when continuous driers are used.

If the average and range of moisture content of the veneer produced are to be satisfactory, the green veneer must be separated into groups so that all the veneer in any group has approximately the same drying characteristics.

Veneer should be separated according to species, thickness, heartwood or sapwood condition, and initial moisture content.

Separate drying schedules are developed for each class of veneer being processed. These schedules indicate the atmospheric conditions to be maintained in the drier. The usual continuous drier has facilities for controlling dry-bulb temperature over a wide range. The time of exposure in the drier is also subject to rather close control. The conveyer system, belt or roller, which moves the veneer through the drier is normally equipped with a variable-speed drive that permits the time in the drier to be established with accuracy.

Fig. 8-9. Chain-belt conveyer, high-temperature veneer drier. (*Courtesy Proctor and Schwartz Company.*)

The third drier variable, relative humidity, is subject to much less precise control. The moisture in the drying atmosphere is controlled through manipulation of the exhaust-stack damper. If a low relative humidity is desired, the damper is fully opened so that the moisture picked up during the drying process is exhausted and is replaced with relatively dry air drawn into the drier from outside. If a high relative humidity is required, the damper is closed, and the moisture-laden air is continuously recirculated in the drier.

The amount of control over temperature, time, and atmospheric moisture content available in high-temperature driers is adequate to permit uniform drying provided that materials with unlike drying characteristics are not mixed and dried together.

5. Drier Operations. In the operation of continuous veneer driers adequate control must be exercised over the drying operation if relatively uniform moisture content is to be achieved. Bethel [3] has pointed out that there are seven variables subject to control in the normal veneer-drying operation. These may be conveniently divided into two groups, as follows:

Variables in the drying operation:
1. Drying temperature
2. Time in the drier
3. Relative humidity

Variables in the material being dried:
1. Initial moisture content of green veneer
2. Species
3. Thickness
4. Heartwood or sapwood condition

Air velocity is an important variable where it is subject to control. Most continuous driers are equipped with fixed-speed fans. Where variable-speed mechanisms are installed on the drier, air velocity may be manipulated to improve control over the drying operation.

The basic objectives of the veneer-drying operation are to obtain a product with average moisture content appropriate to the use that is contemplated and with range in moisture content as small as is feasible.

C. The Veneer Industry

A large variety of wood, both foreign and domestic, is utilized in the manufacture of veneers. A very high percentage of veneer produced in the United States is manufactured from Douglas-fir. A number of other softwood species are also used. The hardwood-veneer industry utilizes redgum, blackgum, tupelos, yellow-poplar, oak, birch, maple, and, in smaller quantities, many other hardwood species. In considering the industry as a whole it is necessary to note that a distinction is made between the hardwood- and softwood-veneer industries because of the marked differences in utilization of their products. In the hardwood-veneer industry a further distinction is made between (1) face veneers, (2) commercial veneers, and (3) veneers for containers.*

1. Face-veneer Manufacturers. These usually limit their activities to the production of sliced, sawed, or stay-log fancy face veneers from carefully selected logs, burls, crotches, and stumps. Such materials are usually handsomely figured and are used as facings for plywood panels employed in the fabrication of furniture and pianos and for interior decorative paneling.

There are more than 50 manufacturers of face veneers scattered through the Ohio and Mississippi River valleys and adjacent territories, the Lake states, and in large cities along the Atlantic seaboard.

The manufacturer of face veneer cuts his raw materials as received and builds up sizable inventories, which are stored in well-ventilated warehouses. After a flitch is cut and the veneers dried, the sheets are bulked in sequence, and the flitch is given an identifying number. Samples, normally consisting of three sheets, one taken from near the top of the flitch, one from about the middle, and another from near the bottom, are marked with the identification number of the flitch, total square-footage content of veneer, and the

---

* The discussion of the hardwood-veneer industry is based in part on "The Veneer Industry," Educ. Ser. 2. The Veneer Association, Chicago, Ill.

number of pieces included in the package. These samples are then sent to salesmen who sell the entire flitch based upon these samples. The importance of preserving sheet sequence of face stocks now becomes evident. The buyer, upon examining the sample sheets, can determine the degree of uniformity of figure in the entire flitch and its suitability for his needs. A uniform figure is essential to the manufacturer who matches veneers to produce highly figured panels. A prospective buyer can also determine length, width, and thickness of sheets, variation in color, and smoothness of cut from properly selected samples of this kind.

In normal times some 80 or more kinds of face veneers are offered to the trade. Importation of exotic woods was greatly curtailed during the Second World War, and many stocks have long since been exhausted; several new species, however, have been introduced in the postwar era.

2. Commercial-veneer Manufacturers. Commercial veneers, or "utility" veneers, are those used for cross bands, cores and backs of plywood panels, and concealed parts of furniture such as drawer bottoms, case and mirror backs, and dust sealers. These are rotary-cut and are made from several species of domestic hardwoods. Manufacturers of commercial veneers are usually located near adequate supplies of suitable timber. Beech, birch, maple, and basswood veneers are produced in large quantities in Wisconsin, Michigan, and New York. Redgum (including sapgum), yellow-poplar, cottonwood, tupelo, sycamore, and oak veneers are made in many southeastern plants from the Ohio River region to the Atlantic and Gulf coastal plains.

Unlike the manufacturer of face veneers, the manufacturer of commercial veneers cuts on order, in which species, thickness, sheet size, and grade requirements are specified. A few plants, however, restrict their production to sheets of standard size. Most commercial veneers are sold to furniture, piano, or radio manufacturers who make their own plywood or to plywood plants that build stock panels or manufacture panels to customer specifications. In several instances a veneer plant is but a part of a large mill manufacturing plywood panels, in which case the entire output of veneer is consumed in the fabrication of panels within the mill.

3. Container-veneer Manufacturers. A wide variety of cheap veneers suitable for crates and packing cases, hampers, fruit and vegetable baskets and kits, meat, butter, and picnic dishes, cheese boxes, drums and hoops, and other similar items are classed as container veneers. Plants manufacturing them are usually established proximate to adequate supplies of cheap timber and suitable markets.

4. Softwood-veneer Manufacturers. Most softwood veneer is produced on the West Coast and in the Rocky Mountain region. In this area plants produce veneer from Douglas-fir, Sitka spruce, western hemlock, balsam fir, western larch, ponderosa pine, sugar pine, western white pine, and redwood. While the great bulk of this veneer is rotary-cut, some sliced softwood veneer is also produced.

Much of the softwood veneer produced in the West is shipped to softwood-plywood plants to be incorporated into softwood plywood. Some of it is used in western mills as center stock and cross-banding for panels fabricated with hardwood faces. Domestic and imported hardwood face stock is used for this purpose. Philippine lauan is extensively used as face stock over soft-wood interior plies. In the southeastern United States, baldcypress and some of the southern yellow pines are used to a very limited extent in the production of center stock for use with hardwood face veneers. Some softwood veneer is shipped from western veneer mills to hardwood-plywood plants in the East where it is also used for interior plies.

II. PLYWOOD

Plywood is a term applied to glued wood constructions built up of veneer layers so that the grain of each layer is at a right angle to that of the adjacent layer in the assembly. This method of assembling wood components is referred to as cross-banded construction. It is this cross-banded construction that distinguishes plywood from laminated wood, where the grain of successive layers is parallel. The cross-banded construction of plywood yields several advantages over the use of solid wood. The most significant advantages of plywood over solid wood are that dimensional stability is improved, strength properties are modified, and wood characteristics are rearranged to maximum advantage.

The simplest plywood construction is the three-ply panel. In this construction the center ply is referred to as the core, or center ply. The term *center* is more commonly used to refer to a veneer ply and *core* to refer to a laminated lumber core or to the core of a particle board. The outer plies are designated *faces* or *backs* depending upon the grade of the veneer. In three-ply construction, the grain direction in the center, or core, is at a right angle to that of the faces or backs.

Plywood constructions may consist of any odd number of plies. In panels having more than three plies the layers between the center, or core, and the face or back are known as *cross bands*.

With respect to its various strength properties and to changes in dimension with changes in moisture content, wood is anisotropic in character. That is, its properties are not the same in each grain direction. For example, wood shrinks and swells with changes in moisture content about 25 times as much radially as it does longitudinally and approximately 50 times as much tangentially as longitudinally. In plywood constructions the shrinkage and swelling in length and width are approximately equal. They are greater than the normal longitudinal dimensional change but substantially less than the expected radial or tangential dimensional change. It should be noted that no improvement in dimensional stability is achieved in the direction of the panel thickness. In a similar way the strength properties of the wood in a plywood panel are redistributed in the cross-banding operation. For example, as a result of the redistribution of cleavage properties, plywood panels can be nailed near their edges without splitting.

The rearrangement of wood properties in a cross-banded construction means that this construction is by its very nature stressed at the glue line. It is very important, therefore, that a plywood panel construction always be so designed that the stresses are as nearly balanced as possible. It is for this reason that flat plywood panels are invariably constructed with an odd number of plies. The matching plies on each side of the center, or core, must balance each other, or the panel will warp or twist when subjected to a change in moisture content. The paired plies are usually matched with respect to species, grain direction, type of veneer cut, original moisture content, and thickness. The exception is in the case of the outer plies, where grade and appearance considerations frequently dictate that some measure of unbalance must be tolerated.

One of the very important advantages of plywood over solid wood is that plywood manufacture provides an opportunity to redistribute wood characteristics. Expensive cabinet woods may be cut into very thin sheets and used only as face veneers, while less expensive woods make up the bulk of the panel. In the case of single-species panels the best quality clear veneer is used for faces, while the more defective wood goes into center stock and crossbanding.

A further advantage of plywood over solid wood is that it can be used in the form of thin panels with large areas that can be fabricated into curved and irregular surfaces not feasible with solid wood.

A. Manufacture of Plywood

The details of plywood manufacture vary over a broad range. Softwood-plywood plants differ in many respects from hardwood-plywood plants. Practices used in large factories are likely to be different from those used in small ones. In general though, while details differ, the basic operations are common to all plywood-manufacturing enterprises. They consist essentially of (1) preparing the stock for gluing, (2) mixing and spreading glue, (3) pressing, and (4) finishing.

1. Preparation of Veneer Stock for Gluing. The amount of work required to prepare veneer stock for gluing depends for the most part on the quality of panel being produced. In the manufacture of low-priced container plywood little work is required on the dry veneer before gluing. In contrast, factories producing high-priced, cabinet-quality plywood may invest as many man-hours in stock preparation ahead of the glue spreader as in all other operations combined. This operation in any plywood plant consists essentially of converting sheets of dry veneer into plies that conform to the appropriate specification with respect to size and grade. In the manufacture of high-grade cabinet plywood the operations performed to prepare stock for gluing are grading and matching, redrying, dry-clipping, jointing, taping or splicing, inspecting and repairing, and sizing.

a. Grading and Matching. The grading and matching operation is most important for the preparation of face plies. The dry veneer suitable for the faces required is selected and marked for dry-clipping. If a one-piece face is required, only sheets of veneer large enough to produce a whole face are selected. If multiple-piece faces are used, the veneers destined to make up a face may have to be matched to meet the appropriate specification.

In certain species a premium is paid for faces selected for color. In the case of birch, for example, faces may be designated selected white or selected red. Selected white faces are made up exclusively of sapwood, and in the preparation of such faces all heartwood must be removed. Similarly, selected red birch is all heartwood. In addition to selecting either all sapwood or all heartwood, it is required that the pieces making up a face be reasonably matched at the joints for color and grain. This is to minimize contrast between adjacent sheets of veneer so that the joint will be inconspicuous.

If an unselected grade is specified, sapwood and heartwood may be mixed in the same face, but it is usually still a requirement that color and grain be reasonably matched at the joints.

With highly figured face veneer the individual sheets may be assembled to accentuate deliberately the contrast at the joint. Slip-matching is done by placing successive sheets of veneer from a sliced flitch adjacent to each other, thus producing a pronounced repetitive figure. Book-matching results when alternate sheets are flipped, giving a symmetrical figure. In this type of matching half the sheets will have the loose side up. Where very elaborate designs are required, as in the case of faces produced with burl, crotch, or butt veneers, matching requires a considerable amount of care, skill, and experience.

In the manufacture of standard stock panels from either softwoods or hardwoods, dry veneer must be checked for grade even though matching may not be required.

b. Redrying. Thin hardwood face veneers are sometimes dried to a moisture content above that required for pressing so that they can be shipped and handled safely. Such veneers must be redried to a suitable moisture content before gluing. This operation is also resorted to when veneers pick up moisture in transit or storage after the original drying. Redrying is accomplished ordinarily by means of platen driers, consisting of a series of smooth, flat, steam-heated platens which alternately close and open to allow moisture to escape from the surface of the wood. The progressive platen driers, already described, have also been successfully employed for redrying.

c. Dry-clipping. Following the grading and matching operation the veneer is clipped to size and squared up for assembly into plies. Two types of clippers are commonly used in dry-clipping operations. The single-sheet clipper is very similar in design to the shear-type green clipper. When very expensive cabinet woods are being worked up, a foot-powered shear clipper is sometimes used. In the processing of less expensive woods, where speed of operation is important for economic reasons, the clipper is powered for electrical, pneumatic, or hydraulic operation. This type of clipper, as the name implies, is used to clip single sheets of veneer.

To process flitches of veneer produced on a slicer or stay-log, a jointing clipper or book clipper is employed. This is also a shear-type clipper but differs from the single-sheet clipper in that it is designed to clip stacks of veneer in a single cut. A stack of veneer is placed on the clipper table and compressed firmly in a tight book through the use of a clamp extending the length of the table just behind the knife. Because the veneer is clamped in place before cutting, the jointing clipper produces a square-edged joint suitable for splicing or taping without further jointing. The principal disadvantage of the jointing clipper as compared with the single clipper is that it is much slower and is, therefore, appropriate only when veneer can reasonably be clipped in batches.

d. Jointing. The jointing operation has the objective of producing a square, straight edge on the veneer so that two sheets can be taped or spliced together and provide an inconspicuous joint line. Two types of jointers are commonly used in plywood plants: the traveling-head jointer and the fixed-head jointer.

In the traveling-head jointer, books of veneer are clamped on the jointer table in much the same manner as in the jointer clipper. A multiple-knife cutterhead is then passed along the veneer book taking a light cut and producing a smooth, square, straight edge. Some traveling-head jointers are equipped with two cutterheads. The first takes a heavy roughing cut and the second a light finishing cut.

The fixed-head jointer is similar in basic construction to the familiar hand-fed lumber jointer. A cutterhead is mounted in the jointer table. On the top of the jointer table, powered chain feed belts are mounted. The books of veneers to be jointed are fed into the jointer between the feed belts and then carried automatically over the cutterhead, where a smooth, square joint is produced. These jointers are equipped with feed belts moving in both directions simultaneously so that a book of veneer can be passed through one direction to joint one edge, turned over, and returned to joint the other edge.

If the veneer being processed in a jointer is to be spliced, the adhesive is commonly applied in the jointing operation. With a traveling-head jointer the glue may be simply hand-painted on to the smooth edge of the book after the cut or it may be applied with a grooved, rubber-covered spreader roll that follows the cutterhead. In the case of the fixed-head jointer a glue applicator roll is mounted in the table behind each cutterhead.

e. Taping and Splicing. When the matcher selects two pieces of veneer to be joined, he butts them together along their dressed edges and holds them firmly in place with gummed tape or with a film of glue spread along the joint. This operation is known as *splicing* and may be done manually but is accomplished more efficiently on specially designed machines known as *splicers,* two general types of which are now in current use.

In a tape-splicing machine (Fig. 8-10) two sheets of veneer to be spliced are fed through a paired series of live rolls, the shafts of which are angled in such a manner that the two pieces of veneer are butted firmly together as they pass across the table. Suspended above the table is a roll of gummed tape. This tape is passed over a moistening device and under the rolls to the joint. An electrically heated roll sets and dries the tape while it is pressed firmly against the veneers. If in the manufacture of a panel three or more pieces of veneer are required to attain sheet width, a workman, the off-bearer, slides the pair of taped pieces back across the table to the operator who, in turn, tapes a third piece to the sheet. In this way sheets of large widths consisting of several individual pieces are readily made. Tape may be either of cloth or paper. It is invariably applied to the "finish side" of the sheet, later to be sanded off the panel during finishing operations.

The tapeless splicer (Fig. 8-11) is a more recent development and permits the joining of individual sheets of veneer without the use of tape. Its chief

Fig. 8-10. Taping hardwood veneer. (*Courtesy Hardwood Plywood Institute, Arlington, Virginia.*)

Fig. 8-11. Splicing veneer on a tapeless splicer. (*Courtesy American Forest Products Industries, Inc.*)

advantage over the tape machine lies in the fact that there is no tape to sand off the faces of the panels after their manufacture or to be left in the glue line. There is also a notable saving in tape costs. Tapeless splicers are somewhat similar in basic design to the tape machines already described. The veneer sheets with glue applied to the edges are carried over an electrically heated strip which cures the glue. The conveyer consists of chain belts or live rolls so arranged as to press the squared, straight edges tightly together as the adhesive sets. Most frequently the adhesive is applied to the edges of the veneer in the jointing operation. However, most tapeless splicers are equipped with glue applicators if it is desirable to apply the glue at the splicer. When animal glue is used in splicing, it is common practice to apply the glue at the jointer and formaldehyde at the splicer.

f. Inspecting and Repairing. Following the taping or splicing operation, the assembled face plies are inspected to assure that they comply with applicable grade standards and that the joints are well made. Minor defects identified in this inspection operation can sometimes be repaired and the ply used. If the defects are too great, the sheet is sent back, clipped, and reprocessed. Where high-priced cabinet woods are being converted into face stock, patching is very carefully performed by hand. Repair of this kind requires much experience and skill on the part of the workman.

In the softwood-plywood industry defects are commonly repaired by patching. Patches take a variety of shapes including circular, oval, elliptical, and lenticular. Some patches have a dog-bone shape.

The traditional patch of the Douglas-fir-plywood industry was for many years the familiar lens-shaped boat patch. This patch was bevel edged and was glued into a bevel-edged hole cut in the veneer to remove the defect. This type of patch is still used in the industry but it has been supplanted in many mills by the Raimann patch. Here the defect is cut out, and an oval or dog-bone shaped patch is inserted in a single operation. This patching technique is used not only with faces but also with cross-banding and center stock.

2. Preparation of Lumber Core. The core stock for thick plywood panels is commonly made up from dimension stock into a lumber core. The dimension stock is frequently produced in the plywood plant from kiln-dried, factory-grade hardwood lumber. The most common procedure is to cut the lumber board into the proper lengths for core stock on a crosscut saw, eliminating major defects. These short pieces are then ripped to random widths between 1 and 4 in. on a straight-line ripsaw. Again unacceptable defects are removed. The narrow strips are then edge-glued to an appropriate width to produce the core. A variety of equipment is used to glue up lumber core stock. The oldest type of machine used for this purpose is the glue reel. This consists of a series of screw clamps mounted on a shaft or a chain conveyer. The core stock is spread with glue and clamped into place in the reel or carrier. Either the glue is cured cold or the carrier is equipped to move into a heated chamber to accelerate glue curing.

A variety of machines are now available for the manufacture of lumber core stock that are much more rapid than the glue reel or clamp carrier. Some of these utilize hot plates to speed glue curing, while others make use of high-frequency alternating electrical currents. The principles of high-frequency glue curing are discussed in section 6.

3. Adhesives. A number of adhesives may be used in the manufacture of plywood. The choice of adhesive to be used in any specific situation depends upon a number of factors. Relative costs must be considered, and, other things being equal, the manufacturer is likely to select the least expensive adhesive that will give the required performance. The performance of a glue in use is, of course, of paramount importance. A plywood panel is no better than its glue bond. If a panel is to be used in exterior exposure, it must be highly moisture-resistant and, in addition, it should be resistant to deterioration caused by fungi, mold, bacteria, and the like. Plywood that is not to be subjected to exterior exposure but is to be used where it may be occasionally wetted or exposed to a high equilibrium moisture content should have a moisture-resistant bond that is also durable in the presence of microorganisms under conditions favorable to their growth. Where the plywood is manufactured for very short-term use, as in the case of one-trip crates or boxes, or where it is to be used only in dry locations, the glue need not be resistant to moisture or to attack by microorganisms.

Ease of use during manufacture is also an important criterion in selecting an adhesive. If an extender is to be used, the glue should be compatible with the extender. Such characteristics as ease of mixing, extended working life, long permissible assembly time, tolerance of veneer moisture, etc., may sometimes result in substantial savings in manufacturing costs that easily compensate for slightly higher mix cost.

Adhesives used in plywood manufacture are either proteins of plant or animal origin or they are synthetic resins. The most commonly used plywood adhesives are reviewed here.

a. Casein Glues. Casein glues, as previously indicated, were developed commercially during the First World War, although the use of casein as an adhesive has been traced back to the Middle Ages. Casein, a derivative of skimmed milk, is obtained by adding a quantity of an organic acid to milk and heating the mixture to about 100°F. This procedure sours the milk and permits rapid separating of the curd and whey. The curd is thoroughly washed and dried and then reduced to the consistency of wheat flour. Proprietary casein glues are offered to the trade as dry powders consisting of a standardized mixture of casein, hydrated lime, and certain alkaline solutions or salts. There are numerous glues of this sort, each compounded for some specific use; e.g., some will permit of long assembly periods, while others may be more highly water-resistant. In use they are merely mixed with water in proportions recommended by the manufacturer.

A second group of casein adhesives, the *wet casein mixes*, are made by the plywood manufacturer by mixing raw casein powder with water, lime, and other ingredients to make a satisfactory adhesive for his particular needs.

The casein glues are suitable for many purposes. Many formulations are relatively water-resistant, but joints made of them weaken materially if subjected to prolonged immersions in water; hence this type of adhesive is not a suitable bonding agent for boat hulls or for construction continuously exposed to excessive high humidities. Casein is an organic material which will deteriorate in the presence of certain fungi unless fortified with suitable toxic agents. Casein glues are somewhat more versatile than most wood-bonding adhesives. They may be mixed and spread at considerably below normal room temperatures. They also exhibit good gap-filling characteristics and therefore are capable of producing good bonds on rough surfaces, even when the glue line itself is excessively thick. Because of their alkalinity they are especially suited for bonding oily woods such as teak and the southern yellow pines.

b. Soybean Glues. Soybean glues made their appearance about 1923, and by 1926 manufacturers of structural Douglas-fir plywood were using them in ever-increasing quantities. The best formulations are very water-resistant, and their cheapness is another point in their favor. Soybean glues are composed of soybean meal in admixture with hydrated lime, caustic soda, sodium silicate, and a waterproofing ingredient. Plywood manufacturers procure the various ingredients from reputable sources and blend them in accordance with schedules they themselves have developed.

Soybean glues have none of the tacky properties exhibited by casein but more nearly resemble wet plaster or cement in their flow characteristics. They are readily applied, however, in mechanical spreaders and may be formulated for either cold-pressing or hot-pressing.

The most significant development in soybean glues since the Second World War has been the development of the "no-clamp" gluing process. In this process a conventional cold-press load is built up and placed under pressure. Instead of being restrained in retaining clamps, as in conventional cold-pressing, the plywood load is kept under pressure in the press for 15 minutes. It is then removed from the press and allowed to cure undisturbed for about 30 minutes. During this curing period the gel strength of the glue increases, and at the end of this period the load may be disassembled and redried or sized. According to Hill,[11] "the gel strength of no-clamp glues develops rapidly for two reasons. First, the glue is formulated at a water ratio lower than that of regular glues. Second, adequate, constant pressure is applied during the entire pressure period which causes a rapid displacement of water from the glueline to the wood."

c. Blood Glues. Soluble dried blood, an important by-product of abbatoirs, has been used with varying degrees of success as the basic constituent of a highly water-resistant glue and as an extender for other adhesives such as soybean, casein, and some of the synthetic resins. Blood glues are prepared in either hot- or cold-press formulations for the production of interior-type softwood plywood. Modern blood glues are closely related in method of formulation and use to the soybean glues. They are dispersed in a caustic soda solution with lime added to improve the moisture resistance. The blood glues have an advantage over synthetic resins for interior plywood in that they set very fast. A recent development in the use of blood glues by the softwood-plywood industry is the replacement of soybean by blood in the formulation of no-clamp adhesives.

d. Synthetic-resin Adhesives. Synthetic-resin adhesives were used first commercially in European plywood plants, and it was not until 1935 that Tego, a phenol-formaldehyde resin in dry sheet form, became available in commercial quantities in the United States. This was followed by several other types of resin adhesives. During the Second World War the production of synthetic-resin adhesives was greatly expedited, and today there are many formulations of the several basic types available to the trade. Their acceptance and utilization by the woodworking industries of America have become so general that old-line, well-established manufacturers of animal, starch, casein, and soybean glues are now marketing many resin compositions under their own trade or brand names.

The synthetic-resin adhesives are of either the *thermosetting* or the *thermoplastic* types. Thermosetting adhesives are those which, when cured under heat and pressure, "set" or harden to form films of great tenacity and strength. The reactions during the curing cycle of such adhesives are irreversible; i.e., upon total cure, repeated applications of heat will in no way soften the bonding matrix. By contrast, thermoplastic adhesives are those which remain soft until cooled. The reaction is reversible to the extent that upon reheating the resin will again soften, only to harden once more as it is cooled.

There are two important classes of thermosetting resin adhesives currently employed in the American plywood industry: the phenol-formaldehyde series and the urea-formaldehyde group. Resorcinol-formaldehyde and melamine-formaldehyde resins are suitable for bonding wood, but because of their higher costs their use by the plywood industry is very limited.

As previously indicated the first phenol-formaldehyde resin used in the United States was marketed in dry sheet form. Upon close examination one finds that it actually consists of a very thin sheet of tissue-like paper suitably coated on both sides with uncured resin. When it is placed between two layers of wood and hot-pressed, moisture released from the wood softens the resin and subsequently facilitates its penetration into the interstices of the elements being bonded. Phenol-formaldehyde-resin adhesives are now available in either liquid or powder form. The liquid resins may be spread as received or diluted with solvents and mixed with extenders in accordance with manufacturer's recommendations. The powder forms are prepared for use by mixing them with water, alcohol, acetone, or combination of these solvents, in proportions recommended by the manufacturer. Some few formulations call for a catalyst, but usually it is not necessary to employ one with resins of this sort.

Phenol-formaldehyde-resin adhesives may be used without additives or they may be extended or filled. When a highly moisture-resistant bond is desired, the resin is used straight. When an interior-type bond is sought, the resin is not infrequently extended.

Urea-formaldehyde-resin adhesives were developed to produce a resin cheaper than the phenolic resins, one capable of curing at a somewhat lower temperature to avoid blistering problems and to achieve a colorless glue line. The earliest formulations of this type required curing temperature of about 200°F, but shortly thereafter the so-called "cold-setting" ureas made their appearance. For the manufacturer unable or unwilling to obtain hot-press equipment, this is an adhesive of the resin type that can be used with the physical facilities at his disposal. It should be clearly borne in mind, however, that the term "cold-setting" is a relative term, as, in the restricted sense at least, there is no really cold-setting resin. The polymerization, or total curing, of any resin film is merely a time-temperature relationship; thus resins that will cure at room temperature in 8 to 16 hours will also completely polymerize in a matter of minutes at higher levels of heat. Polymerization is a chemical reaction whose rate is determined by temperature in the presence of a catalyst. Like most reactions, below a certain temperature, the rate is so sluggish as to be impractical. On this basis, 70°F is usually considered to be the minimum practical temperature for using present-day synthetic-resin adhesives.

The bulk of all urea-formaldehyde-resin adhesive used in plywood manufacture is sold in liquid form. The liquid adhesive is sold either as a low-solids (about 50 per cent solids) or as a high-solids (about 70 per cent solids) resin sirup. Small quantities of urea-formaldehyde-resin adhesive are spray-dried and sold as powders. In the large quantities used in the plywood industry, liquid adhesives have the advantage of lower cost, ease of bulk handling, and ability to be transported in tank cars or trucks. Spray-dried powders are used where good storage properties are particularly sought.

Urea-formaldehyde resins may be fortified with melamine resins to achieve an improvement in moisture resistance. In the manufacture of hardwood plywoods, formulations consisting of 50 per cent urea resin and 50 per cent melamine resin are used for the production of the more weather-resistant types.

Resorcinol-formaldehyde-resin adhesives have found very limited use in the plywood industry. They have performance properties similar to those of the phenol-formaldehyde resins with the added advantage that they can be cured at room temperatures. They are so expensive, however, that it is not economically feasible to utilize them in the plywood-pressing operation. Resorcinol-formaldehyde-resin adhesives have been used to a limited extent in combination with phenol-formaldehyde resins to reduce the curing temperature. They have also been used to edge-glue lumber core that is to be used in highly moisture-resistant panels.

Melamine-formaldehyde-resin adhesives have a limited use in plywood fabrication. They exhibit, except for color, many of the physical characteristics of the phenolics, particularly with respect to heat resistance and durability. They produce a very highly moisture-resistant panel. Melamine-formaldehyde-resin adhesives are more expensive than either the urea or the phenolic types, and this limits their use. The principal utility of the melamine-formaldehyde-resin adhesives in the plywood industry is as a fortifier of urea resins, as previously indicated.

e. Extenders. Extenders are materials combined with the primary adhesive to reduce the cost of the mix. They are usually inexpensive materials that are compatible with the glue and reduce the wet-mix cost. Quite commonly they reduce the quality of the glue mix, but not below the level required to meet the specification of joint quality. The degree of extension of a glue mix is usually denoted by expressing the dry weight of the extender as a percentage of the dry weight of the resin. For example, a glue mix that included 100 lb of 50 per cent solid urea resin and 50 lb of dry extender would be said to be extended 100 per cent. The most common extender for urea-resin adhesives is wheat flour. Rye flour, vetch flour, and soluble dried blood are commonly used in Europe to extend urea resins. Foaming agents are sometimes introduced into the glue mixture for the purpose of using gas or air as an extending agent.

f. Fillers. Fillers are materials that are added to resin-glue formulations to modify the working properties of the mixture. Walnut-shell and pecan-shell flour are commonly added to synthetic-resin adhesives to control the flow properties of the glue. They reduce the flow and minimize the possibility of bleed through and starved joints.

4. Adhesive Mixing. Adhesives used in the manufacture of plywood are usually mixed in large dough-type mixers. The procedures appropriate to the mixing operation vary with the type of adhesive being used. The manufacturer of the adhesive usually provides detailed mixing instructions.

It is important to use care in preparing adhesive mixes to ensure that the correct proportions of ingredients are maintained. It is generally considered good practice to weigh all ingredients entering a mix. Where liquid ingredients are piped to the mixer, volume metering is sometimes found to be more satisfactory.

The quantity of glue mixed at any one time depends upon the properties of the material. A glue with a very short working life must be prepared in small batches. Even when a glue remains usable over a long storage period, if its spreading properties change substantially with time, it may be desirable to prepare small batches and mix frequently.

Where the same mixer is used to prepare several types of adhesive, it is necessary to exercise special care in cleaning the mixer between batches. Residues from the first glue, even in traces, may seriously affect the properties of the second mix.

5. Spreading. In the manufacture of plywood glue is normally applied to the veneer through the use of a spreader (Fig. 8-12*A*). Basically, the glue spreader commonly used by the plywood industry consists of two power-driven rolls to which a supply of the adhesive is provided. The rolls are grooved, and the grooving pattern used depends upon the type of glue being applied. Steel rolls are usually used for spreading casein and soybean glues. Other common plywood glues are usually applied with rubber-covered rolls. The amount of glue that is applied to the veneer in the spreader is regulated by means of a steel doctor roll paired with each spreader roll. The adhesive is supplied to a reservoir underneath the spreader from the glue mixer. A pump distributes the resin to the two pairs of spreader and doctor rolls. The amount of glue spread on the veneer can be controlled by adjusting the pressure of the doctor roll on the spreader roll.

The amount of glue spread is usually described in terms of pounds per thousand feet of glue line. In the hardwood-plywood industry it is customary to express the amount of glue spread as pounds of glue per thousand square feet of single glue line. In the softwood-plywood industry the practice has been to define spread on the basis of pounds of glue per thousand square feet of double glue line.

Spreads vary depending upon the type of glue being used, the species and quality of veneer, and several other factors. It is consistent with good gluing

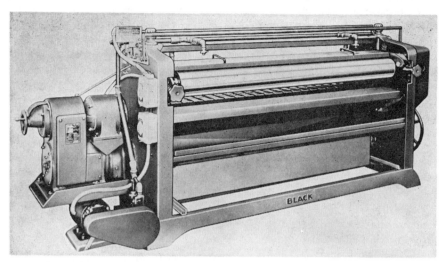

Fig. 8-12*A*. Glue spreader. (*Courtesy Black Brothers Company, Inc., Mendota, Illinois.*)

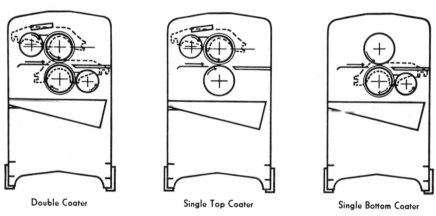

| Double Coater | Single Top Coater | Single Bottom Coater |

Fig. 8-12*B*. Roll arrangement: left, double coater; center, single top coater; right, single bottom coater. (*Courtesy Black Brothers Company, Inc., Mendota, Illinois.*)

practice to use no more than the required minimum of an adhesive in making a joint.

The elapsed time between the beginning of a spreading operation and the application of full pressure upon the glue lines of a plywood panel or batch of panels pressed simultaneously is known as the *assembly time,* or *period*. The rapidity with which an adhesive begins to "set up," or cure, is of primary importance to the manufacturer of plywood. If there is an appreciable hardening of the glue on the surfaces of the wood before full pressure is applied to the joint, flow of the bonding agent is retarded, thick glue films result, and in consequence adhesion is greatly reduced. Similarly, if the glue is too thin and is not permitted to thicken somewhat before pressure is applied, excessive "squeeze-out" and excessive absorption of the resin by the wood occurs, and a starved joint results. It is readily apparent that careful control over assembly is mandatory if high-grade plywood is to be produced.

Assembly time is largely a function of temperature. Generally, the higher the temperature of the wood in the air, the shorter the permissible assembly time for a given glue. It should also be noted that in some cases there is a minimum assembly time as well as a maximum one. The glue can be forced out of the joint if the assembly time is too short with certain adhesives. Some formulations of phenol-formaldehyde adhesives require that a minimum assembly time limitation be observed.

6. Pressing. In wood-gluing operations the parts to be glued are subjected to pressure for the purpose of ensuring proper alignment of components and an intimate contact between the wood and the glue. The adhesive is allowed to cure or at least partially cure while the assembly is under pressure. Pressing may be accomplished at room temperature or at elevated temperatures. The former is known as *cold-pressing*, the latter as *hot-pressing*. Cold-pressing is used with casein, most soybean, and some of the urea-formaldehyde-resin adhesives. Hot-pressing equipment is used to cure blood, some soybean, some urea-formaldehyde, and all phenol-formaldehyde adhesives.

Virtually all plywood is produced in hydraulic presses. These presses apply pressure to the glue line at a magnitude of from 150 to 250 psi. The higher pressures are used with dense woods such as birch, maple, elm, and walnut, while the lower pressures are used with low-density woods such as Douglas-fir, yellow-poplar and cottonwood.

Hydraulic presses depend upon the action of one or more hydraulically operated rams to lift the lower platen, thus closing the press and applying the pressure.

In charging a cold press the panels are stacked on a strong retaining board. When thick panels are pressed, each assembly is usually placed upon a *caul board,* usually a piece of three-ply, $\frac{3}{8}$-in. plywood with waxed surfaces, or a sheet of aluminum. If thin panels are made, two to five or occasionally more assemblies may be placed between cauls. Caul boards are used to assure panel flatness and to guard against injury to the adjacent panel faces while under pressure. A stack of panels large enough to fill the press to capacity is placed on I-beams resting on the lower platen. A top retainer and second set of I-beams are placed on top of the stack. As the press is closed, retainer clamps are adjusted over the beams' flanges and taken up as pressure is applied (Fig. 8-13). Once full pressure is attained the clamps are given a final adjustment, whereupon the press is opened and the bundle removed. The retaining clamps are left on until the adhesive is thoroughly set up, a period which may vary from 8 to 24 or more hours, depending upon the type of bonding agent used. The bundle is then broken and the panels sent to the inspection and finally to the finishing department.

In the no-clamp process, which was previously discussed, the retaining clamps are not used. In this process the plywood assembly remains in the press under pressure for about 15 minutes. After this pressing period, the pressure is removed, and the assembly is taken from the press and stored for a period of several hours while the curing continues.

Fig. 8-13. Bale of cold-pressed plywood being removed from a press. (*Courtesy Hardwood Plywood Institute, Chicago, Illinois.*)

Hot presses are of the multiplaten type (Fig. 8-14). Each of the platens is heated, using steam, hot water, or hot oil. In practice one or two panels are placed between aluminum caul boards. This assembly is then placed between the platens of the press. It is customary to assemble a complete press load at the spreader before beginning to load the press. Because the platens are heated, it is necessary in hot-pressing to load and close the press as quickly as possible to avoid precuring the glue joint before full pressure is applied. With small presses of 8 to 10 openings the panels can be loaded by hand with sufficient speed. Where presses are larger various aids to loading and unloading are built into the pressing installation. The simplest such aid is a hydraulic elevator on the loading side of the press. On the largest presses completely automatic loaders and unloaders are commonly installed.

Pressing times in the hot press vary from 2 to 30 minutes depending upon the thickness of the stock, the distance from the platen surface to the deepest glue line, the type of adhesive, the curing temperature, and the degree of cure desired. With a hot press it is possible to bring about almost complete curing by extending the pressing time. Practically, most manufacturers use a pressing cycle that will permit sufficient cure to enable the panels to be handled in unloading. These panels are then stacked after unloading and allowed to finish curing for several hours before being further processed. This practice is known as "hot-stacking."

Temperatures of 230 to 350°F are commonly used in hot-pressing.

Fig. 8-14. Large hot-pressing operation. Two spreaders feed press. Automatic loaders and unloaders speed up operation. (*Courtesy American Forest Products Industries, Inc.*)

In recent years radio-frequency heat has been used to cure synthetic-resin adhesives. It is a well-known physical phenomenon that when an alternating electric current oscillating in the radio-frequency range is applied to a dielectric material, the material will be heated. This principle is employed in curing glue joints with high-frequency current.

There have been several installations of hot presses designed to cure plywood glue bonds with high-frequency current. While these applications have been technically successful, the economics of energy utilization have discouraged a general acceptance of this method as a substitute for the more conventional multiplaten hot press.

The principal application of high-frequency gluing in the manufacture of plywood has been in the production of lumber core stock. Several edge-gluing machines have been produced that effectively utilize high-frequency current to cure glue bonds in the manufacture of lumber core.

High-frequency techniques have also found some application in the curing of plywood glue bonds in the manufacture of curved or molded plywood.

7. Molded Plywood. Plywood made in the form of flat panels in conventional hot- or cold-press operations can be bent to conform to curved surfaces of single curvature. This is practical where the panel is thin and where the use is such that the panel is fastened firmly and permanently in the curved position. Where thick plywood must be used in a curved form, or where it is not feasible to fasten the panel into the formed shape and hold it there in use, molded plywood is customarily used.

The most common procedure for manufacturing molded plywood is to

Fig. 8-15. Molding a TV cabinet. Panels are cured with high-frequency current. (*Courtesy Hardwood Plywood Institute, Chicago, Illinois.*)

use a press, either hot or cold, that has male and female platens of the desired shape. These platens may be metal and may be heated to serve as a hot press. They may be made of wood lined with a conductor to serve as an electrode, and the curing can be accomplished through the use of a high-frequency generator (Fig. 8-15). Where pieces of a particular shape or form are likely to be large, it is commonly most economical to have specially formed hot plates built. Where shapes are frequently changed, wooden forms and high-frequency curing may be more desirable.

Panels with single or double curvature, such as those used in aircraft skins, hulls of small boats, or in some of the most modern furniture designs, cannot be made with ordinary pressing equipment. Thus molding techniques have been devised which make use of fluid pressures.

In one molding process, veneers interleaved with film glue or veneers coated with an uncured resin are formed over a male mold of wood, metal, or ceramic material and temporarily held in place with wire staples, tape, or by other suitable means. Such an assembly is then placed in a large flexible rubber bag from which the air is immediately expelled. As the vacuum is drawn, the bag presses firmly against the veneers, causing them to conform to the contour of the die. The deflated bag and its contents are then placed in a large steel autoclave where steam, hot water, or both are admitted under pressure. Adequate fluid pressure is brought to bear in this way over the entire assembly, and the heat cures the resin. At the completion of this cycle the molded item is removed, and the operation repeated.

A somewhat similar process employs the use of a female mold. In this instance veneers are laid up along the inner face of the mold and covered with a rubber blanket that is attached to the die. The air is then expelled beneath this covering in order to press the veneers firmly into place. The assembly is then handled in a manner similar to that described above. There are several variations of these two methods for molding panels of unusual shape and contour. In one of these, veneers are laid up in a female die, and a flexible bag placed within the assembly is inflated to bring the veneers into proper contour. Heat may be supplied either from within the bag, from some heating element surrounding the mold, or by the use of high-frequency equipment. In recent years many ingenious methods have been developed at individual plants to mold small plywood panels or parts for use in particular products.

8. Redrying Plywood. In cold-gluing moisture is added to the wood adjacent to the glue line. Unless this moisture is allowed to equalize, later drying of the plywood will result in a number of defects, such as warping, checking, open joints, and glue-line failures. It is therefore necessary that the moisture content of plywood be equalized throughout the panel and also be brought to a level suitable for the subsequent use of the product. It is ordinarily assumed that 10 to 15 per cent moisture in the finished plywood is satisfactory for service in the open air, while for interior use in heated buildings the moisture content of plywood should be 7 to 8 per cent, and somewhat lower for the product intended for furniture manufacture.

On the basis of experiments in kiln-drying aircraft panels, the U.S. Forest Products Laboratory recommends a constant temperature of about 120°F and a constant maximum relative humidity which will permit the stock to dry down to the moisture content desired, with minimum injury to the material and maximum convenience and economy of operation. Temperatures in excess of 120°F will decrease the drying time but are prone to lower the quality of the plywood by increasing warping, checking, and opening of joints.

The length of time required to dry plywood depends largely on its moisture content and the total thickness of plywood, as well as the number and thickness of the constituent plies and the type of wood. Panels dried from very high to very low moisture content should be dried more slowly, to prevent warping and other drying defects. The actual drying time with three- and five-ply panels ranges from 8 to 24 hours. In a few instances mechanical veneer driers and hot-plate presses have been used for redrying thin plywoods with relatively high moisture content. This method has the advantage of speed but involves the use of expensive equipment.

Plywood glued in hot presses does not require redrying after gluing, but good practice calls for a conditioning period to provide for equalization of moisture content. In some shops it is common practice to sponge or spray the panels as they are removed from the press in order to restore surface moisture.

9. Finishing Panels. Unless all plies were trimmed to exact size before gluing, the panels are trimmed after redrying. In some mills the first straight edge is obtained with a hand jointing machine, the panels then pass over a variety saw, set to produce the desired width, and finally through a trim saw for trimming the ends. In a plant equipped with power-driven sizing saws having two saws for cutting both edges in one operation, the hand jointing and the variety-saw operations are eliminated. A power-driven trim saw may be set at a right angle to the sizing saws, so that sizing and trimming of panels are accomplished in one operation.

The finishing of plywood frequently involves the sanding of faces to remove tapes, rough spots, and unevenness in thickness. Great care should be exercised to sand both faces evenly, as the difference in thickness of the faces would destroy the symmetry of construction and lead to distortion with changes in moisture content; furthermore, faces sanded too thin will not finish well and are likely to show the glue line. For sanding, either the endless belt or the drum type of machine may be used, the final cleaning of high-grade plywood usually being done by means of the belt sander.

After sanding, the panels undergo a final inspection and grading. The panels containing surface imperfections, such as checks, splits, or pitch pockets, are sent to the patching room, where these defects are removed and sound pieces of veneer inserted. When this is expertly done, such patched panels are fully as strong and serviceable as sound, unpatched plywoods. The graded plywood is stored piled solid in heated conditioning rooms until needed. It is customary to lay the panels with the faces toward each other when they are crated for shipment.

B. Special Plywoods and Laminates

In addition to conventional plywood described in the foregoing pages, there are a number of special plywoods and veneer laminates which deserve mention.

1. High-density Wood. It has been known for some time that by compressing wood it is possible to improve many of its physical-mechanical properties. Practical application of this principle, however, has been achieved only since the development of the thermosetting synthetic-resin adhesives.

Normal pressures used in gluing wood, ranging from 75 to 200 psi, are accompanied by compression not exceeding 10 per cent of the original thickness of the plywood assembly. If, however, veneers are impregnated with a synthetic resin and then bonded under pressures ranging from 500 to 2,000 psi, considerable compression results, with, of course, a corresponding increase in density. Material of this kind (which may be laid up with either cross or parallel laminations), developed at the U.S. Forest Products Laboratory, was originally designated as compregnated wood and later as *compreg*.

At present compreg and several closely related materials are manufactured under a variety of trade names such as Pluswood, Superpressed Plywood, Improved Wood, and Tegowood.

Compreg is made by impregnating thin sheets of dry veneer with a water-soluble, specially prepared, resin-forming formulation. Following impregnation, the sheets are removed from the bath, laid up, and hot-pressed. Heat has a plasticizing effect upon lignin, and wood treated in this manner will compress much more readily, for a given unit pressure, than one untreated. Varying the intensity of pressure allows the resin-treated wood to be compressed from one-third to one-half of its original dimension and its specific gravity increased from approximately 1.0 to 1.40. Compreg may be made in thin sheets or heavy blocks of considerable thickness. Shrinkage or swelling with atmospheric changes is negligible. In fact, tests at the U.S. Forest Products Laboratory revealed maximum dimensional change of only 3.6 per cent after a 50-day immersion in water. It is also highly resistant to mild acids, alcohols, and many other solvents and is not permanently spotted by such materials. Compreg takes on a high luster from the press platens and needs no further finishing. Since the resin permeates the whole panel, surface scratches or mars may be sanded or burnished out without additional treatment. Compreg has excellent mechanical properties except for impact strength, as the large amount of resin present has some small embrittling effect.

The full utility to which compreg may be put is yet unknown. As a panel it is well suited for bar tops, car tables, furniture, and specialty flooring. During the Second World War it was used for airplane propellers, bases for parachute flares, radio antennae masts, dies, small tools, jigs, and in various parts of aircraft. Tests * by the United States Navy have revealed that compreg made of cottonwood is a suitable decking material for aircraft. Compreg picker sticks used in place of those made from hickory are giving satisfactory service in a number of southern cotton mills.

Tegowood, another densified product, differs from compreg in that thin sheets of veneer are bonded under high pressure with Tego film glue. This material does not appear to have as good dimensional stability or strength characteristics as compreg but is suitable for many purposes.

Impreg is another resin-treated wood developed by the U.S. Forest Products Laboratory. Like compreg it is made of resin-impregnated veneers. After sheet impregnation, the veneers are dried, and the resin cures within the individual sheets. These are then laid up in the conventional manner and hot-pressed with a resin adhesive at normal bonding pressures. The resulting product is a superior plywood with good stability and greatly improved resistance to fungi, insects, fire, and weathering.

Impreg has found more commercial use than any of the densified wood products just described. In recent years it has been extensively used for the manufacture of die models and pattern stock (Fig. 8-16). Its dimensional stability gives it a substantial superiority over solid lumber for this purpose.

* Anon. U.S. Navy Installs Newly-developed Dual Purpose Decking. *Wood Prod.,* **52**(6):20–22. June, 1947.

The automobile industry has been a leading user of impreg pattern stock.

Another development in compressed plywood construction is the *differential density material*, in which a plywood product has different densities throughout its length. This is achieved by using veneer sheets of different lengths and assembling them in such a manner that several more veneer

Fig. 8-16. Die block of automobile fender made with Honduras mahogany impreg veneers. (*Courtesy Hardwood Plywood Institute, Chicago, Illinois.*)

sheets are placed at one end or in the middle of the assembly. Since the shorter veneer sheets are cut to different lengths, the number of veneer layers, and therefore the actual combined thickness of the veneer sheets, varies throughout the length of the plywood. The resulting assembly, consisting of veneers and synthetic-resin glue, is compressed between parallel platens to exactly the same thickness, with the result that a plywood material with density varying throughout its length is obtained. Such material could be used in aircraft propellers, where the part nearest to the hub requires high density while the tips must be as light as possible.

2. Veneer and Plywood Combinations. Various plywood and laminated products can be made by combining veneers or plywood with fiberboard, paper, or nonwood materials such as cloth, asbestos, metal, and foamated plastic materials.

An example of veneer-paper combination is a product constructed of kraft paper and veneer. Kraft paper is somewhat stronger across the grain than thin veneer and therefore may be used to reinforce it. The paper is used either as a core, between layers of veneer, or for outside surfaces. Such veneer-paper combinations are used for packing cases, mirror backs, and cabinet partitions. Papreg and Kimpreg, phenolic-impregnated papers, are also suitable facing material for both veneers and plywood.

For packaging purposes substantial quantities of plywood are produced that consist of face and back plies of heavy-cylinder kraft paper glued to thick veneer centers. This material has also been used for dust bottoms and case backs in low-priced furniture.

A well-known example of the cloth-veneer combination is Flexwood. This is a veneer product used for interior wall finish. Flexwood consists of a sheet of thin hardwood veneer, $\frac{1}{64}$ to $\frac{1}{80}$ in. thick, laminated under heat and pressure with a waterproof adhesive to a coarse cotton fabric. The sheets are then processed in a flexing machine, and the result is a flexible material with a hardwood surface. Flexwood is manufactured in strips 18 to 24 in. wide and 8 to 10 ft long; it can be applied to any surface, flat or curved, with a permanent adhesive, much as wallpaper is hung.

Another example of cloth-and-veneer combination is a special type of conveyer belt where the woven fabric supplies flexibility while the veneer produces the desired surface and a degree of stiffness. A similar product is used for guard cages around machinery.

Another product is *woven wood laminate*. In this process the veneer is coated and impregnated with resin varnishes. It is then cut into strips of the desired width and woven into sheets of any size. The woven sheets of veneer are then bonded with some stiffening material under heat and pressure. The resulting sheets are flexible and can be cut with ordinary shears. When mounted on plywood or other solid backing, these panels can be used for desk tops, card tables, bars, walls, and ceilings. They have a glass-smooth surface, unaffected by moisture, alcohol, and most common acids, and resistant to cigarette burns. Various designs are achieved by combination, in weaving, of woods of various color and grain.

In certain specialty products metal sheets are combined with veneer sheets to produce panels. Some constructions bond metal faces and backs to plywood panels. These panels are useful as fire doors or where the mechanical properties of metal are desired on the surface, and the light weight, high strength-weight ratio, and low cost make wood a desirable core material. They are used in the construction of elevators and railroad passenger cars.

Panels are also being produced with metal foil used as a ply immediately under the face. Because of the high heat conductivity of metal, panels thus constructed are resistant to burning. For example, if a lighted cigarette is inadvertently placed upon the mahogany surface of such a panel, the foil cross band will conduct the heat away from the lighted end so fast that the mahogany face will not reach charring temperature.

Veneers or thin plywood bonded to thick, lightweight cores of foamated thermoplastic resins give promise of being suitable materials for insulation and acoustical boards and for lightweight panels for many special uses.

Plywood is now available prefinished in stock sizes. In the simplest case, the plywood panel is subjected to a finishing operation which may include staining, sealing, and application of a top coat. In some cases the panel may be V-grooved prior to finishing to simulate milled lumber paneling. Some patterns are grooved to uniform widths and others to random widths. Prefinished plywood panels require no further finishing after installation. As a result of economic pressure and foreign competition, less expensive decorative panels with printed reproductions of natural wood grain are also available.

C. Utilization of Plywood

In recent years plywood has gained such popularity and its use has become so varied and diverse that only a general summation of its utilization is presented.

1. Plywood in Construction. Perhaps the greatest field of application for plywood is in construction. The ever-growing popularity of plywood in the construction field is largely due to the advantages it offers in comparison with solid lumber. Some of these advantages have already been mentioned, but they are repeated here to emphasize their importance in a building material. Of particular significance are (1) the equalization of strength properties along the length and width of the panel achieved by alternating the direction of grain in successive plies; (2) greater resistance to end checking and splitting, thus permitting nails and screws to be driven closer to an edge than is possible with solid wood; (3) reduction and equalization of shrinkage and swelling; (4) lightness, which makes handling of plywood on scaffolding easy; (5) ease of cutting into any desired shape; (6) applicability for construction of curved surfaces (Fig. 8-17); (7) good nailing and screw-holding capacity; (8) availability in large sizes cut to exact dimensions; (9) adaptability for sheathing purposes; (10) high range of decorative effects possible.

Of the species used for structural plywood, Douglas-fir is by far the most important. Douglas-fir plywood suitable for construction purposes is made in two types, *interior* and *exterior*. The interior-grade type represents the larger amount of production and consists of plywood with some degree of moisture resistance; it is built to retain its original form and practically all its strength when exposed to occasional but thorough wetting and subsequent normal drying. The exterior type of Douglas-fir plywood is intended to retain its original form and strength when wet and dried repeatedly and otherwise subjected to the elements; it is, therefore, suitable for permanent exterior use.*

---

* Each type, i.e., interior and exterior, is available in several grades. For complete information consult the latest Commercial Standards for Douglas-fir plywood, available

Douglas-fir plywood is extensively used as a roof and wall sheathing and for floors and subfloors (Fig. 8-18). Although on the basis of square-foot cost plywood is somewhat more expensive than some other forms of construction materials, its availability in large panels cut to exact dimensions allows labor savings which tend to equalize the costs. As a subfloor material plywood has the added advantage that large unbroken surfaces minimize the extent of cracks; it also offers a good nailing surface for flooring. The same advantages apply to the use of plywood as sheathing material, but in

Fig. 8-17. Industrial building roofed with arched stressed-skin Douglas-fir plywood panels, each spanning 16 ft. Panels supported on glue-laminated beams and wide-flange steel posts. (*Courtesy Douglas-fir Plywood Association, Tacoma, Washington.*)

order to secure a greater degree of rigidity and strength, the panels must be properly braced and securely nailed to the frame. According to the U.S. Forest Products Laboratory, a rigidity far superior to anything possible with nailing can be obtained by gluing plywood to a frame.

Douglas-fir plywood is peculiarly suited for concrete forms, and many millions of feet of it are consumed annually for this purpose. Plywood concrete forms are easy to construct and easy to remove; they are strong, yet offer a degree of flexibility which allows construction of curved surfaces. Plywood forms produce even concrete surfaces that require little labor to smooth them off. Douglas-fir plywood is equally well adapted to small concrete forms such as sidewalks and curves of residence foundations, and to gigantic projects such as skyscrapers, dams, or bridges. Finally, the same

through the National Bureau of Standards, U.S. Department of Commerce, Washington, D.C.

forms may be reused several times, and then when finally dismantled the plywood still has a salvage value for rough construction.

Douglas-fir plywood, as well as other types of plywood, finds a wide range of application in interior construction. Plywood is occasionally used as lining to be covered with plaster but more commonly as wallboard used as a combination lining and decorative base; the panels are nailed directly to the studding and joints. When used as a decorative base, plywood offers many finishing possibilities. Panels with hardwood face veneers can be

Fig. 8-18. Douglas-fir plywood is used extensively as roof and wall sheathing and for floors and subfloors. (*Courtesy Douglas-fir Plywood Association, Tacoma, Washington.*)

selected according to architectural requirements. If desired they may be stained, painted or enameled, or covered with wallpaper and fabric by first pasting a base of muslin or light paper felt to cover the cracks and take care of slight changes in dimensions that may take place in the walls. Plywood is very well adapted for the modern wall forms requiring curved surfaces and for the fabrication of nooks, partitions, cantilever shelves, etc.

Other household building uses for plywood include the finishing of attics and basements, secondary partitions, closets, and storage rooms. Plywood finds wide application in garages and farm buildings, such as poultry houses and dairy-barn stalls, as well as in office buildings, warehouses, and other industrial buildings. Plywood is a favorite construction material for interior and exterior sheathing, subflooring, and roofing for exposition buildings and similar temporary structures which require speed of construction, maximum safety, high decorative possibilities, and low costs.

In connection with constructional uses of plywood, reference must be made to its utilization in prefabricated houses. Whether made of wood, steel, or other materials, all prefabricated housing systems use a panel as the structural unit. The basic idea of all these systems is that the most difficult part of construction should be done by factory mass-production

Fig. 8-19. Rift-grain oak plywood paneling in St. Raphael's Cathedral, Madison, Wisconsin. (*Courtesy Hardwood Plywood Institute, Chicago, Illinois.*)

methods, by which it can be done more cheaply and efficiently, thus offering to the public cheaper and at the same time more scientifically designed housing. A number of model prefabricated houses utilizing plywood as the principal construction material have recently been built throughout the country.

In addition to the beauty of the natural wood grain and figure plywood is now produced with a variety of textured surfaces which extend its utility as a decorative interior panel material (Fig. 8-19). Some of the textured surfaces consist of grooves or striations machined into the surface ply or embossed on the face. A variety of decorative panels are produced from woods like Douglas-fir and pine that have springwood and summerwood quite different in density. Rotary-cut faces of such woods are subjected to an abrasive treatment that removes some of the soft springwood material and produces an attractive, irregular surface. As stated before, such panels are available factory prefinished.

Parquetry flooring is now being produced from hardwood plywood (Fig. 8-20). Small squares of ⅜-in. three-ply plywood are tongued and grooved

Fig. 8-20. Plywood-block flooring being laid in mastic. (*Courtesy Hardwood Plywood Institute, Chicago, Illinois.*)

on all four sides. This plywood may be laid with a mastic or cement in the same manner used to lay tile flooring.

2. Plywood in Furniture and Cabinetwork. Plywood is widely used in furniture and cabinetmaking (Fig. 8-21). Some of the principal uses are for tops, table leaves, case backs, drawer fronts and bottoms, mirror backs, foot and head panels in beds, radio cabinets, piano cases, sewing-machine cabinets, fixtures, and seats and backs of auditorium chairs, school and theater chairs, and church pews. In fact, plywood is used in all parts of furniture except legs, rungs, frame rockers, and similar solid parts. Furniture plywood is usually made to order in accordance with the manufacturer's specifications as to matching and size. Some furniture manufacturers prefer to buy three-ply panels to be used as a stock on which face veneers are glued at the furniture factory, so that any desired matched effect can be produced as needed. Cabinetwork, such as cupboards, closets, bookcases, and medicine chests, provides another large outlet for plywood.

3. Industrial Uses. To meet the engineering requirements of light weight with maximum strength, there are on the market a number of metal-faced plywood products. Such panels are used in truck bodies, boats, hospitals, and restaurants; more recently such panels have been used in the construction

Fig. 8-21. Buffet constructed with plywood panels. (*Courtesy Drexel Furniture Company, Drexel, North Carolina.*)

of streamlined trains. Plywood covered with heat- and acid-resistant resins is available for use for table tops, office equipment, and wall paneling. Freight-car construction promises an extensive field of use for Douglas-fir plywood in the form of car lining, while railroad passenger cars have long offered a market for plywood, for decorative as well as structural purposes.

Plywood made with water-resistant glues is extensively used in all classes of watercraft (Fig. 8-22). In small pleasure boats plywood is used as a structural and decorative material, while its use for interior paneling in large ocean liners is rapidly increasing. Mobile-home and travel-trailer building consumes millions of square feet of this product (Fig. 8-23). The automobile industry still consumes quantities of plywood, although much less than formerly, for floor boards, trunk shelves, and upholstering frames. During the Second World War enormous quantities of plywood were used in airplane and glider construction.

Fig. 8-22. Molded plywood used in small-boat construction. (*Courtesy Hardwood Plywood Institute, Chicago, Illinois.*)

Fig. 8-23. Plywood in mobile-home construction. (*Courtesy Hardwood Plywood Institute, Chicago, Illinois.*)

163

Millwork is another important outlet for plywood, one of the largest items being door panels of all kinds; much plywood is also used in trunks, traveling boxes, packing cases, and fruit and vegetable baskets and boxes.

An unusual but quite extensive market for Douglas-fir plywood is in the motion-picture industry and on the legitimate stage, for construction of stage sets. Toys, woodenware and novelties, bird houses, Ping-pong table tops, patterns, bulletin boards and signs, educational and industrial exhibits, and amusement devices are but a few other items on the rapidly expanding list of uses for plywood.

A substantial amount of plywood is used in the construction of outdoor signs. Many highway marker signs are now manufactured from plywood. Exterior grades are used for these purposes, and the sign blanks not uncommonly are made of plywood and have a plastic or fiber overlay to provide a good painting surface. Export trade provides an important outlet for large quantities of exterior and interior grades of plywood.

D. Plywood Production: Industrial Status

Plywood is manufactured in practically every state in the Union. The majority of softwood plywood is produced on the Pacific Coast while the bulk of the hardwood plywood manufactured in the United States is produced east of the Mississippi River. In recent years a substantial amount of plywood with hardwood faces and softwood interior plies has been produced on the West Coast.

Table 8-1 gives the production of softwood plywood annually for the years 1924 to 1959 inclusive. These data indicate the remarkable growth of this industry in 36 years. The softwood-plywood industry began in the state of Washington, and that state led in the production of softwood plywood until 1953. After the conclusion of the Second World War, production of softwood plywood in Oregon and California increased very rapidly, while production in Washington leveled off. In 1959 Oregon produced about 65 per cent of the softwood plywood, Washington approximately 20 per cent, and California about 14 per cent, with lesser amounts being manufactured in Montana and Idaho.

The hardwood-plywood industry is made up of a great many small factories distributed widely over the eastern United States. The hardwood-plywood industry increased its production after the Second World War but not nearly as spectacularly as did the softwood-plywood industry, in spite of the fact that American consumption has increased sharply. This is due to the fact that very substantial quantities of hardwood plywood have been imported into the United States since 1950. In 1959 domestic production of hardwood plywood was 960 million square feet. During this same year hardwood-plywood imports reached a new high, with a volume of 1,318 million square feet, representing 59 per cent of United States hardwood-plywood consumption. The principal source of imported hardwood plywood in 1959 was Japan, which shipped into the United States 801 million square feet. The other major source was the Philippine Islands, whose imports for the same year were 214 million square feet. During 1959 production of hardwood

Table 8-1. Annual Production of Softwood Plywood, 1924 to 1959 [a]

Year	Production, sq ft [b]	No. of plants end of year
1924	125,000,000	10
1925	153,000,000	12
1926	173,000,000	12
1927	206,000,000	13
1928	276,000,000	14
1929	358,000,000	16
1930	305,000,000	17
1931	235,000,000	17
1932	200,000,000	17
1933	390,000,000	17
1934	384,000,000	17
1935	480,000,000	17
1936	700,000,000	20
1937	725,000,000	21
1938	650,000,000	21
1939	950,000,000	23
1940	1,200,000,000	25
1941	1,620,000,000	31
1942	1,782,000,000	31
1943	1,430,000,000	30
1944	1,440,000,000	30
1945	1,200,000,000	31
1946	1,395,000,000	33
1947	1,630,000,000	40
1948	1,871,000,000	45
1949	1,899,000,000	55
1950	2,553,652,192	68
1951	2,866,951,594	77
1952	3,049,740,489	86
1953	3,670,433,623	94
1954	3,903,781,385	100
1955	5,075,189,352	111
1956	5,239,810,964	121
1957	5,459,874,284	118
1958	6,339,514,220	127
1959	7,827,938,790	141

[a] SOURCE: Douglas-fir Plywood Association.
[b] Production figures are given in terms of $\frac{3}{8}$-in. three-ply stock.

plywood among the various regions was: South Atlantic, 37 per cent; New England and Middle Atlantic, 18 per cent; Pacific, 18 per cent; East South Central and West South Central, 15 per cent; and East North Central and West North Central, 12 per cent.

SELECTED REFERENCES

1. Anon. Faces and Figures. Reprint from *Homefurnishing Arts*, Spring–Summer number. The Veneer Association, Chicago, Ill. 1935.
2. Anon. Proceedings: Conference on Radio Frequency and Its Applications in Gluing Wood. *Wash. Univ. Col. Forestry, Bull.* 2. 1946.
3. Bethel, J. S. Drying and Conditioning Veneer. *N.C. State Col., Tech. Bull. No.* 4, pp. 52–56g. 1950.
4. Bethel, J. S. Hot Pressing. *N.C. State Col., Tech. Bull. No.* 4, pp. 31–36. 1950.
5. Bethel, J. S., and R. M. Carter. Techniques of Hardwood Plywood Quality Control. Proceedings, Forest Products Research Society, pp. 162–169. 1950.
6. Bethel, J. S., and R. J. Hader. Hardwood Veneer Drying. *N.C. State Col., Tech. Bull. No.* 7. 1952.
7. Bethel, J. S. Kontrola Kvaliteta U. Proizvodnji Ukocenog Drva. Sumarkski List. 1952.
8. Hanchett, D. J. Resorcinol Adhesives for Durable Wood Bonding. *N.C. State Col., Tech. Bull. No.* 4, pp. 44–51. 1950.
9. Harrar, E. S. Veneers and Plywood: Their Manufacture and Use. *Econ. Bot.,* **1**(3):290–305. 1947.
10. Hayward, P. A., et al. American Douglas Fir Plywood and Its Uses. *U.S. Bur. Foreign and Dom. Com., Trade Prom. Ser.* 167. 1937.
11. Hill, Robert F. Soybean Glues. *N.C. State Col., Tech. Bull. No.* 4, pp. 20–23. 1950.
12. Hill, Robert F. Melamine-formaldehyde Resins. *N.C. State Col., Tech. Bull. No.* 4, pp. 24–25. 1950.
13. Johnson, H. H. Veneering. The Macmillan Company, New York. 1928.
14. Knight, E. V., and M. Wulpi. Veneer and Plywood. The Ronald Press Company, New York, 1927.
14a. Kollman, F. Furniere, *Lagenhölzer und Tischlerplatten.* Springer Verlag, Berlin, 1962.
15. Macy, P. A. Vegetable Glues. *N.C. State Col., Tech. Bull. No.* 4, pp. 26–28. 1950.
16. Marra, A. A. Trouble Shooting: The Behaviors of Glue. *N.C. State Col., Tech. Bull. No.* 4, pp. 73–76. 1950.
17. McCormack, P. H. Polyvinyl Glues for Woodworking. *N.C. State Col., Tech. Bull. No.* 4, pp. 57–58. 1950.
18. McLaglen, C. F. Casein Glue. *N.C. State Col., Tech. Bull. No.* 4, pp. 11–16. 1950.
19. Perry, T. D. Modern Plywood, 2d ed. Pitman Publishing Corporation, New York. 1948.
20. Perry, T. D. Wood Adhesives. Pitman Publishing Corporation, New York. 1944.
21. Perry, T. D., and M. F. Bretl. Hot-pressing Technique for Plywood. Reprint from *Am. Soc. Mech. Engin. Trans.* January, 1938.
22. Phinney, H. K. Phenol-formaldehyde and Resorcinol-formaldehyde Resin Glues. *N.C. State Col., Tech. Bull. No.* 4, pp. 37–43. 1950.
23. Rinne, V. J. The Manufacture of Veneer and Plywood. Kuopion Kansallinen Kirjapaino, Kuopio. 1952.
24. Tutt, R., Jr. Animal Glues in Woodworking. *N.C. State Col., Tech. Bull. No.* 4, pp. 1–10. 1950.
25. Wood, A. D., and T. S. Linn. Plywoods: Their Development, Manufacture and Application. Chemical Publishing Company, Inc., New York. 1943.

CHAPTER 9

WOOD FURNITURE

The manufacture of furniture represents one of the oldest forms of wood utilization. A fallen tree provided a natural bench. A piece of the same log, flattened on top, made a convenient table. Because wood was easy to cut and shape with the most crude hand tools, it was early fashioned into a more comfortable chair or a more utilitarian table. The design and construction of wood household furniture soon became an art, and the most skilled of the furniture artisans became famous for the beauty of their design and the skill of their handwork. Some of the early furniture craftsmen published collections of their best designs, and these have been copied ever since. Chippendale's "Gentleman and Cabinetmaker's Director," Hepplewhite's "Cabinetmaker and Upholsterer's Guide," and Sheraton's "Cabinetmaker and Upholsterer's Drawing-book" are classical examples of these early furniture specifications. Original productions of these and other master craftsmen have been saved in museums and faithfully copied over the years.

In the early days of wood furniture manufacture the individual craftsman worked in his shop creating one piece of furniture at a time with painstaking care. Elaborate carvings were the hallmark of fine furniture, and these were accomplished laboriously by hand. Furniture making was an apprentice trade.

With the advent of the industrial revolution came the idea of separation of function in the manufacture of wood furniture as in the manufacture of many other products. One individual or group of individuals designed an article, and another group produced it. And among the producers one man manufactured part A in quantity and another part B, etc. Still other workmen assembled the completed article from the groups of parts previously manufactured. Thus mass production found a place in the building of furniture and with it was introduced the idea of interchangeability of parts, mechanization of work, systematic plant layout, organized production control, and quantitative quality control. Frederick Litchfield in his "Illustrated History of Furniture," [11] published in 1893, made the following somewhat nostalgic comments on the change then occurring in the manufacture of furniture:

167

As the manufacture of furniture is now chiefly carried on in large factories, both in England and on the Continent, the subdivision of labor causes the article to pass through different hands, in successive stages and the wholesale manufacture of furniture by steam, has taken the place of the personal supervision by the master's eye, of the task of the few men who were in the old days the occupants of his workshop. As a writer on the subject has well said, the chisel and the knife are no longer in such cases controlled by the sensitive touch of the human hand. In connection with this we are reminded of Ruskin's precept that "the first condition of a work of art is that it should be conceived and carried out by one person."

Even in this time of organized mass production the finest and most expensive furniture manufactured in the United States has much handwork in it. The great majority of wood furniture is, however, produced on the production line, and it is with this kind of operation that this text is most concerned.

Wood was one of the first materials used in furniture construction. While many other materials have been introduced into the manufacture of household furniture, wood is still the principal component by a very wide margin.

Wood has many things in its favor for use in furniture: (1) it is easily worked with hand and machine tools; (2) it may be satisfactorily fastened together with adhesives, dowels, nails, and screws; (3) it is strong for its weight, enabling the fabrication of strong parts possessing agreeable proportions without incurring excessive weight; (4) damaged wood usually is easily repaired; (5) it is comparatively noiseless when struck; (6) it is a poor conductor of heat and therefore does not feel very hot or very cold when touched, although its temperature may be considerably above or below body temperature; (7) no other material possesses such natural beauty, with variations in figure and color that blend, although the shades of color and the contours of pattern are seldom, if ever, identical; (8) pieces of wood that are not, in themselves, ornamental may be given attractive finishes; and (9) no other material, as it grows old in service, enhances its beauty by taking on a darker color and rich appearance like the patina of old wood furniture.

While wood has many advantages as a furniture material, it is not perfect for this purpose. In its natural state, it has two very objectionable features: (1) it absorbs or releases moisture with changes in atmospheric temperature and humidity, and (2) it shrinks and swells with changes in moisture content. These two properties, working in unison, are the basic causes of most of the problems encountered when wood is used as a furniture material. They account for many of the defects developing in the seasoning of raw material; they are responsible for the splitting and checking of large surfaces, for the failures of many glue joints between veneer and core stock in panels or between structural members, and for the sticking of doors and drawers in humid weather. Elimination of these faults would remove the principal criticism of wood. These effects can be ameliorated, but not removed, by drying wood to the average moisture content it will assume while in service; by giving surfaces moisture-resistant coatings where possible; by using cross-banding veneer between center stock and face veneer of panels; and by using moisture-resistant adhesives, some of which impregnate thin layers of veneer, thereby reducing their tendency to shrink or swell with changes in atmospheric conditions. The use of modified woods that have high dimensional stability is a distinct possibility, but high cost has precluded their intensive use for furniture up to the present time.

Wood, of course, enjoys a record of centuries of excellent service, and this is a powerful influence in maintaining its popularity; but in order to perpetuate this esteem, every effort must be made to better its behavior in service. If such improvement is not forthcoming, wood may yield its position to more satisfactory materials that may be developed.

I. MANUFACTURE OF WOOD HOUSEHOLD FURNITURE

The manufacture of wood household furniture embraces a broad spectrum of industrial operations. At one extreme is the cabinetmaker who works alone or with a few helpers producing custom-made furniture to order. At the other extreme is the large modern factory that produces furniture on a mechanized production line. The industry includes all variants between these two extremes. The size and degree of sophistication of a particular furniture-manufacturing enterprise may be determined by a great many factors. The amount of capital available to the owner is an obvious limiting factor. The quality of furniture being produced will also be a determining factor. Low-priced and medium-priced furniture can be produced in the large quantities required to justify mechanized production lines. These types of furniture, too, are adaptable to machine operations. Very high-priced furniture does not sell in sufficient volume to permit the efficient use of mass-production techniques. This type of furniture, too, requires so much skilled handwork that the value of mechanized production lines cannot be realized. It is true, nevertheless, that the furniture industry in the United States has increasingly adapted modern production-line techniques to the task of manufacturing high-quality furniture. This explains, at least in part, the fact that good furniture is accessible to almost all American families, whereas in many other parts of the world really good furniture is usually available only to the wealthy upper classes. The use of modern woodworking machinery has greatly reduced labor requirements, time, and manufacturing costs, thereby making good furniture available in large quantities at reasonable prices, and it has made possible a parts-fabricating precision that cannot be equaled by human hands. Modern furniture manufacture consists largely of fabricating individual parts in quantity and assembling them into nearly identical pieces of furniture. It is difficult, however, for a furniture factory to standardize on any particular design, especially in the higher grades of furniture. The factory must be equipped to produce a varied line of goods.

A. Factory Processes

To complement the following description of wood household-furniture manufacture, the flow sheet shown in Fig. 9-1 has been prepared. It covers the usual processes encountered in most factories making living-room, dining-room, and bedroom furniture. The flow sheet does not cover every detailed step in the manufacture of such furniture, but it does give the principal operations. Likewise, it cannot be said to apply to a particular factory, because it represents a composite of practices, some of which are not in vogue at all plants.

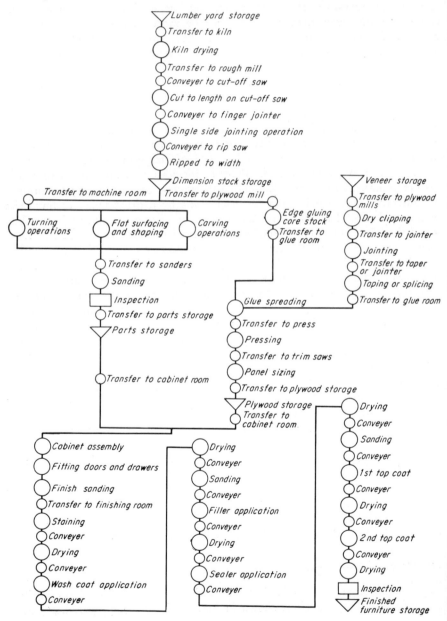

Fig. 9-1. Process chart for typical small furniture factory.

A typical wood-furniture-manufacturing operation might be roughly divided into the following parts: raw-material processing and storage (lumber yard, veneer and plywood storage, etc.) ; rough-mill operations; machine-room operations; cabinet-room operations; finishing; packaging and shipping.

1. Raw-material Processing and Storage. The principal raw materials of the furniture plant are lumber, veneer, and plywood. Such other components as glues, finishing materials, hardware, etc., must also be used, but their handling before use is essentially a conventional warehousing problem. Wood, as a raw material, is brought to the factory principally in the form of lumber. Usually lumber is purchased by grade and air-dry. Some mills, however, buy some of the lumber used in interior parts as mill-run and then segregate into grades at the furniture plant. The chief standard grade of hardwood lumber consumed by furniture factories is the combined grade of No. 1 Common and Selects. Other grades commonly used are Firsts and Seconds and No. 2 Common. The principal thicknesses of lumber purchased are $\frac{4}{4}$, $\frac{5}{4}$, $\frac{6}{4}$, and $\frac{8}{4}$ in.

Many factories purchase hardwood dimension stock in order to avoid costly freight charges on useless material and to reduce the amount of equipment needed in the plant. Hardwood dimension stock is available to the furniture industry in four forms: (1) rough, green, or air-dried blanks of standard sizes; (2) kiln-dried and surfaced blanks of standard sizes; (3) semimachined parts; and (4) finished parts. The first type of hardwood dimension stock necessitates kiln-drying, with its attendant degrade, and almost complete machining at the factory. Stock of the second type has been kiln-dried and also requires less machining. Furniture factories in metropolitan centers and in areas remote from substantial timber resources are increasingly using semifinished and finished dimension. When the latter material is used, the furniture manufacturing operation commences with the cabinet room.

Lumber is delivered to the plant by truck or rail and is stacked in stickered piles in the lumber yard to await use. It is separated by species, grade, and thickness to facilitate future use. When it is needed in the factory, it is taken to the kiln for drying. The kiln-drying operation was described in detail in Chap. 7. In some furniture centers manufacturers prefer to have lumber custom-dried. In so doing, they avoid the necessity of operating kilns.

The furniture factory may purchase all its required plywood components manufactured to order and cut to size. Some factories have their own plywood departments that produce the company's plywood requirements from purchased veneer. Where this is the case, the plywood is manufactured as described in Chap. 8. A not uncommon practice is for the furniture plant to fabricate its high-grade cabinet plywood for such items as table and dresser tops, drawer fronts, side panels for case goods, etc., and to purchase its lower-grade components such as backs for case goods, drawer bottoms, dust bottoms, and similar items.

2. Rough-mill Operations. The rough mill of a furniture factory performs essentially a dimension-stock plant. In the rough mill thick lumber may be resawed into thinner stock before proceeding to the cutoff saw, although it is usually purchased cut to the correct rough thickness. At the cutoff saw, a skilled operator cuts the lumber into lengths suitable for various furniture parts. It is his duty to convert the greatest possible percentage of each board or plank into usable sizes as well as to eliminate defective portions. He must know the permissible defects in every furniture part in order not to waste raw material by trimming too much.

Pieces produced at the cutoff saw are conveyed to a self-feeding, straight-line ripsaw for further removal of defects and dimensioning. The ripsaw operator also must exercise great care in avoiding excessive ripping and waste of raw material.

In some factories a roughing planer or finger jointer is used to plane one surface during the cut-up operation. Sometimes the lumber is processed through the rough-planing operation before the cutoff saw, and sometimes this planing occurs between the cutoff and ripsaws.

Stock for legs and heavy panels must frequently be glued up before it can be used. The gluing operation is essentially the same as that described for lumber core stock in Chap. 8.

3. Machine-room Operations. In the machine room the stock from the rough mill is shaped, machined to final dimensions, and subjected to such other operations as boring, mortising, tenoning, and dovetailing, as required to prepare it for fitting in the cabinet room. All parts that are to be exposed in the finished piece of furniture are sanded to prepare them for finishing operations.

Parts such as posts in case goods that are square or rectangular in cross section are surfaced on a finish planer or a molder and then sanded. Panels, either solid or veneered, are surfaced and drum-sanded. If they are exposed parts in the finished furniture, they are given a further sanding on a belt sander.

Parts that require shaping may be processed on a variety of machines depending upon the nature of the shaping operation. Where single curves are required, the shaping may be accomplished with a band saw. Curved drawer fronts, for example, are commonly band sawed out of a heavy laminated blank. They are then either surfaced or veneered.

Straight components that require profiles to be machined on them are processed through a molder. Profiles on the edges of table tops or bed headboards may be prepared on a single- or double-spindled shaper. An automatic shaper for small parts is equipped with a revolving turntable to which a form of suitable contour is fastened. This form is so designed that several small pieces may be cut at each revolution of the table. Air-operated clamps hold the pieces while they are being shaped by a motor-driven cutterhead suspended from an arm which follows the contour of the pattern or form attached to the revolving turntable.

Other items may be put through a double-end tenoner for the purpose of making tenons on both ends; parts requiring mortising are sent to mortisers of various kinds. Drawer sides and ends may be produced on a dado machine, tenoner, or molder, where the grooves are cut which will hold the drawer bottom. In high-quality furniture drawer fronts, sides, and backs are sent to a dovetailing machine for dovetailing operations on the ends.

Pieces requiring boring for dowels are sent to multiple-spindle boring machines where several holes are to be drilled. Irregularly shaped pieces are cut on scroll saws and jig saws, and pieces for mitered joints are cut on an adjustable miter saw.

Panels requiring open scrollwork can be sent to a router for cutting out the openings. The panel in this instance will be clamped to a jig having

Fig. 9-2. Church furniture. This type of furniture is characterized by much expensive hand carving. (*Courtesy Hardwood Plywood Institute, Chicago, Illinois.*)

grooves of the proper contours on its underside. A steel pin is set in the router table directly beneath the cutting spindle, and this pin extends upward into the grooves of the jig.

Parts that are circular in cross section, such as chair legs, the legs of beds, spools, round posts, etc., are machined on one of the variety of lathes available for furniture manufacture. The profile lathe holds the work between a very slowly revolving headstock and tailstock while a cutterhead equipped with knives of the desired shape revolves at high speed. The contour of the knives is the counterpart of the shape to be turned. Another type of automatic profile lathe uses a back knife. This lathe holds the material between a high-speed revolving headstock and tailstock while a knife of the desired shape operates on the work from the back, moving in and out, as determined by the contour of the work, governed by a master cam. Other specialized lathes are used for fluted and spiral patterns.

Intricate shapes in relief are prepared in carving operations. In very high-priced furniture carving may be done entirely by hand, utilizing the services of a highly skilled carver. Most carving is now accomplished on multiple-spindle carving machines. The original carving is produced by hand and then used as a pattern in the multiple carver. These machines may be equipped with as many as 24 spindles. In using such a machine, the operator moves a tracer over the contour of the hand-carved pattern, and each carving spindle cuts into a blank piece, duplicating the artist's work. The product of the multiple-spindle carver is usually completed by hand by master carvers on a single-spindle carver.

Some types of furniture parts are converted to the proper shape by bending. Here the wood is usually steamed at a temperature of about 212°F for about 1 hour per in. of thickness. After the wood has been steamed, it is placed in a bending form, restrained in the proper shape, and dried. The parts coming from the steam-bending operation are prepared for finishing by machining and sanding.

All material that is to receive a high-quality finish on exposed surfaces is sanded before leaving the machine room. The various kinds of sanders include, among others, the drum, stroke, or belt, hand block, molding, variety, disk, oscillating spindle, and special turning sanders.

4. Cabinet-room Operations. The cabinet room of a furniture factory is its assembly department. Here all the parts previously prepared are collected and assembled into the piece of furniture. While the machine room is equipped with machines of all types, the cabinet room is equipped with jigs, fixtures, and clamps. Large case goods such as dressers, chests, buffets, and the like are assembled in case clamps. Drawers are assembled separately in drawer clamps. Head- and footboards of beds are put together in bed clamps. The principal activities of the cabinet room are assembly and fitting. It is here that the interchangeability of parts pays off. If the individual components have been properly manufactured, they will fit easily together with well-fitted, inconspicuous joints. Where necessary, cold-setting glues are used to assemble pieces; glue joints are set in clamps, or in some instances nails or screws are used to provide gluing pressure. Drawers are properly fitted, and doors are hung. Where glass is incorporated into the piece of furniture, glazing is accomplished in the cabinet room. Hardware, such as hinges, drawer and door pulls, locks, etc., is fitted to the piece. It is the responsibility of the workers in the cabinet room to make sure that the item of furniture is ready for finishing. If machined parts do not fit precisely, they must be modified if possible by the cabinet worker or sent back for rework or replacement if modification on the spot is not feasible. Assembled pieces of furniture are given a final, prefinishing inspection.

5. Finishing Operations. Furniture is finished to achieve two purposes: (1) to enhance the beauty of the article, and (2) to give the article protection and thus increase its useful life. A suitable finish may enhance the beauty of the piece of furniture by improving its color, emphasizing or deemphasizing the grain and figure, and providing a smooth, mirror-like surface. There are an infinite number of furniture-finishing schedules that might be used in furniture-manufacturing operations. A very large number of finishing materials are available to the designer and the manufacturer in selecting the finishing schedule that is appropriate. It is beyond the scope of this text to consider the subject of furniture finishing in great detail.

Fig. 9-3. Baby furniture. Manufacture of furniture of this type constitutes a separate branch of the furniture industry. (*Courtesy Hardwood Plywood Institute, Chicago, Illinois.*)

It will suffice to indicate the general nature of the finishing operation and to note the more important types of finishing materials. In general the basic operations in furniture finishing are (1) staining or coloring, (2) filling, (3) sealing, and (4) applying protective film or top coat.

   a. Staining or Coloring. In general the surfaces of wood are colored either by painting or staining. Painting involves covering the surface of the wood with an opaque coating which usually consists of a pigment of some kind in a resinous base. Some very low-priced furniture is painted to cover up unattractive wood and to reduce finishing costs to a minimum. Some very expensive furniture is also painted, where this type of finish is appropriate to the design, as, for example, in the case of French Provincial reproductions. In general, though, the coloring of wood is accomplished with stains. Stains consist of dyes or pigments dissolved or suspended in a liquid medium. Stains are used to improve the color contrast in the natural grain and figure of wood; to improve the uniformity of color, particularly in wood that has a naturally variegated color; and to change the color of inexpensive woods to make them look like more expensive woods.

Stains are classified, according to the type of carrier or vehicle, as oil, water, and non-grain-raising. Oil stains are inexpensive and easy to use. They are used in finishing some very low-priced furniture and in small workshops but rarely in large factories producing the better grades of furniture. The principal disadvantages of oil stains are that they do not make the color of the wood uniform, they bleed badly, tend to be soluble in top coats, and are not very light-fast.

Water stains were widely used in the furniture industry in the past, but are little used now. Water stains produce good colors, and, in fact, they permit some color effects that cannot be duplicated with other more widely used stains because of the depth of penetration that is normally achieved. The principal disadvantage of a water stain is that it raises the grain and produces a rough surface on the wood. Another disadvantage of water stains is that they are very slow to dry.

Non-grain-raising stains are dyes dissolved in fast-drying solvents. They are referred to in the industry as N.G.R. stains. These stains are currently the most widely used true stains in the furniture industry. They normally contain, in addition to the dye and solvent, a small quantity of a coal-tar diluent which functions to minimize grain-raising. N.G.R. stains may be obtained from colors that compare favorably with water stains for brilliance. They generally dry rapidly and do not bleed badly.

A group of coloring materials now widely used in the furniture industry in lieu of conventional stains are the equalizers, toners, and pigmented wiping stains. Equalizers, as the name implies, are used to equalize the normal nonuniformity of color in the natural wood. They are essentially dilute pigmented lacquers. Toners are similar to equalizers in composition but are used in place of bleaching to tone down the natural color of the wood. Bleaching is a very expensive and time-consuming operation, and it has been largely eliminated in the furniture industry with the advent of toners in the finishing process.

Pigmented wiping stains are widely used as substitute for the more conventional stains. They consist of finely divided pigment dispersed in a hydrocarbon solvent. The more common solvents are naphtha and neutral spirits. These stains must be wiped off after application to achieve uniform color. Since these stains contain pigment rather than dye, they have much more tendency to cover wood grain than do N.G.R. stains, and thus have a uniforming effect.

b. *Fillers.* Fillers are used, as the name suggests, to fill up open pores in the surface of the wood so that the final finish presents a smooth surface. Fillers are more important in finishing coarse-textured woods than fine-textured woods. Long before the development of modern fillers, furniture craftsmen filled the pores with a heavy shellac and then sanded the surface smooth. The early furniture fillers used linseed oil or japan drier as the binder and accordingly were slow to dry. With the development of high-speed, conveyerized finishing systems, it became necessary to develop fast-drying fillers. Present-day fillers dry in a matter of hours and in the case of quick-flash fillers in a matter of minutes.

A filler consists of a finely ground, transparent, colorless pigment carried in an oil binder. In addition, the filler contains colors and dyes to give it the desired color and a drier to speed up the filler-drying operation. Fillers

Fig. 9-4. Modern dining-room furniture. (*Courtesy Hardwood Plywood Institute, Chicago, Illinois.*)

are applied to the surface of the wood, rubbed into the pores of the wood, and then the excess is wiped off. The filler is then allowed to dry before the finishing operation is completed. When wood of very coarse texture is being finished, it is sometimes necessary to double fill in order to assure a good smooth surface.

c. Sealer. Whereas the stain and filler are used in the finishing operation to enhance the appearance of the wood, the sealer is used to protect the wood. It provides a moisture barrier and protects the surface from discoloration due to contaminants; it also provides some protection against wear. In addition, the sealer provides a substantial base for the top coat.

Among the basic materials used as sealers are shellac; nitrocellulose lacquers; and alkyd, vinyl, and various other synthetic-resin varnishes. Sealers are normally sprayed on in modern finishing operations, although they can be brushed on. A good sealer should be tough and should bring out the brilliance of the color coats. It should also increase the appearance of body in the finish.

d. Top Coat. The top coat is primarily a protective coating, but to be satisfactory it must also add to the appearance of the finish. The top coat gives body to the finish; it also provides the surface sheen. A good top coat enhances the color, grain, and figure of the original wood and emphasizes the brilliance of the color coats. A satisfactory top coat must be transparent and free of any haziness.

Ideally, to protect the wood surface, the top coat should be tough and flexible and should adhere to the surface of the furniture. It must be hard enough to resist imprinting and scarring and have reasonable resistance to damage from water, beverages, foods, cleaning agents, hot dishes, etc. The modern top coat also should be adaptable to production-line finishing; i.e., it must spray satisfactorily and dry rapidly. The principal materials used for top coats are lacquer, oleoresinous varnishes, and synthetic-resin varnishes.

e. Finishing Systems. In addition to the principal finishing materials just described, there are a number of other materials used in finishing to achieve the quality of finish desired. Wash coats of shellac or lacquer may be used between stain and filler to stiffen the fibers of the wood so that they may be sanded smooth. Glazing stains are used ahead of the sealer to provide greater depth to the finish.

Interspersed between the applications and drying of finishing materials are extensive sanding operations. The final gloss is achieved by rubbing the top coat with fine abrasives and polishing with wax. Very expensive finishes may involve a sequence of top coats, with sanding and rubbing between coats. As previously indicated, there are a great many combinations of finishing materials and operations possible. Inexpensive finishes usually involve few operations, while expensive finishes require a great many separate processes.

Hager [5] gives the following sample finishing schedules as examples of low-cost, medium-cost, and high-cost finishes:

Low-cost finish:
1. Finish sand furniture to prepare for finishing
2. Stain to desired color
3. Apply coat of sealer
4. Sand sealer coat
5. Apply a flatted top coat

Medium-cost finish:
1. Finish sand furniture to prepare for finishing
2. Stain to desired color
3. Apply light coat of sealer (called a wash coat) to lightly seal the fibers of the wood
4. Fill pores (open-pore wood)
5. Apply coat of sealer

6. Sand sealer coat
7. Uniform color on different parts of the piece by shading light parts to match the darker parts
8. Apply finish coat
9. Sand finish coat
10. Apply second finish coat
11. Rub finish coat
12. Polish finish coat

High-cost finish:

1. Hand sand furniture to prepare for finishing
2. Stain to desired base or undertone color
3. Impregnate wood with waterproofing material
4. Fill pores of wood (open-pore wood)
5. Apply coat of glazing sealer
6. Sand glazing sealer
7. Apply coat of glaze
8. Brush and blend glaze to achieve mellowing of color and to intensify highlights and shadows
9. Apply coat of sealer
10. Sand sealer coat
11. Apply finish coat
12. Sand finish coat
13. Hand pad color into finish coat
14. Add distinguishing marks and treatment to finish
15. Apply second coat of finish
16. Hand rub finish
17. Clean up rubbing
18. Hand wax finish

f. Mechanized Finishing Systems. In the fully mechanized furniture-manufacturing operation the finishing process is accomplished while the pieces of furniture are moving on a continuous conveyer system. The conveyer system must be carefully designed, with spray booths, drying ovens, wiping, sanding, and rubbing stations placed so that the operations can occur at the proper time and drying can be accomplished between finishing operations. The proper operation of a mechanized finishing system combines a properly designed conveyer, finish application and drying apparatus, and suitable finishing materials.

B. Wood Lost in Furniture Manufacture

The manufacture of wood furniture entails a great deal of loss in raw material. Some furniture parts contain less than 50 per cent of the wood necessary for their manufacture. Much of this loss is unavoidable because of (1) artistic shapes required in solid parts; (2) trimming of veneer for panels, particularly when flitches of sawed or sliced veneers are purchased; (3) matching of veneer, especially when unusual designs are sought; and (4) inability to use the entire piece of lumber purchased under regular grading rules because of losses from defects, saw kerf, shavings, and the impossibility of converting all the remaining clear material into usable cuttings of required sizes. The purchase of hardwood dimension stock or completely machined parts by the furniture factory does not solve the wood-loss problem. The same losses are merely transferred to the hardwood dimension-stock factory. The furniture factory, however, does not have to pay freight on useless material when it buys hardwood dimension stock and machined parts.

One operator quotes an average loss of approximately $33\frac{1}{3}$ per cent of the rough lumber taken from the yard and an average loss of about 12 per cent in air-seasoned hardwood dimension stock. The loss of material in green hardwood dimension stock averaged more than 12 per cent. The loss from hardwood dimension stock, while usually considerably less than the loss from lumber, occasionally reaches 25 per cent.

Much loss of raw material can be prevented by (1) proper piling in the storage yard, (2) better kiln-drying equipment and more skillful kiln operators, (3) more careful handling and elimination of unnecessary handling, (4) reduction of spoilage by obtaining better equipment and skilled employees, (5) elimination of oversized blanks for parts, and (6) using extreme care in handling veneers.

C. Some Features of Good Furniture Construction

Much of the quality in furniture lies in unavoidably concealed construction points, such as the type of adhesive used, the fit of a mortise-and-tenon joint, the kind, number, and placement of dowels, the moisture content of the various wooden parts at the time of assembly, and the internal stresses existing in the lumber when it leaves the dry kiln. Rarely can any of these things be ascertained by examining a completely assembled and finished piece of furniture. On these points, the purchaser generally has to rely on the reputation and integrity of the manufacturer. There are several visible evidences of good craftsmanship characteristic of high-grade furniture, on which the purchaser can estimate quality.

1. Very few nails will be found in good furniture.
2. The sides of drawers will be dovetailed into both the front and back panels.
3. Drawer bottoms will have substantial thickness and will be dadoed into the sides and ends of the drawer.
4. The drawer will slide easily and will be treated with a moisture-retardant coating.
5. Center slides on the bottoms of drawers will be fastened to front and back cross rails and glued to the bottoms of the drawers.
6. Bedroom and dining-room furniture will have thin panels between the drawers for dustproofing.
7. The back of a cabinet will be fitted into grooves in the sides rather than being fastened with nails.
8. The legs of chairs, beds, tables, and cabinets will set squarely on a level floor.

9. Corner blocks, glued and screwed to the rails, will strengthen joints at the points of leg attachment on chairs and tables. Absence of corner blocks or substitution of nails for screws in corner blocks is evidence of inferior construction. Corner blocks are not required when bentwood legs are socketed into the rim of a chair seat.

10. The back posts of chairs will be of bentwood or, if sawed from lumber, they will be cut in such a manner that the lower portion will contain a sufficient amount of longitudinal grain so that the bottom of the post cannot be broken off easily.

11. Rungs or stretchers between the legs of chairs will be in sufficient number to give ample bracing.

12. Table tops will be attached to the legs and rails in such a manner that they are permitted to shrink or swell without splitting the base.

13. Carved furniture will have the carvings actually made in solid wood rather than having carved wood blocks or molded compounds glued to the surface of the wood.

14. Veneered panels of a cabinet will have adjacent pieces of face veneer matched, and no piece of face veneer will be spliced to attain the required length in high-quality furniture.

15. Joints will be inconspicuous, indicating that care was exercised in sanding, finishing, and polishing.

16. Fine furniture will possess no glossy or shiny finish.

D. Veneered versus Solid Furniture

Purchasers often hesitate in making a choice between veneered and solid wood furniture, largely because of unfamiliarity with furniture construction and the characteristics of these two types. Briefly stated, either may be highly satisfactory or highly unsatisfactory. Construction methods and the quality of materials in the structure are all-important.

Usually the following advantages are ascribed to furniture of solid wood:

1. Surfaces cannot loosen, blister, or peel off, as may be the case with improperly constructed veneered surfaces.

2. Surfaces can be sandpapered heavily or even planed off when such furniture is to be refinished.

3. Solid wood can be carved, which is impossible in veneered construction unless special provisions have been made.

4. If chips are broken off, the same kind of wood is exposed.

5. Solid wood, when worn down, does not change in figure or appearance.

6. The owner has personal satisfaction in knowing that the article is constructed throughout of the kind of wood visible at the surface.

The advantages of veneered construction in furniture may be summed up as follows:

1. In some respects, a plywood panel is stronger than a solid piece of wood of the same thickness.

2. The use of cross-banding or plywood reduces the tendency of panels to shrink, swell, warp, split, and check.

3. Curved surfaces made of laminated wood are superior in strength to those made of solid wood, are less wasteful of raw material, and may be veneered to give a continuity of figure not attainable by the use of solid wood.

4. Highly figured woods, often cross-grained and therefore unsuited to use in thick sizes because of the resulting uneven shrinkage, can be used for thin face veneers.

5. Because of the thinness of highly figured face veneers, several consecutively cut pieces can be matched to produce symmetrical figures that are impossible of attainment in solid wood.

6. Valuable woods used for face veneers would make furniture prohibitive in cost if used in solid pieces.

7. Veneered construction permits better utilization of wood. More people may enjoy the beauty of rare woods of limited supply when these are used as face veneers, and cores of panels may be made of softer, lighter, and cheaper wood than can be used for solid wood construction.

SELECTED REFERENCES

1. Beaver, W. M. Equalizers, Toners and Pigment Wiping Stains. *N.C. State Col., School of Forestry, Tech. Rpt. No.* 6. 1956.
2. Beaver, W. M. Oil, Water, N.G.R. and Spirit Stains. *N.C. State Col., School of Forestry, Tech. Rpt. No.* 6. 1956.
3. Bethel, J. S. Quality Control in Furniture Factories. Proceedings, Cost and Production Division, Southern Furniture Manufacture Association. 1954.
4. Bethel, J. S., and R. J. Hader. Dimensions and Tolerances for Machined Furniture Parts. *Jour. Forest Prod. Res. Soc.,* 4(5):365–370. 1954.
5. Hager, J. A. Evaluating Furniture Finishes. Proceedings, Forest Products Research Society. 1949.
6. Johnson, A. P., and M. K. Sironen. Manual of the Furniture Arts and Crafts. A. P. Johnson Co., Grand Rapids, Mich. 1928.
7. Kelsey, C. B. Furniture, Its Selection and Use. National Committee on Wood Utilization, U.S. Department of Commerce. 1931.
8. Koehler, A. The Identification of Furniture Woods. *U.S. Dept. Agr., Misc. Cir.* 66. 1926.
9. Korstian, C. F. The Economic Development of the Furniture Industry of the South and Its Future Dependence upon Forestry. *N.C. Dept. Conserv. and Devlpmt., Econ. Paper* 57. 1926.
10. Lenaeus, G. A. Sealers for Wood Furniture. *N.C. State Col., School of Forestry, Tech. Rpt. No.* 6. 1956.
11. Litchfield, F. Illustrated History of Furniture. Tuslove and Hanson, London. 1893.
12. Pattou, A. B., and C. L. Vaughn. Furniture, Furniture Finishing, Decoration, and Patching. Frederick J. Drake & Company, Inc., Wilmette, Ill. 1938.
13. Prak, A. L. Finishing Room Conveyors. *N.C. State Col., School of Forestry, Tech. Rpt. No.* 6. 1956.
14. Quigley, R. I. The Function and Use of Pigmented Toners on Wood Furniture. *N.C. State Col., School of Forestry, Tech. Rpt. No.* 6. 1956.
15. Quigley, R. I. Top Coats for Wood Furniture. *N.C. State Col., School of Forestry, Tech. Rpt. No.* 6. 1956.

16. Stevens, W. C. The Practice of Wood Bending. *Forest Prod. Res. Rec.* 10. Department of Scientific and Industrial Research, London. 1936.
17. Stevens, W. C. Machinery and Equipment for Bending Wood. *Forest Prod. Res. Rec.* 25. Department of Scientific and Industrial Research, London. 1938.
18. Stevens, W. C., and N. Turner. Methods of Bending Wood by Hand. *Forest Prod. Res. Bull.* 17. Department of Scientific and Industrial Research, London. 1938.
19. Thomas, F. H. Fillers and Filling. *N.C. State Col., School of Forestry, Tech. Rpt. No.* 6. 1956.
20. Upson, A. T., and A. O. Benson. Production and Use of Small Dimension Stock in the Chair Industry. Association of Wood Using Industries, Chicago, Ill. 1923.
21. Vanderwalker, F. N. Wood Finishing: Plain and Decorative. Frederick J. Drake & Company, Inc., Wilmette, Ill. 1936.
22. Waring, R. G. Principles of Mill and Paint-shop Practice. The Bruce Publishing Company, Milwaukee. 1930.

CHAPTER 10

SHINGLES AND SHAKES

The origin of shingles is obscured in the beginnings of history. There is little doubt that handmade shingles, comparable to present-day shakes, were among the first wood products manufactured by man. It has been surmised that primitive man probably learned early that certain straight-grained woods could be split with sharp stones, long before steel implements were in use. The earliest European writings mention shingles and shakes. One definition from an old English dictionary, which is comprehensible in modern English, is as follows: "Shyngles, wyche be tyles of woode such as churches and steples be covered with. (Huloet.)" * In terms of modern usage, a shingle is a small, thin piece of wood with parallel edges, thinner at one end than the other, commonly used for covering the sides and roofs of buildings. Present-day shingles are sawed. Shakes are split or rived boards that are the same thickness at both ends, although they may be tapered in splitting by reversing the block each time a shake is split off. Modern shakes are sometimes sawed with a taper on one side or machined on the top side to simulate a split surface. In early times shingles were shaved, rived, or split with a broadax or froe after the logs had been cut into bolts of the proper length. Sometimes they were tapered at one end with a drawknife so they would lie flat like modern shingles. Thus early shingles were more like modern shakes. Early American settlers brought the art to this country, and they found species that were well suited to shingle making. Shingles were easy to make compared to the laborious methods of manufacturing lumber then in use, and they soon became a standard building material.

The beginnings of the western redcedar shingle industry in the Pacific Northwest followed closely after trapping, which was the first industry in the region. It has been reported that Hudson's Bay Company trappers cut shingles for a living after they became too old for the rigors of the trapline. These shingles were shipped to California or around Cape Horn to the eastern states. The following extract quotes an interesting record of an early shipment of shingles:

* Quoted in The Century Dictionary and Cyclopedia. The Century Company, 1911.

184

That shingles were a staple article of commerce along the North Pacific coast over a hundred years ago is well established by a reference to the logs of Hudson's Bay trading schooners, one example of which will suffice. The schooner "Columbia," the first vessel launched on the Columbia River, at Vancouver, Washington, was dispatched to the Fraser River in 1830, under command of William Ryan. Leaving Fort Vancouver on July 9 she arrived at Fort Langley on the Fraser on August 15. The captain's entry for August 17 reads: "Received on board 16 casks salmon, 12 bdls. shingles, 13 bales beaver, 16 bales salmon." [2]

By 1845, shaved shingles were being produced in commercial quantities in the Pacific Northwest, but full-scale production did not develop until the introduction of the block shingle machine from Canada and the arrival of the Great Northern Railroad at Seattle.

A. Species Used for Shingles

The variety of species that have been used for shingles in the United States parallels closely the depletion of supplies of suitable timbers as lumbering moved westward. The Pilgrims found the heartwood of virgin white pine highly durable and of excellent properties for shingles, far better than the heartwood of second- and third-growth trees of today. The John Howard Payne house, immortalized in the song "Home, Sweet Home," is a favorite example of the durability of shingles. Built in 1660, the siding of white pine shingles is still intact and the roof was only recently re-covered with redcedar shingles. Later, northern and southern white-cedars were used, and in the development of the middle western states, durable white oak and chestnut shingles and shakes were common. Northern white-cedar, from the northeastern and Lake states and from Canada, is still the source of small quantities of shingles. In the Gulf states it was early discovered that cypress possessed the qualities desired for shingles, and this species is still used to a limited extent. As population moved westward and the large stands of western redcedar and redwood were discovered, the national source of shingles moved to the West Coast, and redcedar soon dominated the shingle market. Redwood, partly because the wood is more valuable for other uses, has never been developed as a major source of shingles, although limited quantities are sold in California and occasionally shipped to other states.

Locally, numerous species have been and are still being used to a very limited extent. Southern pine in the South, sugar pine in the Sierra regions of California, hemlock and spruce in the East, and occasionally Douglas-fir in the West are among many that could be mentioned. Port-Orford-cedar possesses ideal properties, but short supplies and more profitable uses have limited this species to local consumption. Alaska yellow-cedar falls in a somewhat similar category. In some parts of the country, small operators are still manufacturing shingles from hardwoods for local use.

1. Western Redcedar. At the present time western redcedar accounts for approximately 95 per cent of all shingles produced in the United States. The principal commercial stands are located west of the Cascade Mountains in Washington and Oregon, but the species is also found in commercial quantities in some parts of Idaho and Montana and east of the Cascades in Washington. Extensive stands are also found in British Columbia, Canada. In 1959, production of western redcedar shingles was 2,302,064 squares.* During the same year, production of western redcedar shakes was 1,298,000 squares. There is some duplication in these data, since 592,410 squares of shakes was processed from shingles. In 1959, there were 168 mills producing shingles and/or shakes, according to the Bureau of the Census.[11] While western redcedar dominated the shingle industry, the small fraction of non-redcedar shingles was produced from redwood, bald cypress, northern and southern white-cedars, and small quantities of other species cut for local use. Western redcedar dominates the field so completely that this chapter will deal primarily with the shingle industry of the Pacific Northwest.

The production of shingles and shakes has remained relatively stable for the past 20 years, fluctuating from year to year somewhat in response to the level of activity of the home-building industry. In 1945, production of red-cedar shingles and shakes amounted to about 3 million squares; in 1954, it was about 3.8 million squares; and in 1958, it was about 3.2 million squares. At present, production is limited primarily by lack of timber supply rather than lack of demand.

B. Qualities Desired in Wood Used for Shingles

Wood for use in the manufacture of shingles should have the following properties: (1) durability; (2) freedom from splitting in nailing; (3) dimensional stability, i.e., low ratio of tangential to radial shrinkage and minimum shrinkage in all planes; (4) light weight; (5) good insulating properties; (6) adequate strength; (7) straight, even grain for ease of manufacture; (8) ability to take stains and paint; (9) ability to resist abrasion; and (10) pleasing appearance. Availability of sufficient supplies of logs or bolts, at a reasonable cost, is also an important factor.

Western redcedar, redwood, cypress, and northern and southern white-cedars all possess the desired properties to a high degree when compared with other species commonly used for lumber and other purposes. Western redcedar is somewhat superior to the other shingle species on most counts, which is one of the reasons for its dominance of the shingle field. The availability of larger stands of redcedar has also played an important part. As matters now stand, suitable supplies of logs of all shingle species, including redcedar, are in short supply. Unless research can develop treatments for shingles made from other woods that will make them equal or superior to redcedar, this species will continue to be the predominant source of wood shingles.

C. The Manufacture of Shingles

Most cedar logs are purchased on a specialized cedar-log market. The cost of logs is a dominant factor in the production costs of shingles, but

* For the definition of this term see p. 194.

prices so fluctuate that specific figures will not be given. Some of the reasons for this are differences in log grades and scaling practices in the United States and Canada, differences in logging costs, fluctuations in Douglas-fir production (upon which cedar-log supplies depend to a large extent), log inventories at various times in the year, and government price and quota controls.

A small percentage of logs destined for small shingle mills located away from water transportation are bolted by hand in the woods. Shingle bolts are split segments of a log, ranging from 48 to 54 in. in length, depending on the length of the shingles to be manufactured. By far the greatest number of logs are cut to standard log lengths in the woods and handled with conventional logging equipment prior to delivery to water transportation. After arrival at the shingle mill, the logs are usually dumped in a log pond for ease of sorting. They are then cut to convenient lengths for elevation into the mill, which is usually accomplished by means of a conventional log haul. Most shingle mills are of two-story construction with the shingle machines on the second level. Steam-driven drag saws were formerly used to cut the logs in the pond, but power-driven chain saws are becoming popular because of their greater speed and efficiency. On reaching the log deck the logs are cut into 16-, 18-, or 24-in. lengths, depending on the length of the shingles to be manufactured. Deck saws are up to 10 ft in diameter and are operated at a rate of speed that will ensure a smooth, accurate cut. These blocks are then quartered, either by sawing or splitting, and reduced still further, if necessary, to make them of suitable size for the shingle machines. Sawing is more common than splitting, probably because of the lower cost of sawing equipment. Splitting is more desirable because of the saving in saw kerf, and since the splits tend to follow the plane of the rays, more true radial surfaces are obtained. This results in a higher proportion of vertical-grain shingles. Splitters, operated by either steam or pneumatic power, may be either vertical or horizontal and may have three or more knives. Some mills use a one knife splitter and resplit as many times as necessary to obtain the desired size of blocks. Multiple-knife splitter heads are more common. Some prefer three-knife to four-knife splitters because of the larger proportion of wider shingles that can be cut from larger blocks. Sawyers working on a piece-work basis prefer large blocks because fewer need be placed in the machine, and it is, therefore, cutting shingles a greater percentage of the time. The blunt, wedge-shaped knives usually enter the blocks only about 1 in. The blocks are brought to the splitter by some form of mechanical conveyer. Some mills use a plunger with a ball-bearing head to elevate the block above the conveyer chains in preparation for splitting. This device permits the operator to turn the block to whatever position will provide the maximum yield of high-grade material. Shadow guides are commonly used to indicate the exact positions that the knives will strike when the splitter head is lowered.

After quartering, in most Canadian and a few American mills, the blocks are passed to a specialized saw table, equipped with a vertically oriented circular saw called a *knee bolter*. Here the bark and sapwood and certain surface defects are removed. Most American mills dispense with this operation because the bark and sapwood are removed from each shingle at the time it is edged in the normal process of manufacture. The blocks are then moved by conveyer from the splitter or the knee bolter to the shingle machines (Fig. 10-1), where sufficient quantities are stock-piled within convenient reach of the sawyer. Virtually all machines now in use in the red-cedar shingle industry are of the vertical type, although a few of the older horizontal machines are still in use in other regions. The vertical design makes it possible for one operator to edge and grade the shingles as they are cut from the block. In this way the sawyer is constantly aware of the

Fig. 10-1. Shingle machine showing carriage and setworks. (*Courtesy Sumner Iron Works, Everett, Washington.*)

quality of the shingles being produced, which permits him to adjust the block when necessary to stay within specification limits for edge-grain shingles. It may be necessary to adjust the block several times to keep the cutting face as closely as possible in the plane of the wood rays, or at right angles to the growth rings. Each such adjustment results in the loss of some material, called a *spalt*, which is the slice of wood that must be removed to reorient the block.

Shingle machines are equipped with clamping jaws that hold the blocks rigidly in place on a reciprocating power-driven carriage (Fig. 10-1). The blocks are placed in the machine in the vertical position and are oriented in such a manner that the shingles are cut from the radial surface that was exposed by splitting. The carriage, which is actuated by an eccentric crank from a geared drive, passes the block back and forth through the saw at the rate of about 35 strokes per minute. On each return stroke of the carriage, pawls engage the carriage feed rolls, tilting the carriage, so that a tapered shingle is cut on each forward stroke. The butt, or thick end, of the shingles is cut from the top of the block on one stroke and from the bottom on the next stroke. Machines called *double-butting machines* make two strokes before the carriage is tilted. This gives two butts at the top, then two at the bottom, instead of alternating as with single-butting machines. The saws used are approximately 40 in. in diameter, from 8 to 10 gauge at the center, tapering to 16 to 19 gauge at the rims, and have about 80 teeth.

As each shingle comes off the saw the sawyer picks it up and places it on a *springboard* (sometimes called *shingle jointer,* or *knot saw*) for edging, or jointing (Fig. 10-2). The springboard is a small hinged table having a guide

Fig. 10-2. Shingle sawyer trimming edge of a shingle with his right hand as he takes the next shingle from the saw with his left hand. (*Courtesy Redcedar Shingle Bureau, Seattle, Washington.*)

for the butt of the shingle that is at a true right angle with the edging saw. After the shingle has been placed in the proper position on the springboard, the latter is pushed down, bringing the edge of the shingle into contact with a circular saw. This trims away any bark or sapwood that may be present or merely joints the edge at right angles to the butt. The shingle is then turned over and the process repeated in order to make both edges parallel and to remove any defects that may be present on the second edge. Defects in the center as well as the edge can be removed with this saw, making one or two acceptable narrower shingles from a larger shingle that would be rejected or reduced to a lower grade. While the edging is in process, the sawyer grades the shingle, then tosses it into the proper grade chute, where it passes to the shingle packer, usually located on a lower level. Rejected shingles are kept separate.

Packers are stationed beside bins fed by the sawyer with a packing frame near at hand (Fig. 10-3). The shingles are regraded by the weaver as they are packed as extra assurance that no mistakes are made. Enough of each grade is placed in the packing frame to produce a standard-size bundle. One half of the butts are at one end of the bundle and the other half at the other, with the thin tips overlapping in the center. The bundle is bound by *band sticks*, one each at the center of the bundle above and below, the ends being bound under pressure with metal straps or bands. This method of packaging assures flat shingles, even during the drying process. The packer

Fig. 10-3. Shingle packer binding off a bundle of shingles in a packing frame. The various grades of shingles come from sawyer at upper level through chutes at rear. (*Courtesy Redcedar Shingle Bureau, Seattle, Washington.*)

then places the bundles on a conveyer that takes them either to storage or to the drying kilns. At this stage the shingles are in the green condition, with the wood usually well above the fiber saturation point. Unless cargo shipment is contemplated, the shingles must be dried. Repressing, accomplished in a horizontal-press frame, is necessary after kiln-drying to take up the slack caused by shrinkage.

1. Manufacturing Methods in Canada. Operating techniques in Canadian mills are similar to those in the United States, with some minor exceptions. Machine speeds are limited by law to 34 strokes per minute, whereas there is no limit in this country. It has been found, however, that 36 strokes per minute is about the maximum that is consistent with satisfactory grade recovery and minimum waste. The output per machine in the United States is approximately $4\frac{1}{2}$ squares per hour; in Canada the rate is $3\frac{1}{2}$ squares per hour. Several other factors besides carriage speeds are involved. The sawyers, commonly employed in Canadian mills, work less rapidly than

American workers. Also, Canadian mills concentrate on the production of the high grades. This reduces output because the blocks must be turned more often and the machines are idle a greater part of the time. The lower rate of production has tended to offset lower wage rates in Canada. Returns to the owners are, however, greater because of the larger proportion of higher-grade shingles produced. The average yield of shingles per M bd ft log scale is approximately 25 per cent higher in the Pacific Northwest than in British Columbia, or 12½ and 10 squares, respectively.

D. Seasoning and Storage

Green shingles are usually shipped by water or by truck, since freight rates on these carriers are based on volume and not on weight. Cargo, or water, shipments declined sharply during the Second World War, amounting to less than 10 per cent of total shipments in 1944, and this form of transportation has still recovered only a small part of its original volume.[14] Truck shipments of green shingles between Oregon and California have increased since then, but it is doubted if they have offset losses in cargo movements. This means that approximately 90 per cent of all redcedar shingles are dried before shipment in order to reduce transportation charges. Small quantities of shingles are still air-seasoned, but the major portion is kiln-dried. Western redcedar requires mild drying schedules because the wood is subject to collapse and honeycombing. Overdrying of shingles to reduce shipping costs was one of the practices that created prejudices against wood shingles prior to the establishment of minimum standards. Satisfactory drying schedules have been developed through research, and most operators now recognize the importance of proper seasoning. Savings in freight costs are not the only advantages of drying. Properly kiln-dried shingles assure uniform moisture content at the time of laying, which equalizes shrinkage and virtually eliminates cupping and warping after the shingles are laid. To preserve these advantages it is desirable that shingles be stored under conditions which provide sufficient circulation of air and which avoid exposure of outer layers to rain or the direct rays of the sun.

E. Staining and the Application of Wood Preservatives

The heartwood of redcedar is highly durable, and neither stains nor preservatives are required to obtain long, satisfactory service from a roof covered with untreated, all-heartwood shingles. Evidence of this is found in the fact that by far the largest proportion of shingles is not stained or otherwise treated. Color, however, is sometimes desired, and the life of shingles containing sapwood can be prolonged by the use of appropriate wood preservatives. Shingles may be stained by dipping or by pressure methods in commercial plants or by hand on the job. They may also be stained in place, on roofs or sidewalls, but the results are less effective. Painting, although satisfactory for shingles on sidewalls, is not generally recommended for shingled roofs.

The advantages of stained shingles are: (1) they provide a choice of colors or combinations of colors; (2) stains of good quality contain linseed oil, which reduces shrinking and swelling during wet and dry periods; (3) preservative treatments can be combined with staining to prolong the life of shingles that contain sapwood; and (4) stains reduce the rate of mechanical wear of the surface of shingles. Mechanical wear is obvious only after long periods of time, but shingles eventually become thinner as a result of attrition by sand and dust particles and of leaching and mild hydrolysis of the surface layers of the wood from exposure to rain. Proper prestaining, followed by restaining of the exposed butts, will help to prevent deterioration. The additional cost of stained shingles is small compared with the cost of laying the roof. It is highly desirable, therefore, that they be bought from a reliable dealer or, in the case of on-the-job applications, that only the best ingredients be purchased. Additional information on this subject can be obtained in selected references 7 and 10.

The preservation of shingles is not practiced extensively but will, no doubt, become more common now that substitute species that are less durable are coming under consideration for use as shingles. For years, creosote was the standard preservative for shingles, some companies developing "decolorized" fractions of creosote for use with light-colored stains. Some of the newer preservatives, such as pentachlorophenol and copper naphthanate, may well find a place in this field. The disadvantage of the standard grade of creosote has been the impossibility of using it in combination with light colors.

Preservatives are best applied by pressure methods, but reasonably effective results can be obtained either by dipping or by the use of the conventional hot-and-cold-bath method of applying wood preservatives. Preservatives applied to a roof or sidewall after it has been laid cannot be expected to be effective in the prevention of decay; the preservative will not reach inaccessible areas under and between the shingles where decay is most likely to become established. A more complete discussion of the subject of preserving shingles can be found in selected reference 10.

F. Classification of Shingles

The establishment in 1931 of quality standards for the manufacture of wood shingles by the Bureau of Standards of the Department of Commerce marked a turning point in the elimination of earlier practices that had been detrimental to the future of wood shingles. The Redcedar Shingle Bureau and many shingle manufacturers gave full support to the project, and between 85 and 90 per cent of the shingles now manufactured conform to these standards. The standard, as revised, is known as Commercial Standard 31–52 and covers the No. 1 grades of redwood and cypress shingles, as well as those manufactured from redcedar. The grade marks of accepted association inspection agencies assure conformance to this standard. This applies, as well, to shingles manufactured in Canada for the American market.

1. Length and Width of Shingles. Most shingles are packed and sold in random widths, ranging from 3 to 14 in. This leaves only length and thickness as the criteria for classification. Standard lengths are 16, 18, and 24 in. Ten per cent of the shingles in any one shipment may be 1 in. either over or under the specified length. *Dimension* shingles are an exception to the general rule, since they are jointed to definite widths.

2. Thickness of Shingles. The thickness of shingles is measured at the butt when green and designated by the number of butts required to make up a specified dimension. For example, 4/2 in. means that four butts are required to equal 2 in. When length and thickness are combined there are three standard sizes of shingles: 16 in.—5/2 in.; 18 in.—5/2¼ in.; and 24 in.—4/2 in. Nonstandard specialty items, either thicker or thinner than the above, are produced in small quantities.

3. Grades. Specifications for shingles are given in Commercial Standard 31–52 * and in the grade rule books of the various shingle associations.[7] These should be studied with care by all prospective purchasers and users of shingles. The grades for the several lengths of shingles are somewhat similar. The following summary of the grades of 16-in. shingles is presented as an example:

> *No. 1 Shingles.* No. 1 redcedar shingles represent the best grade that is manufactured. These shingles are intended primarily for roof construction, where the shingles should lie flat and tight and where there must be complete protection from rain water driven by high winds.
> No. 1 shingles must be 100 per cent edge or vertical grain, 100 per cent clear, and 100 per cent heartwood. None of these shingles should be wider than 14 in. and none narrower than 3 in. There should be 20 courses of shingles at each end of the bundle. Not more than 10 per cent of the combined width of the shingles laid side by side (running inches) in any shipment may be less than 4 in. in width.
> *No. 2 Shingles.* Shingles of No. 2 grade must be clear or free from blemishes for three-fourths of their length as measured from the butts. A maximum width of only 1 in. of sapwood is permissible in the first 10 in. Mixed vertical and flat grains are allowed. No shingles shall be wider than 14 in. and none narrower than 3 in. Not more than 20 per cent of the running inches in any shipment may be less than 4 in.
> *No. 3 Shingles.* No. 3 shingles must be 8 in. clear or better and may contain sapwood. No shingle shall be wider than 14 in. and none narrower than 2½ in., and not more than 30 per cent of the running inches in any shipment may be less than 4 in. wide. Knotholes up to 3 in. in diameter are permitted in the upper half.

Undercoursing grade. A fourth grade of shingle, known unofficially as No. 4 but properly designated Undercoursing grade, is produced by many mills and is recognized by the industry through the Redcedar Shingle Bureau. As the name implies, it is intended solely for the undercourse of double-coursed shingle sidewalls. Since the undercourse is always completely covered, the grade permits certain defects throughout the area of the shingle. Many mills mix their No. 3 grade shingles with their Undercoursing grade, producing a grade known as Special Undercoursing. The Undercoursing grades were produced as a result of the increased popularity of double-coursed sidewall construction.

---

* U.S. Department of Commerce, Bureau of Standards. Wood Shingles (Red Cedar, Tidewater Red Cypress, California Redwood). Commercial Standard, CS 31–52.

The following commercial designations are generally recognized in the American shingle market:

16 in.—5/2 in.	Perfects, or XXXXX
18 in.—5/2 in.	Eurekas *
18 in.—5/2¼ in.	Perfections
24 in.—4/2 in.	Royals

* Manufactured to a limited extent in Canada and marketed in small quantities in the United States.

No. 2 and No. 3 shingles are sometimes designated as 8-, 10-, or 16-in. clears, based on the number of inches of wood, measured from the butt, that is clear of defects.

4. *Unit of Measure*. The unit of measure used in marketing shingles is the *square*, or *square pack*, which is defined by the Bureau of Standards as "a unit providing sufficient shingles for the coverage of an area of 100 sq. ft. when the shingles are laid at any specified exposure to the weather." For roofing, the specified exposures depend upon the pitch of the roof. For 16-in. shingles an exposure of 3¾ in. is recommended for roof pitches of 1/8 or 1/6, and 5 in. for roof pitches of 5/24, 1/4, 1/3, and 1/2. For 18-in. shingles the corresponding exposures are 4¼ in. and 5½ in., and for 24-in. shingles they are 5¾ in. and 7½ in. For sidewall shingles recommended exposures for 16-in. shingles are 6 to 7½ in. for single course and 8 to 12 in. for double course. For 18-in. shingles single course exposures are 6 to 8½ in. and double-course 8 to 14 in. For 20-in. shingles the corresponding exposures are 8 to 11 in. and 12 to 16 in. Bundle sizes are determined by the number of courses of shingle butts at each end of the bundle. When a bundle is 20/20, it means that there are 40 layers, or courses, of shingles overlapping under the band stick, with 20 courses of butts on each side of the sticks. Dimension shingles are sold by the piece, or net count, but 1 square, when laid according to specified exposures, will cover 100 sq ft.

G. Application of Shingles

The application of shingles, although based on a good deal of research as well as practical experience, is an industrial art rather than a technical subject. Authoritative handbooks are available that give detailed instructions and information on the proper methods of laying shingles on roofs and sidewalls, recommended exposures, special adaptations for hips, ridges, valleys and flashings, types and sizes of nails to use, overroofing, double-coursing, staining and painting, and numerous other techniques essential to the building trades.[6,7] Such information is beyond the scope of this text, but it is recommended that everyone who is interested in wood utilization obtain a good handbook for ready reference. Some general remarks may, however, be of interest.

No. 1 shingles should always be used for the roofs of primary buildings, where maximum protection from the weather and long life are essential. The lower grades can be used for the roofs of secondary structures, such as home garages and numerous farm buildings, but anything less than the best grade of shingle is an unwise investment for the roof of a primary structure. The properties of the No. 1 grade—all-vertical grain to provide greater strength and minimum shrinking and swelling, 100 per cent heartwood to provide maximum durability, and freedom from defects—assure minimum maintenance costs and maximum life. No. 2 shingles are standard for sidewalls and well suited for roofs of secondary buildings, for overroofing over old shingles, and for double-coursing under No. 1 shingles. No. 3, the economy grade, gives good service when used in proper locations, such as undercoursing, overroofing of farm buildings, or sidewalls of garages, summer cabins, and farm buildings.

Strict adherence to the recommendations of the accepted trade associations with respect to exposures and spacing, types and sizes of nails, and other details cannot be overemphasized. They are designed to assure the user of maximum service. Digressions from these accepted standards usually result in false economies. Users of shingles should follow the instructions proposed by the Redcedar Shingle Bureau,[7] or other reliable sources.

Overroofing, a relatively recent development, has numerous advantages over reroofing when a roof is in need of replacement. Among these are the following: (1) it provides double insulation, with consequent savings in fuel and greater comfort in summer; (2) it can be done at any time, in the rainy season as well as during dry periods; (3) it is less costly, as it eliminates the labor cost of removing the old shingles; (4) it adds strength to the roof without appreciably increasing the dead load (shingles weigh less than 2 lb per sq ft); and (5) the litter of old shingles, always a nuisance to remove, is eliminated. In reroofing barns and feed sheds it excludes the possibility of nails and other dangerous debris falling into haymows or feed bins.

Shingles should not be wetted or soaked prior to installation. It was long believed that wetting and consequent swelling would ensure that the shingles would not be laid too close together. Proper lateral spacing of dry shingles will accomplish the same purpose more efficiently, and the laying of wet shingles is detrimental to the construction of a good roof. Shrinkage stresses between nails often cause splits and checks as a shingle dries. The equilibrium moisture content attained in the dealer's storage shed in the area in which the shingles are to be used is usually the most suitable for laying the shingles. It should be remembered that the spacing recommendations of the appropriate trade associations make the proper allowance for shrinking and swelling.

1. The Life of Shingles. No. 1 shingles, if properly applied, can be expected to have a life of 35 years under average climatic conditions on steep roofs; 25 years on roofs with a one-fourth pitch; and a life in excess of the life of the average structure when used as sidewalls. The above expectancies can be considered as averages; there are numerous records of roofs that have lasted in excess of 75 years. An example is quoted as follows: " 'Lowland,' the magnificent mansion of the late George Abbot James, at Nahant, Massachusetts, was built and roofed with redcedar shingles in 1872 and is found to be in perfect state of preservation." [*] The life expectancies

---

[*] The Encyclopedia Americana. Americana Corporation, New York. 1947.

of No. 2 and No. 3 shingles depend to a large extent on the proper selection of the uses to which these lower grades are to be put. No. 2 shingles, when used on sidewalls, will match or even exceed the life of No. 1 shingles used on the roof of the same building. No. 3 shingles will give service commensurate with their cost if used where exposure conditions are not too severe. Even flat-grained shingles have been known to last as long as 25 years on roofs.

2. *Advantages of Wood Shingles.* Among the more important advantages of wood shingles are (1) superior insulating properties; (2) long life; (3) resistance to wind, snow, and hail damage; (4) suitability in nearly all climates; and (5) attractive appearance. Wood in general has far better insulating qualities than most other building materials, and western redcedar is high in efficiency in this respect when compared with other woods used for construction purposes. The thermal conductivity of shingles as normally laid up is only about one-third that of asphalt roofing. In other words, wood shingles, on an average, provide three times as much insulation as asphalt shingles. The life of wood shingles compares favorably with, and usually exceeds, that of composition roofings. Wood shingles are less subject to damage by hail and high winds, and their performance is little affected by climatic conditions; they serve equally well in both hot and cold climates. To most people, wood shingles are more attractive in appearance than most competitive materials. Weathering and slight variations in the natural color of the heartwood of redcedar and the soft texture that results from the broken lines of shingles laid in random widths gives a harmonious effect not found in the stereotyped patterns of most composition roofings. The colors of stained shingles, since the stains are absorbed by the wood, are more soft and natural than the harsh colors of many composition materials.

The outstanding advantage of redcedar shingles is their long life. In a questionnaire distributed by the Bureau of Standards to 148 field offices of the Home Owner's Loan Association and Federal Housing Administration in 48 states and the District of Columbia,[16] information was requested on the length of life of wood and asphalt shingles. The results showed that the average life of wood shingles in all states was approximately 20 years, as compared with about 14 for asphalt shingles. It was further shown that climate had a marked effect on the life of the asphalt product, whereas it made much less difference in the case of wood shingles. In 21 states south of a line corresponding roughly to the 40th parallel of latitude, the average life of asphalt shingles was only 12 years. North of this line the average was 16 years. This was not a scientifically planned experiment, but the results are indicative of the relative lives of the two roofing materials.

Among lesser advantages, it is claimed that damages to shrubbery and the dangers attending snow and ice avalanches from roofs are reduced on all but the steepest roofs when wood shingles are used, that condensation is not a serious problem under shingle coverings, and that noise transmission is far less than with most other roofing and sidewall substitutes. The disadvantages of wood shingles are higher initial cost and their inability to meet building-code restrictions that have become established throughout the country.

H. Character of the Redcedar Shingle Industry

Shingle mills are of two types: "straight" shingle mills and "combination" mills. The former produces shingles exclusively; the latter, both shingles and cedar lumber. Not many sawmills in the Pacific Northwest combine the manufacture of cedar with other species such as Douglas-fir and hemlock. Most companies prefer to sell their cedar logs on the open market. A few large, integrated operations retain their cedar logs in order to obtain the added profits that can be derived by converting the logs in their own shingle mill or sawmill. By and large, however, the cedar industry is one of intensive specialization.

Shingle mills vary in size from 1- to 25-machine installations. There has been a trend in the shingle industry toward fewer plants with more machines.

The shingle industry is characterized by instability and intermittency of operation. Many of the small operations are insecurely financed, and an average of 20 to 30 per cent of all mills are shut down most of the time. Idle mills are most numerous from December to March, but a rather large number is closed in all months of the year. Unfavorable weather conditions in the logging areas cause most of the winter shutdowns. Other factors contributing to the unevenness of production are local log shortages, insufficient railroad cars, scarcity of labor, competition for labor from more stable industries, labor disputes, plant breakdowns, mill fires, slack orders, and insolvency caused by inadequate financing. At no time is production up to full capacity, even when the demand for shingles is high. The same conditions exist in Canada.

I. Marketing of Shingles

The history of marketing shingles is an excellent example of the damage to a product that can result from a lack of specifications and inspection standards and a strong trade association to enforce them. In the early 1930s, the quality of shingles in the United States was at a very low level. A high proportion of the output consisted of improperly dried, thin, flat-grained shingles. It is generally believed that this condition influenced materially the establishment of the asphalt-roofing industry on a competitive basis. It may also have had an influence on the establishment of fire ordinances and underwriters' codes that are inimical to the interests of the wood-shingle industry. Loose, cupped, and warped shingles are known to contribute to the ignitability of roofs and to the spread of fires. The first impact of competition from other types of roofing materials caused reductions in the prices of shingles; the second result was a marked improvement in the quality of shingles. During the same period, Canadian shingles commanded a premium on the American market because of the more rigid inspection to which they were subjected.

1. Markets. Shingles are used in all parts of the United States, but restrictive legislation has played an important part in the pattern of distribution. Consumption is limited to a large extent to rural and urban areas where discriminatory legislation has been less severe than in many large

cities. Consumption is not proportional to density of population, largely because of the building-code restrictions just mentioned. A vigorous educational program conducted by the Western Redcedar Shingle Bureau has resulted in some modification of restrictive building codes. Shingles are much
more extensively used in building on the West Coast than in other sections
of the country, both because of proximity to the source of supply and because building codes in this region have been historically much less restrictive.

2. Competitive Factors. In spite of their many advantages, wood shingles have lost ground steadily to asphalt shingles and roll roofing since
shortly after the First World War. The principal reasons are quoted as follows: "Among the factors operating against wood shingles and in favor of
other roofings are (1) the location of the respective industries in relation to
markets, and differences in the distribution systems; (2) building regulations; (3) the relative costs of the several roofings; (4) fire-insurance rates;
and (5) trade promotion and salesmanship. All these play an important part
in the ultimate choice of the consumer and tend to create an expanding
market, particularly for asphalt roofings, and a contracting market for wood
shingles." [14]

Competition from other types of roofing materials such as asbestos shingles, some of the older metal roofings, tile, and slate is not a serious threat
to wood shingles because these products occupy a higher price level or their
weight results in shipping-cost penalties. Aluminum siding and roofings are
being sold at competitive prices and are being used rather extensively for
barns and other farm buildings at the time this is written. Thin aluminum
sheeting has certain obvious shortcomings for use as roofing and sidewalls
but, if sold at a low enough price, it could offer serious competition to wood
shingles. More serious from the standpoint of competition is an aluminum
shingle that provides for a 5-in. exposure and carries a guarantee of 60 years.
Aluminum bevel siding, prime-coated for any color of house paint, is also
available. Plastic shingles of all colors, which are fused to each other and to
the roof boards or shingles, are marketed by a large company. They have an
expected life of 40 years.

3. Building Codes. Restrictive building regulations date to the period
prior to the acceptance of commercial standards and specifications for wood
shingles. Before and just after 1930, a large part of the domestic shingle
production consisted of flat-grained shingles that cupped, warped, and split,
thus contributing to fire hazard. Many of these restrictions remain in force,
even though conformance to Commercial Standard 31–52 and its revisions
has resulted in a "vastly improved shingle which practically eliminates all
danger from brands and sparks on roofs." [14] In spite of determined efforts,
and some success, by the Redcedar Shingle Bureau to combat restrictive
codes, the sentiment against wood shingles continues to grow in some parts
of the country. Even when restrictions are not exclusively prohibitive, increased fire-insurance rates are a further handicap to wood shingles. Actually,
though wood shingles have a lower ignition point than mineral-coated asphalt
shingles, they do not contribute as much fuel to a fire as do asphalt materials
when once ignited. Evidence tends to show that many restrictive building
codes are predicated on prejudices of previous years and that many of these

objections are not valid in the case of wood shingles manufactured in accordance with present-day specifications.

4. *Future Trends.* In discussing the competition of composition shingles with wood shingles it should be remembered that this conflict concerns only one of the many wood-using industries of the country.* As a matter of fact, the composition-roofing industries utilize large quantities of wood fiber, either in the form of paper or matted fiber, as a base that is impregnated with asphalt. Snoke [9] reports that both groundwood fiber and chemical-wood fibers are used in the manufacture of felts for asphalt-prepared roofing. Usually the two are used together, the former in amounts up to nearly 19 per cent, the latter up to 14 per cent, depending on the formulation. In some formulations the felts may contain as much as 32 per cent of wood fibers. The other components, depending on the type of roofing being made, consist of rags, up to 75 per cent; jute, up to about 15 per cent; and wool, rayon, silk, hair, and kapok in lesser quantities. Recent research has shown the way to the use of mechanically fiberized wood residue for this purpose in even larger proportions, a process that can play an important part in better and more complete wood utilization. Neither must it be forgotten that domestic wood shingles must face the competition of Canadian importations. The exact extent and form of this competition are unpredictable because of quotas, tariff restrictions, and trade agreements that may change rather rapidly in the years to come.

The demise of the wood-shingle industry has been predicted from time to time for many years. The Stanford report forecast a substantial reduction in shingle and shake production from 1952 to 1960. Production figures for 1959 fail, however, to confirm this prediction. It seems safe to assume that in spite of market problems and building-code restrictions wood shingles and shakes will continue to enjoy a substantial popularity because of attractive appearance, long life, and low maintenance cost, and that production is, in the long run, more likely to be limited by supply of raw material than by product demand.

J. The Manufacture of Shakes

Shake making was an important business in colonial and pioneer times. Shakes have returned to popularity in recent years because of the pleasing architectural effects that are possible with them. Shakes are more costly than shingles and are most commonly used where the effect justifies the added cost.

Handmade split shakes are usually 25, 31, or 37 in. long and vary in width from 5 to 18 in. They are usually from $\frac{1}{2}$ to $1\frac{1}{4}$ in. in thickness. Some are tapered, but if ease of production or a particularly rough effect is desired, they are left the full thickness at both ends. Numerous mechanical methods, including both sawing and special machining, have been devised. Among these are (1) resawing thick split shakes from corner to corner, resulting in two shakes having a rough, split surface and a flat, sawed back;

---

* There is, however, a definite parallel between the building-code restrictions governing the use of wood shingles and similar regulations against the use of wood for other purposes.

(2) running a heavy-sawed shingle through a special machine head that imparts to the surface the rough appearance of a split shake; and (3) passing a heavy shingle under a heated pressure roll that is embossed with the impression of a split shake. Various methods of grooving shingles to simulate the appearance of a split surface have also been devised. In 1959, about 36 per cent of the production of redcedar roofing was in the form of shakes. Fifty-four per cent of the shakes produced were manufactured directly from logs and bolts. The remainder were processed from purchased shingles. At the present time, the manufacture of shakes is almost entirely limited to the Pacific Coast states and to certain areas of the Rocky Mountains. Tray shakes, used for fruit boxes, once a common commodity in California, have all but disappeared.

SELECTED REFERENCES

1. Anon. B.C. Leads All Canada in Production of Sawn Lumber, Shingles, Ties and Lath, Says Statistical Bureau Report. *Brit. Columbia Lumberman*, **31**(9):99. 1947.
2. Anon. 1. How Red Cedar Shingles Are Made; 2. Western Red Cedar Shakes; 3. Development of the Shingle Industry. *Timberman*, **33**:7. 1932.
3. Anon. Lumber Industry Facts: 1960. National Lumber Manufacturers Association, Washington, D.C. 1961.
4. Anon. Tariff on Canadian Lumber Cut: Shingles Enter Free. *West Coast Lumberman*, **47**(12):64. 1947.
5. Anon. 1. Cedar Shortage Hits Shingles; 2. Shingle Methods Need Revising. *Timberman*, **47**:3. 1946.
6. Gilmore, W. J., H. R. Sinnard, and E. H. Davis. Roofs and Exterior Walls of Red Cedar Shingles. *Oreg. State Col. Ext., Bull.* 540. 1940.
7. Grondal, B. L., and W. W. Woodbridge. Certigrade Handbook of Red Cedar Shingles. Red Cedar Shingles Bureau. 1942.
8. Kirkland, B. P. Forest Resources of the Douglas Fir Region. Joint Committee on Forest Conservation, Portland, Oreg. 1946.
9. Snoke, H. R. Asphalt-prepared Roll Roofings and Shingles. *Bldg. Mater. and Structures Rpt.* BMS7u. U.S. Department of Commerce, Bureau of Standards. 1941.
10. U.S. Department of Agriculture, Forest Service, Forest Products Laboratory. The Preservative Treatment and Staining of Shingles. Report No. R761. Revised, June, 1935.
11. U.S. Department of Commerce, Bureau of the Census. Current Industrial Reports: Red Cedar Shingles. 1959.
12. U.S. Department of Commerce, Bureau of the Census. Industry and Product Reports: Shingle Mills. 1960.
13. U.S. Department of Commerce, Bureau of Foreign and Domestic Commerce. Commercial Standard CS 31–52: Wood Shingles. 1952.
14. U.S. Tariff Commission. Report to the United States Senate on Red Cedar Shingles. Report No. 149, 2d ser. 1942.
15. U.S. Tariff Commission. Red Cedar Shingles. War Changes in Industry Series. Report No. 8. 1945.
16. Waldron, L. J., and H. R. Snoke. Roofing in the United States: Results of a Questionnaire. *Bldg. Mater. and Structures Rpt.* BMS s7. U.S. Department of Commerce, Bureau of Standards. 1940.

CHAPTER 11

WOOD CONTAINERS

I. COOPERAGE

Cooperage is a collective term descriptive of a variety of wood containers made of assemblies of staves and heads securely held together with wooden, iron-wire, or strap-steel hoops. The term cooperage is also applied to an establishment where barrels and similarly designed containers are assembled.

Barrels, casks, and kegs with leakproof joints used for packaging liquids, foods processed in liquids, semisolids, and occasionally heavy solids are classed as *tight cooperage*. By contrast, *slack cooperage* containers are designed for packaging, storing, and transporting dry materials. Slack barrels are manufactured to conform to a wide range of quality standards. The best barrels are provided with siftproof joints and are especially well suited for packaging finely divided solids such as flour, dyestuffs, and chemicals.

Large tanks and vats, silos, flumes, and standpipes fabricated in place with profiled staves bound together with steel bands of from $\frac{3}{8}$ to 1 in. in diameter are commonly regarded as forms of heavy cooperage. Container woodenware such as pails, tubs, churns, firkins, piggins, kits, and kanakins, fashioned with staves and hoops and fitted with special tops and/or bottoms, constitute other forms of coopers' ware.

In recent years a lightweight, cylindrical, plywood barrel with wooden or metal tops and a pair of head hoops has become a popular container with many shippers.

A. Historical

The origin of the wooden barrel antedates the recorded history of mankind. Archaeologists have chronicled that barrels were made by both the ancient Babylonians and Egyptians, while numerous passages in the Old Testament attest to their common usage at least several centuries before the birth of Christ. When the barrel first acquired its bilge (central bulge) is also a matter of speculation, although there is abundant evidence that curved staves were used prior to 525 B.C.

Historians agree that the precursor of the stave barrel was a short section of a hollowed log the ends of which were covered with animal skins. Such containers were patently heavy, cumbersome, and of limited capacity; and their transportation was obviously difficult. One may conjecture that the ancient cooper, in an effort to produce a more efficient vessel, conceived the idea of fashioning together a number of hand-carved staves into a cylindrical structure of much less bulk and proportionately greater capacity. His hoops, probably leather thongs, not only served to avert leakage along the joints but also may have sealed checks in the staves themselves that developed in seasoning.

During the period of the Crusades (1096–1291), the barrel was introduced from the Holy Land into Europe, and by the fifteenth century several standard sizes of barrels were used as universally accepted scales of measure. In 1423, English lawmakers decreed that the capacity of a hogshead would be standardized at 63 wine gallons. Later, by other imposed statutes, the capacity of a wine barrel was fixed at $31\frac{1}{2}$ gal; those for vinegar, eels, and herring at 30 gal; those for honey and codfish at 32 gal. In a similar manner a barrel of gunpowder was required to weigh 100 lb; one of soap, 265 lb; and one of candles, 120 lb.

The barrel was also intimately associated with early American history. Columbus, in preparing for his great westward quest into the vast unknown, provisioned his ships with food and drink packaged in cooperage. Later, when the Pilgrims founded their colony at Plymouth in 1620, they not only brought the bulk of their food and supplies in coopers' ware, but also had the foresight to include among their numbers a journeyman cooper, John Alden, hired expressly for the purpose of plying his trade in the New World.

Export was the very lifeblood of the American colonies, and the demand for cooperage was tremendous all along the Atlantic seaboard. Codfish, whale oil, molasses, and rum from New England, whisky from Pennsylvania, and naval stores, tobacco, and rice from the Carolinas and Virginia were among the many commodities packaged in barrels for consignment abroad. In due course of time European merchants began to recognize the superior qualities of American-made cooperage, and vast quantities of staves and headings were purchased by importers in England, Ireland, Spain, and Portugal.

Time has witnessed many changes in the American cooperage industry from the colonial era of hand-rived staves to the highly mechanized, mass-production methods of the twentieth century. The discovery of petroleum in Pennsylvania during the middle of the nineteenth century and the subsequent exploitation of the oil fields created an unprecedented demand for tight cooperage, and it was during this period that the first power-driven, stave-cutting equipment was made available to the cooper. Thereafter machines of various types rapidly replaced the various laborious and time-consuming hand operations, and within a few years the industry became totally mechanized.

No historical review of cooperage would be complete, however, without mention of the curved stave. This structure embodies the principle of the double arch,* and in the more than 2,000 years that it has been employed in

* Longitudinally and transversely curved.

the assembly of barrels, no one has been able to improve upon the basic design in terms of increased mechanical efficiency. In a properly raised barrel, the projected planes of all stave joints intersect along the longitudinal axis of the structure, with each stave serving as a key. Thus, when an external blow is applied, the shock is instantly transmitted to all parts of the assembly, and the damaging effect at the point of contact is greatly minimized. In fact, a blow of sufficient magnitude to fracture a stave in a cylindrical barrel ordinarily can be absorbed by a bilge barrel having staves of comparable length and thickness.

B. Tight Cooperage

Tight barrels are designed primarily for packaging liquids such as vinegar, fruit juices, sirups, alcohols and alcoholic beverages, organic solvents, mineral and vegetable oils, paints and chemicals, as well as meats, fruits, and vegetables processed in brine or other liquids. They are also used for packaging semisolids such as lard and vegetable shortenings, preserved fruits, resins, waxes, pastes, and greases. They make excellent shipping containers for the transportation of such highly dangerous commodities as gunpowder, corrosives, poisons, and inflammables.

As it will be determined later, there are several kinds of tight barrels. The details of their construction are somewhat variable, as each type has been designed for a specific use. Those having capacities of less than 30 gal are known in the trade as *kegs*, whereas those which exceed a capacity of 60 gal are usually designated *hogsheads*. The terms *tierce* and *cask* are usually applied to heavy barrels of great capacity used for packaging fish, wines, and a few other commodities.

1. Manufacture of Tight Cooperage. The component parts of a tight barrel are staves, heads, and hoops. These may be manufactured at separate and independent plants and sold to coopers for assembly into barrels in accordance with customer specifications, or they may all be manufactured and assembled within the walls of a single facility. It has also been a common practice among a number of the larger cooperage plants to operate captive stave and heading mills in areas adjacent to adequate supplies of suitable timber.

a. Staves. Woods permitted in the manufacture of staves and heads for tight cooperage under the Interstate Commerce Commission regulations include white oak, chestnut oak, red oak, black cherry, beech, sweet birch, yellow birch, sugar maple, Douglas-fir, and Scandinavian pine. Logs selected for stave manufacture are cut into bolts of predetermined length and then reduced to flitches which will produce quarter-grained staves not to exceed 6 in. in width. In stave production the flitches are rough peeled and then fed into a stave-bolt equalizer to square their ends and to assure staves of equal length. A common type of equalizer (Fig. 11-1) is a machine composed of a pair of circular saw blades mounted on a common drive shaft seated in a heavy metal frame. In some operations, however, equalizing is done after the staves have been ejected from the saw.

At one time tight staves were made by forcing roughly hewed blanks between a pair of heavy, curved, stationary knives. The products thus formed were known as "bucked staves," and their manufacture was laborious and wasteful. The stave bucker has long since given way to the *barrel* or *drum saw*, and virtually all tight staves are now manufactured using equipment of this sort. The barrel saw (Fig. 11-2) is composed of a cylindrical steel drum with peripheral saw teeth at one end, mounted on a power-driven shaft seated in a heavy metal frame. Staves are produced by placing a flitch on a manually operated carriage which runs on a track parallel to the longitudinal axis of the saw. With each forward stroke of the carriage the flitch is forced into the whirling blade. At the completion of the stroke a stave drops into a cradle within the cylinder and is automatically ejected when the carriage is returned to its original position. Barrel saws are manufactured in several diameters, as the staves cut on any one of them are essentially segments of a cylinder equal in diameter to that of the saw blade used.

Fig. 11-1. Stave-bolt equalizer. (*Courtesy John S. Oram Co., Cleveland, Ohio.*)

Another stave saw is the *bilge saw* (Fig. 11-3). This differs from the barrel saw in that the blade has the configuration of a conventional barrel, and the carriage track is similarly curved. Staves cut with this saw have a "built-in" bilge, but owing to the prevalence of short grain, which is invariably present in the region of the bilge, they are usually weaker and more readily broken than those of comparable dimensions produced on conventional barrel saws. Accordingly, bilge staves are rather restricted in use, and their production is usually limited to lengths of 24 in. or less.

Green staves, as they are ejected from the saw, require seasoning before further machining and their eventual assembly into barrels. It is common practice at many mills to crib-pile staves in the open or in well-ventilated drying sheds for several weeks or even months before they are kiln-dried. At other facilities they are bulk-stacked and kiln-dried without preliminary air-seasoning. For best results, staves should be dried to a moisture content of from 10 to 12 per cent. Below this level many woods become too brittle and are incapable of withstanding rough handling to which staves are frequently subjected. On the other hand, barrels assembled with air-dried staves commonly dry out and develop open joints while in storage and thus become unsuitable for many uses.

When staves are thoroughly seasoned they are ready to be jointed. The jointing operation shapes and bevels stave edges in such a manner that

Fig. 11-2. Barrel saw. (Saw blade depicted as transparent in order to show position of stave ejector.) (*Courtesy Baxter D. Whitney & Sons, Winchendon, Massachusetts.*)

Fig. 11-3. Cutting pine staves with bilge saw. Note workman at left equalizing a stave after it leaves the saw. (*Courtesy U.S. Forest Service.*)

uniformly tight joints are assured along the entire length of all contact faces in an assembled barrel. This operation is accomplished on a machine known as a stave jointer (Fig. 11-4). Its principal component is a large, power-driven, vertically oriented metal wheel with a dished face in which jointer knives are locked into position at regular intervals. In operating the jointer a laborer selects a rough stave, clamps it on a horizontal cradle in front of the wheel, and with the use of a manually operated foot treadle brings it into contact with the whirling blades which shape and bevel an edge. At

Fig. 11-4. Stave jointing. (*Courtesy Louisville Cooperage Company, Louisville, Kentucky.*)

the completion of this operation, the stave is turned end-for-end, and the second edge is similarly profiled. Machines of this type are built in several sizes in order to accommodate staves of different standard lengths. The jointer wheels, depending upon size, barrel diameter, and the kind of staves being manufactured, are furnished with two, four, six, or even eight knives, and are driven at speeds ranging from 450 to 650 rpm. Jointed staves are classified in terms of differences in width between their mid-sections and their ends. Thus half-bilge, three-quarter bilge, or seven-eighths bilge staves are those in which the width of their mid-sections exceeds that of their ends by $\frac{1}{2}$ in., $\frac{3}{4}$ in., and $\frac{7}{8}$ in., respectively.

b. Headings. The headings of tight barrels are made from the same species of wood used in stave production, although they are frequently manufactured at plants engaged only in the fashioning of this sort of stock. Bolts and flitches selected for headings are cut into blanks 1 to $1\frac{1}{4}$ in. in thickness and then, like staves, are thoroughly seasoned before they are worked up into finished goods. Upon their removal from the kiln, the blanks are delivered to a finishing mill, where they are dressed to a predetermined thickness and then tongued and grooved or dowel-jointed. Following these preliminary operations, a workman, known in the trade as a matcher, selects a number of pieces and assembles them into a "square" from which a one-piece circular heading is cut. In the assembly of dowel-jointed headings the seams are commonly caulked with dry rush or cattail leaves, called flags.

Fig. 11-5. Turning a head for a tight barrel. (*Courtesy Pioneer Cooperage Company of Illinois, Chicago, Illinois.*)

When flagging is not used the joints are commonly edge-glued. Well-trained matchers make every reasonable effort to assemble squares that will produce "goosenecks" * of minimum size in the head-cutting machine.

Figure 11-5 depicts a machine of this sort. In its operation a square is manually positioned between a pair of revolving disk-like clamps and turned against a high-speed, concaved saw and a knife. The saw produces a heading of circular configuration with a beveled edge, while the knife, working behind the saw, carves a second and opposing bevel from the opposite side. Thus the finished heading, when automatically ejected from the machine at the end of the cutting cycle, is furnished with a double-beveled, or wedge-shaped, rim, which, as will be shown, gives the assurance of a tight fit in the

* Trade term applied to waste trimmings from a head-cutting machine.

assembled barrel. Heading manufacturers bale their wares in sets of 20 pairs for shipment to cooperage houses.

c. Hoops. Only hoops of steel are used in the fabrication of tight barrels. Because they must be made in several sizes and with varying degrees of flare to conform to a variety of barrel styles offered to consumers, hoops are usually made at the cooperage where machines are available for coiling, riveting, and flaring. Hoop steel ranges from 1¼ to 1¾ in. in width, and from 16 to 19 Birmingham gauge in thickness; it is usually supplied in the form of flat strips, or occasionally in coils.

Fig. 11-6. Raising tight cooperage. (*Courtesy U.S. Forest Service.*)

2. Tight-barrel Assembly. The initial step in the assembly of a barrel is known as raising. First a workman selects two temporary, heavy steel truss hoops (a head hoop and a bilge hoop) of appropriate diameters and positions each of them on a device known as a setup (Fig. 11-6). Next he selects a number of staves of random width but of predetermined length and contour, and by trial and error of manipulation fits and refits them in the setup until he has a snug basal assembly. The number of staves used in raising a barrel is important only in that no fewer should be used than the minimum number stipulated by Interstate Commerce Commission regulations for barrels of specific capacities. Once a barrel has been raised it is removed from the setup, with the two truss hoops still in place, and put in a tunnel where it is steamed for about 30 minutes at 200 to 210°F. Steaming softens the wood sufficiently to enable the cooper to draw the tops of the staves together without breakage. This operation, known as windlassing, is accomplished with a cable and windlass. While the assembly is in the windlass, a second pair of temporary truss hoops is fitted at its top, and the structure is then ready for firing.

Firing should not be confused with charring, as it is essentially a baking process using dry heat, and has a threefold purpose. First, it drives from the staves moisture absorbed during the steaming treatment just described; secondly, it stress-relieves the wood fibers in the vicinity of the bilge; and finally, it "sets" the staves in such a manner that they do not spring back

Fig. 11-7. Charring bourbon whisky barrels. (*Courtesy U.S. Forest Service.*)

appreciably when hoops are removed. In practice a barrel is ordinarily held in a firing chamber for about 15 minutes at temperatures ranging from 400 to 500°F. Progressive firing tunnels are used in some plants where barrels move through them on conveyers at predetermined rates of travel.

After barrels have been fired and cooled to room temperature, they are leveled and trussed. Leveling is essentially a tamping procedure used to bring the staves into perfect transverse alignment. Once a barrel is properly leveled it is placed in a trusser. This is a machine with several power-driven arms used to drive the truss hoops down over the sloping walls of the structure, thus giving the assurance of uniformly tight joints.

At this point barrels destined for use in the aging of bourbon whisky are internally charred in accordance with government regulations (Fig. 11-7). This was done at one time by kindling a small fire within the structure. In modern cooperage operations, however, charring is done with oil or gas flames, since they are more easily controlled. Some manufacturers employ a bung protector during this procedure. Charring appears to be a necessary step in the manufacture of bourbon whisky barrels, for it is believed that certain impurities in the distillate are absorbed by the carbonized surface during the aging process. A bung protector is a water-saturated pad firmly affixed to the mid-section of the stave selected to receive the bung and is used to prevent excessive char at the point where the bung hole is to be drilled.

Fig. 11-8. Heading-up machine. (*Courtesy Pioneer Cooperage Company of Illinois, Chicago, Illinois.*)

Whisky-barrel headings also must be charred, and particular care is exercised to avoid burning or charring of their beveled rims. The charring of these components is a separate operation and is done just prior to the time the headings are seated in a barrel.

When barrels have been leveled and trussed, they are ready to receive their headings. In a series of three operations, which the cooper calls *working off*, a barrel is placed between two adjustable chuck rings in a multipurpose device known as a *howeling, crozing,* and *chamfering* machine. Small knives much like diminutive carpenter's gouges cut a pair of semicircular grooves, *howels*, one at either end on the inside of the barrel, 1 to 1¼ in. from the rims of the staves. A second pair of knives carves a V-shaped groove, the *croze*, in each of the howels. The configuration of the croze is such that the double-beveled edge of a heading may be firmly seated to produce a strong, watertight joint. Finally, a third set of knives bevels the inner faces of the staves between the howels and the rims of the staves. This portion of the barrel is known as the *chamfer* or *chime*. It not only facilitates emplacement of the headings but also safeguards against leakage which could develop if small pieces sheared off the ends of staves during periods of rough handling.

Headings are inserted and properly seated in the croze using a heading-up machine (Fig. 11-8), after which the barrels are placed, sanded, and smoothed to improve their general appearance. From the sander they are taken to a hooping machine, a device similar to a trusser, where the temporary truss hoops are replaced with permanent heading, chime, and bilge hoops fabri-

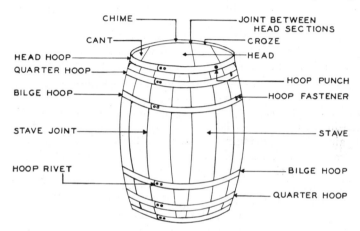

Fig. 11-9. Component parts of an assembled barrel.

cated of strap steel. Special metal hoop fasteners driven into the staves are also used to keep hoops in their proper position.

A finished tight barrel usually carries six hoops (Fig. 11-9). The medial pair, the bilge hoops, are driven first, then the quarter, or chime, hoops, in the order noted. Where maximum strength is requisite a fourth pair of hoops may be added. Known as eighth hoops, they are seated between the head and quarter hoops. Nearly all tight barrels used in export trade, where rough handling and hard usage are virtually inevitable, are furnished with eight hoops.

To prevent leakage or to guard against contamination of many liquid commodities, it has been found expedient to coat the inner walls of these containers with an inert substance such as paraffin, silicate of soda, or glue. Coatings of this sort are not required for all types of tight cooperage. The use of any one of them is governed by the quality of the package and the nature of the commodity it is to receive. Paraffin linings are used primarily for water-based liquids such as brine, beverages, fruit juices, sirups, and vinegar. Glue is somewhat water-soluble and should be restricted to products containing alcohol solvents, shellac, lacquers, paints, and numerous chemicals. Silicate of soda imparts neither taste nor odor and makes an excellent liner for barrels packaging food products such as lard, molasses, vegetable oils, processed meats, and sausage casings, or industrial commodities such as waxes, white oil, greases, and numerous chemicals, including caustics.

3. Kinds of Tight Barrels. The Associated Cooperage Industries of America, Inc., recognizes, without respect to their capacities, eight standard classes of tight barrels, the principal features of each of which are indicated below.*

a. Bourbon Barrel. This, the king of the tight barrels, is also known as the whisky barrel. It is used for storing and packaging distilled spirits, and, when internally charred, for aging bourbon whiskies. Bourbon barrels are assembled with choice quartersawed staves and headings of white oak heartwood. Staves are equalized to 34, 35, and 36 in. in length, and when thoroughly dried they must be ¾, ⅞, and 1 in. in thickness, respectively. No single stave may exceed a width of 5½ in. A maximum diameter of 20⅞ in. has been established for headings, which, when dried and dressed, shall not be less than 1 in. thick. Ordinarily six strap-steel hoops are used in the assembly of these barrels, but an additional pair of eighth hoops is commonly placed on those destined for overseas shipment.

b. White Oak Barrel, Sap Clear. Quartersawed staves and headings are also used in the assembly of barrels of this class. Minor defects in the wood that impair neither the imperviousness nor the serviceability of these containers are permitted. No liner is required even when packaging such commodities as wines, vinegar, fruit juices, and liquid chemicals.

c. White Oak Barrel, Tight Sap. Barrels in this class are raised with quartersawed white oak staves and headings, but sapwood is permitted provided the springwood pores are closed (occluded with tyloses). With appropriate linings, the barrels are suitable containers for such industrial products as shellac, varnish, oils, alcohol, dimalt, soft drink sirups, paints, lacquers, and foods processed or stored in brine. This barrel is also used without a lining to package meats, fruit juices, and several beverages.

d. White Oak Barrel, Sound Sap. Since the staves and heading in barrels of this class may include any amount of sound, bright sapwood, the barrels must be lined to prevent leakage. They are used for a host of commodities including liquid chemicals, vegetable and mineral oils, lard, greases, vinegar, cider, solvents, turpentine, denatured alcohol, catsup, sauerkraut, meats, and condensed milk.

e. Red Oak Barrel. This barrel is manufactured with staves and headings of red oak; bright, sound sapwood is permitted. Inasmuch as the springwood pores of most red oak timbers are more or less open (relatively free of tyloses), linings are necessary in barrels used for packaging fluids and meats, vegetables, and fruits processed in liquids. This barrel is used without a lining for lard, greases, and other semisolids.

f. Gum Barrels. Barrels in this category are made of sweetgum (redgum), although blackgum and tupelo staves and headings may be substituted when agreed upon by the purchaser. Unlined gum barrels are used for holding semisolids, dry chemicals, powdered dyes, and heavy sirups; and lined bar-

* For complete specifications covering tight barrels the reader is referred to The Wooden Barrel Manual, issued by the Associated Cooperage Industries of America, Inc., St. Louis, Missouri.

rels are suitable for liquid chemicals, industrial oils, tanning compounds, foods processed in liquids, and similar commodities.

g. Ash Barrel. With the exception of pumpkin ash, which is not an acceptable cooperage material, the staves and headings of this barrel are made from any of the commercial ash timbers. Lined ash barrels are used for packaging the same general classes of commodities listed for the red oak barrel. Unlined, the barrels may be used for processed meat, fish, and meat products.

h. Douglas-fir Barrel. This barrel is manufactured on the West Coast. The wood of Douglas-fir usually imparts objectionable tastes to most foodstuffs with which it comes into direct contact, and linings must be used on those barrels destined to receive products for human consumption, such as salmon, processed fruits, and meats. Unlined barrels are suitable for many industrial liquids and semisolids.

4. Plywood Whisky Barrels. Whisky barrels of 50-gal capacity made with 14 interchangeable laminated staves and plywood headings have been used sparingly during the past several years. The staves are fashioned from a four-ply parallel laminate composed of white oak heartwood rotary-cut veneers; the headings are turned from five-ply white oak panels of conventional design.

5. Beer Barrels. The beer barrel is an unusually sturdy structure distinguished by a very pronounced bilge. It is restricted in its use to the packaging of malt beverages, and its staves and heading, which are appreciably thicker than those in the whisky barrel and its allies, are manufactured with machines specifically designed for their production.

A government statute fixes the capacity of this container at 31 gal, but also permits the manufacture of half-, quarter-, and eighth-capacity barrels as well. The full barrel is assembled with 16 to 21 staves of white oak heartwood. Each stave is 31 in. in length and $1\frac{3}{4}$ in. in thickness at its extremities; it must be also $1\frac{7}{16}$ in. in depth in the region of the bilge. Headings are also required to measure $1\frac{3}{4}$ in. in thickness. These heavy components, which so characterize the beer barrel, are not only capable of absorbing severe impact loads without damage to the structure, but also provide excellent thermal insulation for the brew while it is in transit.

C. Slack Cooperage

Slack barrels are used for packaging, storing, and shipping products of both farm and industrial origin. Because of their efficiently engineered construction, they can often withstand greater abuse than other types of wooden containers and are especially well suited for overseas shipments of goods where rough handling is inevitable.

Slack barrels are manufactured in several types and sizes. Each type is designed for a specific use and in consequence will vary from others in the details of its construction. The weight of a commodity to be packaged frequently governs the selection of staves and headings to be used, as well as the type and number of hoops required. In packaging foodstuffs the kind of wood used is often of major concern, since that of several species will impart objectionable tastes, odors, or discolorations. Powdered ingredients such as dried milk and eggs, spices, expensive chemicals, and the like, call for moistureproof, siftproof containers. Such barrels are assembled with top-quality tongued and grooved staves, one-piece headings, and strap-steel hoops. Barrels raised with plain-jointed staves, multiple-piece headings, and hoops of wood, wire, or beaded steel are used for packaging chinaware, nails, un-shelled oysters, dried fish, meat and poultry, vegetables, fruits, and hard-ware. When used with a paper-bag liner they are excellent containers for many finely divided and even powdered materials. Because of the number of slack-barrel styles available, a potential user should seek the advice of a competent cooper. He is thoroughly qualified to recommend the best and cheapest container that will not only meet the consumer's needs but will also comply with the rules and regulations of the Interstate Commerce Commis-sion, as well as those covering export trade.

1. Manufacture of Slack Cooperage. Like the components of a tight barrel, those of slack barrels are commonly manufactured at separate facili-ties and shipped to cooperage houses for assembly.

a. Staves. A large number of timbers, including sweetgum (redgum), beech, birch, maple, oak, ash, elm, sycamore, hackberry, cottonwood, willow, basswood, blackgum, numerous pines, Douglas-fir, and western hemlock, have been employed in the production of slack staves, which, incidently, are produced by either sawing or slicing.

In the production of sawed staves tree-length logs are cut into bolts which are in turn further reduced to flitches. The flitches are rough-peeled, equal-ized, and then reduced to staves of any desirable thickness using a barrel saw.

Sliced staves, on the other hand, are made by repeatedly driving a flitch against a knife ground to the arc of a circle of predetermined diameter. When staves are cut in this manner, the flitches are given a preliminary heating treatment which softens the wood and thus increases the shearing efficiency of the knife. In practice, flitches are either (1) stacked on cars and transferred to steam tunnels, where they are steamed for at least 24 hours with 80- to 110-lb steam, or (2) boiled in open tanks for several hours. They are then immediately barked and equalized, and, while still hot, deliv-ered to stave-cutting machines.

The slicing machine, or *stave cutter* as it is also known, consists essentially of a *tumbler,* which is a device used to hold a flitch in position for cutting, and a *knife.* The flitch is placed into the tumbler in such a way that it will produce a quartersliced stave when driven lengthwise across the knife. Machines of this sort are capable of 150 or more strokes a minute and thus can produce 40,000 staves in an 8-hour day when manipulated by a skilled operator. Most manufacturers of slack staves claim that the smoothest and brightest staves are produced from steamed flitches. Over- or undercooking, however, invariably results in blanks of inferior quality. Both sawed and sliced stave blanks are then carefully seasoned, much in the manner of tight staves, then jointed, and finally bundled for shipment to a cooperage.

Hardwoods are especially well adapted for the production of sliced staves. Softwoods, on the other hand, are usually made on barrel and bilge saws because of the frequency with which they shatter under the knife following a heat treatment.

b. Headings. One-piece and multiple-piece headings are used in the assembly of slack barrels. These are turned from preassembled headblanks in a manner similar to that described for tight headings, using the same group of woods employed in fabricating slack staves.

c. Hoops. Years ago wooden hoops were used exclusively in the assembly of all slack barrels. As these containers came into general use, however, it was found necessary to improve upon the strength of many of them and to provide others with tighter, and even siftproof, joints. This led to the use of iron-wire and beaded-steel hoops, and now it is only when strength and exceedingly tight joints are of minor concern that slack barrels are assembled with only wooden hoops.

American elm, because the physical and mechanical properties of its wood are affected but little by prolonged boiling or steaming, a necessary step in hoop manufacture, was, and still is, preferred to all other species. At the present time about 95 per cent of the wood hoops produced are fashioned from this species.

Wood hoops are made by sawing or slicing tapered splines from planks of suitable dimensions. These splines are fed into an intricate machine which planes them, points one end, and laps * the other. They are then placed in a steam tunnel or an open vat of boiling water for several hours to facilitate coiling operations, after which, and while still hot and wet, they are fed into a hoop coiler. The hoop coiler is a device which laps the two ends and then fastens them together with rivets or staples to form a finished hoop of prescribed diameter. The hoops are then permitted to dry before they are gathered into bundles for consignment to an assembly plant.

Standard wood hoops are 5 to 8 ft in length, $1\frac{3}{8}$ in. in width, and $\frac{9}{32}$ in. in thickness at one edge and $\frac{5}{32}$ in. at the other. This trapezoid-like, cross-sectional configuration gives the hoops a slight flare which ensures a snugger fit against the tapering walls of the barrel.

d. Head Liners. Head liners are thin strips of elm 12, 14, and 18 in. in length, $\frac{1}{2}$ in. deep, and $\frac{3}{16}$ in. in thickness, which are nailed or stapled to the chamfer to prevent headings from slipping out of the croze. Two such strips are used for each head on all barrels assembled with wood hoops when the net weight of their contents will exceed a weight of 250 lb. They are also used on barrels furnished with beaded-steel hoops when the materials to be packaged exceed 600 lb. When deemed advisable, the liner may be made to extend completely around the circumference of the head.

2. Slack-barrel Assembly. The assembly of a slack barrel follows the same general pattern of that of a tight barrel, previously described. A number of selected staves are placed in the raiser, then steamed and windlassed. With temporary hoops in place, the barrel is then fired, leveled, trussed, crozed, and chamfered. Finally, the headings are set and the permanent hoops driven into position. Hoops may be of wood, iron-wire, beaded-steel bands or combinations of any two of them. If tongue and grooved staves are used,

---

* The lap is in reality a scarfed or feathered end to which the pointed tip is fastened when the hoop is coiled.

the matching operation is performed in machines designed for the purpose, but otherwise the barrels are raised in the usual manner.

3. Kinds of Slack Barrels. The Associated Cooperage Industries of America, Inc., recognizes nine types of slack barrels. Each has been designed for specific uses, and all, when manufactured to Association standards, meet minimum specifications of the Interstate Commerce Commission and all regulations pertaining to export-shipping containers.

a. No. 1 Barrel. This is the finest of the slack barrels and is manufactured exclusively with No. 1 grade staves and headings. When the contents of the barrel must "breathe," plain-jointed staves and headings are used. Tongued and grooved staves and "T & G," or butt-jointed and glued, headings are specified for barrels when siftproof joints are required or when protection against moisture or contamination is of paramount importance. As an added measure of safety a paper-bag liner is also commonly employed. The No. 1 barrel is used principally for packaging dried milk and eggs, dry paint products, soap chips and powders, sugar, salt, spices, dry chemicals, lime, and similar commodities.

b. No. 2 Barrel. Barrels in this class are assembled with No. 2 grade staves and headings. The staves may be either edge-jointed or tongued and grooved, but as a general practice all headings are edge-jointed. Carrying four beaded-steel hoops, the No. 2 barrel is an exceptionally strong structure which is used primarily as a shipping container for many kinds of hardware, including chain, valves, hinges, and plumbing supplies, and for low-grade chemicals.

c. Meat-and-poultry Barrel. This barrel is also assembled with No. 2 staves, but as a rule supplied with but one heading. It is furnished with two wooden head hoops and four wire hoops, one pair of which is placed immediately below the head hoops and the other near the bilge. This barrel is an excellent and sturdy shipping container for dressed turkey, chicken, and duck, and for cuts of beef and pork.

d. Glass-and-pottery Barrel. Other than the fact that this barrel is furnished with two heads, it is identical to the meat-and-poultry barrel described above. As its name implies, it is used as a shipping container for glass and ceramic goods.

e. Fish Barrel. This is a barrel used for packaging and shipping fish, crabs, and other seafoods. Assembled with No. 2 plain butt-jointed staves, it is available with one or two heads and with various combinations of wire and wood hoops, as dictated by consumers' specifications.

f. Potato Barrel. The potato barrel is fabricated with dead-cull staves and a single No. 2 heading of either pine or hardwood. It is furnished with one of three combinations of hoops, viz.: (1) six wood hoops; (2) four wood and two wire hoops; or (3) two wood and four wire hoops.

g. Apple Barrel. Like the potato barrel, this container is also assembled with dead-cull staves and one No. 2 heading. It is usually bound with six wood hoops.

h. Vegetable Barrel. This is a No. 2 barrel with plain butt-jointed staves and headings and six wood hoops.

i. Nail-and-spike Kegs. The keg is made in a variety of capacities and used for packaging nuts, bolts, nails, screws, spikes, rivets, and other types of small, bulky metal products. Southern yellow pine is used almost exclusively for both the staves and headings. It is available either with four beaded-steel hoops or with two steel head hoops and two iron-wire bilge hoops.

4. Tobacco Hogsheads. No discussion of slack cooperage would be complete without mention of the tobacco hogshead. This is a straight-walled container with a capacity of approximately 1,000 lb of compressed leaf, and more than 600,000 of them are needed each fall to handle the tobacco harvest in North Carolina and South Carolina alone.

Leaf tobacco purchased at auctions held in the fall and early winter throughout the tobacco belts of the Carolinas, Virginia, and Maryland is packed in hogsheads * and dispatched to processing plants for redrying in

Fig. 11-10. Tobacco hogshead: *a*, loaded barrel ready for storage; *b*, removing the two mats. (*Courtesy A. E. Wackerman.*)

order that it may be stored for protracted lengths of time without deteriorating. Once dried, the leaf is repacked and the hogsheads taken to nearby well-ventilated storage sheds where the tobacco is permitted to age for at least 2 years. At the conclusion of the aging cycle the hogsheads are taken to a stemmery where the tobacco is removed once more and steamed and stemmed.†

For many, many years the tobacco hogsheads were made with southern yellow pine or oak staves and heads bound together with wood or metal hoops. These containers received exceedingly rough handling in their circuitous travels from warehouse floors to the final processing plants. Damaged staves, heads, and hoops were counted in the thousands, and many hogsheads had to be virtually rebuilt after each time they were emptied. Such damage was regarded as inevitable, and cooperage shops were a major department of every tobacco factory. But when labor and material costs began to rise appreciably, the tobacco manufacturers began to search for a better container, and in the late 1930s a prefabricated hogshead of Douglas-fir plywood staves and heading made its appearance. It was quickly accepted by

* Georgia leaf is usually baled in burlap.

† Process wherein the tough, fibrous, midrib (stem) of the leaves is separated from the blade.

several of the large manufacturers and today has all but replaced the old-style container.

The Douglas-fir plywood hogshead is composed of two interchangeable mats that, when properly mated, form the cylindrical wall of the container. Each mat is composed of 15 three-ply staves that are 54 in. long, $42\frac{7}{32}$ in. wide, and $\frac{5}{16}$ in. in thickness. In the assembly of a mat, spacers of 20-gauge metal are temporarily inserted at each joint, and the staves are then bound together with five 20-gauge, equidistant galvanized-iron bands $1\frac{1}{2}$ in. wide and $72\frac{5}{8}$ in. in length. Male hinges are riveted at one end of each band, and female hinges are similarly attached at the other. A pair of liners (small strips of $\frac{7}{16}$-in. plywood, $1\frac{1}{4}$ in. wide and $4\frac{5}{8}$ in. long) are nailed on the inner face of each stave, one at either end, to provide adequate seating for the heads.

The heads of tobacco hogsheads are turned from $\frac{7}{16}$-in. three-ply Douglas-fir plywood and measure 46 in. in diameter. An oak spline, $\frac{3}{4}$ in. by $1\frac{1}{4}$ in. by 11 ft $9\frac{1}{4}$ in., in the form of a hoop, is nailed to each head piece. When the head is properly seated, this band fits snugly against the faces of the liners and protects the ends of the staves against breakage when a heavy, fully packed barrel is upended.

To assemble the barrel, two mats are placed side by side, and steel pins are inserted into one set of mating hinges. The two remaining edges are then drawn together and similarly pinned. The shell is then upended and the bottom heading carefully driven into position. When the barrel has been packed with leaf, the three top pins from one hinge row are removed and the top heading inserted, after which the hinges are again drawn together with a ratchet and the pins replaced. To empty this type of container, it is only necessary to remove the pins from one hinge row.

Plywood hogsheads have several distinct advantages over the older type of containers:

1. They weigh only two-thirds as much (about 100 lb) as the old style barrel and hence effect a material saving in shipping costs.

2. Inasmuch as they are engineered containers, damage from rough handling is only a fraction of that previously experienced, and since all parts are interchangeable, any defective component may be easily and readily replaced; thus a long service life is assured.

3. When tobacco is consigned to West Coast and Canadian tobacco plants, the emptied barrels can be knocked down and the mats and headings returned to the shipper for reuse. The old-style barrel, on the other hand, had little salvage value outside the tobacco-producing centers, and its disposal was always a problem since the costs involved in dismantling, returning, and reassembling exceeded the price of new containers.

D. Status of the Cooperage Industry

The wooden barrel was once the nearly universal storage and shipping container for a host of farm and industrial commodities. New and improved methods of merchandising, distributing, and packaging goods have had a profound effect on the use of barrels, and in some instances barrels have been completely replaced by other types of containers.

1. *Slack Cooperage.* The slack barrel is still used in quantity for packaging such products as powdered milk, seafoods, meat, poultry, tobacco, dry chemicals, soap, powders, detergents, nails, spikes, rivets, bolts, and a few other agricultural and industrial commodities. The ever-increasing use of fiber cartons, fiber drums, and sacks made of cotton, paper, and burlap, however, has resulted in the loss of many markets. According to the Stanford Research Institute's recent report, "America's Demand for Wood, 1929–1975," slack-barrel production decreased from 55 million barrels in 1929 to 28 million in 1952. This decrease in production is expected to continue, and it is predicted that no more than 17 million units will be required in the year 1975.

2. *Whisky Barrels.* Between 17 and 20 million new white oak barrels have been used annually for the storage and aging of whiskies since the repeal of the Eighteenth Constitutional Amendment. Despite several successful patents for processes to expedite aging, some of which even eliminated the use of barrels, a Federal statute decrees that all such spirituous liquids manufactured in the United States shall be aged in new white oak barrels. Hence, so long as this law remains unchanged there is little likelihood that competitive containers will enter the market or that a major curtailment in whisky-barrel production will occur.

3. *Beer Barrels.* Prior to the enactment of the Volstead Act in 1919, the bulk of all malt beverages produced in the United States was packaged and distributed in barrels and kegs of oak. However, when the brewing of beer and ale again became legalized in 1933, brewers were unable to procure the then-conventional beer barrels in the quantities required and out of sheer necessity turned to the metal drum, with some misgivings on the part of several of them, to supplement their needs. However, this lightweight, stainless-steel, double-walled container quickly proved to be the equal, if not actually the superior, of its wooden counterpart, for not only did it require much less maintenance but it also cost only half as much. New methods of merchandising also have had their impact on the use of barrels per se, and currently over 50 per cent of all malt beverages are packaged in nonreturnable bottles and cans. Moreover, about 80 per cent of all beer and ale that is barreled is now put up in the metal drum, and it appears that it will be only a matter of time before the old wooden beer barrel disappears completely from the American scene.

4. *Wine Barrels.* During the past several years the demand for barrels to be used for the storage of wines has increased rather than diminished. On the other hand, their use in distributing and marketing such spirits has fallen sharply. Large volumes of wine are now being marketed in bottles and jugs, while bulk shipments are made in ever-increasing quantities in glass lined tank cars and tanker trunks.

5. *Oil Barrels.* In the early beginnings of the petroleum industry all crudes were shipped to refineries in wooden barrels. By 1865 wooden tank cars were put into service, and in 1872 the first underground pipe lines from oil wells to refineries were put into operation. Production of the oil barrel reached its peak in the year 1907 when over seven million new barrels were placed in service. Shortly thereafter, and largely as the result of a rapidly expanding automotive industry, new methods of merchandising and packaging petroleum products were introduced. Bulk shipments of gasoline and oil to distributors were made in steel tank cars, while many other petroleum products were packaged for the retail trade in small, standardized metal containers. In 1934, 6 million steel drums, 44 million pails and buckets, and 535 million sealed tins of from 1 qt to 5 gal capacity were needed to package petroleum products marketed by service-station operators and other retailers. Wooden oil barrels are seldom encountered today in the domestic petroleum markets, although appreciable quantities of them are still used for export shipments of lubricating oils, greases, and waxes.

6. Pork Barrels. This form of tight barrel was once used in prodigious quantities by meat packers as a shipping container for cuts of beef and pork and even for poultry. During the past several years, however, it has been largely replaced by the wooden case and wire-bound box, and as a result the current requirements for barrels of this sort, even though larger quantities of meat products are being produced than ever before, stands at something less than 150,000 units annually.

7. Future of Tight Cooperage. Competitive containers such as steel and fiber drums, bulk shipments in tank cars or tanker trucks, and the use of pipe lines have had a marked effect on tight-cooperage production, which is currently 78 per cent below 1929 levels. The future outlook for tight cooperage, however, is much better than that for slack cooperage, as it has been predicted that by 1975 production will have increased by at least 35 per cent above present levels.

E. Container Woodenware

Woodenware is a collective term applicable to several articles of turnery and also to such wooden containers as pails, kits, buckets, tubs, and piggins. Since these structures are assembled with staves, hoops, and headings and used in handling liquids and semisolids, they are also regarded as forms of tight cooperage. All these items have about the same general configuration; yet each one of them has certain specific features which serve to distinguish it from the rest. A pail, for example, is assembled with strap-iron hoops and a wire bail (handle) and is widest at its top. It may or may not be furnished with a cover depending upon customer specifications. A kit, by contrast, has the shape of an inverted pail. The staves are held in place with either iron-wire or strap-iron hoops, while the heads are seated in grooves near the upper rim, much like the croze in a conventional barrel. A kit is also furnished with a wire bail. A bucket (also known as a kanakin) has the same configuration as a kit, but is assembled with hoops and a handle of wood, and with a cover that slips over the rim rather than one that is seated in the staves. Tubs are larger pails with two side handles, usually without headings. Large, straight-walled containers made of staves and wooden hoops and used for packaging "wheels" of cheese are also known as tubs. A piggin is a wooden pail with one of its staves extending beyond its rim to form a handle.

This structure obtained its name from farmers who used it to carry feed to their hogs and is still in use in certain rural areas of the South.

Production of woodenware in the United States had its inception during early colonial times in New England, where containers such as pails and kits were in demand for packaging salted fish. Soon containers of this sort were being used for handling pickles, jellies, butter, lard, maple sirup, printer's ink, and many other commodities. By 1850, woodenware plants had become established in the Lake states, where there was an abundance of white pine and red spruce. The first plant in the Pacific Northwest was established in 1921 and produced both pails and kits assembled with Sitka spruce staves and headings.

As with barrels, the general decline in the use of container woodenware in recent years has been greatly influenced by new innovations in merchandising and distribution and by the introduction of many new, inexpensive, attractive, and serviceable lightweight containers of glass, fiber, and metal.

F. Tanks and Vats

Tanks and vats assembled with massive tongue-and-groove staves and headings bound together with steel rods fitted with couplings are regarded as forms of heavy cooperage. Both the staves and the individual pieces composing the headings are produced from lumber. The staves are made by passing planks of predetermined dimensions through a machine which joints and tongues and grooves their edges, and profiles their faces in such a manner that their cross-sectional configuration is in the form of an arc from a circle of specified diameter. But unlike barrels, which are assembled in great numbers at a cooperage, each tank or vat is usually a carefully designed and engineered structure for a specific use, with its assembly undertaken at that point where it is to be permanently located.

White oak, Douglas-fir, baldcypress, and redwood are the timbers principally used in the heavy cooperage industry. The growing scarcity of tidewater baldcypress, once the prime favorite of tank manufacturers, has resulted in the greater use of redwood in recent years. Lumber of this species is readily available in large, clear sizes, and this, coupled with its comparative lightness in weight, excellent stability, and natural resistance to decay and chemical corrosion, makes redwood timber an excellent processing or storage vessel for a wide variety of industrial products.

For storing corrosive materials, wood tanks are generally more satisfactory than similar containers fabricated of metal. The petroleum industry, for example, has determined that it is much more economical to store corrosive sour oil crudes in redwood tanks than in those erected with any other type of structural material. Wood tanks are used in the pulp and paper industry for storing cold sulfite cooking liquor and other processed or processing materials. Brine processing of onions, olives, cauliflower, pickles, and tomatoes is usually accomplished in paraffin-lined tanks of from 2,000 to 500,000-gal capacity. Wooden tanks employed in the fermentation and aging of malt beverages are standard equipment in nearly every brewery. Similar vessels are also to be found in distilleries and wineries for housing large inventories of cider, vinegar, fruit juices, wines, and raw whiskies. Water is less likely to freeze in a wooden tank than a metal one because of the excellent thermal-insulating characteristics of wood. This is one of the principal reasons why railroads used wooden boiler-water storage tanks along their rights of way during the era of the steam locomotive. Batteries of wooden extractor tanks and vats may be observed in most tanneries, while wooden dye vats are common in textile plants. In fact many industries have found the wooden tank and vat to render many years of satisfactory service at

Fig. 11-11. Battery of redwood tanks used for storage of corrosive sour-oil crudes. (*Courtesy California Redwood Association, San Francisco, California.*)

very nominal cost. One winery reports that it has several wine tanks of redwood that have been in continuous use for more than 90 years and are still in a state of excellent repair.

II. BOXES AND CRATES

The volume of wood consumed annually in the fabrication of boxes and crates is exceeded only by that used in general construction, and for each of the past several years 15 per cent of the lumber cut has been needed to meet the agricultural and industrial demands for box shook * and crating stock. In addition to the nearly 700 facilities currently producing such items from lumber, about 30 per cent of all container-grade veneer manufactured eventually reaches the market in the form of sheathing for crates, paper-faced boxboards, container plywood, and wire-bound boxes.

* Term applied in the trade to the components of a knocked-down box.

A conventional box is generally conceived to be a container with six lumber faces held firmly together with nails or other metal fastenings. A crate, in contrast, consists of an open framework of boards or dimension lumber (sometimes both) so constructed that framing members serve as the structural load-carrying components of the assembly. Nevertheless, the distinction among some forms of boxes and crates is not always readily apparent. For example, lightweight structures with solid ends and open-slatted sides, tops, and bottoms, used for packaging produce and fruits from farms and orchards, respectively, are designated crates. Yet somewhat similar open containers, the components of which are held together with staples and wire, are known in the trade as wire-bound boxes. And again, crates used for heavy goods such as machinery and the larger household appliances are often sheathed with veneer, plywood, or thin boards, and in this form are exceedingly box-like in appearance. A large box used for packaging such items, often referred to as a packing case, would be composed of solid sides and ends of lumber fastened together without benefit of a primary structural framework.

Battens and cleats are often nailed to the ends and in the corners for reinforcement, and cleats are sometimes attached to the bottoms of boxes to facilitate mechanical handling. Internal blocks and braces are employed whenever it becomes necessary to keep objects from tilting or shifting while in transit, but these inserts are not regarded as integral components of a box.

Considerable time and money have been expended during the past decade or so to improve upon the structural characteristics of boxes and crates. In fact, during the early stages of the Second World War the amount of damage to, and the spoilage of, war material in transit and at supply bases in the combat areas reached appalling proportions. These losses, many of which were traceable to inefficiently designed and poorly constructed containers, led the military establishment and several civilian agencies to institute crash programs of research aimed at developing better box-and-crate designs and improving packaging practices. It was also during this period that a number of lightweight but exceedingly sturdy wooden containers were created for use in the airborne shipment of goods.

A. Species Used

Practically all but the heaviest and most valuable American woods are used in the fabrication of boxes and crates. A preference is generally shown for cheap, readily available lumber, known for its good machining characteristics, high strength-weight ratio, resistance to splitting, and ability to hold nails. Woods of the paler hues are also desired, since better color contrasts are obtainable with the inks and other marking materials used to indicate shipper, consignee, order number, and other information required by contract or law.

Members of the Division of Packaging Research at the U.S. Forest Products Laboratory in Madison, Wisconsin, have divided the commercially important native wood into four groups, using as a basis those properties deemed important in box construction (see Table 11-1).

Table 11-1. Various Commercially Important Native Woods and Their Properties Rating for Nailed and Lock-corner Wood Box Construction [a]

Species	Comparative bending strength [b]	Comparative shock resistance [b]	Nail withdrawal resistance [c] End, lb	Edge (radial), lb	Flat (tangential), lb	Tendency to split when nailed	Dry weight (15% M.C.), lb/cu ft	Ease of drying	Ease of working
Group I:									
Aspen	64	65	137	194	204	L	27	M	M
Basswood	61	53	138 [d]	199 [d]	194 [d]	L [d]	26	M [d]	Easy [d]
Buckeye	58	53	[d]	[d]	202	[d]	26	M	Easy
Cedar	68	69	118	192	273	L	25	M	Easy [d]
Chestnut	67	69	172	258	196	L	30	M	Easy
Cottonwood	60	65	132	191	291	L [d]	27	M [d]	M
Cypress	83	76	144	266	200	L	33	Easy	M
Fir (true firs)	72	62	87 [d]	176 [d]	[d]	L [d]	27	Easy	M
Magnolia	78	81	[d]		284	L	36	D	M
Pine (except southern yellow)	74	74	168	270	226	[d]	29	Easy	Easy
Redwood	82	66	106	221	207 [d]	[d]	29	D	M
Spruce	68	66	143 [d]	205	223	L	28	Easy	Easy [d]
Willow	61	102	[d]			L	28	D	Easy
Yellow-poplar	76	76	162	212		L	30	M	Easy
Group II:									
Douglas-fir	82	75	183	273	296	M	32	Easy	D
Hemlock	76	70	138	245	253	M	30	M	M

Species									
Southern yellow pine......	100	102	237	330	376	M	39	Easy	D
Tamarack..............	84	83	d	d	d	M	38	M	M
Western larch..........	97	107	180	299	319	M	39	D	D
Group III:									
Ash (except white)......	83	112	d	d	d	M	39	D	D
Soft elm..............	88	142	236	344	339	L	37	D	D
Soft maple............	80	101	280	333	338	M	35	D	D
Sweetgum.............	85	106	192	292	278	L	36	M–D	M
Sycamore.............	74	79	d	d	d	L	36	D	M
Tupelo................	82	78	233	376	345	L	36	d	M
Group IV:									
Beech................	102	134	358	495	460	H	44	M	D
Birch................	89	152	331	473	451	H	42	M	D
Hackberry............	76	145	d	d	d	M	37	M	D
Hard maple...........	103	134	376	488	437	H	41	D	D
Hickory..............	134	283	d	d	d	H	50	D	D
Oak.................	100	130	316	418	433	L	45	D	D
Pecan................	109	157	d	d	d	H	53	d	D
Rock elm.............	106	190	d	d	d	d	44	D	d
White ash............	109	136	385	455	452	M	43	M	D

LEGEND: L = low; M = moderate; D = difficult; H = high.

[a] Adapted from U.S. Department of Agriculture, Forest Service, Forest Products Laboratory. Nailed and Lock-corner Wood Boxes. Report No. 2129. 1958.

[b] In comparison with the average of all species listed.

[c] Average direct withdrawal resistance for 1 sevenpenny cement-coated nail driven to a depth of 1¼ in. and pulled at once.

[d] Complete data not available and thus no value given.

Group I consists of lightweight woods particularly noted for their resistance to splitting and the facility with which they may be seasoned and machined. All of them possess moderate strength and satisfactory nail-holding capacity.

The woods in group II are harder, heavier, and stronger, and exhibit better nail-holding capacity than those in the first group. On the other hand, they are somewhat more difficult to machine and season, and are much more prone to splitting.

Group III woods show about the same physical and mechanical properties as those from group II, but are much less inclined to split.

Those in the last group (IV) exhibit superior strength and nail-holding capacity, but they are the most difficult to season and machine, and exhibit the greatest tendency to split in nailing.

B. Classes of Wood Boxes

There are currently three general classes of wood boxes available to consumers, viz.: (1) the nailed wood box, (2) the cleated plywood box, and (3) the wire-bound box.

1. Nailed Wood Boxes. Boxes in this category are available in hundreds of sizes and capacities, each one of which is an adaptation of one of eight standard styles. They are ordinarily appreciably heavier, size for size, than either the cleated plywood or wire-bound boxes and generally offer considerably more resistance to puncturing, crushing, and corner smashing. Formulas are available for determining the thicknesses of their component parts when their over-all dimensions have been determined and the anticipated weight of their contents is known.

a. Style 1 Box. This box (Fig. 11-12) is composed of plain sides, top, bottom, and ends, all without benefit of cleats or other reinforcement. It is structurally the weakest of the eight standard box styles, yet it is a wholly satisfactory shipping container for many kinds of merchandise. It is exceedingly easy to construct and requires less storage space than other boxes of similar dimensions owing to the absence of cleats. Its principal limitations are that (1) the sides, top, and bottom cannot be stagger-nailed to the ends; (2) nailing the sides into the end grain of the ends is unavoidable; and (3) end-splitting is always a threat because the grain of the sides and ends is essentially parallel.

b. Style 2 Box. This box differs from the one described above in that four cleats are nailed to each of the two ends. The vertical pair are called *through cleats* (Fig. 11-12), the horizontal pair are *fillers*. The use of end cleats permits stagger-nailing of the top, bottom, and sides to the ends. Side-grain nailing can be thus effectively increased, with better joint strength resulting. The cleats also reinforce the ends and provide a recess for shipping tickets as well as handholds for lifting. Style 2 boxes when properly assembled can be used for merchandise weighing from 800 to 1,000 lb.

c. Style 2½ Box. This box is similar to the Style 2 container except that the filler cleats are seated in recesses cut into the through cleats (Fig. 11-12).

This design is commonly used with group IV woods since the through cleats carry a portion of the load when the top is attached.

d. Style 3 Box. In this box the filler and through cleats are miter-jointed (Fig. 11-12). Both Style 2½ and Style 3 boxes have the same advantages as the Style 2 box and are commonly used for the same purposes.

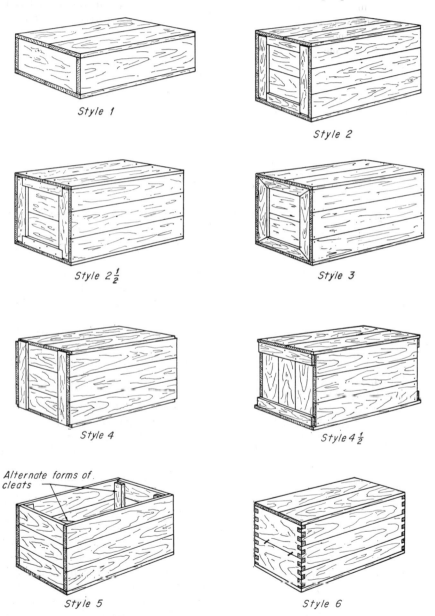

Style 1

Style 2

Style 2½

Style 3

Style 4

Style 4½

Alternate forms of cleats

Style 5

Style 6

Fig. 11-12. Styles of wood boxes. (*Courtesy U.S. Forest Products Laboratory.*)

e. Style 4 Box. Only two through cleats are used on the ends of this box (Fig. 11-12). In attaching the cleats, nails are used which are longer than the combined thickness of end board and cleat, so that they may be clinched on the inside after they have been driven. This box is of somewhat similar construction to the four previously described but should never be used for gross weights in excess of 500 lb. The two cleats provide protection against end-splitting of the end boards, and when the structure is placed on its side, the cleats may be used for handholds.

f. Style 4½ Box. This is a relatively new member of the box family and differs from a Style 4 box only in that its ends are constructed with vertically aligned boards (Fig. 11-12) reinforced with a pair of clinch-nailed horizontal cleats.

g. Style 5 Box. In the assembly of this box a cleat is nailed at each of the inside corners (Fig. 11-12). In some instances cleats of triangular cross section are used. Because these structures are placed on the inside, there is less likelihood of their being knocked off in handling. Like the Style 1 box, the Style 5 box requires less storage space but offers no handholds and is often difficult to handle.

Fig. 11-13. Diagrammatic sketch of cleated plywood box.

h. Style 6 Box. Boxes in this class are similar to those of Style 1 except that their sides and ends are made with locked corners (Fig. 11-12). Bottoms and tops are nailed to the sides in the usual manner.

2. Cleated Plywood Boxes. Figure 11-13 is a diagrammatic sketch of a typical cleated plywood box. Note that the ends, top, bottom, and sides are made of one-piece plywood panels reinforced with through and filler cleats and that the package, when sealed for shipment, is bound with stapled steel straps. Boxes of this sort are light in weight and offer excellent resistance to mashing at the corners, diagonal distortion, and puncturing. They are essentially dustproof, easily handled, and difficult to pilfer while in transit. While they can be made to hold loads up to 1,000 lb, they are ordinarily of much smaller stature and are especially well suited for airborne shipments. For maximum joint strength three-way corner construction (see page 230) is recommended.

3. Wire-bound Wood Boxes. Boxes of this sort are lightweight, prefabricated shipping containers made occasionally with thin boards, but much more commonly with sheets of container-grade veneer. Mortised cleats are attached to the edges of the sides, top, and bottom in a manner shown in Fig. 11-14, after which they are assembled into four-piece mats using two or more strands of binding wire. When the mats are squared, the cleats lock into position, and the endpieces, which are usually reinforced with one to four battens, are inserted and nailed to the inner faces of the end cleats. To close or seal such a box, it is merely necessary to bring the two ends of each

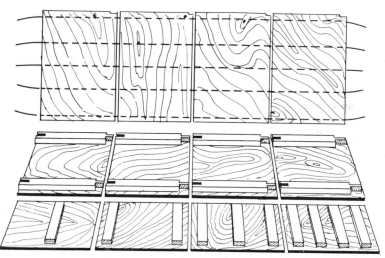

Fig. 11-14. Diagrammatic sketch of knocked-down, wire-bound box. Note four styles of end panels available.

strand of wire together and twist. Some manufacturers use binding wires which have loops fashioned at either end. In such a case the box is sealed by passing one loop through the other and then bending one of them.

Although light in weight, this is an unusually sturdy container and is being used in ever-increasing numbers for shipment of farm produce and for dressed meat and poultry.

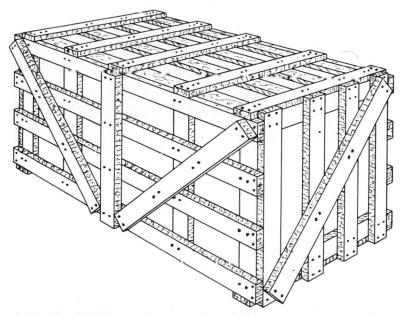

Fig. 11-15. Open, three-way-corner, diagonally braced crate.

C. Wood Crates

There are no general standardized designs for crates such as there are for boxes, although several styles have been standardized for packaging items by individual shippers and by fruit packers' associations and cooperatives. The open, three-way-corner, diagonally braced crate, an example of which is shown in Fig. 11-15, is commonly used for shipping stoves, plumbing fixtures, bicycles, plate glass, and a host of other moderately heavy, durable goods. The three-way corner eliminates end-grain nailing and is recommended whenever maximum joint strength is required, since end-grain nailing is only about half as efficient as side-grain nailing. Box-type crates are used to provide maximum protection of goods against weathering, salt-water spray, dust, and rough handling while in transit or storage. Aircraft parts including engines, machinery, piano, and other heavy, often expensive, items are commonly shipped in such containers. The sheathing may consist of single sheets of veneer, plywood, or lumber. A typical lumber-sheathed crate is illustrated in Fig. 11-16.

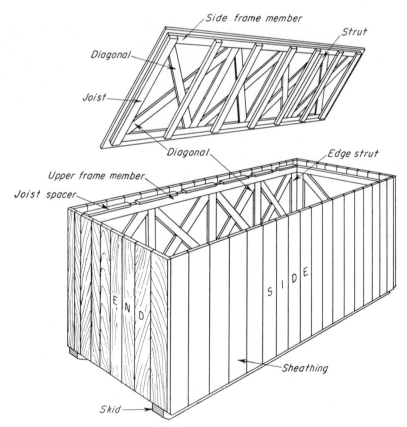

Fig. 11-16. Lumber-sheathed crate. (*Courtesy U.S. Forest Service.*)

D. Nails Used in Box-and-crate Construction

Smooth, bright, round-shanked, iron-wire nails of the box, clout, common, cooler, and sinker types are those most commonly employed in the assembly of boxes and crates. Spirally grooved and annular-grooved nails are also occasionally employed because of their superior holding characteristics; however, their use for such purposes is greatly restricted because of their higher costs.

The holding power of round-shanked nails of any of the types enumerated above may be materially improved by coating them with a resin or dipping them in a slurry of portland cement. When thoroughly dry, such coatings add to the friction between wood and nail, with the result that nail-withdrawal resistance may be increased by as much as 50 per cent. Even better tenacity is claimed for such nails when they have been etched for 7 hours in a 10 per cent aqueous solution of monoammonium sulfate.

Nails made of galvanized iron, cadmium-plated brass, copper, aluminum, and stainless steel show no appreciably better holding capacity than those made of iron, but they are recommended for use when rust and corrosion must be avoided at all costs.

Box and common nails are used most frequently for boxes and crates fabricated with lumber. Both types are of identical length in any nail class; both feature diamond points; but box nails are of smaller diameter, class for class. When clinching is practiced, nothing is gained by coating or etching either type prior to use.

Clout nails are largely restricted in use to fastening cleats and battens on veneer and plywood faces of wire-bound and plywood boxes. The basal portions of these nails are flattened much like a duck's bill and are especially well suited for machine driving and clinching. It should be added, however, that machines are in general use for driving nails of any sort, but particularly when the container is of a standard design and assembled in large numbers.

Blunt-pointed nails are recommended when the denser hardwoods are used, since they are much less apt to split boards than those with sharp points. For an excellent comprehensive treatise dealing with nails, nailing practices, and nailing patterns, the reader is referred to U.S. Department of Agriculture, Handbook No. 160, "Nailing Better Wood Boxes and Crates."

E. The Place of Wood Boxes and Crates in Agriculture

Boxes and crates of many styles have become nearly universal shipping containers for a variety of products from farm and orchard. In some instances, they have completely replaced the slack barrel, while in a few others, they themselves have been replaced by sacks, baskets, hampers, and fiber cartons.

More than 70 million two-compartment nailed boxes are required annually to package the nation's orange crop. Another 12 million are needed for other citrus fruits.

The apple producers of the Pacific Northwest were the first to replace the slack barrel with boxes in the shipment of apples. The box, now standardized at 2,173.5 cu in., has since been adopted by producers in several apple-growing centers in the United States. A pear box similar to the standard apple box is used by Washington, Oregon, and California growers.

Cherries, peaches, plums, prunes, grapes, and other soft fruits are packaged in shallow wood containers known in the trade as lugs. Open-slatted crates are preferred by many growers producing lettuce, cauliflower, melons, celery, and tomatoes.

Cheese, butter, and egg producers are other large consumers of agricultural boxes and crates.

F. Industrial Boxes and Crates

Approximately 60 per cent of all wood boxes and crates are used for shipping the products of American industry. Such containers range from large, heavy, box-like crates used to package castings, forgings, and machinery often weighing several tons, to small, compact yet sturdy cases required for delicate and sensitive scientific instruments, which not infrequently weigh but a few ounces. Containers such as these are seldom standardized, but rather are assembled in accordance with customer specifications or with railway and Interstate Commerce Commission regulations.

Wood boxes constructed in accordance with specifications acceptable to the Interstate Commerce Commission are used in large numbers for transporting explosives and dangerous chemicals. Wood boxes and lumber-sheathed crates, because they afford maximum protection in transit against pilferage, contamination, breakage, spoilage, or corrosion of their contents, are used in vast numbers for shipment of goods overseas. Millions of board feet of lumber are required annually for crating farm and industrial machinery, aircraft and automotive parts, and heavy electrical goods such as motors, generators, transformers, and massive insulators.

The American fishing industries ice-pack much of their catch in wood boxes for rapid delivery to inland consumers. Some meat packers are again using boxes for bulk shipments of pork and lard, while several canneries that had discarded the box in favor of fiberboard shipping containers have recently adopted some of the newer lightweight shipping boxes because of the added protection afforded their merchandise while in transit.

Large volumes of low-grade lumber are consumed each year in the manufacture of a number of lightweight but sturdy containers that collectively are known as returnable boxes. In this category are such familiar items as the bottled-milk crate, carrying and shipping cases for bottled carbonated and malt beverages, boxes for bottled mineral water, banana and bread boxes. Some of these containers are doing yeoman service, and records at one brewery reveal that many of their cases are still serviceable after more than 1,200 round trips.

G. Status of the Industry

Nearly 15 billion board feet of lumber was required to fabricate the boxes and crates used in the peak year of the Second World War. But with the nation's return to a peacetime economy in 1946, consumption had dropped to about 4 billion board feet. During the past several years paperboard containers have captured some of the markets formerly held by the box and crate; yet modest increases in the use of the wooden containers in some areas have partially offset such losses. About 4.5 billion board feet of lumber is currently required to meet the needs for boxes and crates, and it is anticipated that the demand will increase to about 5.3 billion board feet by 1975 despite rapidly increasing competition from the paperboard items.

The continuing popularity of the wood box and crate is due principally to the activities of the National Wood Box Association and its regional affiliates. All these agencies, working closely with state and Federal agencies, have rendered many invaluable services to shippers, carriers, and consumers of goods by coordinating packaging specifications, improving designs, and promoting their general use through educational and trade-promotion programs.

Increased competition from the paperboard containers has also been noted by the manufacturers of wire-bound and plywood boxes, baskets, and hampers. Nevertheless, and in spite of further market losses in some areas, it is expected that the consumption of container-grade veneer will rise from the current 3 billion square feet to 4.7 billion square feet by 1975.

III. VENEER BASKETS

The use of veneers in basket manufacture had its beginning in 1866 when James Kirby of Benton Harbor, Michigan, made a $\frac{1}{3}$-bushel basket for the peach growers of southern Michigan. About this same time A. W. Wells & Company of St. Joseph, Michigan, also introduced a basket made of sliced veneers for the use of the orchardists in his area. Both these commodities were immediate successes; others were quickly placed on the market; and all of them became serious competitors of the hand-woven splint baskets then in common use. There are currently many kinds and capacities of standardized veneer baskets and hampers available to both farmer and orchardist, as the veneer-basket industry now has plants operating near every major fruit- and vegetable-producing area in the United States.

In the early days veneer baskets were fashioned by hand, and their production in most plants was limited to a few hundred daily. The inventive genius of several basket manufacturers, however, soon led to the development of numerous mechanical devices that have long since replaced the slow, laborious hand operations. Modern basket factories are highly mechanized, with the largest producing more than 2,000 dozen finished items daily.

A. Species Used

The bulk of all veneer used in the manufacture of baskets is produced by rotary cutting. A preference is shown for those woods that are flexible yet tough, moderately heavy, odorless, light in color, and able to retain nails, brads, staples, and wire stitching. Species used in quantity include beech, American elm, the hard and soft maples, black and yellow birches, sweet-gum, tupelo-gums, basswood, yellow-poplar, northern black cottonwood, and some of the ashes. Quantities of eastern spruce and white pine are also used in some areas.

B. Classes of Veneer Baskets

Six classes or kinds of veneer baskets are currently available to consumers, viz.: (1) berry boxes, (2) till baskets, (3) hampers, (4) round-stave baskets, (5) splint or market baskets, and (6) Climax or grape baskets.

1. Berry Boxes. These familiar containers, used in distributing and marketing strawberries, currants, raspberries, blueberries, and similar fruits, are available in several styles. The American box, which is used in greater quantities than any of the others, is easily identified by two scored and folded sheets of veneer that make up the four sides and double bottom stapled to a reinforcing spline-like rim. The Hallock box features a raised bottom and is most commonly used in the Pacific Northwest and in the Lake states. A stitched tray has found much favor with fruit producers in Missouri and Idaho. The metal-rim box, widely used throughout the United States, is second only to the American box, which it closely resembles (Fig. 11-17). In recent years a berry box has also appeared on the market made of paperboard. All berry boxes are manufactured in dry-measure capacities of ½ pt, 1 pt, and 1 qt.

2. Till Baskets. Till baskets are square or rectangular containers in the form of an inverted frustum of a pyramid. They are used principally in marketing soft fruits such as apricots, peaches, plums, and prunes. In their fabrication two predimensioned sheets are fed into a machine which, in one continuous operation, scores and folds them over an anvil to form the basket and then finishes them off by attaching a wood or metal rim. Containers such as these are made in four capacities ranging from 1 to 4 qt dry measure.

3. Hampers. Hampers are flat-bottomed baskets in the form of an inverted frustum of a cone. They are fashioned with solid wooden one- or two-piece bottoms and tapered veneer staves securely stapled to two to four wood hoops (splines of veneer), the exact number depending upon their size. There are currently nine standard sizes of hampers with a capacity range of ⅛, ¼, ½, ⅝, ¾, 1, 1¼, 1½, to 2 bu.

4. Round-stave Baskets. As the name implies, these baskets are made with precut staves of veneer and are supplied in two distinct styles. The first of these, the *round-bottom basket*, features continuous staves which form both the sides and bottom of the container. In the other, the *tub basket*, the structure is assembled with short, straight staves attached to a solid lumber or plywood bottom.

Both types are fitted with wood hoops, a pair of wire handles, and, when specified, a top (Fig. 11-17). Tub baskets carry three hoops; round-bottom baskets two or three. When only top and middle hoops are used, the bottoms of these baskets are definitely rounded, but if a third hoop (basal) is attached, the sides of the basket are straighter and the bottom appreciably flatter. The three-hoop structure is commonly referred to as a bent-bottom basket in some localities. Round-stave baskets are made in the same nine sizes as listed above for hampers.

5. Splint, or Market, Baskets. These are rectilinear structures woven with veneer splints and furnished with a rigid handle. Originally made with

Fig. 11-17. Types of veneer baskets and boxes: *top row,* tub baskets; *second row,* berry boxes; *bottom row,* Climax boxes. (*Courtesy U.S. Forest Service.*)

a diamond weave, the market basket now comes in a square-braid weave and is available in six standard sizes of 4-, 8-, 12-, 16-, 24-, and 32-qt capacity, dry measure. Its use is largely restricted to the agricultural areas of the central United States.

6. Climax, or Grape, Baskets. Climax baskets superficially resemble oblong till baskets but are readily distinguished by their rounded ends, continuous walls, solid wood bottoms, and wire or wood handles. By Federal statute, the 1-qt Climax is restricted in use to the packaging of mushrooms. The 2-, 4-, and 12-qt baskets are used for grapes and a few other fruits and vegetables (Fig. 11-17).

C. Basket Standardization

Prior to the Standard Container Acts of 1916 and 1928, respectively, baskets were manufactured in many styles, shapes, and alleged capacities. These Acts not only established the capacity for each container but also reduced the number of containers in each of the several categories. Thus, while there were 44 berry boxes of various styles and capacities on the market prior to the enactment of these Acts, today such boxes are limited to three capacities (see page 235). Similarly, till baskets were reduced from 44 to 4; round-stave baskets from 36 to 9; hampers from 75 to 9; and Climax baskets from 30 to 4. The Bureau of Agriculture Economics has been responsible for the administration of these Acts and also for the elimination of many odd-capacity containers, several of which were so fabricated that they create a false illusion with respect to their actual size and capacity.

Basket standardization has resulted in numerous benefits to all but the unscrupulous. The consumer is assured of a full measure of the commodity he purchases. The producer of the commodity pays less for the containers when they can be manufactured by mass-production methods. Finally, the problems of the manufacturer are simplified to the extent that he can standardize his equipment for the production of a comparatively few items and concentrate his efforts upon the production of those items he is required to stock in quantity.

IV. PALLETS

A pallet is a low, sturdy platform on which materials in process of manufacture, or finished goods, may be stacked in order to expedite their handling, movement, and storage with the use of mechanical fork lifts and/or hand trucks. While pallets are not containers in the ordinary sense, a multitude of them serve as carriers in the shipment of unitized loads of goods and hence are included within the scope of the subject matter of this chapter.

Pallets first came into use in the early 1920s with the introduction of mechanical lifts. In 1938 the Navy began experimenting with them when it initiated its warehouse and port terminal expansion program, and it was the service-wide adoption of pallets that really triggered the revolution in the mechanical handling of goods on pallets. The military establishments were quick to recognize the many savings in time and dollars and initiated palletization programs which during the Second World War called for more than 90 million pallets. With the return of a peacetime economy many of the American industries turned to the use of pallets. A total of 23 million pallets were manufactured in 1950 to take care of industrial demands, and 5 years later production had risen to 43 million units. It has been reliably estimated that the demands for pallets will continue to increase at a nominal rate during the next few years, and that more than 51 million will be produced in the year 1975.

Several national organizations have made significant contributions to the improvement and standardization of wood pallets. Among these, the work of the National Wooden Pallet Manufacturers Association has been outstanding. Originally a division of the National Wood Box Manufacturers Association when it came into being in 1946, it is now an active, aggressive, independent trade association with a membership of more than 85 companies.

Among its many endeavors, this association has been able to demonstrate beyond all reasonable doubt that by installing carefully planned and properly executed palletization programs, manufacturers, warehouse and storage firms, carriers, receivers of semifinished and finished goods, and other businesses in which the movement of goods plays a major role can anticipate substantial savings in dollars. In fact, 16 major advantages are claimed for palletization programs. These are enumerated below:

1. Palletization materially increases the amount of goods that can be stored in a given area by utilizing the air space as well as the floor space.
2. The incidence of accidents to personnel decreases when mechanical lifts replace manual labor.
3. Palletization effects savings in handling costs by as much as 80 per cent; statistics reveal a national industry-wide average of from 40 to 45 per cent.
4. The rapidity with which goods may be handled is reflected in a saving of man-hours; overtime wage payments can be usually eliminated.
5. Palletized goods in storage or in transit can be readily ventilated if need be.
6. Palletization greatly simplifies inventory controls.
7. Damage to goods is minimized when they are mechanically handled in palletized shipments.
8. Unitized pallet loads of goods require only one or two labels; thus appreciable savings in materials and labor are effected.
9. Goods which are strapped or glued to pallets in unitized bonded loads are much less subject to pilferage.
10. Palletization permits uniform stacking and placement of goods and stocks, thus giving the assurance of unobstructed floors and aisles of uniform width and accessibility.
11. A pallet is a subfloor to which steel strapping can be attached with facility if and when required.
12. Pallets can often eliminate the need for costly conveyer systems, since loading and unloading can be done at any point accessible to the handling equipment.
13. Palletization practically eliminates production holdups and bottlenecks of goods in process.
14. The necessity for taking production workers off their jobs "to lend a hand," in the movement of goods is unnecessary in palletized operations.
15. Palletization not only cuts loading and unloading time of carriers by as much as 50 per cent but has also made it possible for one railway car or trailer truck to do the work of two or even three. Thus not only are demurrage charges much less frequent, but more efficient use of carrier equipment is possible, which in turn is reflected in increases in revenue to the carrier operators themselves.
16. Pallets also effect material savings in the loading and unloading of steamships. This permits a faster turn-around of the ships and a substantial saving in shipping costs.

There are three basic principles which must always be considered in the proper handling of all goods: (1) within practical limits, loads of goods to be handled should be as large as possible; (2) all goods should be handled a minimum number of times; (3) mechanical equipment should be used rather than expensive and relatively inefficient manual labor.

Palletization programs must be tailored to fit specific operations. Consideration must always be given to the four elements concerned with commodity movements: (1) fixed paths, (2) variable paths, (3) distance, and (4) frequency. Other factors must also be considered. For example, in operations concerned with the movement of lightweight goods in small quantities over short distances or from fixed origins to fixed destinations the use of pallets might well be questioned. On the other hand, it may well be stated as a corollary that the need for pallets and mechanical lifting devices increases in direct proportion to the increase in the amount of goods to be moved, and to the distance, frequency of such movement, and the variability of the paths which the goods must traverse.

A. Classes of Pallets

Pallets are grouped into three general classes: (1) *expendable,* (2) *general purpose,* and (3) *special purpose.* Before discussing these structures, however the reader is urged to examine Fig. 11-18 in order that he may familiarize himself with their component parts and the nomenclature used in their designation.

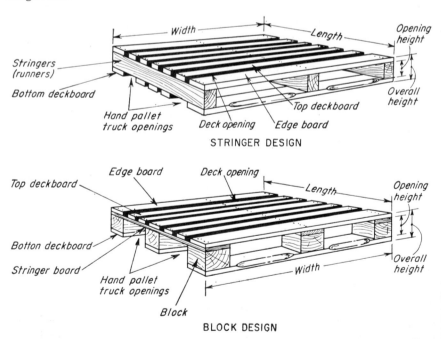

Fig. 11-18. Component parts and structural features of pallets. (*Courtesy National Wooden Pallet Manufacturers Association, Washington, D.C.*)

Expendable pallets are for the most part one-use, nonreturnable items used for shipping. Known also as *shipping pallets,* the details of their construction are largely governed by considerations of cost.

General-purpose pallets, by contrast, are reusable structures designed and constructed to give years of service under the repeated abuses of hard handling. They are most commonly used for the movement of materials and goods in process from one operation to the next, and for the warehousing and shipment of semifinished and finished products of American industry.

Special-purpose pallets, as the name implies, are those designed to handle a particular kind of load or to perform a special service. A few of them are covered by patents.

Pallets in each of these three classes are assembled using a number of standardized designs, styles, and constructions, each of which will now be described in turn.

B. Pallet Designs

There are currently two basic designs used in the fabrication of pallets. The first of these, the *two-way pallet* (Fig. 11-19a), permits the entry of fork lifts or hand trucks on only two opposite sides. The other, the *four-way pallet,* is so constructed that lifting equipment may enter from any one of its four sides. There are, however, two modifications for this design. In one, the *four-way, notched stringer* (Fig. 11-19b), provision is made for the entrance of fork lifts on all four sides, but hand trucks are permitted to enter on only two opposite sides. The other, the *four-way block design* (Fig. 11-19c), may be entered on all four sides by both types of lifting equipment. Yet another modification of the four-way block design permits diagonal entry of lifting equipment at each of the four corners. Sometimes designated as the *eight-way block pallet,* its use is largely restricted to warehouse activity.

C. Pallet Styling

The number and nature of the decks incorporated in the fabrication of a pallet dictate its style, two basic forms of which are recognized. Those constructed with a single deck are designated *single-faced pallets* (Fig. 11-19d). *Double-faced pallets* feature top and bottom decks and are furnished in two substyles. One of these is the *reversible pallet* (Fig. 11-19e), so-called because both decks are of identical construction. When the top deck is composed of more members than the one on the bottom, the pallet is *nonreversible* (Fig. 11-19f).

D. Pallet Construction

Three principal features of construction are also used to classify pallets. When the ends of the deck boards lie flush with the sides of the outer pair of stringers, or rows of blocks, the structure is known as a *flush-stringer pallet* (Figs. 11-19a–c, e, f). If, however, the outer stringers or rows of blocks lie inboard of the top deck boards but flush with those in the bottom deck, the

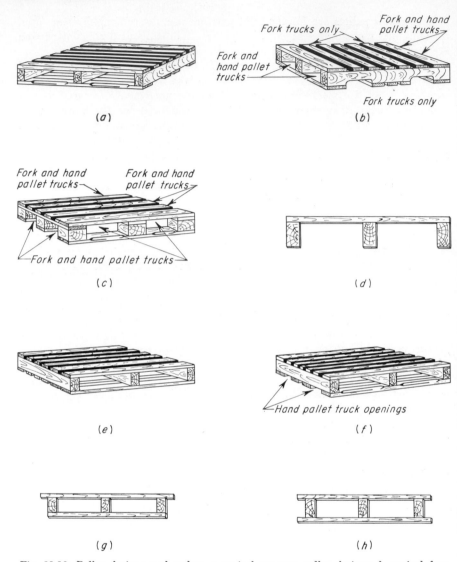

Fig. 11-19. Pallet designs and styles: *a*, typical two-way pallet design; *b*, typical four-way, notched stringer design; *c*, typical four-way block design; *d*, single-faced pallet; *e*, double-faced reversible pallet; *f*, double-faced nonreversible pallet; *g*, single-wing pallet; *h*, double-wing pallet. (*Courtesy National Wooden Pallet Manufacturers Association, Washington, D.C.*)

structure is termed a *single-wing pallet* (Fig. 11-19g). *Double-wing pallet* construction features stringers and blocks which lie inboard of both upper and lower deck boards (Fig. 11-19h).

E. Pallet Type Designations

For the convenience of both producer and consumer, the National Wooden Pallet Manufacturers Association has assigned a simple numerical designation to each of six types of pallets, in which various combinations of style and construction are embodied. These are summarized below. Each type, however, is available in any of the standard designs (see page 239), the choice being the customer's prerogative (Fig. 11-20).

Designation	Basic Features
Type 1	Single-faced, nonreversible
Type 2	Double-faced, flush-stringer or block, nonreversible
Type 3	Double-faced, flush-stringer or block, reversible
Type 4	Double-faced, single-wing, nonreversible
Type 5	Double-faced, double-wing, nonreversible
Type 6	Double-faced, double-wing, reversible

F. Expendable Pallets

As previously indicated, expendable pallets are usually one trip items. Cost and strength are two primary considerations in their fabrication. Three general forms are in common use. The *no-block pallet* (Fig. 11-21a), which resembles a panel of a crate, is a two-way structure. Because of the limited height of its openings, it is usually handled with a saber-bladed fork lift.

The two others are made with blocks rather than stringers in order to conserve both weight and timber. The Econ-O-Mee pallet is a reversible affair with three deck boards above and below. The Sturd-Dee pallet is a nonreversible structure with five top deck boards. The details of construction of these two expendables are shown in Fig. 11-21b and 11-21c, respectively.

G. General-purpose Pallets

Pallet types 1 to 6 are examples of general-purpose pallets. Since they have already been described (Fig. 11-20), further discussions are unwarranted.

H. Special-purpose Pallets

These pallets, unlike the general-purpose pallets, which are covered by specifications, may be of any style or design and may incorporate any construction regarded as necessary for satisfactory performance. Included in this category are reusable lightweight pallets used for such bulky commodities as cases of empty cans, electric-light bulbs, and packaged paper products; and heavy-duty pallets for such commodities as steel sheets, rolls

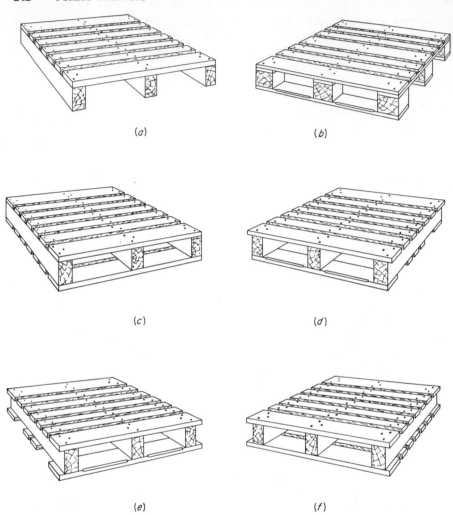

Fig. 11-20. Types of pallets: *a*, type 1 pallet; *b*, type 2 pallet; *c*, type 3 pallet; *d*, type 4 pallet; *e*, type 5 pallet; *f*, type 6 pallet. (*Courtesy National Wooden Pallet Manufacturers Association, Washington, D.C.*)

of newsprint, and reels of wire. A unique structure, the Take-It-Or-Leave-It pallet (Fig. 11-21*d*), permits fork lifts to enter above or below the load. It is commonly used in palletized warehouses when the stored commodities are destined for shipment without pallets. Others are designed specifically for handling drums and kegs.

Cargo cribs and crate-and-bin pallets are in reality cribs or boxes, crates, and bins mounted on pallets. They are available in demountable and rigid forms and are used for shipping and distributing machined parts, tires, coarse granulates, orchard products, and many other items of unusual shapes.

Three patented pallets are worthy of mention here. The *metal-cap* pallet (Fig. 11-21*e*) has the ends of its deck boards and outer pair of stringers bound with metal straps to prevent loosening and failure of the fastenings. The *lifetime* pallet (Fig. 11-21*f*) features tongue-and-groove joints between deck boards and outer stringers which are further reinforced with angle irons to prevent racking. The *clinch-tite* pallet is made with lighter holes bored in the stringer above the deck boards in order to clinch the nails within the stringer after they have been driven (Fig. 11-21*g*).

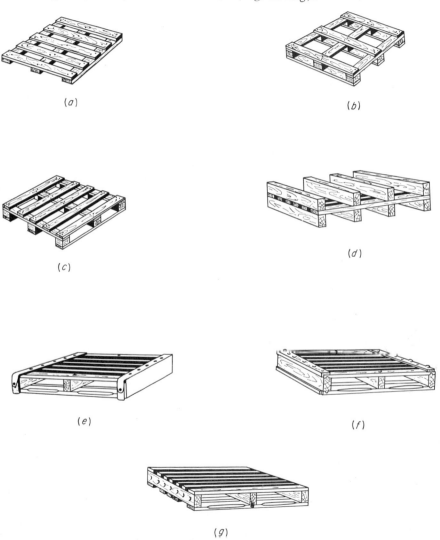

Fig. 11-21. Expendable and special-purpose pallets: *a,* no-block pallet; *b,* Econ-O-Mee three-deck-board pallet; *c,* Sturd-Dee five-deck-board pallet; *d,* Take-It-Or-Leave-It pallet; *e,* metal-cap pallet; *f,* lifetime pallet; *g,* clinch-tite pallet. (*Courtesy National Wooden Pallet Manufacturers Association, Washington, D.C.*)

I. Woods Used in the Fabrication of Pallets

Lumber used in the fabrication of pallets is required to be sound, free of decay, and free of all knots whose average diameter is greater than one-third the width of the piece in which they occur. Mineral streaks, stains, pinworm holes, and mild seasoning checks are permitted. Boards with wane that is not more than one-fourth the thickness and one-sixth the width of the piece on which it occurs are also acceptable. Each board and stringer, however, must be square-edged and free from any combination of defects that would adversely affect its strength and ability to hold fastenings.

Species of wood regarded as acceptable for the manufacture of boxes and crates are equally suitable for the fabrication of pallets. In fact, it is customary for pallet manufacturers to employ the same four species groupings outlined in Table 11-1 (see page 224), since those properties of importance in box-and-crate-construction are essentially the same as those for pallets.

J. Grades of Pallets

Two grades of pallets are recognized, each of which is based upon the moisture content of its deck boards. Grade AA pallets are those in which the moisture content of deck boards made from woods in groups I, II, and III does not exceed 20 per cent, or 25 per cent for group IV woods. Stringers and blocks of AA pallets may be in any stage of dryness.

Grade A pallets are those in which no restriction in moisture content is made for any of its component parts.

K. Structural Details

1. Deck Boards. All deck boards must be surfaced on one side and not less than $^{25}\!/_{32}$ in. in thickness. Both upper- and lower-edge deck boards must be at least $5\frac{1}{2}$ in. wide; those inserted between them may range from a minimum of $3\frac{1}{2}$ in. to a maximum of $8\frac{1}{2}$ in. in width. It is a common practice to provide the outer deck boards of the lower deck with a chamfer (Fig. 11-18). This should be at least 12 in. long, cut on an angle of 35° with the face, and not less than $\frac{1}{4}$ in. from the lower edge of the board.

2. Stringers and Blocks. Two stringers or six blocks are required on all pallets with deck boards that do not exceed a length of 24 in. Three stringers or nine blocks are required for pallets carrying deck boards of from 25 to 48 in. in length. If loads are apt to exceed 2,000 lb on pallets over 48 in. long, special-purpose structures should be considered.

The minimum cross-sectional dimensions of a stringer after surfacing have been set at $1\frac{3}{4}$ by $3\frac{3}{4}$ in., whereas the width and length of blocks usually conform to the width of the outside deck boards and stringer boards, respectively. Stringer boards are used on block pallets and usually conform to the same thicknesses of deck boards.

3. Fastenings. An average pallet will be composed of 36 joints, and it has been reliably estimated that 80 per cent of all pallet failures can be traced to improperly used or applied fastenings. A number of different types of fastenings are used for wood pallets; the most common are (1) nuts and bolts, (2) drive screws (helically threaded nails), (3) annular ring nails, and (4) common nails.

Common wire nails are occasionally used on expendable pallets where clinching is possible, but they are totally unsuited for warehouse and returnable pallets, where almost constant racking and torsion forces cause them to loosen and back out, thus destroying the effectiveness of the joint.

Bolted construction is superior to all other forms; it is also the most expensive. Carriage and car bolts have been found to be equally satisfactory in most instances, but for enduring and permanent joints Tenuts or their equivalent are recommended.

Drive screws provide the second most satisfactory type of joint. No. 6, flathead, $2\frac{1}{2}$-in., diamond-pointed screws with a minimum of four flutes are recommended for attaching $2\frac{5}{32}$-in. deck boards to stringers and blocks. An alternate fastener, the annular ring nail, should be of No. 10 wire gauge, $2\frac{1}{2}$ in. in length, diamond-pointed, and carry 20 annular ring threads to the inch.

L. Preservatives for Wood Pallets

The use of certain wood preservatives is recommended for pallets destined for prolonged inside storage as well as for out-of-door and cold-storage uses. Preservatives not only provide protection against the ravages of decay, attacks by powder-post beetles and termites, and discolorations occasioned by mold and blue-stain fungi, but they also have a stabilizing influence on the wood and thus reduce the number and severity of defects traceable to changes in moisture content such as warping, checking, splitting, and nail back-outs.

Three types of oil-borne, water-repellent preservatives are recommended by the N.W.P.M.A. for pallet treatment: (1) copper naphthenate (0.5 per cent metal), (2) pentachlorophenol (5 per cent solution), (3) zinc sulfonate (5 per cent solution). All three of these are equally acceptable for all industrial applications. Only those pallets treated with zinc sulfonate are permitted to be used in handling foodstuffs.

M. Gluing and Strapping of Bonded Unit Loads

1. Gluing. Unitized pallet loads are sometimes bonded together with glue or glued paper or cloth strips to provide maximum protection while in storage and transit. In the bonding operation it is also desirable on occasion to glue the bottom layer of the load to the pallet itself. Glues should be avoided, however, when their use might damage the palletized material, or when surfaces are uneven thus providing poor bonding areas.

2. Strapping. Steel straps fastened to the pallet members are also used to stabilize unitized pallet loads. These are used in several ways depending on the nature of the unit packages and the manner in which they are stacked. Over-the-load straps are used when vertical security of the bonded items is required; horizontal straps secure individual layers of items. Such straps may be round or flat; the size is governed by the gross weight of the load or by the weight of each of its component layers.

N. Pallet Maintenance

Properly designed returnable pallets can be expected to provide many trips of trouble-free service under normal conditions of use. By exercising a few precautionary measures extensive damage and repair can be largely avoided. A few common-sense rules in the use and handling of pallets and pallet-lifting equipment should be strictly enforced. Early repairs of minor damage are much less expensive than major reworks. Regular inspection and repair go hand in hand, and when properly exercised will pay good dividends.

SELECTED REFERENCES

1. Anderson, L. O. Nailing Better Wood Boxes and Crates. U.S. Department of Agriculture, Handbook No. 160. 1959.
2. Anon. American Veneer Packages. Lumber Buyer's Publishing Co., Chicago, Ill. 1941.
3. Anon. Minimum Standard Specifications for Warehouse, Permanent or Returnable Wooden Pallets. National Wooden Pallet Manufacturers Association, Washington, D.C. 1954.
4. Anon. Pallets and Palletization. National Wooden Pallet Manufacturers Association, Washington, D.C. 1954.
5. Anon. The Wooden Barrel Manual. The Associated Cooperage Industries of America, Inc., St. Louis, Mo. Revised, 1951.
6. Blew, J. O., Jr. Preservatives for Wood Pallets. U.S. Department of Agriculture, Forest Service, Forest Products Laboratory Report No. 2166. 1959.
7. Carey, L. C. Containers for Fruits and Vegetables. *U.S. Dept. Agr. Farmers' Bull.* 1821. 1939.
8. Coyne, F. E. The Development of the Cooperage Industry in the United States, 1620–1940. Lumber Buyers' Publishing Co., Chicago, Ill. 1940.
9. Hankerson, F. P. The Cooperage Handbook. Chemical Publishing Company, Inc., New York. 1947.
10. Heebink, T. B., and E. W. Fobes. Hardwood Pallet Manufacturing. U.S. Department of Agriculture, Forest Service, Forest Products Laboratory Report No. 2132. 1958.
11. Johnson, R. P. A. Wood Tanks. U.S. Department of Agriculture, Forest Service, Forest Products Laboratory Report No. R-1285. 1945.
12. Sowder, A. M. Timber Requirements of the Cooperage Industry. Unnumbered report, The Forest Survey. U.S. Department of Agriculture, Forest Service. 1942.
13. Stanford Research Institute. America's Demand for Wood, 1929–1975. Palo Alto, Calif. 1954.
14. Stern, E. G. Manufacture of Tight Plywood Cooperage. Proceedings, Forest Products Research Society, Vol. 1. 1947.
15. U.S. Department of Agriculture, Forest Service, Forest Products Laboratory. Nailed and Lock-corner Wood Boxes. Report No. 2129. 1958.
16. Wackerman, A. E. Douglas Fir Revolutionizes the Tobacco Hogshead. *Barrel, Box and Packages,* **47**(1):3–7. 1942.
17. Walters, C. S. The Illinois Container Industry. *Ill. Agr. Expt. Sta., Bull.* 534. 1948.
18. Warner, J. R., and D. R. Cowan. Wood Pallets. *Lake States Forest Expt. Sta., Sta. Paper* 43. 1956.

CHAPTER 12

WOOD COMPOSITION BOARD

From the viewpoint of total utilization of the forest resource, those industries that are relatively indiscriminate in terms of the properties of the wood raw material they use are of increasing importance. High-quality lumber and plywood are prized for the figure, grain, color, and unique mechanical properties that are inherent in the structure of the wood as it occurs in the living tree. As the wood industry uses an increasing volume of small, second-growth material, the proportion of the total volume of forest resource that is adaptable to these uses becomes smaller. If the costs of these materials are to be kept at a reasonable level, it becomes more and more necessary to develop those wood-utilization processes that can use wood in a variety of forms and in large quantity, and in which grain, figure, texture, and color are relatively unimportant.

Fig. 12-1. Pulpwood in a concentration yard awaiting debarking. Slabs and edgings appear in foreground and roundwood in background. Much wood previously regarded as waste is now used in production of composition boards. (*Courtesy American Forest Products Industries, Inc.*)

247

In general the wood industries of this type have fallen into one of two categories. The first of these is the chemical industry, which uses wood as a chemical raw material. The second category includes those wood industries that reduce the wood to small pieces or particles and then reassemble them into a useful form. The largest of the first industrial group is the pulp and paper industry, which is described in Chap. 17. The second group, whose activities are based upon the reduction of wood to small components and its reassembly into special forms, is represented by the wood-composition-board industry.

Composition boards can be classified as fiberboard or particle board, depending upon the nature of the basic wood component. Fiberboards are produced from mechanical pulps, while particle boards are manufactured from small portions of whole wood in the form of splinters, chips, flakes, or shavings. Fiberboards are designated as insulation board or hardboard depending upon the density. Table 12-1 gives a comparison of the three types of composition board.

Table 12-1. Comparison of Types of Composition Boards

Board type	Process	Final board density	Basic wood element	Type of binder	Board thickness
Insulation board	Wet	Low: 10–26 lb per cu ft	Coarse, non-chemical fiber	Natural or none	$\frac{1}{2}$–1 in.
Hardboard	Wet Semiwet Dry	High: 50–80 lb per cu ft	Coarse, non-chemical fiber	Natural or adhesive	$\frac{1}{10}$–$\frac{1}{4}$ in.
Particle board	Dry	Medium: 25–50 lb per cu ft	Small piece of whole wood	Adhesive	$\frac{3}{8}$–$1\frac{1}{2}$ in.

The composition-board industry is one of the most rapidly expanding industrial complexes in the United States. The Stanford report [21] points out that "about 16 times as much hardboard was used in 1953 as in 1929 [and] insulating board has also shown a healthy 700 per cent increase during the same period."

The Technical Committee, Forest Products Research Society, Wood Composition Board Division,[22] reported that hardboard production increased from 957,080,000 sq ft in 1948 to 1,498,193,000 sq ft in 1956. Particle board, which was not made in the United States prior to the Second World War, was produced at a level of 570,000,000 sq ft in 1956 according to the same report.

The fiberboard and particle-board industries utilize the residues from other woodworking plants in large measure and accordingly provide opportunities to reduce the cost of other products and expand the development of completely integrated wood industries. The Technical Committee, Forest Products Research Society, Wood Composition Board Division, in its 1958 status report, indicated the principal sources of raw material for hardboard and particle-board plants as shown in Table 12-2.

Table 12-2. Source of Raw Materials for Composition-board Plants [a]

Principal raw material	Number of plants		
	Insulation board	Hard- board	Particle board
Pulpwood........................	13	11	9
Sawmill or plywood-mill residues....	4	8	21
Furniture or millwork manufacturing residues........................	..	..	22
Waste paper.....................	2		
Bagasse.........................	2		
Mixed or unknown raw material.....	2	2	11
Total plants reported...........	23	21	63

[a] SOURCE: Technical Committee, Forest Products Research Society, Wood Composition Board Division. 1958 Status of the Composition Board Industry. *Forest Prod. Jour.*, 9(2):53–56. 1959.

I. TYPES OF PROCESSES AND PRODUCTS

Composition boards, as was indicated, may be classified according to the type of basic particle into *fiberboards* and *particle boards*. Where the basic particle is essentially pulp, consisting of individual fibers or small clumps of fibers, the resulting product may be classed as a fiberboard. Where the basic particle is a chip, flake, or splinter consisting of small units of wood in its original form, the resulting product may be termed *particle board*. Within these two broad categories a wide variety of products are manufactured and sold.

A. Fiberboards

1. Wet Process. The present-day hardboard industry in the United States developed from a process invented by William H. Mason during the 1920s. The Masonite product was the prototype of the wet-process hardboard. In this process the wood raw material is chipped. The chips are then loaded into a high-pressure cylinder called a "gun." Each gun has a capacity of about 260 lb of chips. After the gun has been charged, it is subjected to a steam pressure of 400 psi for ¼ minute. High-pressure steam is then introduced into the chamber, and the pressure is increased to 1,000 psi and continued for from 1 to 1½ minutes. At the end of this steaming cycle, the pressure is suddenly released by means of a quick-opening, hydraulically operated valve. The resulting explosion produces a mass of wood fibers. According to Boehm and Harper [2] the explosion operation performs two functions: "First, under the influence of high steam pressure, moisture, and high temperature, the wood undergoes a hydrolytic reaction which breaks down the lignocellulose bond. . . . Secondly, the sudden release of the hydrolyzed chips to atmospheric pressure tears them apart to produce a characteristically brown, fluffy fiber." The fiber thus produced is refined, washed, and then felted into a mat on the wire of a Fourdrinier machine, where some of the water is mechanically removed. The felted mat is cut to length and then loaded into a multiplaten press on a screen. The board is pressed under carefully controlled conditions of temperature, moisture con-

Fig. 12-2. Mat formation in wet-process fiberboard operation. As fiber-water slurry flows on to moving wire, water is removed first by vacuum and then by roll-pressing. (*Courtesy Hardboard Association.*)

tent, and pressure to form a hardboard. The partially hydrolyzed lignin serves as a natural bonding agent, and the resulting product is a reconstituted wood product. The screen used in this wet process permits the escape of steam during the pressing operation. A board thus produced will have the impression of the screen on the back.

Another wet-board process used in this country and abroad utilizes the Asplund defibrator to convert the wood chips to pulp. This is a continuous process, as contrasted with the batch-type explosion process utilized in the production of Masonite hardboard. In the Asplund defibrator the chips are fed continuously from a hopper via a screw feed into a preheater, where they are heated in steam under pressures of from 125 to 175 psi. In the preheater, the lignin is partially hydrolyzed to facilitate fiberization and ultimate bonding in the board. From the preheater the hot chips are again moved via a screw feed into the space between two grinding disks. One disk is stationary and the other rotates. The chips are introduced into the tapered space separating the two disks and forced to the outside to contact the grinding surfaces. The chips are converted into fiber as they pass through the narrow clearance between the grinding surfaces of the fixed and rotating disks. Pulp produced in this manner may be converted into wet-formed hardboard in the same manner as described for the Masonite hardboard.

Still a third process for producing fiber from chips for the production of hardboard utilizes Bauer refiners. These machines are revolving-disk mills that reduce chips to fibers through attrition. As in the case of the Asplund defibrator the chips are heated in steam to accomplish partial hydrolysis

Fig. 12-3. Bauer refiner. (*Courtesy Hardboard Association.*)

before they are introduced into the refiner. Some Bauer refiners are single-disk mills where the chips are reduced to fiber as they pass between a revolving disk and a stationary plate. Double-disk mills accomplish the reduction to fiber as the chips pass between two disks revolving in opposite directions. This is a two-step process as contrasted with the one-step Asplund process. After the pulp is prepared, the felted mat may be formed on a Fourdrinier machine or on a cylinder board machine as a continuous wet process.

It is quite common to combine the two processes. The Asplund defibrator is used to produce the fiber and the Bauer mill to refine it.

Some wet-process boards are formed in a batch-forming process. Instead of using a Fourdrinier or a cylinder machine that forms a mat on a continuously moving screen, in the batch-forming process a measured quantity of pulp slurry is introduced into a deckle box. A suction applied to the bottom of the deckle box removes most of the water and forms the mat. The deckle box in this system has dimensions comparable to those required in the finished board. Mats formed in this way may be further reduced in moisture content through a precompression operation. The batch-forming process was developed by Chapman Manufacturing Company of Corvallis, Oregon.

2. Dry Process. The dry process may be distinguished from the wet process in that the pulp is conveyed in air in the former and in water in the latter. In the wet process the mat has a much higher moisture content when it enters the press than is the case in the dry process. Where the mat is formed from a water-dispersed pulp, it may be put through a continuous drier, similar to a Coe veneer drier, before pressing to reduce the moisture content.

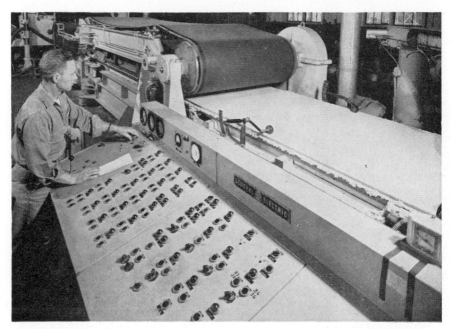

Fig. 12-4. Mat formation in dry-process fiberboard operation. (*Courtesy Hardboard Association.*)

Alternatively, the pulp may be dried before the board is formed and an air-felting operation used to prepare the mat. A dry-process board may be introduced into the multiplaten press between chrome-plated cauls without the use of the bottom screen, since the amount of moisture to be driven off in the pressing operation is much less than it is in a wet-process mat. Such boards have two smooth surfaces, rather than the screen-marked back typical of the wet-process board. Where the moisture content, temperature, and pressure are properly controlled, dry-process boards may be produced with a natural ligneous bond. In some dry-process boards, however, small quantities of synthetic-resin adhesive are introduced into the mix to serve as a binder.

Virtually all the new installations since the Second World War have been dry-process plants. These plants do not have the large water requirements of wet-process plants; hence, they can be built in locations where the wet process would not be possible. The dry process makes economically feasible installations of smaller capacity than would be practical with the wet process.

3. Treated Hardboard. Most hardboards are treated during manufacture to increase their moisture resistance and dimensional stability and to improve some mechanical properties. Synthetic resins added to the pulp furnish improve many of the moisture and strength properties of the board. The precise effect of these treatments upon the physical properties of the board depends upon the type and amount of resin used.

A process known as oil-tempering is widely used throughout the hardboard

Fig. 12 5. Precompression between heavy rubber belts reduces mat thickness to 3 in. (*Courtesy American Forest Products Industries, Inc.*)

industry to achieve the same purpose. This involves saturating the finished board in drying oils such as tung oil, tall oil, soybean oil, and linseed oil. The impregnated boards are then heated at about 300°F in a kiln for several hours. The tempering process must be carefully controlled to achieve complete polymerization. Failure to do this may result in a board with properties inferior to the untreated board.

B. Particle Boards

Particle boards differ from hardboards in that the particles are larger units that exhibit many of the characteristics of the original wood. The particles are produced by cutting or breaking wood into chips, slices, shavings, or splinters, rather than by reducing wood to fibers or clumps of fibers by breaking down the lignin bond between cells. Hardboards depend for their within-board cohesiveness upon the mechanical felting of the fibers, upon the formation under heat and pressure of a natural ligneous bond, augmented, sometimes, by the addition of a synthetic-resin adhesive. Particle boards, on the other hand, depend almost entirely upon an adhesive additive for within-board cohesiveness, since the particles are too coarse to be susceptible to benefits from fiber-felting. Particle boards, as commercially produced, cover a much broader spectrum of appearance and properties than do hardboards. Particle boards may be generally classified on the basis of particle geometry or manufacturing process employed.

1. Particle Geometry. Particle boards are produced using a wide variety of particle size and shape. In general, particles may be identified in two broad groupings: (1) those that are produced by the action of a cutting knife to form flakes with predetermined geometry; and (2) those that are formed primarily through a breaking or tearing action, sometimes combined with cutting, to form particles of indistinct geometry.

a. Flakes. Flakes are particles produced on special cutting equipment to predetermined dimensions. They are referred to in Germany as engineered particles, as contrasted with the relatively random dimensions of particles in other categories. The properties of a particle board depend to a very large extent upon particle geometry. Accordingly, these properties are quite controllable with flakes, but only subject to limited control where other particle types are used.

b. Shavings. Shavings are the residues from lumber-surfacing operations. These residues are commonly processed through a ball or hammer mill to break them up into small pieces that will interlock during mat formation.

c. Splinters. Splinters are small, rectangular particles of irregular size and shape. They are commonly produced by processing solid wood through a knife or hammer hog, or a chipper, and then further processing the hogged product through a hammer mill.

d. Layered Boards. Some types of particle-board processes permit the production of boards having different types of particles on the surface than in the center of the board. The best known of the three-layered boards is that produced by the patented Swiss Novopan process. This board has a splinter or milled-flake center with thin-flake surfaces on the face and back. The United States Plywood Company is the American licensee of this process and markets the boards under the trade name Novoply. The Behr process, also developed in Europe, yields a three-layered, all-flake board, with thick center and thin surface flakes. The Roddis Division, Weyerhaeuser Company, produces this board under license and markets it as Timblend. The Pack River Lumber Company also has developed a patented process for producing a three-layered board with thin surface flakes or "wafers" and heavier interior flakes. The Bähre Metallwerk, KG., Springe, Germany, has perfected a process for producing the equivalent of a three-layered board from a single-particle furnish. In this process the forming machine is, in effect, an air-separator which deposits the fine fractions on the top and bottom of the board and the coarse fractions in the center.

Experimentally, three-layered boards have been produced using coarse splinter particles as the bulk center-ply furnish and fine particles as top and bottom surfaces. This type of board has not been produced in the United States in commercial quantities to date.

2. Screening. Following particle production, the output is usually screened to remove fines and particles that are too large. The oversize particles are returned to the mill for reprocessing, and the fines normally are consigned to the fuel bin. Screening is necessary to ensure some uniformity in particle size. Fines are removed from the product mix for several reasons. They do not contribute to board strength and, if present in too large quantities, they reduce strength substantially. Fines demand a very large quantity of adhesive, greatly increasing the cost of the board. If fines are not removed, they may settle to one surface of a board, producing a product that is unbalanced and dimensionally unstable (see Fig. 17-12).

3. Drying. The flakes, shavings, and splinters that make up a particle board must be dry before the board is produced, since all particle boards are dry-formed. If dry waste is used as a raw material, the drying operation may be omitted. When green wood is utilized as a source of particles, a drying operation is required. The general practice in producing particle boards in the United States is to dry particles in a rotary drum-type drier. There are several versions of this type of drier currently in use, but the most common is the direct-fired drier. This type consists of a large, cylindrical drum with flights, or vanes, attached to the inner surface. Hot gases from a furnace burning wood, oil, gas, or coal are forced through the drum at high velocity by an exhaust fan. Particles are introduced into the drum at the gas-intake end. They are carried through the drum by the hot gases until they settle on the vanes inside the cylinder sheath. They are then lifted by the rotation of the drum and dumped back into the stream of hot gases. Dry particles are light in weight and are carried through the drum very rapidly. Wet particles settle quickly out of the gas stream onto the vanes and must be repeatedly returned to the gas stream for drying. They require longer to pass through the drier. One type of rotary drum drier has an entering gas temperature of 700 to 1200°F and a leaving gas temperature of 220 to 400°F, depending upon the size, shape, and moisture content of the entering particles.

Several types of driers depend upon the gradual mechanical movement of particles through a heated chamber. A so-called turbodrier, manufactured in Germany, consists of an upright cylinder with a series of rotating circular shelves. The particles are introduced at the top, make one rotation on the top shelf, and are then dumped, via tilting plates, onto the shelf below. They progress thus through successive stages to the bottom of the drier and are discharged. Hot air circulates over the particles at all levels, and the dumping procedure serves thoroughly to mix the chips, flakes, or splinters. Other driers, utilizing the same general principle, are equipped with vibrating trays installed in a drying kiln in such a manner that the particles entering from the top of the kiln are dumped on the top tray, move to the end of the tray, and are dumped on the end of the second tray, thus progressing over a whole series of trays to the bottom of the kiln, where the particles are discharged. Hot air is forced over the trays, drying the particles.

A variety of suspension driers are produced and used in Europe. These driers usually combine drying with separation of the particles by size. While the various designs differ in construction and operation, the basic principle of all the suspension driers is the same. Essentially, these driers consist of a drying column. Hot gas or hot air is introduced into the base of the column, and the particles to be dried are introduced into the top of the column. The heavy particles fall down through the hot-gas column and are collected on a screen at the bottom, where they continue to dry and are ultimately removed for use. The lighter particles are dried in the hot gas, and as they become lighter in weight, they rise with the gas and are exhausted and collected. Thus the particles are separated during drying into coarse and fine fractions and can therefore be used to produce a layered board.

In all the chip-drying processes, high-temperature drying is accomplished. That is, the drying temperature is above the boiling point of water and in most cases is above the kindling temperature of wood. In one model of the rotary drum drier the temperature of the gas as it leaves the furnace is 1800°F.

4. Mixing. In preparing a furnish for the particle-board lay-up operation it is necessary to measure the ingredients and then thoroughly mix them. The basic ingredients are wood particles and a resin adhesive. Sometimes other additives such as wax or preservative material are also incorporated into the mix. Mixers are of two types, batch and continuous, and the metering equipment required is determined in part by the type of mixer used.

a. Batch Mixing. Batch mixers are of simple design and operation. They are essentially mixing chambers into which the measured quantities of resin and wood are introduced. They are so equipped that the ingredients can be agitated to provide a uniform spread of adhesive over the particles. Most batch mixers resemble the conventional glue mixer. They are chambers with power-driven stirrers in the center. The wood required for a batch is weighed out and loaded into the mixer. The adhesive can be weighed in the same manner, or, if liquid resin is used, a proper volume can be metered out of a storage tank.

b. Continuous Mixing. Continuous mixers, as the name implies, provide for a continuous flow of product mix. Adhesives and wood are metered in, and after proper agitation as the mix passes through the mixer, the wood-glue complex flows out to the mat-forming operation. A number of different kinds of continuous mixers have been designed for this purpose, but a complete description of each is beyond the scope of this book. All of them involve metering in the chips and adhesive on a continuous basis and then providing for a satisfactory distribution of the resin over the wood particles. Liquid adhesives are normally used with continuous mixers, and these can be measured in by controlling flow through an adjustable valve. Wood is metered in through a gravimetric chip feeder similar to the type commonly used in pulp mills. In this device the particles are discharged from a hopper onto a belt. They pass over a special weighing section of the belt designed so that the belt speed is controlled in response to the indicated weight.

5. Board Formation. Particle boards are formed in one of two ways: by flat-pressing or by extrusion. These two methods are quite different, and the boards produced have different properties.

a. Flat-pressing. In the manufacture of flat-pressed particle board, the basic manufacturing operation is generally performed in a multiple-opening hot press, similar to that used in the manufacture of plywood or hardboard. The product mix is measured onto a caul tray either by volume or weight. The mat is several times as thick as the target thickness of the finished board. The mat is customarily prepressed to reduce the mat thickness either by passing it under a series of heavy rolls or by processing it through a single-opening hydraulic cold press. The prepressed mats with their caul plates are

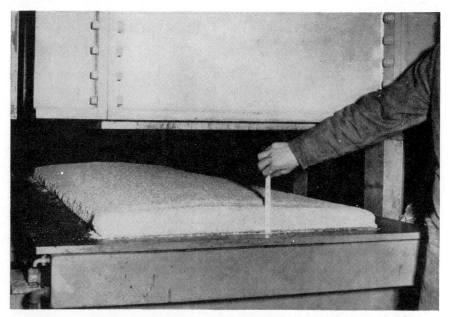

Fig. 12-6. Prepress for particle-board mat. (*Courtesy Onsrud Machine Works.*)

fed into a press loader, from which the multiple-opening hot press is automatically loaded, all openings simultaneously. The board is consolidated under heat and pressure in the press, the pressing cycle being determined by the thickness of the board and the curing properties of the adhesive.

When layered boards are being produced, the mat-forming process must be modified to permit the back, center, and face particles to be metered out separately on the caul tray.

The standard method of producing flat-pressed particle board in a multiplaten press is essentially a batch process. A modification of the flat-press method is *the Bartrev continuous process,* developed in England. In this process the mat is laid on a caul that consists of a stainless-steel endless belt. The mat is conveyed on the steel belt through a high-frequency preheating unit. In this unit the wood-adhesive complex is heated to approximately 70°C. As the mat leaves the high-frequency preheater it moves into the pressing section of the machine. Here a second stainless-steel belt is situated above the mat, and, therefore, the mat is carried through the pressing unit between the two stainless-steel caul conveyer belts. An endless chain of heated platens applies continuous pressure to both sides of the board through the steel belts. The platens are heated by resistance heaters. The platen chain resembles, but on a much larger scale, the chain-feed mechanism on a veneer tapeless splicer or fixed-head veneer jointer. The mat is separated from the moving platen belt by the stainless-steel caul belts. Boards produced by this process resemble conventional flat-pressed boards in properties.

Fig. 12-7. Multiplaten hardboard press viewed from unloading end. (*Courtesy American Forest Products Industries, Inc.*)

b. Extrusion Processes. The board formation and pressing operations in the extrusion processes differ appreciably from those in the flat-press processes, and these differences are reflected in the properties of the boards produced. In all flat-pressed boards, the particles are oriented so that the long dimension of the particle tends to be approximately parallel to the plane of the surface of the board. All extruded boards have particles in the finished product positioned so that the long dimension of the particle is randomly oriented in a plane perpendicular to the face of the board, as contrasted with the flat-pressed board, where the particles are randomly oriented in a plane parallel to the board face.

The various extrusion methods for manufacturing particle board are all continuous processes. The product mix, consisting essentially of wood particles and resin adhesive, is forced between two rigidly mounted heated platens. The original extrusion process was developed in Germany by Otto Kreibaum. *The Kreibaum process* consists of a vertical pressing unit. The product mix is loaded into a hopper, from which it is discharged into the open end of the pressing unit. A reciprocating ram forces the wood-adhesive complex between the heated platens, compressing the particles and forcing an increment of board out of the press. The compressed particles are glued together as the adhesive is polymerized under the influence of the heat from the platens. In this process, boards are sometimes produced which have holes throughout their length to decrease the weight of the board. This is accomplished by extruding the board over metal rods. The rods are heated and contribute to the cure of the adhesive. The use of the center rods makes it possible to speed up the rate of production of thick boards, since the heated rods, combined with the heated platens, reduce the distance that heat must be transmitted through the board to effect a cure of the resin.

A vertical extrusion press is also produced in the United States by the Chipcraft Company. Basically, this process is similar to the Kreibaum process, although it differs substantially in details of design and operation.

The Lane Company, Inc., has produced a horizontal extruder for the manufacture of particle board. In principle the horizontal extruder is similar to the vertical extruder except that the hot platens are horizontally oriented. Again, design and operating details are substantially different.

In an extrusion press the thickness of the board is determined by the spacing of the platens. This is also true of the Bartrev process. In a flat-press process thickness is determined by placing between the platens metal stops that limit the distance between the platens when the press load is under pressure. In the extrusion process density is controlled by adjusting the pressure on the ram or by varying the quantity of product mix introduced ahead of the ram during each cycle. In the flat-press process density is controlled by adjusting the thickness of the mat for a given stop thickness.

In the continuous processes a moving saw cuts boards off to the proper length as the extruded ribbon comes out of the press. Boards from a multi-platen press are trimmed for length and width on a battery of equalizer saws.

II. PROPERTIES OF BOARDS

A. Hardboard

Hardboard is commonly manufactured in thicknesses of $\frac{1}{8}$, $\frac{3}{16}$, $\frac{1}{4}$, and $\frac{5}{16}$ in. Panels manufactured for stock and ultimate sale through retail outlets are produced in standard stock sizes of 4 by 4, 4 by 6, 4 by 8, 4 by 10, 4 by 12, and 4 by 16 ft. Large industrial users who require special sizes in large quantities may buy panels cut to specified size in carload or truckload lots.

Standard hardboard is normally manufactured to a density in the specific gravity range of 0.80 to 1.15. Special high-density boards, sometimes referred to as "super" hardboards, have specific gravities in the range 1.35 to 1.45.

Hardboards vary considerably in physical properties, as might be expected from the variety of processes and the degree of control which may

Fig. 12-8. A variety of hardboard products. (*Courtesy Hardboard Association.*)

be exercised in the operation of a plant to produce the type of material demanded by consumers (Table 12-3).

Table 12-3. Properties of Hardboards [a]

Property	Untreated hardboard	Treated hardboard
Specific gravity...............	0.88–1.04	0.95–1.12
Tensile strength parallel to surface, psi.....................	1,000–3,000	4,000–5,500
Modulus of rupture, psi.........	3,000–6,000	7,500–10,500
Modulus of elasticity in bending, psi.......................	400,000–800,000	800,000–1,000,000
Compression strength parallel to surface, psi..................	2,000–4,000	4,200–5,300
Water absorption, 24-hr immersion, by wt.................	9–18	5–13
Tensile strength perpendicular to surface, psi.................	125–350	250–400

[a] SOURCE: U.S. Forest Service.

B. Particle Boards

Particle boards vary more than hardboards in properties. According to Lewis [12] the important factors determining the properties of a particle board are: (1) type and size of particle; (2) type of manufacture (extruded or flat-press); (3) kind and amount of resin; (4) distribution and orientation of particles; (5) density; (6) quality of manufacture (effectiveness of resin-spread, felting, and uniformity of formation); (7) moisture content during pressing; and (8) postmanufacturing treatment. Because of the wide variation in properties represented by commercial particle boards, it is not feasible to be very specific in reporting upon these board properties.

1. Flat-pressed Particle Boards. Flat-pressed boards are manufactured in thicknesses from ¼ to 2 in. Thickness above 1 in. is rare, however, since it is usually more economically feasible to laminate two 1-in. boards than to manufacture a 2-in. board. Flat-pressed boards are not usually manufactured in the hardboard thicknesses since they cannot compete with hardboard in price. Table 12-4 gives a summary of properties of flat-pressed particle

Table 12-4. Summary of Flat-pressed Particle Board [a]

Property	Medium-density board	High-density board
Specific gravity.................	0.4–0.8	0.8–1.2
Modulus of rupture, psi.........	1,500–8,000	3,000–7,500
Modulus of elasticity, psi.......	150,000–700,000	400,000–1,000,000
Tensile strength parallel to surface, psi....................	500–4,000	1,000–5,000
Tensile strength perpendicular to surface, psi.................	40–400	275–400
24-hr water absorption per cent by wt......................	20–75	15–40
Thickness swelling, per cent, after 24-hr soak at 70°F	20–75	15–40

[a] SOURCE: U.S. Forest Service.

board based on tests performed on commercial boards representing a considerable variation in design. Accurate values for any particular board can be obtained only by test. The values reported in Table 12-4 are for particle boards as they are produced and not as core stock in plywood, where additional plies have been added.

2. Extruded Particle Boards. Because of the orientation of particles across the board in extruded products, these boards have low bending strength and are rarely used in the form produced. This type of board almost invariably has additional veneer plies added to make up a plywood-type panel. After they have been glued up into plywood panels, the original low strength is unimportant. They must, however, have sufficient initial bending strength to withstand stresses imposed in handling prior to secondary gluing. Again, because of the particle orientation and method of manufacture, extruded boards have different dimensional properties than flat-pressed boards. Spring-back, or thickness-swelling, is a phenomenon of board length in an extruded board and of board thickness in a flat-pressed board. Because extruded boards are rarely used as manufactured, a summary of physical properties for this type of board has little meaning.

III. FUTURE DEVELOPMENTS IN FIBERBOARD
AND PARTICLE-BOARD MANUFACTURE

As has been previously indicated, the fiberboards and particle boards, which are the subject of this chapter, represent one of the most rapidly expanding segments of the wood-products industrial group. While fiberboard has been manufactured in the United States for about 40 years, this industry's most rapid growth has been since the Second World War. The entire particle-board industry in this country has developed in this latter period. Consequently, developments come with rapidity. As new factories are built and older plants are expanded, improvements are made in manufacturing techniques.

Much research is under way to determine the factors influencing board quality and to determine ways of making better boards more economically. The adhesives used early in the manufacture of particle board were developed for other uses and adapted to this new product. More recently the adhesive industry has developed a whole new spectrum of adhesives specifically designed for board manufacture, and all the major adhesive companies are continuing research directed toward the production of improved binders.

Despite the extensive research efforts already expended, much additional research is required. Accepted standards of board quality are already avail-

Fig. 12-9. Sample of flat-pressed particle boards showing variation in particle geometry. (*Courtesy American Forest Products Industries, Inc.*)

able for hardboard. A whole gamut of Federal specifications, commercial standards, A.S.T.M. standards, and tentative standards describe board properties and standard methods of test. No comparable sets of standards and specifications have been developed and accepted for particle board, although much work has been undertaken to develop the basis for such standards. It is reasonable to assume that standards for particle board, acceptable to the manufacturing industry and consumers, will eventually be developed.

SELECTED REFERENCES

1. Anon. Wood Composition Boards: General Characteristics, Uses and Manufacturing Processes. Report, Pacific Power and Light Company. 1955.
2. Boehm, R. M., and J. S. Harper. Application of Hardboard in Veneered Panels. Proceedings, Forest Products Research Society. 1951.
3. Clark, J. A New Dry Process Multi-ply Board. *Forest Prod. Jour.*, **5**(4):209–213. 1955.
4. Connelly, T. J. The Kreibaum Process for Extruded Core Board. *Forest Prod. Jour.*, **5**(4):21A–23A. 1955.
5. Crafton, J. M. Extruded Resin-bonded Wood-particle Board. *Jour. Forest Prod. Res. Soc.*, **4**(5):231–233. 1954.
6. De Pau, R. T. Small Plant Set-up for Insulation and Hardboard Manufacture. Proceedings, Forest Products Research Society. 1950.
7. Elmendorf, A. Manufacture of Hardboard Out of Wood Waste. *Jour. Forest Prod. Res. Soc.*, **2**(4):7–9. 1952.
8. Hofstrand, A. D. Relationship of Specific Gravity to Moduli of Rupture and Elasticity in Commercial Hardboard. *Forest Prod. Jour.*, **8**(6):177–180. 1958.
9. Johnson, E. S., A. A. Carlyle, L. B. McGee, and R. A. McLean. Wood Particle Board Handbook. Report, North Carolina State College. 1956.
10. Larmore, F. D. Influence of Specific Gravity and Resin Content on Properties of Particle Board. *Forest Prod. Jour.*, **9**(4):131–134. 1959.
11. Lewis, W. C. The Hardboard Industry in the United States. *Jour. Forest Prod. Res. Soc.*, **2**(4):3–6. 1952.
12. Lewis, W. C. Testing and Evaluating Procedures for Building Boards. *Forest Prod. Jour.*, **6**(7):241–246. 1956.
13. Locke, E. G., and K. G. Johnson. Fiber Products as a Factor in Utilization. *Forest Prod. Jour.*, **5**(3):153–156. 1955.
14. Lowgren, U. Wood Pulping by the Asplund Defibrator. Proceedings, Forest Products Research Society. 1948.
15. Markwardt, L. J. Report on Progress in Development of Testing Methods for Fiberboards. U.S. Department of Agriculture, Forest Service, Forest Products Laboratory Report No. 2105. 1958.
16. Marra, G. G. Particle Boards: Their Classification and Composition. *Forest Prod. Jour.*, **8**(12):11A–16A. 1958.
17. Miller, H. C. L. Chipcore: Its Characteristics and Production, Dry Process Panel. *Jour. Forest Prod. Res. Soc.*, **3**(5):149–152. 1953.
18. Mottet, A. L. The Manufacture of General Purpose and Decorative Pressed Wood Boards by a Dry Process. *Jour. Forest Prod. Res. Soc.*, **4**(5):224–226. 1954.
19. Roddin, H. High Temperature Drying of Chips. *Forest Prod. Jour.*, **9**(9):5A–6A. 1959.
20. Risheii, C. A. Fundamental Economics of the Particle Board Industry. *Forest Prod. Jour.*, **7**(1):6–9. 1957.

21. Stanford Research Institute. America's Demand for Wood, 1929–1975. Palo Alto, Calif. 1954.

22. Technical Committee, Forest Products Research Society, Wood Composition Board Division. 1958 Status of the Composition Board Industry. *Forest Prod. Jour.*, **9** (2) :53–56. 1959.

23. Textor, C. K. Wood Fiber Production with Revolving Disk Mills. Proceedings Forest Products Research Society. 1948.

24. Turner, H. D. The Effect of Particle Size and Shape on the Strength and Dimensional Stability of Resin Bonded Wood Particle Panels. *Jour. Forest Prod. Res. Soc.*, **4** (5) :210–223. 1954.

25. Turner, H. D. Wood Handbook. U.S. Department of Agriculture, Forest Service. 1955.

26. Walton, A. T. How Insulating Board Products Are Made by the Simpson Logging Company. *Jour. Forest Prod. Res. Soc.*, **1** (1) :91–94. 1951.

27. Williston, E. M. An Analysis of the Panel Products Field. *Forest Prod. Jour.*, **8** (2) :72–79. 1958.

28. Williston, E. M. Fibreboard and Particle Board, Vols. 1–6. Food and Agriculture Organization of the United Nations, Rome, Italy. 1958.

CHAPTER 13

WOOD FLOUR

Wood flour is nothing more than wood per se that has been finely divided and screened to desired particle size by mechanical means. En masse, the finest grinds superficially resemble wheat flour; the coarser grades of such material, however, are distinctly fibrous in character. Wood flour was first produced on a commercial scale in Norway in 1906, but is now being processed in a number of countries including Canada, Denmark, England, France, U.S.S.R., Germany, the Netherlands, Sweden, and the United States. Approximately 85 per cent of the current total output is consumed as a filler in the formulation of explosives, linoleums, and plastics and other molded products.

A. Raw Materials

The majority of the consumers of wood flour insist upon a product that is light in color, light in weight, fluffy, highly absorptive, and essentially free from resin. Such specifications preclude the use of woods which are highly resinous, deeply colored, or in the high-density ranges. About 75 per cent of the output of wood flour in the United States is traceable to the woods of the three commercial soft pines.* The residual 25 per cent is constituted largely of spruce, hemlock, balsam fir, pond pine, cottonwood, aspen, beech, birch, and maple. Small quantities of Douglas-fir flour are used in phenolic molding powders and in electrical insulation. When color is not of major concern, willow, yellow-poplar, and even redwood can be and have been used.

B. Preparation of Wood

Wood flour produced in the United States is made by pulverizing dry sawdust (8 to 10 per cent moisture content) and planing mill shavings obtained in the remanufacture of kiln-dried lumber. The supply of raw material is further augmented by such solid-wood residues as slabs, trimming, and

* Eastern white, western white, and sugar pines.

edgings. To be of use, however, such material must be dry and free from bark, dirt, and any other undesirable foreign or extraneous matter. Solid wood destined for pulverization is first reduced to small chips or splinter-like pieces using mechanical chippers or hogs. It is also necessary at times to run planing-mill shavings through a hog before their final reduction. The sawdust, shavings, and partially divided solid-wood residues are passed over a vibrating screen which separates the coarser material from that which is of a size suitable for pulverization. The former is retained on the screens but is eventually returned to a hog for further reduction in particle size. The acceptable stock, that which passes through the screens, is conveyed directly to a pulverizing mill.

C. Manufacture of Wood Flour

In the early days, wood flour was obtained by screening sawdust. Yields were low (about 4 per cent on screens of 80 mesh and finer), and demands not infrequently exceeded available supplies. Some flour was also obtained by screening sander dust. Such material usually contained quantities of the abrasive, which, if declared objectionable by the consumer, was removed by air separation. Small quantities of wood flour are still produced in these two ways, but the bulk of it is now manufactured using one of four general types of pulverizing mills.

1. Attrition Mills. The oldest mill of this type is the *stone mill,* an adaptation of the ancient burrstone grinder used in making grist. It consists of two superimposed grindstones, 40 to 60 in. in diameter, enclosed in a wooden housing. The upper stone, which is stationary, is provided with a central opening through which the screenings are delivered to the grinding faces. The inner (or lower) face of this stone is slightly dished and features a number of short radial grooves that permit the coarser particles to work their way in between the grinding faces. The opposing inner (or upper) face of the lower stone, which is power-driven, is ungrooved and perfectly flat. As the wood becomes pulverized by the grinding action of the two stones, the particles migrate to the periphery of the stones and ultimately fall into the bottom of the housing. The fineness of the finished product is controlled by the spacing of the stones. In this type of mill the initial size of the raw material must first be reduced to that of sawdust. To prevent overheating, which may result in charring or even combustion, small amounts of water or steam are introduced between the stones. Very fine wood flour can be produced by this method. Stone mills are common abroad, especially in Scandinavia, but are seldom used in this country because of their relatively high power consumption.

Modern attrition mills have many improvements over the old stone mills and are of two general types, *double* and *single attrition.* The double attrition mills consist of two metal grinding disks, revolving in opposite directions, enclosed in an iron housing. The solid lower disk is covered with grinding plates of hard steel provided with numerous V-shaped depressions with sharp cutting edges. The upper disk is made of spokes to which grinding plates, similar to those on the lower disk, are attached. These plates, however, do not extend all the way to the center of the upper disk, leaving an opening in the center. Wood fed into the upper disk gradually finds its way through the spokes to the grinding surface, and the wood flour is forced out at the periphery of the disks and is deposited at the bottom of the enclosing case.

Single attrition mills are equipped with a single revolving disk similar to the upper spoke disk of the double attrition mills, and the wood material is pulverized by a rubbing action of this disk against the sides of the housing, which are provided with a suitable grinding surface. Attrition mills are generally equipped with a pneumatic collecting system consisting of a manifold connected to the suction side of a fan which blows the flour to the sifting room.

2. Hammer Mill. A hammer mill is composed of a horizontal, power-driven shaft fitted with a series of steel disks separated by spacers. Attached to each disk near its periphery are a number of free-swinging hammers. The hammer system in turn is housed in a cylindrical metal chamber furnished at the top with a feed-in opening and with a screen of a prescribed mesh below. The interior wall of this cylinder is also commonly provided with a grinding surface. This assembly is enclosed in a two-piece metal casing. Screenings fed into this mill are pulverized by the impact of the rapidly revolving hammers forcibly throwing the particles of wood against the grinding surface of the cylinder. Air separators (described later) are used in collecting material of 60 mesh or finer. The collection of various sizes of particles is controlled by air speed.

Hammer mills are somewhat lower in first cost but are considerably higher in maintenance charges than attrition mills of a corresponding capacity. Hammer mills are also objectionable because of their tendency to permit slivers of comparatively small diameter, but of considerable length, to escape through their screens. The attrition mills will produce a larger tonnage of wood flour with less horsepower. The capacity of different grinding equipment varies greatly with the type, size of the grinding surface, power used, fineness of the product desired, and the type of wood.

3. Beater Mill. The beater mill is an upright, or vertical, adaptation of the hammer mill. The hammers have been replaced by a number of free-swinging beater arms staggered along the vertically oriented drive shaft. When in motion, these arms whirl the screenings, which are let in at the top of the cylinder, and beat the material until it is small enough to pass through a basal screen into the collection system.

4. Roller Mills. Some wood flour is also produced by crushing in roller mills. This is accomplished by passing the raw material between either (1) a moving roller and a stationary metal bedplate, or (2) two or more pairs of steel rollers surfaced with teeth of varying degrees of fineness. In the first type of mill pulverization takes place at the line of contact between the roller and bedplate; in the second mill the paired rollers are arranged in vertical banks, and the particles of wood are pulverized as they pass vertically down through each pair of rollers to the collection system at the base. A patented process in which the wood is impregnated with urea resin to increase its brittleness prior to pulverization was issued in 1944.

The particles of wood flour as they come from the grinders vary in degree of fineness.* The coarser grades are separated by means of separators, called sifters. However, if stock finer than 60 mesh is desired, the grading is frequently accomplished by means of an air separator based on the principle that for any given air velocity a point is reached at which the air resistance of a particle of a given size is balanced by its weight, and all larger particles settle. The speed with which the air passes through the separator therefore determines the fineness of the product. Fast currents will carry larger particles, while the slower ones will pick up only fine dust. Thus, by regulating the speed of air, the grade of the resulting product can also be regulated without the use of screen. The big advantage of an air separator is that even the finest flour collects in the enclosing cyclone dust collector, with hardly any loss.

Finely ground wood flour will absorb considerable moisture (4 to 6 per cent) from the atmosphere. Ordinarily no attempt is made to redry it at the processing plants, as consumers prefer to readjust moisture content in accordance with their particular requirements. Wood flour is usually packed in 100-lb jute sacks, using screw-compression-type packers.

The presence of air-borne wood dust in wood-flour manufacturing plants creates a constant fire hazard. Under certain circumstances the presence of such dust in the air may create conditions favoring explosive combustion. Such conditions exist when the concentration of wood dust is such that each particle is surrounded by enough air to allow complete combustion and when spacing of the particles is close enough to allow propagation of flame from one particle to another. Either greater or lesser concentration of particles will not cause explosion. The fire hazard in wood-flour plants can be considerably minimized by the observance of the following points: (1) installation of an efficient dust-collecting and air-conditioning system; (2) elimination of dust accumulations on rafters, open shelves, corners, and other poorly accessible parts of the building; (3) use of completely enclosed, nonsparking motors and switches, and elimination of all sources of static by proper grounding; (4) installation of sprinkling systems and nonsparking floors; and (5) enforcement of no-smoking rules and all other general safety rules.

The cost of milling and screening depends upon the type of mill employed and the grade of flour produced. Equipment capable of milling 1 ton of 40-mesh flour per hour varies between $7,500 and $15,000. The cost of similar machines to produce finer grades is not substantially greater, but their output per hour for grades up to 100 mesh varies inversely with the degree of fineness required. That is to say, a mill that will yield 1 ton of 40-mesh flour can produce only ½ ton of 80-mesh material in the same time interval. Mill output varies as the square of the ratio of the meshes for particle sizes above 100 mesh; i.e., it takes four times as long to grind a ton of 200-mesh material than it does to produce an equal weight of 100-mesh stock.

Wood flour is usually sold by the ton, and is currently bringing from $18

---

* The sizes of wood flour most commonly used are 40, 60, and 80 mesh. The finest commercial grade of wood is about 150 mesh, although sizes as fine as 350 to 400 have been attained with pulverizing equipment.

to $55. Raw materials make up about 40 per cent, and processing and overhead 60 per cent, of the production costs.

D. Classes of Wood Flour

There are three principal classes of wood flour currently available to the consumer, the grading of which is based upon degree of fineness, color, and resin content.

1. Nontechnical Wood Flour. The bulk of the wood flour currently produced in the United States is classed as nontechnical wood flour. It is manufactured to less rigid specifications than technical flour, described below, and not infrequently consists of a mixture of both hardwood and softwood species. Large volumes of this grade of material are consumed in the manufacture of linoleum, dynamite, moldings, artificial wood carvings, and plastic wood.

2. Technical Wood Flour. This material is manufactured to rigid quality standards and commands a high price. Specific consumer requirements relate to its absorptive properties, resin content, weight, color, and fineness. Flour coarser than 60-mesh screenings is unacceptable. Most of this class of flour produced in the United States is made from sawdust and shavings of white pine. When other woods are used in its preparation, no mixing of the various species is permitted. Technical wood flour is used principally as a filler in resinoid plastics.

3. Granularmetric Wood Flour. This product is a special grade of technical flour in which the wood particles must conform to a specific weight and size. It is difficult to produce within the tolerances allowed and hence is very expensive.

E. Uses of Wood Flour

The principal uses of wood flour fall into the following categories: (1) fillers, either as extenders to reduce the amount of more expensive ingredients or as modifiers of physical properties; (2) absorbents; and (3) abrasives.

As a filler, wood flour finds its greatest application in the manufacture of linoleum, particularly of the inlaid type, and to a lesser degree in the battleship and printed grades. The linoleum industry normally consumes more than 50 per cent of the wood flour used in this country. In addition to acting as an extender, wood flour also functions as a carrier for colors throughout the thickness of the linoleum sheet.

Large quantities of wood flour are utilized as filler in the manufacture of plastics, composition flooring, artificial wood, moldings, insulating brick, and similar products. Although the use of wood flour in the plastics industry may be motivated by considerations of economy, the principal reason for its use lies in the ability of this material to modify favorably the physical properties of plastic materials by increasing impact strength, reducing brittleness and shrinkage, and providing opacity to transparent or translucent plastics.

Wood flour is also utilized as an extender in glues used in woodworking and plywood plants, especially with the synthetic adhesives.

Other examples of the use of wood flour involving modification of physical properties include its utilization in foundries as an antibinding agent, in ceramics to effect porosity in burned-out materials, in electrical equipment to increase insulation, and in special types of paint to obtain sound-deadening properties. Wood flour is used decoratively in "oatmeal" and "velvet" wallpapers, where colored grains of wood flour sprinkled over the surface provide desired design and texture.

The most important use of wood flour as an absorbent is in the manufacture of dynamite. It is added to nitroglycerine together with oxidizing agents, such as sodium nitrate and ammonium nitrate, primarily to reduce sensitivity to shock. Wood flour and other vegetable fiber have largely replaced *Kieselguhr* (a fine earth) as an explosive absorbent because the combustion of organic material during an explosion contributes to the force of the blast.

Wood flour is used as a mild abrasive and as an oil absorbent in the cleaning of furs, and as an abrasive in certain types of soaps especially recommended for use by mechanics. Square-cut, relatively coarse hardwood wood flour is extensively used in the jewelry trade for cleaning silverware.* The following is a partial list of other products in which wood flour is an important ingredient: artificial wood, ash trays, door knobs, handles, molded furniture parts, ornaments, and novelties.

F. Status of the Industry

During the past several years wood flour has enjoyed expanding markets and a number of new uses; yet most of the tonnage produced is still consumed as one of several ingredients used in the formulation of other commodities. Thus it is reasonable to assume that the future of the wood-flour industry will depend, in no small measure, upon the continuing growth and economic health of the major consumers of its products. Moreover, there appears little likelihood that wood flour will lose its volume markets to other organic or mineral fillers in the foreseeable future, for in the final analysis relatively low production costs and the highly desirable physical properties of wood flour preclude serious competition from such substitute materials.

Prior to 1940 imports averaged 20 per cent of the total tonnage of wood flour consumed in the United States. During the Second World War, however, foreign supplies were cut off, and the domestic producers enlarged their facilities in order to keep pace with the demands. There is now adequate production capacity to provide both current and projected future industrial requirements; except in the event of a general economic decline, the industry should remain in a stable condition for many years to come.

---

* This type of wood flour can be produced more advantageously with a rotary knife cutter, since such a cutter takes sawdust and shavings equally well and produces sharp, clear-cut particles. A 30-in. cutter will produce from 750 to 1,000 lb of wood flour per hour.

SELECTED REFERENCES

1. Anon. Fillers: Cotton Flock and Wood Flour. *Mod. Plastics,* **17**(2):53. 1939.
2. Anon. Wood Flour as Filler for Rubber Flooring. *India Rubber World,* **99**(6):61. 1939.
3. Aries, R. S. Wood Flour Production. *Amer. Lumberman,* (3282):213–221. 1944.
4. Bowen, P. P. Manufacture and Use of Wood Flour. Proceedings Forest Products Research Society, Vol. 2, pp. 276–279. 1948.
5. Buckley, E. H. Wood Flour Production in Canada. Unnumbered reprint, *Canad. Lumberman.* May, 1952.
6. Cosgrove, G. F. Wood Flour and Its Uses. *Hardwood Rec.,* **62**(2):24. 1926.
7. Emmett, J. A. Machines and Methods for Making Wood Flour. *Wood-Worker,* **56**(12):26–28. 1938.
8. Emmett, J. A. Wood Flour Production: Possibilities as a Profitable Means of Waste Utilization. *Timberman,* **39**(9):91–92. 1938.
9. Friedlander, C. J. Woodwaste and Wood Flour. *Wood,* **2**(7):296. 1937.
10. Gangloff, W. C. Fillers in the Plastic Industry. *Chem. Indus.,* **53**(4):512–514. 1943.
11. Harrison, J. D. B. Wood Flour. Forest Products Laboratories, Dominion of Canada Forest Service. 1940.
12. Hartman, I., and J. Nagy. Inflammability and Explosibility of Powder Used in the Plastics Industry. *U.S. Bur. Mines, Rpt. Invest.,* RI-3751. 1944.
13. Jahn, E. C. Wood Flour Market Expanding. *West Coast Lumberman,* **66**(7):40. 1939.
14. Reineke, L. H. Wood Flour. U.S. Department of Agriculture, Forest Service, Forest Products Laboratory Report No. 1666–9. 1956.
15. Stamm, A. J., and E. E. Harris. Chemical Processing of Wood. Chemical Publishing Company, Inc., New York. 1953.
16. Steidle, H. H. Wood Flour: Its Manufacture and Principal Uses. *Amer. Lumberman,* (2079):54. 1942.
17. Studley, J. D. Sawdust, Wood Flour, Shavings and Excelsior. U.S. Department of Commerce, unnumbered report. 1936.
18. Thomas, J. O. The Manufacture of Wood Flour. *Wood,* **4**(5):207. 1939.

CHAPTER 14

SAWDUST AND SHAVINGS

Wood residues are unavoidable in all types of primary and secondary wood processing. Economic utilization of these residues, including sawdust and shavings, is one of the major factors in more profitable utilization of the raw material used in the manufacture of wood products. The most common use is for fuel at the point of origin. However, there are a number of other outlets for sawdust and shavings, and some of these are described in this chapter.

Most markets for sawdust and shavings are specialized and frequently local in character, and profits are marginal. Because of this, and in spite of large total unused quantities, it is not uncommon to find local shortages of sawdust and shavings of the right kind at times of greatest consumption. The main reasons for this are as follows: (1) shaving and sawdust production, as well as their greatest consumption, is seasonal in many localities; (2) users of sawdust and shavings are becoming increasingly more exacting about the quality and composition of sawdust best suited for their purposes, and this makes the average run of sawmill and factory stock of little commercial value; (3) because of the low margin of profit on sawdust and shavings, it is seldom profitable to supplement local output by shipping such materials from the regions where they are available in large volumes.

I. SAWDUST

Between 10 and 13 per cent of the total content of the log is reduced to sawdust in milling operations. The percentage of sawdust produced depends largely on the average width of the saw kerf and the thickness of lumber sawed. Mills using band saws, and those cutting thick stock, produce less sawdust than mills whose output consists mostly of 4/4- and 5/4-in. lumber.

Sawdust is commonly classified as softwood, hardwood, mixed softwood, and mixed softwood and hardwood. The bulk of the sawdust is used green. In most cases it has not been found economical to dry it. Some dry sawdust, however, is produced from machining air-seasoned and kiln-dried lumber. Green sawdust is used mostly as fuel at or near the point of production.

Green hardwood sawdust is used in fairly large amounts for meat smoking, and both green softwood and green hardwood sawdust are used in increasingly large amounts for agricultural purposes as soil conditioners and as mulches. The sawdust sold by dealers for industrial purposes is either air- or kiln-dried.

Demand for sawdust for industrial uses is becoming so exacting that sawdust is usually graded for size. Frequently sawdust is also separated by species, since for some uses any woods that may discolor the product with which they come in contact must be carefully excluded. For this reason the sawdust of one species is usually worth more than that of a mixture from several woods. Hardwood sawdust is either sifted or is sold in various approximate size grades, designated by the types of machinery from which it results, such as sander dust, resaw dust, etc. Sifted hardwood sawdust is designated by mesh sizes, from 8 to 40 mesh. The sifting may be done either at the plant or by the dealers in their storage places. Softwood sawdust is seldom sifted.

The sawdust business is so competitive and the margin of profit so small that excepting for a few of the most progressive concerns, sawmills and woodworking plants find it unprofitable to sell direct to consumers. The sawdust, therefore, is usually handled by dealers who contract with the producers for the output of a mill or factory. Small city woodworking plants are frequently willing to give their sawdust gratis to reduce the fire hazard and to save the space that it would otherwise occupy. In the case of larger plants, the dealers usually contract for a definite number of cars for a stated period of time. Dealers in the large cities usually maintain storage sheds and screening plants. The sawdust is taken directly from the cars to mechanical screens, bagged automatically, and stored by species and grades.

A. Limitations in Utilization of Sawdust [14]

The profitable handling of sawdust is limited to a large extent by several factors.

1. Handling Charges. The margin of profit on sawdust is so small that all handling must be reduced to a minimum; this factor prevents many sawmill operators from engaging in the sifting and grading of sawdust necessary for its direct disposal to the consumer.

2. Cost of Transportation. This is the chief limiting factor in the marketing of sawdust. With the low prices which sawdust brings on the market, freight rates control the distance from which this product can be profitably shipped; ordinarily shipping of sawdust by rail is limited to a radius of less than 100 miles.

3. Lack of Uniformity of Product. This factor makes many present and prospective consumers of sawdust decide to use other materials. Many industrial uses require sawdust of uniform quality in regard to species, grade, moisture content, and freedom from foreign substances; all these requirements can seldom be guaranteed either by the producers or by the distributors.

4. Quantity of Sawdust Available. The demand for sawdust is seasonal, more being required in the fall and the winter, i.e., at the time of least activity of sawmills. On the other hand, the use of sawdust for wood flour, alcohol, wood briquettes, and other products requires installation of specialized equipment which must be kept busy throughout the year. Unless large and dependable supplies of a given quality can be guaranteed, the installation of such equipment is not justified.

5. Physical State of Sawdust. The small size of its particles and the bulkiness of sawdust in large quantities place definite restriction on its chemical and mechanical utilization.

B. Uses of Sawdust [14]

For convenience of discussion the uses of sawdust may be grouped as follows: (1) sawdust as fuel, (2) sawdust in its natural condition, (3) sawdust in the manufacture of other products, and (4) sawdust in the manufacture of chemical products.

1. Sawdust as Fuel. The utilization of sawdust as fuel is discussed in the chapter on Wood Fuel (see Chap. 15). The observation is in order, however, that even though the largest single use of green sawdust continues to be for fuel, the total amount of sawdust so used is steadily declining. This is due to the replacement of steam-powered plants, which used waste wood for fuel at the point of its production, by plants that use electric power, generated elsewhere or by internal-combustion engines.

2. Sawdust in Its Natural Condition. The uses of sawdust in its natural state are so numerous that only a few major ones can be enumerated. Many of these depend on its high absorptivity, its bulk, and even its chemical composition.

a. Sawdust Mulches.[2,4,6,7,10] Use of sawdust for agricultural purposes has increased rapidly in recent years. It is used either as a soil conditioner or as mulch. Most soils are in need of organic matter because the usual crop rotation seldom maintains soil humus at a desirable level. Fine wood residues provide inexpensive plant material that can be used to supplement other sources of humus.

Wood itself contains only small amounts of chemicals valuable as fertilizers. The principal organic compounds of wood that are of agricultural interest are cellulose, the pentosans, and lignin; of these lignin is considered to be the most valuable. When sawdust is added to soil, the cellulose and the pentosans are attacked most rapidly by microorganisms. Lignin, its degradation products, and the residue of microorganisms tend to remain in the soil for a considerable time as constituents of humus, thereby improving the physical condition of the soil. Lignin also reduces leaching and, therefore acts as a storehouse of nutrients.

When undecomposed sawdust is mixed with soil, a temporary harmful effect on crops may be observed. It manifests itself in the yellowing of plants indicating depletion of the available soil nitrogen. This is brought about by decomposition of wood particles by bacteria and fungi, a process which requires nitrogen in excess of that provided by the sawdust. This nitrogen deficiency effect of sawdust seldom extends beyond the first season if n more than 3 or 4 tons of dry material per acre are added to the soil. Addition of a larger amount may result in nitrate depression over several years. To overcome this initial effect of the sawdust, it is recommended that about 24 lb of nitrogen, corresponding to 115 lb of ammonium nitrate, be added per ton of dry wood. Ultimately the nitrogen assimilated by the microorganisms is released as they die. It is also possible that phosphate deficiency may be brought about by sawdust addition. Information on this subject, however, is not at present available.

Although most kinds of sawdust are acid, unless the sawdust is applied to lime-requiring crops, the acid is of minor importance. In the case of acid-requiring plants, such as blueberries and azaleas, the resulting acidity is beneficial.

When used as bedding for animals and poultry, sawdust acts as an absorbent for liquid manure, which contains 90 per cent of the total nitrogen in manure. By adding 50 lb of super phosphate per ton of dry wood, the nitrogen in the liquid manure is fixed in such a form that it does not readily evaporate.

Sawdust also may be composted with farm manure, legume waste, and other residues high in nitrogen. In place of these, chemical nitrogen, limestone, and phosphate can be added. Under conditions of adequate moisture, sawdust compost should be ready for use in 3 to 6 months. Inoculation of sawdust composting material with a cellulose-decomposing fungus, *Coprinus ephemerus*, is advocated by S. A. Wilde.[15]

Sawdust is considered to be an excellent mulch for fruit orchards, blueberry and strawberry gardens, and for vegetable and flower gardens, provided adequate nitrogen is supplied to the soil. A substantial increase in yield has been reported for diverse agricultural crops.

b. Stable and Kennel Bedding. Sawdust as well as shavings are extensively used as bedding for animals, especially for cattle at large dairy farms. For this purpose white pine and spruce are considered the most desirable and bring the highest price. In kennels eastern redcedar sawdust and shavings are preferred because of their reputed effect as an insect repellent. In a study conducted by the School of Forestry at the University of Minnesota it was determined that hardwood sawdust and chips have only about half the absorptive capacity of straw.*

c. Absorbent. Sawdust is widely used in butcher shops, fish markets, machine shops, garages, factories, and warehouses as floor covering. Any kind of sawdust may be used for this purpose, though clean, light-colored material is preferred.

d. Leather Manufacture. Sawdust is used in the leather industry to the extent of more than 1,000 tons annually. Its greatest use at tanneries is in the manufacture of patent leathers and uppers, for moistening the hides for stretching. The requirements of tanners vary greatly in regard to the species and the fineness of sawdust desired. In all cases, however, the requirements

* Minnesota Forestry Notes. School of Forestry, St. Paul, Minn. 1958. Also see reference 7.

are very exacting as to freedom from splinters, grease, and any foreign material that might impart color to the leather. The difficulties experienced in obtaining sawdust of uniform quality and free from discoloring substances have caused many leather manufacturers to turn to other methods of moistening hides. White pine and other lightweight, light-colored, nonstaining woods are preferred.

e. Conditioning of Furs. Sawdust is used extensively for moistening the skins when they come to the dressing plant and also in cleaning the pelts to remove all grease and dirt, to give the hair a light, fluffy appearance, and to restore luster. The operation is carried on in drums in which the pelts are tumbled about with sawdust. Carefully screened, kiln-dried, fine hard maple sawdust is generally used by furriers, though limited quantities of other hardwoods are also used by some fur dressers, who by experience have found that a certain kind of sawdust is especially suited for a given type of fur. The sawdust used for removing grease and dirt is mixed with some cleaning ingredient which cuts the grease while the sawdust absorbs it.

f. Cleaning, Drying, and Polishing Metals. Several hundred tons of sawdust are used annually for removing oil and grease from metal and metal products such as cutlery, metal objects turned in lathes, and aluminum wares. The finished products are first dipped into kerosene or other solvent, and the solvent is then dried with sawdust; kiln-dried hard maple sawdust is preferred. Sawdust is also used in the plating industry for drying and polishing finished products removed from the plating solution. Likewise, dry sawdust is frequently used to clean heavy machinery of grease and oil.

g. Packing. Sawdust is frequently used as a packing medium, especially in the shipment of grapes and other fruit, as well as in packing fragile articles such as glass, china, and canned goods. For packing fruit coarse, granular sawdust is preferred because it is light-colored, odorless, and tasteless, though sawdust from redwood, white fir, and even Douglas-fir is also used. About 4,000 tons of sifted sawdust is consumed annually by the grape packers of California alone.

Other uses of sawdust in the natural state include *curing of freshly poured cement.* From 1 to 2 in. of wet sawdust from a nonstaining species is spread for a period of 10 days to 2 weeks, to allow a slow curing of the cement. Sawdust is also used *in icehouses,* though this use is falling off sharply owing to electric refrigeration and the manufacture of artificial ice; *for stuffed toys and pin cushions; in nursery practice* for "heeling" the stock for a short time prior to shipment; *in insect bait,* as a carrier for arsenic and other poisons to combat chewing insects such as cutworms; *in meat and salt smoking,* for which purposes green hickory sawdust is preferred, although maple, oak, walnut, and mahogany are also extensively used.

h. Fill Insulation. Dry sawdust, along with other shredded and granulated organic and inorganic materials, can be used as a fill insulation by blowing or pouring it into the desired space. Sawdust and shavings compare favorably with other types of fill insulation in heat-insulation properties. The principal objections to the use of mill-run sawdust are that it is too damp, therefore, conceivably contributing to conditions favoring decay, and also that it is nonuniform. In the absence of authoritative information on sawdust as an insulating material, local building codes tend to place restrictions on its use in residences.

3. Sawdust in the Manufacture of Other Products. Quantities of sawdust and shavings, principally white pine, are utilized in the manufacture of *wood flour* (see Chap. 13); sawdust is also an important ingredient of a number of composition products, some of which are described below.

a. Composition Flooring. This is made usually of caustic magnesia cement with various amounts of sawdust as a filler. Though different kinds of sawdust are used for this purpose, medium fine (20 to 40 mesh) maple sawdust is generally preferred. The dry ingredients, consisting of sawdust and pigment, are thoroughly mixed and a solution of magnesium chloride is added. The mixture is then spread like cement. Usually two coats are used, the top layer containing a larger percentage of fine material, such as wood flour.

b. Stucco and Plasters. Sawdust is added to bind the mass together, producing at the same time a lighter and more porous product which can be nailed without damage and which has better insulating qualities.

c. Gypsum Compositions. Sawdust is used in a number of gypsum products, such as interior partitions, insulation products, and roofing materials. Sawdust makes gypsum products more porous, decreasing their weight and increasing their insulating qualities; it also softens the finished product to the extent that it can be nailed and sawed; finally, it lessens the cost of the finished product.

d. Clay Products. The use of sawdust in clay products today is practically limited to the manufacture of porous, hollow building tiles. Sawdust is added to make the tiles lighter in weight, more resistant to heat, and capable of holding nails. When the clay is fired the sawdust is burned out of the tile, leaving it in a porous condition. Sawdust is not used in the manufacture of common bricks, flue linings, or clay pipes, because it tends to increase the absorptive capacity of these products.

e. Concrete Products and Artificial Stones. Sawdust is used in the preparation of the so-called "lightweight concretes," which contain as much as one-third to one-half sawdust. The advantages of such concrete are that the product can be sawed, fastened with nails and screws, and polished. This product is said to be very fire-resistant and nonconductive of sound. This type of concrete is used for floors where it is desired to attach wooden members to concrete by means of nails or screws; it has also been suggested for construction of floors in cattle barns.

A number of products called *artificial stones* are made of different binders with sawdust as a filler. They can be colored and when polished resemble marble or stone. Such products are widely used for building purposes in the form of roof, floor, and panel tiles and building stones.

f. Artificial Wood. This is a product made of sawdust, paper waste, casein glue, and limestone or chalk. The ingredients are ground together, moistened with water, and molded. The finished product is said to possess many of the properties of natural wood.

g. Abrasives. Wet sawdust is mixed with coke, salt, and sand in the manufacture of silicon carbide abrasives. When these are fused in an electric furnace, the sawdust burns out first, leaving the rest of the mass in a porous condition, thus providing outlets for escaping gases generated during the fusion; if no such provision were made, these escaping gases would rupture the fusing mass. Any kind of sawdust, dry or wet, can be used.

h. Floor-sweeping Compounds. In the manufacture of sweeping compounds, sawdust is mixed with an antiseptic and cleaning ingredients, such as clean, fine sand, to which essential oils, such as oil of eucalyptus or oil of sassafras, are added to impart a pleasant odor and to absorb the fine particles of dust. Ordinary damp sawdust is probably just as satisfactory, and on concrete floors it is preferred, since it prevents excessive dust.

i. Molded Articles and Plastics. Finely ground sawdust and shavings combined with a suitable binder, such as starch, glue, or aluminum sulfate, are used in making molded ornaments, dolls, novelties, insulators, and similar items.

A relatively low-cost plastic mix, in which a large percentage of the material consists of hydrolyzed dry sawdust, has been developed by the U.S. Forest Products Laboratory. A number of other similar products have also been announced by other research organizations (see page 491). None of them has as yet reached commercial development on an extensive scale.

j. Board Products. A number of processes have recently been announced for the manufacture of board products made entirely of pressed sawdust and other granulated wood waste with the addition of a suitable binder; other types of wallboard are made by combining sawdust with wood pulp or other type of vegetable fiber.

It has been estimated that as much as 2,000 sq ft of board may be obtained from a ton of sawdust. Extreme simplicity of operation, dispensing with the service of skilled technicians, is claimed for one process. In addition to flat boards, a variety of other pressed items, such as containers, disposable dishes, and moldings, can be made of any suitable sawdust-resin mixture. Commercial development of an inexpensive molding process for sawdust and similar wood residue will go a long way in solving the major wood-waste problems now confronting the wood-conversion industries of this country.

4. Sawdust in the Manufacture of Chemical Products. The possibilities of sawdust utilization in the manufacture of chemical products include (1) destructive distillation; (2) steam extraction; (3) production of ethyl alcohol and cattle food; and (4) production of oxalic acid, dyes, and similar products.

a. Destructive Distillation of Hardwood Sawdust. Although a possibility, this technique has not achieved commercial success, largely because of the bulkiness of the material, which necessitates large quantities of it for uninterrupted production, and the difficulties encountered in carbonization of sawdust.

b. Steam Extraction. Steam extraction of turpentine and pine oil from the sawdust of resinous woods in a manner similar to that employed with chipped southern pine stumps is technically quite feasible. It is, however, questionable if at present the yield obtained and the supply of raw material would justify the cost of operation.

c. Ethyl Alcohol and Yeast from Wood. The production of ethyl alcohol and yeast from sawdust and other wood waste is discussed in the section on saccharification of wood.

d. Oxalic Acid. In the past, considerable quantities of oxalic acid were produced by fusion of sawdust with a caustic. Though at present oxalic acid is made by other processes, opportunities for more efficient fusion of sawdust have been demonstrated by Othmer and his coworkers.[9] This method yields up to 75 per cent of oxalic acid on the basis of dry weight of wood; in addition, considerable quantities of acetic and formic acids and methanol are obtained. The approximate yield of chemicals in the experimental runs were as follows:

Material used, lb		*Product formed, lb*	
100	dry sawdust	45.5	oxalic acid
9	sodium hydroxide	11.7	acetic acid
34.7	lime	2.48	formic acid
61.1	100% sulfuric acid	5.5	methanol
		85.0	calcium sulfate (waste)
		3.0	wood oil

By carrying out fusions on several different species of wood, it has been shown that a constant ratio exists between the yields obtained and the amount of cellulose in wood. Since, however, the yields were somewhat higher than could be expected from the cellulose alone, it appears that other constituents of the wood also entered into the reaction. Commercial success of this treatment depends largely on the efficiency of recovery of caustic soda and success in separating the materials resulting from the fusion.

Oxalic acid is an important organic acid, of which some 10 million pounds is used annually in the United States. It is used in large quantities by laundries as an acid rinse; also in the production of celluloid and rayon, in the manufacture of explosives, leather manufacture and dressing, purification of glycerol and stearin, bleaching straw, and in the preparation of certain dyes.

II. SHAVINGS

On the basis of volume, the most important use of shavings is for fuel, either as such or in the form of wood briquettes. Other important uses for shavings include packing of such products as canned goods, paint, and building-stone blocks, clean shavings of nonstaining species being preferred; stable and kennel bedding; lime burning; wood flour; insulation; wood pulp; and in wallboards made of shavings and resins. White pine shavings are in greatest demand; the use of other light-colored woods such as spruce, cottonwood, aspen, and basswood is also quite extensive.

The chief difficulty encountered in the profitable utilization of shavings is their bulkiness, which makes long-distance shipping economically prohibitive. According to the Department of Commerce, 12 to 15 cents per 100 lb is usually the maximum freight rate that can be absorbed by the product. The shavings are either shipped in carload lots or, to decrease their bulk and for convenience of handling, pressed into bales weighing 80 to 240 lb each.

SELECTED REFERENCES

1. Allison, F. E., and M. S. Anderson. The Use of Sawdust for Mulches and Soil Improvement. *U.S. Dept. Agr., Cir.* 891. 1951.
2. Anon. What to Expect from Sawdust Mulches. *Oreg. Agr. Expt. Sta., Agr. Prog.,* 4, No. 1. 1956.
3. Anon. Wood Products for Fertilizers. *Northeast. Wood Util. Council, Bull.* 32. 1950.
4. Bollen, W. B., and K. C. Lu. Effect of Douglas-fir Sawdust Mulches and Incorporations on Soil Microbial Activities and Plant Growth. *Soil Microbial Sci. Soc. of Amer. Proc.,* **21**(1):35–41. 1957.
5. Burks, G. F., and E. E. Behre. Timber Resources Review: Chap. 3, Growth and Utilization of Domestic Timber. U.S. Department of Agriculture, Forest Service. 1955.
6. Frese, Paul (guest ed.). Handbook on Mulches. *Plants and Gardens* (special printing), **13**(1). 1957.
7. Gessel, S. P. Composts and Mulches from Wood Waste. *New Woods Use Ser., Cir.* 33. University of Washington. 1959.
8. McKenzie, W. M. The Effect on Nitrogen Availability of Adding Fragmented Wood to the Soil. *Austral. Jour. Agr. Res.,* **9**(5):664–679. 1958.
9. Othmer, D. F., C. H. Gamer, and J. J. Jacobs, Jr. Oxalic Acid from Sawdust. *Indus. and Engin. Chem.,* **34**:262–273. 1942.
10. Overholser, J. L. Sawdust Mulches for Larger Crops, Better Soils. *Oreg. Forest Prod. Lab., Rpt.* G-5. 1955.
11. Simmons, F. C., and A. R. Bond. Sawmill "Waste" in Maryland. *Northeast. Forest Expt. Sta., Sta. Paper* 74. 1955.
12. Smith, W. R., G. H. Englerth, and M. A. Taras. Wood Residues in North Carolina: Raw Material for Industry. *U.S. Forest Serv. and N.C. Dept. Conserv. and Devlpmt., Resources—Indust. Ser.* 8. 1955.
13. Stanford Research Institute. America's Demand for Wood, 1929–1975. Palo Alto, Calif. 1954.
14. U.S. Department of Agriculture, Forest Service, Forest Products Laboratory. Uses of Sawdust and Shavings. Report No. R1661-1 (revised). 1961.
15. Wilde, S. A. Marketable Sawdust Composts: Their Preparation and Fertilizing Value. *Forest Prod. Jour.,* **8**(11):323–326. 1958.

CHAPTER 15

WOOD FUEL

At one time wood was extensively used in the United States for domestic heating and cooking purposes, as locomotive fuel, and for generating steam and later electricity in industrial establishments.* Per-capita consumption of wood fuel in the United States reached its peak in about 1860 with an average of 4.5 cords. The peak of volume consumption, however, did not occur until the period just before 1880, when the total consumption of wood fuel was estimated at about 146 million cords. The 1952 rate of wood-fuel consumption for domestic use is placed at about 50 million cords, and it is expected to decline by 1975 to about 26 million cords, 17 million of which will be used for fireplaces and 9 million for other heating purposes.[1,12] In spite of this sharp decline, wood is still consumed in larger quantities for fuel than for any other purpose except lumber.

Industrial use of wood as fuel in this country is now confined largely to woodworking industries and is met almost entirely from wood residues. It places little demand on merchantable timber. In many instances this is, at least in part, an incineration operation and is quite inefficient. The diversion of solid-wood residues to other more profitable uses, such as pulp chips, and conversion from steam power to internal combustion engines by small mills have resulted in further displacement of wood as industrial fuel by other types of fuel.

Requirements for wood as a household heating and cooking fuel are related directly to the rate of modernization of domestic heating and cooking equipment. In cities and suburban areas, the use of wood as fuel is now limited mostly to fireplaces. Exceptions to this are found in the sawmilling districts of Oregon and Washington, where sawdust and hogged wood are still used for domestic heating. In rural districts wood has also been steadily giving way to other types of fuel. The number of homes still using wood for central heating and cooking, both in rural and in city areas, is believed to be less than 10 per cent of the total occupied dwellings.

* R. V. Reynolds and A. H. Pierson [11] report that on the basis of timber consumed in the 300 years of American history, fuel has accounted for more than half (estimated at 1,000 billion cubic feet of standing timber) of the total volume cut from our forests and for more than twice as much as lumber; even in 1940 the quantity of timber cut for fuel was second only to lumber. They estimate further that about 75 per cent of this total was hardwoods, 40 to 50 per cent of which was oak.

This trend is expected to continue. The only use of wood as fuel that may be expected to show a small increase is its use in fireplaces. This is due to the inclusion of fireplaces in the growing number of new suburban homes being built. It is estimated that the cordwood consumption of fireplace wood will increase from the present total of about 14 million cords to 17 million by 1975.[12] Estimated trends in consumption of fuel wood for other than industrial uses are shown in Table 15-1.

Table 15-1. Estimated Consumption of Fuel Wood in the United States, 1940 to 2000, Millions of Cords [a]

Year	Rural farm	Rural	Urban	Fire-places	Other	Total
1940	38.2	16.6	9.4	12.0	3.1	79.3
1950	20.6	13.0	5.8	14.0	1.6	55.0
1960	7.5	9.3	2.4	14.0	0.6	33.8
1965	4.4	5.8	2.4	15.0	0.4	28.0
1970	2.9	4.3	2.4	16.0	0.2	25.8
1975	2.3	3.7	2.4	17.0	0.2	25.1
2000	2.5	2.9	2.4	21.0	0.2	29.0

[a] SOURCE: Stanford Research Institute. America's Demand for Wood, 1929–1975. Palo Alto, Calif. 1954.

Of the estimated 50 million cords of fuel wood consumed in 1952 for non-industrial uses about one-third was produced from living commercial trees, primarily in farm woodlots. The remainder was a by-product of sawmills and wood-conversion plants. About two-thirds of the fuel wood cut from living trees is from hardwoods and one-third from softwoods. It is anticipated that in the future the proportion of fuel wood cut from living commercial trees will decline still further. A greater share of the fireplace wood will be a by-product of logging and manufacturing operations.

In Canada also, consumption of fuel wood is rapidly declining. From 1940 to 1950 there was a decrease of 58 per cent in the use of fuel wood, and a further decline of more than 50 per cent of both cordwood and wood residues is predicted by 1980.[3] Estimated consumption of cordwood for fuel in 1955 was 3,610,000 cords, valued at about 34 million dollars. In addition, sawmills produced some 650,000 cords of slabs and edging suitable for fuel and some 900,000 units (a unit equals 200 cu ft) of hogged fuel and sawdust. Estimates for fuel wood consumption in 1980 are placed at 2 million cords of the standing timber and 0.5 million cords of the mill residues.

I. WOOD COMBUSTION

A. Advantages and Disadvantages of Wood Fuel

One of the advantages of wood as a fuel is that dry wood ignites readily and gives a quick, hot flame. This makes wood desirable when quick heat for a short time is wanted. Wood fuel is clean and leaves a small amount of ash (about 1 per cent of the ovendry weight of wood), which has some value as a fertilizer. Where locally available and burned in properly designed furnaces and stoves, wood should be just as economical as, or more so than, other types of fuel.

Some of the disadvantages of wood as a fuel are: (1) wood fuel, except when used as sawdust, must be seasoned, requiring large storage space; (2) because of its bulkiness and relatively low margin of profit, wood can be transported only short distances; (3) wood burns rapidly, necessitating frequent refueling; (4) wood fuel gives a low efficiency if burned in an ordinary furnace; (5) it is difficult to obtain wood fuel of uniform quality, especially in so far as its moisture content is concerned.

Some of the above disadvantages, viz., the inefficiency of burning and the necessity for frequent refueling, can be largely eliminated by the use of equipment designed for burning wood. Examples of such improved equipment are the slow-combustion stoves and furnaces now in wide use in European countries; improved wood stoves and furnaces of American design are now also available. These improved stoves are designed to utilize all the combustible elements, which in the ordinary wood furnace are allowed to escape up the chimney. To accomplish this the wood is burned very slowly, with just enough air admitted to allow a slow distillation of wood gases, which are then mixed with an air current of high temperature and burned under the most favorable conditions by being passed through a zone of high temperature. Such stoves and furnaces are said to have greater efficiency [1] than ordinary burners. The slow-combustion furnaces require attention only every 8 to 24 hours.*

B. Available Heat from Wood

The heat value of a substance is determined by the amount of heat (expressed in British thermal units) produced in burning it to total ash. Since different woods are fundamentally alike in the chemical composition of the wood substance, *at the same moisture content,* the heat value obtained from unit weights of all woods, regardless of the species, is about the same. Exceptions to this are woods containing such combustible materials as resins, oils, and gums; these substances have a high calorific value (approximately 17,500 Btu † per lb), and therefore woods containing appreciable amounts

---

* Hale [5] found that when hard maple was burned in the ordinary furnace, its heat efficiency at 20 per cent moisture was 50 to 60 per cent of the gross calorific value of hard maple at that moisture content; white pine had an efficiency of only 50 per cent; and anthracite coal burned in the same furnace gave an efficiency of 73 per cent.

The most inefficient way of burning wood in dwellings is in fireplaces. "An experiment conducted in Georgia is said to have shown that in order to maintain a prescribed temperature in a certain room it was necessary to burn 10 times as much wood in a fireplace as the amount required when using a good wood stove." [11]

When consumed in the fireplace, wood fuel must be regarded mainly as a luxury, intended to satisfy aesthetic needs of the occupants.

† *Btu*—British thermal unit: the quantity of heat required to raise 1 lb of water 1°F, at or near its maximum density.

of them will have correspondingly high heat values. Tests conducted at the Forest Products Laboratories of Canada indicate that the average *higher* calorific value of hardwoods is about 8500 Btu per lb of ovendry wood,* while that of white pine is 9000, Douglas-fir 9200, and western redcedar 9700 Btu per lb of ovendry wood.[5] By way of comparison, anthracite coal yields 12,900 Btu per lb, bituminous coal 13,200, and crude oil 19,500 per lb.

The higher calorific value represents the total amount of heat given off by wood burned in a calorimeter, or it represents the calculated heat value based on a chemical analysis of the wood. In any fuel containing hydrogen, the calorific value, as determined in a calorimeter, is higher than that obtainable under most working conditions because of the latent heat of the volatilized water formed in the combustion of the hydrogen. This heat is released when the vapor is condensed, but in practice the vapor usually passes off uncondensed. In estimating the higher calorific value of wood on the basis of a chemical analysis, to secure a reasonable degree of accuracy the calculations should be based on an ultimate analysis that reduces the fuel to its elementary constituents, such as carbon, hydrogen, oxygen, nitrogen, ash, and moisture. An ultimate analysis does not reveal how these constituents are combined in the fuel. Fuels of almost identical ultimate analyses produce different heating values when tested in a calorimeter. Although methods other than these two have been used to determine calorific values, the heat values determined in a calorimeter are regarded as the most reliable.

The *lower* calorific value, the available heat from the combustion of wood, is greatly affected by several factors, such as (1) the amount of moisture in the wood, (2) moisture formed in the burning of hydrogen, (3) the amount of air admitted for combustion, (4) heat carried away in dry chimney gases, (5) heat loss because of unconsumed gaseous combustible matter discharged with flue gases, (6) losses due to unconsumed solid combustible matter, and (7) radiation losses from the combustion chamber. Minor losses, usually ignored, include (1) heating of water vapor in the air admitted for combustion, (2) heating of wood substance to the ignition temperature, and (3) the heat required to separate bound water from wood substance.† Table 15-2, prepared on the basis of certain assumptions, shows the effect of moisture on the lower calorific value of wood.

Wood fuel is generally used in the air-dry condition. A comparison of the available heat in green and air-dry wood and their equivalents in coal is given in Table 15-3.

* The U.S. Forest Products Laboratory uses a figure of 8600 Btu as the average higher calorific value for hardwoods.

† For 1 lb of ovendry wood, the loss due to atmospheric moisture (assuming an initial temperature of 62°F, a relative humidity of 80 per cent, 50 per cent excess air, and a flue-gas temperature of 450°F) is about 15 Btu, the loss due to heating the wood to the ignition temperature (assuming that the initial temperature of the wood is 62°F and that the ignition temperature is 472°F) is about 134 Btu, and the loss due to the heat required to separate water and wood substance is about 34 Btu in drying from the fiber-saturation point (30 per cent moisture content, ovendry basis) to the ovendry condition.

Table 15-2. Heat Available from 1 Lb of Moist Wood

Moisture content of wood, %		Available heat, Btu [a]
Ovendry basis	Moist basis	
0	0.00	7098
5	4.76	6701
10	9.09	6341
15	13.04	6011
20	16.67	5710
25	20.00	5432
30	23.08	5176
40	28.57	4718
50	33.33	4322
75	42.86	3529
100	50.00	2934
150	60.00	2101
200	66.67	1546
250	71.43	1149

[a] Based on a higher calorific value of 8600 Btu, an initial wood temperature of 62°F, a flue-gas temperature of 450°F, an initial air temperature of 62°F, and 50 per cent excess air.

Table 15-3. Approximate Weight and Heating Value per Cord (80 Cu Ft) of Different Woods, Green and Air-dry (20 Per Cent Moisture) [a]

Woods	Weight, lb		Available heat, million Btu		Equivalent in coal tons	
	Green	Air-dry	Green	Air-dry	Green	Air-dry
Ash......................	3,840	3,440	16.5	20.0	0.75	0.91
Aspen....................	3,440	2,160	10.3	12.5	0.47	0.57
Beech, American...........	4,320	3,760	17.3	21.8	0.79	0.99
Birch, yellow..............	4,560	3,680	17.3	21.3	0.79	0.97
Douglas-fir...............	3,200	2,400	13.0	18.0	0.59	0.82
Elm, American.............	4,320	2,900	14.3	17.2	0.65	0.78
Hickory, shagbark.........	5,040	4,240	20.7	24.6	0.94	1.12
Maple, red................	4,000	3,200	15.0	18.6	0.68	0.85
Maple, sugar..............	4,480	3,680	18.4	21.3	0.84	0.97
Oak, red..................	5,120	3,680	17.9	21.3	0.81	0.97
Oak, white................	5,040	3,920	19.2	22.7	0.87	1.04
Pine, eastern white.........	2,880	2,080	12.1	13.3	0.55	0.60
Pine, southern yellow........	4,000	2,600	14.2	20.5	0.64	0.93

[a] Data from the U.S. Forest Products Laboratory, except for Douglas-fir and southern pine. Heat value of coal, under similar conditions of combustion, is considered to be 11,000 Btu.

II. TYPES OF WOOD FUEL

Besides cordwood cut expressly for use as a fuel, other types of wood fuel include mill residues, such as edgings, trimmings, slabs, and short lengths; sawdust and shavings; wood briquettes; wood charcoal; and wood gas.

A. Cordwood

Wood fuel is usually sold by the cord. The standard cord is a pile of wood 8 ft long, 4 ft wide, and 4 ft high. Wood fuel, however, is usually sold by the *short cord,* also called *face cord, rick,* or *rank,* which is a pile of wood 8 ft long and 4 ft high, the width varying with the length of pieces, usually 12, 14, 16, 18, 24, or 36 in. It is customary to pile green wood 2 or 3 in. higher than 4 ft to allow for settling and shrinkage.

Although the space occupied by a standard cord equals 128 cu ft, the solid volume of wood is considerably smaller because of the voids between the sticks. It is customary to assume that a well-piled standard cord will contain 90 cu ft of solid wood. Actually, the volume of wood in the cord varies greatly, depending on (1) the care exercised in piling; (2) the size of the sticks (the smaller the diameter of the sticks, the smaller the solid-wood content of the cord); (3) whether the sticks are round or split (split sticks 10 in. in diameter and over reduce the solid-wood volume by as much as 5 cu ft or more per cord; however, a cord of split wood generally has a greater solid content than a cord of *small*—less than 10 in. in diameter— roundwood); (4) the number of crooked and limby sticks in the cord; and (5) bark thickness.

Measurements made at the Lake States Forest Experiment Station on hard maple and yellow birch cordwood 4 ft 3 in. long, cut for a hardwood distillation plant and piled 4 ft 4 in. high, gave the figures shown in Table 15-4.

The same study showed that if sticks 10 in. in diameter and larger were split, the solid content of the cord would be reduced by about 5 cu ft; if sticks more than 16 in. in diameter were split, the solid content of the cord would be reduced by 7 cu ft.

On the basis of a comprehensive study made in New York State, Hoyle [6] reports that one experienced man can cut $1\frac{1}{3}$ to 7 cords of 14-in. wood per 9-hour day, 2 cords a day per man being a representative figure under average conditions. This includes all woods operations such as felling, skidding, sawing, and splitting when necessary. Hoyle's study, however, was made before the power chain saws largely replaced hand tools. It can be assumed that the rate of production under comparable conditions should be higher when power chain saws are used in woods operations.

The cutting of fuel wood into shorter lengths is frequently done with a portable buzz saw run by a gasoline engine or a farm tractor (Fig. 15-1). If the saw is kept busy all the time, two men can cut about 2 cords of fuel wood per hour. Chain saws are now frequently used in place of buzz saws. This does away with the necessity of skidding tree trunks to a roadside.

Table 15-4. Amount of Solid Wood per Cord or Sticks of Different Diameters [a]

Middle diameter of sticks, outside bark, in.	Number of sticks	Cu ft content with bark
5	93	50
6	75	59
7	61	66
8	50	70
9	42	74
10	35	77
11	30	80
12	26	82
14	20	84
16	15	86
18	12	87
20	10	88

[a] Lake States Forest Experiment Station, *Tech. Note* 8. October, 1928.

Fig. 15-1. Portable buzz saw run by a tractor in a Michigan State University woodlot. (*Courtesy P. W. Robbins.*)

B. Sawdust and Shavings and Other Mill Residue

The greatest use of sawdust and shavings is for fuel. Because of their bulkiness they are consumed largely either at the points of production or by industrial plants nearby. The use of sawdust for fuel requires specially constructed grates and a proper mixture of sawdust with larger pieces of wood residue. For better results green sawdust and shavings are frequently combined with about 50 per cent of hogged mill residue.

The heat value of green sawdust depends largely on its moisture and bark content. Since most of it contains 40 to 70 per cent of moisture, the heat value of green sawdust when burned is only about half or less that of ovendry wood. The advantages of sawdust as a fuel for generating power are its cheapness, cleanliness, and steadiness of combustion. The principal drawback to its use as a fuel is its bulkiness, which necessitates considerable handling and a large storage space, the latter being necessary to ensure against a shortage of fuel. It is estimated that about 1 cu ft of hogged wood and sawdust is required to develop 1 horsepower.

In the Pacific Northwest, sawdust is also extensively used for domestic heating and in hotels. Most of the sawdust used for this purpose is unseasoned Douglas-fir, just as it comes from the saw. It is sold by the unit, which equals 200 cu ft of space and contains about 80 cu ft of solid wood. The use of sawdust in the home requires a special attachment which can be installed in any make of coal-burning furnace. A typical attachment contains a hopper from which the sawdust is fed by gravity to the grate (Fig. 15-2). Important features of such an installation are light feeding of sawdust and control of the volume of air admitted, the two ensuring complete and continued combustion. On the Pacific Coast the normal winter consumption of sawdust ranges from $1\frac{1}{2}$ to 2 units per room.

C. Wood Briquettes

The name *wood briquette* is applied to sawdust, shavings, and hogged wood residue compressed together with or without the aid of a binder for use as a fuel. The manufacturing of wood briquettes for fuel has been successful in European countries, especially in Germany and France, for many years; in the United States, however, briquetting attained a degree of success only after equipment was perfected for compressing dry sawdust and shavings without artificial binders.

The major deterrent factors for successful development of the wood-waste briquetting industry in this country are as follows: (1) a large and continuous supply of suitable raw material is required to make operation profitable; (2) production costs must be kept very low in order to compete with other forms of cheap fuel; (3) because of the bulkiness and low margin of profit, the product must be sold largely in the vicinity of producing plants and therefore requires a dependable local market; (4) with the methods now in use, successful briquetting requires dry material, which necessitates special equipment for drying wood waste. All these conditions can be met in only very few localities, viz., the Pacific Coast region, the Inland Empire, and the South. Wood residues used for briquetting are chiefly from coniferous species. However, at least one southern plant is producing wood briquettes from oak sawdust.

The advent of natural gas in the Pacific Northwest will probably limit the expansion of the market for wood briquettes. At the same time, the diversion

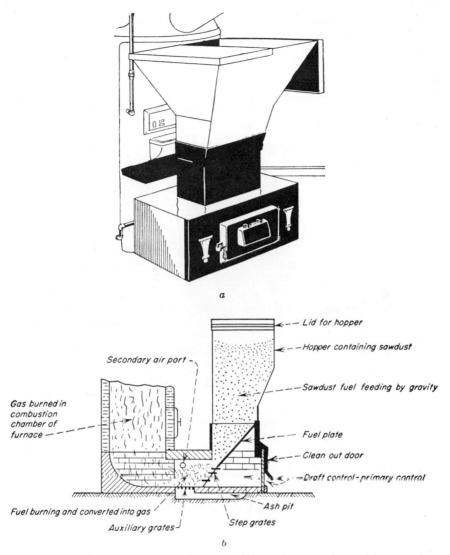

Fig. 15-2. Typical sawdust burner: *a*, conventional attachment to front of heating plant for burning sawdust; *b*, section of sawdust burner attached to furnace. (*Reproduced by permission from the Utilization of Sawmill Waste and Sawdust for Fuel, by J. H. Jenkins and F. W. Guernsey, Forest Service Circular 48, Department of Mines and Resources, Ottawa, Canada.*)

of mill refuse to pulp chips will also reduce the sources of raw material suitable for making wood briquettes.

A number of systems have been devised for briquetting wood residues; all of them depend on compression of wood material. In some systems, binders such as coal-tar pitch, petroleum refuse, and waste sulfite liquor are added to the wood before pressing. A few processes resort to mechanical binding such as encircling the briquettes with wire or the use of a central core made of wire or rope.

The only commercially successful systems, however, are those that produce wood briquettes without any chemical or mechanical binders. In these methods, briquettes are formed under pressure adequate to destroy the natural elasticity of the wood or in some cases sufficient to produce cohesion between wood particles due to the softening effect on lignin of heat induced by friction at high pressure.

Fig. 15-3. Rear view of Pres-to-log briquetting machine. Pres-to-log briquette is shown in upper right corner. (*Courtesy Wood Briquettes, Inc., Lewiston, Idaho.*)

The commercially most successful of the binderless briquette processes is the Pres-to-log. This is the trade name of the pressed-wood product in the form of a cylinder $4\frac{1}{8}$ in. in diameter and $12\frac{3}{4}$ in. long, each weighing about 8 lb. The wood material used in the manufacture of Pres-to-logs must be dry and of uniform size. The wood residue, if green, is first ground and then dried with flue gases to a moisture content of 6 to 8 per cent before it is blown through a conveyer pipe to the machine.

The pressing machine has three principal parts: the pressing mechanism, consisting of a tapering screw auger, and a cam-actuated pressing head; a large revolving disk containing a series of dies; and the hydraulic pressure-holding cylinder (Fig. 15-3). The raw material is blown into a tapering spiral compressing screw, and as it moves forward it is compressed into a constantly decreasing space against the rear end of the pressing head. The head is equipped with a cutting edge and a slot. As the head revolves, the material compressed against it is cut in the form of a continuous spiral, which passes through the slot to the front of the head and is compressed to its final density in the cylindrical opening in the die wheel by the forward movement of the head. The pressure exerted by the screw-auger extension head ranges from 15,000 to 25,000 psi. The material is pressed against the rod of the pressure-holding cylinder, which closes the die opening on the side opposite to the pressure head. When one die opening is filled, the die wheel is rotated so that the next opening comes against the pressing screw. Since intense heat, ranging from 350 to 450°F, is developed during the pressing, the die wheel is cooled by means of an internally circulating water system.

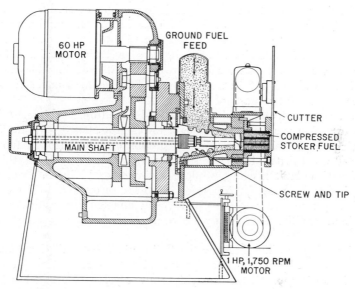

Fig. 15-4. Cross section of Pres-to-logs stoker fuel machine showing method of compressing wood waste into stoker fuel. (*Courtesy Wood Briquettes, Inc., Lewiston, Idaho.*)

It takes about 21 minutes for the disk to be filled. When the first filled die returns to starting position, the briquette is sufficiently cool to be forced out of the die. This is accomplished by forcing into the die holding a cooled briquette the raw material for a new briquette. The pressure on the rod of the pressure-holding cylinder, held against the cool end of the briquette, is so adjusted that the rod gradually recedes, allowing the cool briquette to unload and at the same time exerting sufficient pressure to compress the raw material into a briquette. When the entire briquette is out, the pressure on the rod is released, the briquette is allowed to drop, and the die wheel moves forward to the next position. The pressure cylinder forces the piston rod against the next opening and the operation is repeated. Such a machine has an average capacity of 12 tons of briquettes in 24 hours of operation. A smaller machine capable of producing briquettes of any size from ¼ to 1½ in. in diameter has recently been developed (Fig. 15-4).

Wood briquettes are used for domestic heating in furnaces and fireplaces and in heating and cooking stoves. They are also well adapted for use in diners and lounge cars, automobile trailers, and similar places. The advantages claimed for wood briquettes are their cleanliness, ease of handling and igniting, small volume of smoke, low ash content, and compactness. Pound for pound the heat value of wood briquettes equals that of wood at the same moisture and resin contents. If resins are used for binding, the heat value of briquettes will be correspondingly higher. Since, however, wood briquettes are usually drier than firewood, they will generate more heat per pound of material than firewood, and therefore when used for domestic heating in place of cordwood, a proportionally higher price can justifiably be paid for them.

Wood briquettes treated with chemicals to produce colored flames simulating old driftwood have been developed for use in fireplaces and are generally found on the market as a novelty during the Christmas season. Other special types of wood briquettes include fire lighters and briquettes for automobile tourists' use. Such briquettes are impregnated with inflammable substances to facilitate combustion. Still another type of briquette is made of one part of dry sawdust and two parts of coal dust.

There are several other stoker-size wood-briquetting machines.[9] Of these, a Swiss extrusion machine, the Glomera, forms briquettes $3\frac{1}{2}$ in. in diameter by forcing wood residue by means of pistons or rams through tubes. The compacted wood material tends to separate into disks $\frac{1}{4}$ to $\frac{1}{2}$ in. thick.

An American-built machine produces briquettes $1\frac{1}{8}$ in. in diameter and several inches long by forcing wood particles through 8 to 16 extrusion tubes. The pressure face of the piston has a dimpled center, the imprint of which improves the interlocking between successive layers of compressed wood.

A briquetting machine built by the California Pellet Mill Company forms wood briquettes by forcing wood material with a rotor through tapering orifices in a die cap. All these machines are capable of producing from $\frac{1}{2}$ to 2 tons of wood pellets an hour.

Comprehensive studies conducted in Michigan[4] on the stoker-type wood briquettes indicate that briquettes can be produced at competitive prices with stoker coal and that wood briquettes have a number of advantages over the coal. Most important of these are cleanliness, relative freedom from smoke and clinker formations, and a superior uniformity of combustion. Need for additional storage space, which must be waterproof, is perhaps the principal disadvantage of wood briquettes as a stoker fuel.

D. Charcoal

For use of charcoal as a fuel, see Chap. 19, Carbonization and Destructive Distillation of Wood.

E. Producer Gas

The term *producer gas* is applied to a mixture of combustible and noncombustible gases generated by slow combustion of a fuel in specially designed units called producer-gas units. Different types of fuel have been successfully used in the manufacture of producer gas, among them wood and charcoal. Wood can be used either in the form of small pieces or as a mixture of sawdust with hogged fuel. Sawdust is used green and can be added to the extent of about 50 per cent of the mixture.

The principle of the producer-gas plant involves the conversion of fuel into gas by means of retarded combustion. The fuel is burned in a thick layer in a specially designed chamber called a *generator*,* into which just enough air is admitted to allow complete combustion only in the portion of the fuel bed nearest to the air supply. The oxygen of the air combines with the fuel gases at the source of air supply, forming carbon dioxide gas. This gas passes through the glowing fuel in the rest of the fuel bed, absorbing carbon and forming carbon monoxide. These reactions can be represented as follows: [7]

$$C + O_2 \rightarrow CO_2$$

$$CO_2 + C \rightarrow 2CO$$

These reactions are exothermic and liberate a certain amount of heat in excess of that necessary to maintain the working temperatures of the generator. This excess heat can be utilized for production of more carbon monoxide by admitting water into the generator either in the form of steam or as a fine jet of water. When steam comes in contact with glowing carbon at temperatures above 1000°C, hydrogen and carbon monoxide are formed, thus:

$$H_2O + C \rightarrow H_2 + CO$$

Since this reaction is endothermic, the amount of water admitted into the generator must be carefully regulated so as not to lower the effective temperature within the generator.

Carbon monoxide is allowed to mix with air in the mixing valve, forming a combustible gas. The heating value of wood gas is estimated at 130 to 150 Btu per cu ft of gas. The gas can either be burned directly beneath steam boilers or can be utilized in internal-combustion engines in place of gasoline. If used as a fuel for steam boilers, the gas can be burned without further cleaning; if used in internal-combustion engines, it must be passed through filters to remove tarry and carbonaceous products. The filtered gas is then directed to a mixing valve where it is mixed with air, and the mixture is then passed into the manifold.

Portable producer-gas plants, suitable for automobiles, trucks, and tractors, have been extensively used in Europe, the U.S.S.R., and Australia, when gasoline shortage did much to promote the use of wood gas as a substitute. Because of the low efficiency of the wood-burning gas-producer plants, compared with diesel and gasoline engines, the use of these units in tractors and automotive units is rapidly declining except in the areas where gasoline and oil shortages still prevail.

* Some writers call the generator a *producer* while others confine the term *producer* to the entire producer-gas unit, including in addition to the generator the cleaning and the cooling units.

SELECTED REFERENCES

1. Anon. How to Burn Wood. *Northeast. Wood Util. Council, Bull.* 28. 1949.
2. Cash, J. R., and M. G. Cash. Producer Gas for Motor Vehicles. Angus and Robertson, Ltd., Sydney. 1940.
3. Davis, John, et al. The Outlook for the Canadian Forest Industries. Royal Commission on Canada's Economic Progress, Ottawa, Canada. 1957.
4. Garland, H. Possibilities for the Production of Wood Briquette Stoker Fuel in Northern Michigan. Forest Products Research Division, Michigan College of Mining and Technology, mimeo., Houghton, Michigan. 1950.
5. Hale, J. D. Heating Value of Wood Fuels. Forest Products Laboratories of Canada, Ottawa, Canada. 1933.
6. Hoyle, R. J. Harvesting and Marketing Timber in New York. *N.Y. State Col. Forestry, Tech. Pub.* 49. 1936.
7. Kissin, I. Gas Producers for Motor Vehicles and Their Operation with Forest Fuels. *Imp. Forestry Bur., Oxford, Tech. Commun.* 1. 1942.
8. Rees, L. W. Wood as Fuel. *Minn. Agr. Ext. Div., Univ. Minn., Spec. Bull.* 158. 1932.
9. Reineke, L. H. Briquettes from Wood Waste. U.S. Department of Agriculture, Forest Service, Forest Products Laboratory Report No. 1666-13 (revised). 1960.
10. Reineke, L. H. Wood Fuel Combustion Practice. U.S. Department of Agriculture, Forest Service, Forest Products Laboratory Report No. 1666-18 (revised). 1960.
11. Reynolds, R. V., and A. H. Pierson. Fuel Wood Used in the United States 1630–1930. *U.S. Dept. Agr. Cir.* 641. 1942.
12. Stanford Research Institute. America's Demand for Wood, 1929–1975. Palo Alto, Calif. 1954.
13. U.S. Department of Agriculture, Forest Service. Timber Resources for America's Future. Forest Resource Report No. 14. 1958.

CHAPTER 16

SECONDARY WOOD PRODUCTS

Space does not permit detailed discussions of the numerous lesser products fabricated of wood. A number in this category are of only minor importance in the total national economy, yet at the local level some of these wood-working plants provide the principal source of income for many families in small communities. The few items described on the following pages serve to illustrate the nature and ramifications of such industries.

I. WOOD-CASED LEAD PENCILS

Paleographers are in general agreement that the art of writing, as distinguished from epigraphy (inscriptions on stone and wood), originated more than 5,000 years ago with the Chinese emperor Fu-he. Using a pigmented fluid and brush, he attempted, with considerable success, to record his thoughts and experiences on thin sheets of animal skin. It is also known that the early Romans used a stylus to inscribe their records on waxed tablets of boxwood and on sheets of vellum. Crude pens made their appearance in the seventh century, but the now most universally used writing implement of them all, the wood-cased lead pencil, did not appear until the middle of the sixteenth century. It was about this time that vast deposits of nearly pure graphite were discovered in Cumberland County, England. The soft metal was found ideal for marking sheep to establish ownership identity, and shortly thereafter small rectangular slugs of the metal were fitted between slats of wood to form the first pencils. The Faber Pencil Company was organized in Nuremberg, Germany, in 1760, and is the oldest pencil manufacturer still in existence. In 1876 round leads replaced the rectangular slugs, which by that time were molded with powdered graphite and clay. According to the latest statistics released by the Bureau of the Census, 10,404,000 gross of black pencils were manufactured in the United States during the year 1954; another 1,090,000 gross of colored items were produced in the same period.

A. Pencil Materials

The essential materials used in the manufacture of pencils are graphite, clay, wood, and glue. Graphite is mined in several countries in North America, Europe, Asia, and Africa, but the Ceylon and Mexican graphites are considered best suited to pencil manufacture. Graphite comes in the form of tiny silver flakes and also as amorphous graphite, and the two kinds frequently are blended in making leads. While formerly used exclusively for pencils, only about 7 per cent now finds its way into pencil factories. Suitable clay for the manufacture of pencils is found in New Jersey, Austria, and Czechoslovakia; that coming from Czechoslovakia is reputed to be the best in the world.

The ideal woods for lead pencils are eastern redcedar and southern redcedar, but because of the scarcity of these species, incense-cedar now accounts for almost 98 per cent of the total amount of wood used in this country for pencil making. In Europe, alder is used extensively and is supplemented by imports of African pencil-cedar (*Juniperus procera* Hochst.). The now general use of mechanical sharpeners, however, is widening the range of species that may be used for pencils. Redwood and some woods of the genus *Thuja* are potential pencil woods that have not as yet come into general use.

The preferred woods for pencil manufacture are pleasantly scented, reddish in color, rather brittle, easy to cut with a knife, free from annual rings having hard bands of summerwood, and capable of taking a smooth finish. When alder and the sapwood of the preferred species are used, the wood is often dyed and perfumed so that the final product resembles redcedar in color and scent. Brittleness is an undesirable quality in wood for most uses, but in pencil manufacture it is a desirable property. When the lead, which is rather brittle, is placed in a brittle wood, a flexure of the pencil sufficient to break the lead will usually break the case.

B. Pencil-slat Manufacture

In the days when the eastern and southern redcedars were the two principal American pencil woods, it was a common practice to supplement the log supply with purchases of redcedar fence rails. The split rails and logs upon delivery to concentration yards were placed on wet, mucky soils or in shallow ponds to expedite the decay of the perishable sapwood zone, which was then considered unsuited for pencil cases. This was at best a slow process that required at least 3 or 4 years, but it was generally believed that the heartwood became mellow and brittle during this period, two desirable properties of pencil stock. However, as previously indicated, the bulk of the pencils currently produced in the United States are made from incense-cedar. The practice of permitting the sapwood to rot has been abolished, as such material is now worked up into cheap grades of pencils and penholders. Quantities of penholders are also made of white pine, white birch, basswood, and yellow-poplar.

Logs selected for pencil stock are cut into planks 2⅝ in. in thickness. As a rule the planks are air-seasoned for 3 months and then kiln-dried to a moisture content of 5 to 6 per cent, after which they are ripped into $2\frac{5}{8}$ by $2\frac{5}{8}$-in. strips called squares. The squares in turn are cut into defect-free blocks 8 in. in length, or in multiples thereof, and then reduced to slats of approximately $\frac{3}{16}$ in. in thickness. Finally the slats are fed into a machine which, in one operation, planes them to one-half the thickness of the pencil and cuts eight parallel, equidistant grooves to a depth of one-half the diameter of the lead. When finished in this manner, the slats are packaged and dispatched to the factories for pencil manufacture.

Pencil planks are graded on the basis of the number of defect-free 8-in. blocks they will produce. Three grades are recognized for incense-cedar. No. 1 planks must contain 50 per cent or more of usable blocks; No. 2, 25 to 50 per cent; and No. 3, $12\frac{1}{2}$ to 25 per cent. The principal defect is pocket rot; however, small knots and distorted grain are also causes for rejection.

C. Pencil Assembly and Finishing

The first step in pencil assembly is placing leads in the grooves of a slat that has first been brushed with an adhesive. Following this operation, a second mating slat is placed on top of the leaded one to form a block. A number of these built-up blocks are then put in a frame, pressed under a hydraulically operated head, and while under full pressure, locked in the frame and put aside to allow the adhesive to set.

After the glue has been cured, the blocks are ready to be made into separate pencils. This is done in either of two ways. In the first, the blocks are run through a one-spindle molder that half-shapes them into pencils; then the blocks are turned over and put through the machine again. This method allows the block to be turned end for end in order to get smooth planing, in case there is a slightly cross-grained slat on the second side. The other method forms the completely shaped pencil (whether it is round, oval, hexagonal, or of other shapes) in one passage through a two-spindle molder. Fig. 16-1 shows a two-spindle pencil molder.

The shaped pencils are then sandpapered. Higher-grade pencils are painted, varnished, and burnished. The trimming of the ends is accomplished by two large, rapidly revolving, sandpaper-covered drums set opposite each other a pencil length apart. An endless belt carries the pencils between the two drums. This process removes all varnish on the ends and at the same time reduces the pencils to the same length, usually 7 in.

Fine drawing pencils are made without erasers. A cheap pencil is drilled at one end and an eraser glued into the hole. A better pencil is furnished with an eraser held in a brass or plastic ferrule. To attach an eraser with a ferrule, several steps are necessary. First a shoulder is cut or pressed on one end of the pencil, usually by an automatic machine; the ferrule is fitted on, and the rubber plug inserted, sometimes by an automatic machine and often partly by hand. The eraser is a compound of rubber and an abrasive. To give maximum elasticity and length of life, the rubber erasive material is vulcanized with sulfur.

The last step in the manufacture of a pencil is stamping. Gold leaf, sheet aluminum, and bronze powder are usually used to stamp on the manufacturer's name and trade-mark, and the degree of hardness of the lead. The steel stamping die is generally electrically heated, and it is applied with some pressure, thereby pressing the material into the painted surface. Branded pencils, after careful inspection, are ready to be boxed for shipment.

Pencil manufacture proper is essentially an assembling process, but a few of the larger companies produce their own leads, ferrules, erasers, and slats.

Fig. 16-1. Two-spindle pencil molder. (*Courtesy S. A. Woods Machine Company.*)

II. MATCHES

Phosphorus was discovered in 1670 by the alchemist Brand of Hamburg, Germany. Shortly thereafter it was found that by rubbing a small piece of this element between two sheets of brown paper a flame could be produced which in turn could be used to ignite splinters of wood previously tipped with sulfur. Over 100 years elapsed, however, before the friction match was introduced. The first of these was made in 1827 by the Englishman John Walker, who tipped splinters of wood with a mixture of chlorate of potash, sugar, and gum arabic. Ignition took place when the splinter was drawn rapidly through folds of sandpaper. Walker's matches and several others which appeared shortly thereafter were sold under the name "Lucifers." The first phosphoric matches were invented in 1831 by Dr. Charles Sauria of St. Lothair. He failed to protect his discovery with patents, however, and within a comparatively short time phosphoric matches were being widely manufactured in both Europe and America.

A modern match may be defined as a piece of inflammable material, such as wood, cardboard, or waxed thread,* tipped with a mixture of chemicals that may be easily ignited by the heat of friction.

Matches made in the United States fall into one of two classes: (1) the wood-splint, double-dip, strike-anywhere match; and (2) the wood- or cardboard-splint safety match. The head of the former is composed of a *bulb* and *tip*. The bulb is formulated with a mixture of slow-burning chemicals that are difficult to ignite by frictional means, but which will immediately burst into flame when sufficient heat is generated by the burning tip that is readily ignited by frictional heat produced in striking. Bulb and tip are readily distinguishable by differences in color. Matches of this sort have been manufactured in the United States since 1905.

The safety or strike-on-the-box cardboard or wood-splint matches have heads that contain large quantities of chlorate of potash. Ignition is accomplished by striking the head on the side of a box or folder coated with a composition containing red phosphorus and an abrasive material.

To reduce the fire hazard from discarded matches, splints are impregnated with a weak solution of ammonium phosphate to prevent them from becoming glowing embers after the flames have been extinguished.

The manufacture of matches has been developed to the point where it is a continuous mechanical operation requiring comparatively little labor. In fact, from the time the wood is put into the splint-making machine, the product is untouched by human hands until used by the ultimate consumer.

Two types of wooden match splints are used in this country: the square splint and the round, grooved splint. The latter is more popular than the former, and about 97½ per cent of American wooden match splints are of this kind. The principal species used for the round, grooved splints are the white pines, while aspen is the favored wood for square match splints.

Because most American match factories are located in the eastern half of the United States, many plants using pine match splints obtain their wood in the form of match blocks from mills located principally in Washington and Idaho. Aspen for square-splint matches is usually obtained from points much closer to the match factory. The Bureau of the Census reported that for the year 1954, 26,863 million strike-anywhere matches and 47,545 million safety matches were produced in the United States.

A. Pine Match-block Manufacture

Sawmills usually produce match plank in two thicknesses, each sufficiently heavy so that when air-seasoned or kiln-dried, the lumber will surface to thicknesses of 2 and 2½ in., respectively. The bulk of match plank is air-seasoned from 1 to 2 years, and the balance is carefully kiln-dried. Some prejudice against the use of kiln-dried match blocks exists, based on the

---

* American matches are made with wood or cardboard splints, but small foreign matches, *vestas*, often consist of a splint of cotton thread coated with stearine and paraffin. These splints are smoothed and rounded by drawing them through holes in a metal plate, after which they are cut to length and tipped with igniting materials.

charge that they produce darker colored and weaker splints. If the kiln-drying operation is performed properly, these objections should be eliminated.

Match plank is produced by sawing parallel to the bark of the log in order to obtain straight-grained stock and is selected on the basis of straightness of grain, softness, uniformity of texture, whiteness, freedom from decay, and potential yields of match blocks between knots. Most match-block companies require at least a 60 per cent yield of match blocks from each plank.

After the stock is dried to the proper moisture content, the planks are taken into the match-block factory, where they are crosscut into lengths of

Fig. 16-2. Splint-cutting section of a match machine. (*Courtesy Palmer Match Company.*)

1⅞, 2⅛, and 2⅜ in. The resulting blocks are then carefully inspected for cross-grained areas and other defects. Those not requiring removal of defect are known as *picker blocks*. Defective blocks are sent to a *chopper* who removes the defects with a hatchet. Acceptable blocks from the chopper must be at least 2 in. wide and are known as *chopped blocks*. After a final inspection, the blocks are stored in bins for shipment to match factories.

It is estimated that between 20 to 25 per cent of the annual production of western white pine lumber is used for match plank.

B. Manufacture of Round, Grooved-splint Matches

Both strike-anywhere and safety matches are now being made with the "round," grooved pine splint. The match blocks are placed, with the end grain of the wood up, on a ratchet-driven conveyer that carries them to cutting dies. The feed mechanism for the cutting dies is illustrated in Fig. 16-2. The dies are of a tubular type, descended vertically into the blocks. There are usually 42 or 57 of these dies in one cutting head so that, at each descent, this number of match splints can be cut simultaneously. The head makes about 300 strokes per minute.* On the downward stroke, the dies cut the splints; on the upward stroke, the splints are implanted into holes in an iron plate that is one of a series hinged together to form a continuous belt consisting of 1,600 to 2,400 plates. The perforated area of each plate is about 4 by 16 in., containing 600 or more holes. This belt and the cutting head are geared together in such a fashion that the belt moves forward as the head descends, and as the head rises another set of holes is in position to receive the new splints. The belt travels over large carrier wheels mounted in a suitable framework set on the floor of the factory.

Because the conveyer feeding the match blocks to the cutting dies is set

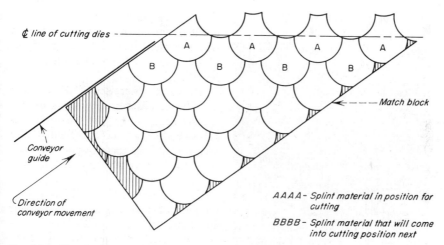

Fig. 16-3. Sketch showing method of cutting round, grooved match splints and utilization of match blocks. Shaded areas are not used. (*Courtesy Palmer Match Company.*)

at an angle with the line of dies, the *face* of the block to be cut moves in a combined forward and lateral manner *with respect to the dies;* the block actually moves in only the one direction determined by the conveyer. With each intermittent movement of the conveyer belt, the cutting face of the block moves, in a direction at right angles to the line of dies, one-half the diameter of a match splint. This is sufficient to cause the face of the block to move laterally, with respect to the dies, so that another row of match-splint material is centered under the dies. When tubular cutting dies are used, the resulting match splints are not cylindrical, but they have one large convex face and three smaller concave faces. Occasional splints will have a flat face because of insufficient wood on those cut from the edges of the match block. Figure 16-3 presents a sketch showing this method of cutting and the utilization of wood in a match block. About 90 per cent of the wood in a shipment of match blocks is converted into splints.

At some plants two lines of match blocks, separated by a partition above the conveyer, are fed into the splint-cutting machine. This procedure fills

* To increase match production in the Second World War, cutting heads were speeded up to 420 strokes per minute.

the ends of the plates in the match-machine belt, leaving an unfilled area in the middle of each plate.

After the splints have been implanted in the belt, the next operation performed as the belt moves forward is that of brushing. Here a revolving brush strikes the splints and removes broken or weak ones and, in addition, removes tiny splinters from the remaining sticks.

Following the brushing operation, the belt is depressed so that the splints are dipped into a vat containing ammonium prosphate to impregnate them for the prevention of afterglow. The splints are then subjected to a blast of warm air for drying.

When the splints have been dried, the belt again carries them through a vat of melted paraffin to provide a flame carrier, the splints being dipped about one-fourth to one-third their length.

The next operation of the process consists of passing the splints over a composition roller that carries material for the bulb of the head of the strike-anywhere match or, in the case of the safety match, material for the complete head.

The splints are then subjected to a current of properly conditioned air for the purpose of setting the composition. In order to give sufficient time for drying, the belt is made to travel up and down through a series of huge carrier wheels.

Fig. 16-4. Charge end of match machine. (*Courtesy Palmer Match Company.*)

To complete the heads of strike-anywhere matches, the partially dried bulbs pass over a second roller that carries the sensitive composition that will form the tips; final drying in a properly conditioned atmosphere follows. For making matches in a 60-minute cycle, the factory must maintain a temperature of about 72°F and a relative humidity of 50 per cent. At the termination of the cycle, the completed matches are pushed out of the belt plates into a conveyer, to be delivered to automatic packing equipment (Fig. 16-4). The time required to make a completed match varies from 40 to 60 minutes, depending on the individual factory, and the daily output per machine is usually between seven and nine million matches.

C. Manufacture of Square-splint Matches

Match splints of this sort are usually made of aspen. Logs are procured in lengths of 8 ft. At the match factory they are cut into 2-ft bolts, then in turn debarked and rotary-peeled in a small high-speed lathe into sheets of veneer $\frac{1}{10}$ in. in thickness. A spur (or set of several spurs) inserted ahead of the knife scores a rotating bolt in such a manner that the veneer comes off as either (1) two 12-in. ribbons or (2) several strips, each of which is equal in width to the length of the matches to be produced.

When the single-score method is employed, the 12-in.-wide sheets are bulked into piles 10 ft in length and 80 to 90 layers in thickness, guillotined to match length, and then fed into the match-splint chopper, which produces a shower of splints with each stroke. The splints thus formed are conveyed to slotted shaker screens which rejects all "shorts" and "longs." The acceptable splints are then immersed in a chemical solution that will prevent afterglow once the flame of the match has been extinguished. From the immersion tanks they pass through a series of drying ovens and then a barrel tumbler, which not only removes superficial splinters but also smoothes and polishes each splint. Finally they are placed in large hoppers from which they are set up for insertion in the plates of the belt of the match machine.

The strips of veneer cut to match lengths are handled a little differently. These are fed into a machine which first impregnates the sheets with afterglow-preventing chemicals, then dries and cuts them into splints, and finally assembles the sticks into trays for the dipping machine.

At the dipping machine, a joggling mechanism and plunger inserts the splints (from either source) into perforated steel plates fashioned into a long endless belt. Once inserted the splints are partially immersed in hot wax and then dipped into a shallow tray containing the chemicals which compose the striking head. The match is now finished. From the tray the belt travels through driers and then onto a table, where punches behind the plates force out a specific number of matches into a box waiting to receive them. The "skillet," or outer cover, is then automatically slipped over the inner container. Square matches are of the safety type.

D. Manufacture of Paper Matches

Paper matches are made into books, or folders, usually containing 20 matches each. Book matches are of the safety type and are made automatically by a machine that slits and dips the cardboard, puts the composition on the cover, binds, and cuts the books apart.

E. Match Packaging

When wood-splint matches are ejected from the plates of the match-making machine, all heads point in one direction. Matches that are sold in small boxes containing about 40 matches are packed with the heads all in one direction, but strike-anywhere matches in the larger household sizes of boxes are packaged so that the match heads in one layer, containing approximately half the matches, point in one direction, while the heads in the second layer point in the opposite direction. This is done as a safety measure. Should the box be thrown on a hard surface, accidentally igniting the matches, only half the contents would be likely to light.

Small amounts of wood-splint matches of the safety type are encased in paper folders with the heads and friction paper concealed. These matches are ignited when withdrawn from the folder and are particularly adapted for use in windy situations.

F. Matchbox Manufacture

The making of matchboxes is incidental to match manufacture. Equipment for this purpose consists of printing presses and box-making machines. Machines for producing household-size boxes make strips, boxes, and slides (the outer cases). A slitter takes large rolls of cardboard and cuts strips of proper width for the various box parts. Protection strips are folded at the proper points; boxes are folded and glued; and the slide machine takes the printed box covers and folds, glues, and sands them automatically.

Small boxes, having a capacity of about 40 matches each, are made in several ways. Some are made entirely of cardboard, similar to the household-size box, and still others are made with paper-covered wooden slides and boxes. These two types are popular in the United States. A paper box may be perforated so that it can be torn in two for easy attachment to smokers' stands and ash trays. In the manufacture of boxes with wood parts, $\frac{1}{30}$-in. veneer is used and the veneer is scored in the proper places for subsequent folding before it leaves the lathe. Three pieces of veneer are required for each box: one for the bottom of the inside box, one for the rim of this box, and one for the slide or outer case. Very thin paper applied with an adhesive holds the parts together. Birch and maple are commonly used in the United States for these parts of wood boxes.

III. TOOTHPICKS

Charles Forster of Strong, Maine, is credited with being the first American to manufacture toothpicks. It is recorded that he conceived the idea of producing them on a commercial basis after he had observed their use by natives in South America. His first products, although fashioned by hand, were an immediate success, and by 1860 he was literally forced to devise means for their mechanical production in order to keep supply in balance with demand.

A. Flat, Round-end Toothpicks

These simple little articles now sustain a prosperous industry that annually consumes nearly 4 million board feet of white birch lumber. Toothpicks of this sort have been made with maple, gum, and aspen, but white birch is preferred because of its whiteness, straightness of grain, strength, and freedom from objectionable tastes.

In modern machine manufacture of flat toothpicks, white birch logs are reduced to bolts of from $10\frac{1}{2}$ to 24 in. in length. Acceptable bolts must be at least 8 in. but no more than 20 in. in diameter and free from knots and red heartwood. Toothpicks are made from veneer, and in the initial stages of processing the bolts may be debarked and then rotary-cut in a cold condition, but the more common practice is to soften the wood by prolonged steam·

ing or immersion in hot water prior to insertion in a lathe. Spurs set in milled recesses of the pressure bar at $2\frac{7}{16}$-in. intervals score the rotating bolt just ahead of the cutting knife so that several ribbons $\frac{1}{20}$ in. thick and $2\frac{7}{16}$ in. wide are produced simultaneously. Good bolts may ordinarily be reduced to 3-in. cores.

Each ribbon of veneer so produced is then fed into a pointing machine that bevels or skives the edges to produce the thin ends required in a toothpick; the veneer is pulled over two knives by means of feed rolls. These knives may be adjusted to appropriate angles.

After passing through the pointing machine, the skived strip is rolled on a winding spool. When the spool is filled, the roll of veneer is slipped off and is taken to the cutting machine.

The toothpick-cutting machine is an automatic device of the punch-press type that cuts and forms the toothpicks. It is operated at a speed of about 2,000 strokes per minute, each stroke forcing two toothpicks through dies in the bedplate of the machine. The dies are so arranged that there is little loss of material, and approximately 154 toothpicks are obtained from each linear foot of veneer.

The formed toothpicks are dried in trays either in a steam-heated, box-type oven having a recirculating air system or in a more modern automatic conveyer drier. In the box-type oven, the drying time is about 2 hours.

When thoroughly dried, the toothpicks are placed in a tumbling barrel or polishing drum with a small amount of powdered chalk or shaved paraffin. The tumbling action wears off the sharp corners, and the chalk or paraffin gives the toothpick a glaze.

Polished toothpicks are placed in an automatic straightening and box-filling machine. This machine assembles the toothpicks in an orderly manner and places them in small boxes ready for closing. A standard packing case holds 1,152 of these small boxes, each of which contains 750 toothpicks.

A complete unit of toothpick machines consists of one veneer lathe, one pointing machine, one veneer-winding spool, six toothpick-cutting machines, one drying oven, and one straightening and box-filling machine. Such a unit requires only about 15 horsepower for its operation, and its hourly output is about 1,440,000 toothpicks. One standard cord of sound white birch wood will yield from six million to nine million flat toothpicks.

B. Circular Toothpicks

Another type of toothpick with a circular cross section is produced in prodigious numbers. The producers of this item, however, have refrained from divulging the manufacturing procedure on the grounds that to do so would be a disclosure of trade secrets.

C. Specialty Toothpicks

Colored and mint-flavored toothpicks fall into this category. Brightly colored "party picks" in red, green, purple, and other hues are used primarily for handling and/or serving condiments and other foodstuffs. One large midwestern manufacturer produces these items from basswood veneer. His process differs from the more conventional knife-cutting method in that several hundred picks are produced simultaneously using die-punching techniques. The green sticks thus obtained are dried, polished, mechanically sorted to eliminate defective pieces, and then immersed in dye vats. After adequate penetration of the dye has been obtained, the picks are again dried and then boxed. This firm currently produces 3,750 million colored sticks annually. In the production of mint-flavored toothpicks, flavoring extract is used in place of dye.

IV. WOOD EXCELSIOR

Excelsior, a term applied to long, thin strands of shredded wood, is of American origin and first appeared on the market in about 1860 under the name of *wood fiber*. Some years later an upholstering firm, in advertising the virtues and superior qualities of the wood fiber used in padding and filling its cushions and mattresses, coined the name "excelsior," by which the material is now universally known. Excelsior, in its modern connotation, also includes shredded newsprint and other paper waste.

A. Manufacture of Excelsior

The principal qualities desired of a wood for excelsior manufacture include freedom from objectionable odors and stain-producing substances, lightness in weight, lightness in color, and the capacity to produce soft but strong and resilient strands. Bolts used in its production must be straight-grained and devoid of bow, crook, twist, knots, and decay. Cottonwood, basswood, yellow-poplar, aspen, and the southern yellow pines are species principally employed.

1. Preparation of Material. Excelsior wood is usually delivered to a mill in the form of 36- to 56-in.-long bolts. These are usually peeled upon delivery and then piled in $4\frac{1}{2}$ by 8-ft racks for seasoning, inasmuch as the wood must be dried to a 16 to 18 per cent moisture content before processing. The drying period may be as much as 12 to 24 months in the North, and 6 to 9 months in the South. When adequately seasoned, the bolts are cut into 16-in. lengths, and those of more than 9 in. in diameter (6 in. in some mills) are split longitudinally in accordance with recommended patterns delineated in Fig. 16-5.

2. Manufacture. The actual manufacture of excelsior is a relatively simple process. Two types of machines are used in its production. In one, the upright machine (Fig. 16-6), a block of wood is placed and held firmly between two corrugated feed rolls. The block is then brought into contact with a moving slide equipped with steel spurs and a knife, the cutting edge of which is generally toothed. The spurs score the wood, and the resulting strands are then cut off by the knife that follows. No spurs are used, however, in cutting very coarse or ribbon-fine grades. The shreds of wood drop into a conveyer at the end of each stroke of the machine and are delivered to a press where they are made up into bales of from 60 to 200 lb. Several upright machines are usually framed together. A typical installation consists of a battery of six to eight machines, a number that may be efficiently attended by one operator.

The horizontal machines (Fig. 16-7) work on the same principle as the upright, except that each one can handle eight blocks of wood at the same

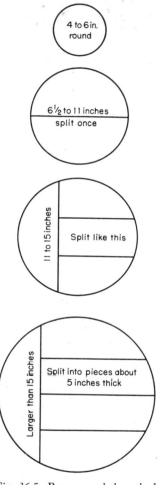

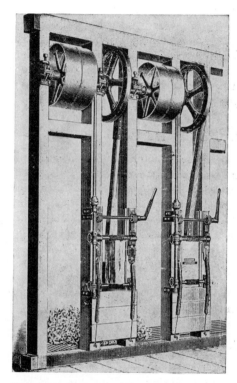

Fig. 16-5. Recommended method of splitting wood blocks for excelsior machine.

Fig. 16-6. Two upright excelsior machines framed together. (*Courtesy Indianapolis Excelsior Machine Company, Indianapolis, Indiana.*)

time. The blocks are fastened directly above slides equipped with oscillating knives or with spurs and knives. The shreds of excelsior drop on a conveyer belt and are carried to the baler. The capacity of an upright machine is 800 to 1,000 lb of the common grade of excelsior in 10 hours, while the horizontal machine can produce 8,000 to 10,000 lb in the same period.

All residues, including the sawdust produced in crosscutting of the bolts

Fig. 16-7. Eight-block horizontal excelsior machine. (*Courtesy Lewis T. Kline, Alpena, Michigan.*)

and the spalts (small pieces of bolts that cannot be entirely utilized in the excelsior machine), are burned as fuel to generate steam for the plant.

The equipment of a typical excelsior-manufacturing plant consists of a barker, cutoff saw, wood splitter, battery of shredding machines, knife and spur grinder, baling device, and scales.

B. Grades of Excelsior and Yield

Excelsior is manufactured in several grades, determined by the thickness and width of strands. These grades are ordinarily classified as follows: (1) *coarse excelsior*—$\frac{1}{20}$ to $\frac{1}{80}$ in. in thickness and $\frac{1}{4}$ to $\frac{1}{16}$ in. in width; (2) *standard*, or *common, excelsior*—$\frac{1}{30}$ to $\frac{1}{100}$ in. in thickness, generally subdivided into several width classes, viz., *fine*, $\frac{1}{26}$ in. wide, *medium*, $\frac{1}{8}$ in. wide, and *coarse*, $\frac{7}{32}$ in. wide; (3) *wood wool*—$\frac{1}{100}$ to $\frac{1}{500}$ in. in thickness and $\frac{1}{32}$ to $\frac{1}{500}$ in. in width. The bulk of production, perhaps 80 to 90 per cent of the total output of excelsior, is of coarse and standard grades.

The average yield of excelsior equals about 1 ton for each cord of 56-in. wood used but varies somewhat with the grade and the quality of raw material.

C. Uses of Excelsior

Formerly the two principal uses for excelsior were as mattress stuffing and as a packing medium. Since 1920, however, its use in the mattress industry has been largely discontinued in favor of cotton linters. Today the common and coarse grades are used principally for protective packing and to some extent in kennel and stable bedding and for stuffing toys; the fine grade, or wood wool, is used largely in low-priced upholstered furniture and in automobile seats and cushions. Recently an insulating material consisting of excelsior and gypsum has been put on the market; excelsior is also used for reinforcing gypsum-fiber concrete in a certain type of building boards. Such boards are low in density, which makes them excellent for heat and sound insulation. At the same time they are strong enough to carry part of the load in structural design. Their rough surface provides a good base for stucco, plaster, or cement. The material may be nailed and sawed and may also be impregnated with fire-resistant chemicals. These uses are, however, of little importance compared with the total production of excelsior. Considerable quantities of excelsior are made into paper-covered pads and wrappers used in packing. Finer grades of excelsior are used to a limited extent in filtering gas and liquid, and the extra fine grades are suitable as absorbent material.

V. WOOD TURNINGS *

Industries classified as wood turners include all plants whose principal activity is the manufacture of round or shaped articles. Wood turning is frequently combined with other processes in the same plant, as in the manufacture and assembly of furniture. Other industries, illustrated by sporting-goods manufacturers and plants making garden or farm tools, may produce turnings for their own consumption, in which case wood turning is integrated with other fabrication processes of major importance to the manufacturer.

A. Location and Raw Materials

Wood-turning industries, like other primary wood-using industries, are concentrated around their supply of raw material. Of particular significance is the relationship of the turning industry to two woods, white birch and hickory, whose principal markets are for specialty turnings and striking-tool handles. Regions and specific areas within them containing forests producing the best quality of white birch and hickory have the greatest concentration of wood turners. New England states, particularly Maine, New Hampshire, and Vermont, the best white birch area, have more wood-turning plants than the combined total in all other states. Tennessee and to a lesser extent Kentucky are the two leading states in the manufacture of hickory handles.

Associated with white birch, and used in ever-increasing quantities for turning as the supply of white birch diminishes, are such northern hardwoods as beech, yellow birch, and hard maple. Turning plants in the southern New England states, New York, Pennsylvania, Ohio, Indiana, Michigan, and the Appalachian Mountain range generally use these denser hardwoods. In the South oak, walnut, cherry, dogwood, and gum are used in the manufacture of important specialty products.

White ash and other ash species have been universally used for handles for garden and farm tools and for baseball bats. Plants manufacturing ash squares and turnings are located in the commercial ash regions of the states east of the Mississippi River. More baseball bats are made in Kentucky than in any other state.

Several other species of wood used by turners to supplement their local supply of woods desirable for turning are hackberry, yellow-poplar, aspen, persimmon, cottonwood, southern yellow pine, and Douglas-fir. The soft-

---

* This section was written by Professor Roy M. Carter, School of Forestry, North Carolina State College, Raleigh, North Carolina.

woods are used primarily for broom handles, and the hardwoods are used in conjunction with other woods for specialty turnings.

B. Importance of the Industry

Hundreds of small communities and towns in the Northeast are wholly dependent upon the wood-turning industry, and collectively the 450 wood-turning plants in the Northeast are an important contributor to the economy of the states and the region.

Wood-turning plants vary in size from single-family units to factories employing over 200 people. Specialty products and changes in toy and novelty markets necessitate adaptation of machines and methods to new products and procedures. Wood turners in northern New England states possess more native ingenuity than in any other section of the United States and provide the art and skill for machine modification so essential to the success of wood turners.

Although the industry does not use a large volume of wood on a national basis, in New England it uses more hardwood than any other industry. As a primary user of forest products, its consumption of wood represents approximately 1 per cent of the total annual forest drain. Wood turning provides the highest value and volume market for white birch, hickory, ash, dogwood, and persimmon.

C. Types of Products

Wood turners make a greater variety of products than any other primary wood industry. From childhood, when toys and miniature furniture items are desired, throughout our entire life cycle, wood turnings serve our needs. A partial list shows the multiplicity of applications and products (see p. 311).

Many woods not specifically mentioned in the list of turned products may be of importance locally in the manufacture of turnings. For example, in Grand Rapids, Michigan, hard maple is used for furniture dowels, while in the South furniture dowels are frequently hickory, oak, and soft maple, although the South is a large consumer of furniture dowels produced in New England. Some furniture plants produce dowels for their own consumption from scrap edgings.

The list of turned products does not include the many shapes and sizes of classified products. An indication of the variety may be gleaned from specifications for sizes and diameters of standard dowels, of which there are over 600 separate items. More than 3,000 different shapes, sizes, and kinds of bobbins are manufactured.

D. Manufacturing Processes

Turning stock is obtained by the wood-turning industry from (1) bolts or short logs, (2) squares and rough dimension stock, and (3) conventional factory lumber. In the white birch belt of New England and Canada, the industry uses bolts, predominantly 4 ft long, purchased at the mill or road-

Partial List of Turned Products

Item		Species used	
Major	Classification	Greatest volume	Other principal materials
Dowels	Rods	White birch	Birch, beech, maple
	Furniture	White birch	Dense hardwoods
	Other uses	White birch	Dense hardwoods
Handles	Cutlery	Dense hardwoods	Birch compreg
	Striking tool	Hickory	Almost none
	Garden tool	Ash	Dense hardwoods
	Farm tool	Ash	Dense hardwoods
	Cutting tool	White birch	Dense hardwoods
	Broom	Dense hardwoods	Hardwoods and softwoods
	Saw	Maple, apple	Dense hardwoods
	Brush	Dense hardwoods	Hardwoods and softwoods
	Knife	Dense hardwoods	Imported woods
	Shovel	Ash, hickory	Dense hardwoods
	Bailwood	White birch	Dense hardwoods
Toys	Complete toy	White birch	Hardwoods
	Toy parts	White birch	Hardwoods and softwoods
	Toy furniture	Dense hardwoods	Hardwoods
Novelties	Checkers, chess	White birch	Hardwoods
	Bottle caps	White birch	Hardwoods
	Beads, buttons	White birch	Hardwoods
	Ornaments	White birch	Hardwoods
	Gift shop items	White birch	Hardwoods
Furniture	Lamp parts	Cherry, walnut	Dense hardwoods
	Chair parts	Dense hardwoods	Hardwoods
	Table parts	Dense hardwoods	Hardwoods
	Knobs	White birch	Hardwoods
	Children's	Dense hardwoods	Hardwoods
Textile	Bobbins	Hard maple	Almost none
	Spools, thread	White birch	Birch, beech, maple
	Shuttles	Dogwood	Birch impreg
	Skewers	Birch, beech, maple	Dense hardwoods
Sporting goods	Bats	Ash	Hackberry, hickory
	Golf tees	White birch	Birch, beech, maple
	Golf heads	Persimmon	Birch compreg
	Croquet	Hard maple	Dense hardwoods
	Rackets	Dense hardwoods	Hardwoods, softwoods
Kitchenware	Bowls, spoons	Birch, beech, maple	Mahogany, walnut
	Handles	Dense hardwoods	Plastics
	Rolling pins	Hard maple	Birch, beech, maple
Shoe parts	Lasts	Hard maple	Almost none
	Heels	Hard maple	Hardwoods, plastics
	Shanks	Dense hardwoods	Hardwoods
Office supplies	Paper-roll plugs	Soft hardwoods	Dense hardwoods
	Paper-roll centers	White birch	Birch, beech, maple
	Pen holders	White birch	Birch, beech, maple
	Paper holders	White birch	Birch, beech, maple
Clothespins	Round, flat, spring	White birch	Birch, beech, maple
Bearings	Oil type	Imported woods	Birch, beech, maple
	Oil impreg	Maple, yellow birch	Beech, dogwood

side. A considerable number of plants purchase and log timber tracts, and some manufacturers own land to maintain a supply of reserve timber. In the hickory, ash, and Southern turning-industry area, short logs, 8 ft or less in length, and bolts are obtained in a manner similar to those in New England. A considerable quantity of squares and flat stock is shipped from Canada to New England wood turners to supplement their supply, while in the South squares are obtained from sawmills catering to turning stock. Conventional factory lumber is generally used by turning plants located some distance from the bolt or short-log supply.

E. Bolts and Short-log Conversion

Following the lumber boom in New England from 1870 to 1890, when the large accessible timber was cut, a bolt-length circular sawmill was developed to use the smaller, less uniform timber. To obtain high yields from small, tapered, crooked trees, logs were cut into 50-in. bolts, the ideal length for handling by one man. Long-log lumber mills were replaced by bolt mills designed to cut bolts not over 75 in. in length. The material cut by these mills was then converted into parts readily usable by the furniture, textile, sporting-goods, and toy industries, which were centered in the New England area for many years.

The bolt-length circular sawmill, originally called "snapdragon" and now known as the bolter mill or bolter, incorporates several advantageous features: (1) the saw cuts parallel with the bark to obtain the straight grain desirable for turnings and dimension stock; (2) one man can turn the bolts on the carriage as well as handle them in the yard and in the woods; (3) the yield from bolts averages approximately 15 per cent greater than the yield in lumber from similar quality long logs; (4) bolts are placed on a dogless carriage, and after each cut the bolt is pushed to an adjustable guide fence instead of being pulled toward the saw as in long-lumber mills; and (5) by always sawing parallel with the bark as the bolt is turned, taper in the bolt is confined to the lower quality central heartwood zone. In bolter operations an off-bearer or mechanical conveyer removes the slab or board from the side of the saw opposite from the sawyer. Boards or planks are cut to the thickness of green turning squares, and the cut boards are subsequently ripsawed into squares. If bolt size permits, a flitch may be cut whose thickness will be the width of boards desired after resawing the flitch.

A typical bolter mill includes one or two auxiliary saws, which are essential to obtain maximum production as well as to increase yield. When squares of several sizes are made, a normal practice, one table ripsaw, called a *stripper,* with adjustable fence cuts planks into large squares, and the second ripsaw cuts the edgings from the first ripsaw and boards into small squares. When boards are cut for multiple-dowel, molder-type plants, a twin circular resaw or twin band resaw and stripper saws are used. The flitch, 3, 4, or 5 in. thick, usually has one straight edge which is placed against the twin saw fence producing two boards with each cut. The guide fence and space between saws are adjustable for sawing boards of different thickness. A conveyer or manual return facilitates continuing the sawing of the flitch. Recovery of usable material from edgings and boards cut before flitching is made on the stripper saws.

Daily production of a bolter mill varies from 3,000 to 15,000 bd ft, or 5 to 25 cords, per day, depending upon (1) the diameter and quality of bolts, (2) the size and type of product, and (3) the efficiency and number of resaws. Average production for a six-man crew is around 6,500 bd ft per day. Appropriate job assignments are one bolter sawyer, two stripper saw men, two stackers, and one yard man to keep the sawyer supplied with bolts and to remove stacked squares or flats to the air-drying yard or kiln.

F. Product Processes

Processes used in manufacturing wood turnings may be divided into three basic machine principles in which (1) the wood is turned and the knife is held rigid, (2) the wood is stationary and the knives rotate, and (3) the wood is slowly rotated to make contact with the rotating knives. All wood turnings are made by using one or more of these procedures, with many variations in machines and knives to form the various machined shapes found in wood turnings. The oldest machine principle, turning the wood, is in use today in mountain craft and custom woodworking shops. Our early settlers and craftsmen used a leather thong or rope wrapped around the wood and a wheel which was rotated by a thong extended to a foot pedal. The operator turned the wood with his feet and had his hands free to use cutting tools. A few craftsmen in the southern Appalachian Mountains are still using this method today for wood turnings and potter's wheels.

Turned products made by this first basic machine principle are spools, bobbins, lamp bases, bowls, furniture parts, and many others. All home-workshop lathes use this principle. Examples of holding the wood and making turnings by rotating the knives are single-dowel machines and multiple-dowel molders for manufacturing dowels used for some items under each of the major products in the partial list of turned products. The third basic principle, slowly rotating the wood against rotating knives, is very common and is used to make such typical products as baseball bats, furniture, shoe lasts, and novelties.

1. Dowels. Commercial dowels range in length from 2 to 48 in., with diameters varying from $\frac{1}{8}$ to $1\frac{1}{4}$ in. The combination of lengths and diameters totals 384 dowel types. Added to these are 138 lengths and diameters of dowel pins used in furniture, and over 100 special-use dowel products, thus giving a combined total of over 600 dowel sizes.

Dowels are manufactured either from 50-in.-long bolter-mill squares or from bolter-mill boards, locally termed *flats* or *slats*. Kiln-dried, rough $\frac{3}{8}$ to $1\frac{1}{2}$-in. squares are fed continuously, end to end, through a single-dowel machine which has a revolving cutterhead with interior knives; the outfeed portion of the head is made approximately the size of the dowel to prevent dowel vibration. Infeed rolls force the squares through the center of the cutterhead.

Multiple-dowel slat machines were developed in the late 1930s by using the principle of a double-surface molder. Instead of using straight knives in the cutterheads, knives were machined with a series of 180° circular grooves one-half the diameter of the dowel. The top cutterhead shapes one-half of the dowel, and the bottom cutterhead shapes the other, each removing about $\frac{3}{64}$ in. of wood between the dowels. Knives must be precisely machined and in perfect alignment on each matched cutterhead to produce a round, smooth dowel. Kiln-dried slats 4 to $4\frac{1}{4}$ in. wide and $\frac{3}{8}$ to $\frac{1}{2}$ in. thick will

Fig. 16-8. Shaping hoe handles. (*Courtesy Blanchard Handle Corporation, Chattanooga, Tennessee.*)

make thirteen $\frac{1}{4}$-in. dowels. Accuracy in cutting rough green slats is essential for the efficiency of the multiple-dowel machine as well as to obtain maximum yield in dowels.

Off-bearers of both single- and multiple-dowel machines quickly select the 48-in. clear-length dowels and place those containing defects in racks or on conveyer belts. Defects are subsequently cut out, and the clear lengths are placed in frames for binding into hexagonal bundles, usually 6 to 48 in. long. Bundles are end-trimmed to exact lengths. A bundle contains 500 dowels $\frac{1}{4}$ in. in diameter and only 50 dowels 1 in. in diameter.

The first multiple-dowel machines on record in the turning industry were made in Maine. Now several woodworking-machinery manufacturers have machines especially designed for dowels. Fork, hoe, shovel, and other long handles are made by this method using 75-in. bolts, cutting the dowels green.

2. Bobbins. Textile industries use over 3,000 types and sizes of bobbins for various machines, types of thread, and process combinations. Bobbin users prefer hard maple, although some beech, yellow birch, and dogwood bobbins have been made.

Bobbin producers made split bobbins for many years from cross sections of logs cut to oversize bobbin length. A splitter made clear, roughly square splits for immediate turning and subsequent drying in a tapered dowel form. This process produced straight bobbins, although the yield was usually low.

Fig. 16-9. Wood bobbin-turning machine. (*Courtesy Draper Corporation.*)

Bobbin manufacturing is now more mechanized and uses squares or plank made into dowels. Green, $1\frac{3}{4}$ in., bobbin squares, for example, are cut to a 9-in. length and then hopper-fed to an end-chucking lathe which turns the square against a stationary, step-tapered knife. Green rough turnings are approximately 1 in. in diameter on the small end and $1\frac{1}{4}$ in. on the ferrule end. The $1\frac{1}{4}$-in. diameter extends along the bobbin about $1\frac{3}{4}$ in. and is tapered down to 1 in. Rough bobbins are commonly shipped green to finishers, where they are kiln-dried in wire baskets to a moisture content of 6 to 9 per cent.

Dry shuttle-type bobbins must be machined to an accuracy of $\pm.001$ to .003 in. in order to avoid spindle wobble, to operate in automatic bobbin machines, and to meet the increased precision required in the use of synthetic fibers. At least three and usually five drills are used to finish boring the hole through the bobbin center for various spindle types.

A new, fully automatic bobbin machine is in operation for machining and drilling dry rough bobbins at the rate of 90 to 100 per minute. The machine has 12 motors, each with a self-centering pick-up cup or socket in the center of which is an end-chuck for turning. The bobbin exterior is finished, and the center is semifinished for final drill finishing to fit the spindles of textile machines.

Bobbins receive the highest quality finishing coatings to protect the wood surface from rough treatment. Usually five to seven coats of oleoresinous or synthetic varnish are applied and are baked at temperatures as high as 295°F to obtain a hard, tough protective film.

3. Spools. Thread spools were once handled many times during manufacturing. New processes and machines, developed by the turning industry's ingenuity, have converted the process to one of the most automatic in the entire wood industry. From the time the dried square or dowel leaves the kiln until it is bagged and ready for shipment, the spool is untouched by man. A 4-ft dowel is picked up by a notched set of flanges and carried past saws which cut the spools to length. The blocks are then fed to a tumbler to remove trim ends, dust, and slivers, and finally dumped on a sorter-conveyer belt where cull blocks are removed. From elevated storage bins the blocks are gravity-fed to automatic spool machines, where they are drilled, barreled, and given a notch to hold the thread. The spools are then conveyed to a tumbling drum, an inspection-sorting belt, and a bagging machine, where the filled bags are tied, tagged, and sent to the shipping department.

4. Ax Handles. Hickory bolts, 38 to 40 in. long, are cut for ax handles on a traveling bed, center-mounted on a 36- to 54-in. bolt saw. Bolts are first quartered, then the blocks are taper-cut into flitches for ax-handle blanks measuring $1\frac{5}{8}$ in. thick by $3\frac{3}{8}$ in. wide at the small end, and at the large end $2\frac{1}{16}$ in. thick by $3\frac{5}{8}$ in. wide for single-bit handles. An operator pushes the bolt through the saw to overcome spring tension which enables the saw table to return to its initial position. In some cases green handles are cut with saws to rough handle shape, although most plants dry the handle blanks prior to shaping.

Air-dried or kiln-dried ax-handle blanks are rough cut to size on saw-type, gouge-type, or knife-type copying lathes. The cutting tool, mounted on a traveling head, moves the length of the slowly turning handle blank, and a cam replica of the handle moves the cutting head up and down, forming the handle. Handles are finished by sanding with 2- or 3-grit size abrasive belts, a saw cut is made in the handle eye, and the grip end is trimmed. Handles may be unfinished, although usually waxed, or they may be finished with oil-based sealers and buffed or burnished.

The handles of many striking tools like hammers, picks, grub hoes, etc., are made on lathes with cam devices for shaping the handle.

5. Products Turned before Drying. Wooden handles for containers, called bailwood, are commonly turned green on lathes. The square or dowel is rotated against a fixed knife, which shapes the handle, and a drill through center is made. "Hot-box" drying, conducted as fast as possible, prepares the handle for inserting the bail.

Paper-roll plugs, roughing operations on lasts, bowling pins, and a number of other products are turned in a green condition. Veneer products like spatulas, tongue depressors, steak markers, matches, toothpicks, etc., are cut green and dried before shipment. Baseball bats, bowling pins, and other large-sized turnings are rough-turned green to reduce drying time and reduce degrade.

Fig. 16-10. Sanding ax handles. (*Courtesy Marion Handle Mills, Inc., Marion, Virginia.*)

SELECTED REFERENCES

1. Anderson, I. V. Match Plank Cut from Western White Pine Trees. *Timberman*, **36**:15–18, 36. 1935.
2. Anon. Matches, 2d ed. The Palmer Match Co., Akron, Ohio. 1935.
3. Anon. The Story of the Lead Pencil. American Lead Pencil Co., Hoboken, N.J. 1936.
4. Banton, H. L. Variation of the Physical Properties and Machinability of Woods Native to Maine and Commonly Used in Wood Industries When Impregnated with Chemicals. Unpublished thesis. University of Maine, Orono, Maine. 1946.
5. Carter, R. M. Trends and Problems of the Wood Turning Industry in the Northeast. Vermont Bureau of Industrial Research. Sixth Vermont Wood Products Conference, Northfield, Vt. 1946.
6. Fleming, C. C., and A. L. Guptill. The Pencil: Its History, Manufacture and Use. Koh-I-Noor Pencil Co., Inc. New York. 1936.
7. Lehman, J. W. Products from Hickory Bolts. Southeastern Forest Experiment Station, Hickory Task Force Report No. 6. 1958.

8. Panshin, A. J. The Michigan Wood Turning Industry. Department of Economic Development of the State of Michigan. 1947.
9. Parsons, F. W. A Tale of Yesterday, Today, and Tomorrow. Joseph Dixon Crucible Co., Jersey City, N.J. 1937.
10. Redman, G. P. Short Log Bolter for Furniture Stock. Southern Furniture Manufacturers Association, High Point, N.C. Undated.
11. Smith, W. P. The Striking Tool Handle and Blank Industry in the Tennessee Valley Area. TVA Report No. 204-52. 1952.

PART III

CHEMICALLY DERIVED PRODUCTS FROM WOOD

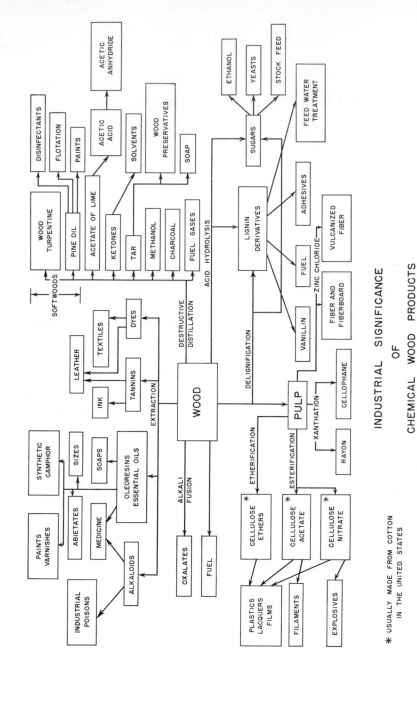

INDUSTRIAL SIGNIFICANCE

OF

CHEMICAL WOOD PRODUCTS

* USUALLY MADE FROM COTTON
  IN THE UNITED STATES

(By permission from Wood Chemistry, Louis E. Wise, ed., Reinhold Publishing Corporation, New York, 1944.)

CHAPTER 17

PULP AND PAPER

The pulp and paper industry * of the United States ranks among the first 10 industries of this country in the value of its products; it employs about half a million people, and its total assets in 1960 were estimated at more than nine billion dollars.

The real beginning of the paper industry of the United States may be dated from 1827, when Henry Barclay of Saugerties, New York, imported from England the first papermaking machine. By 1850 all but two paper mills in the United States were equipped with paper machines. Further expansion of the industry, however, was severely hampered by a scarcity of rags, the only source of raw fibrous material at that time. It was only with the introduction of wood pulp about 1860 that the American pulp and paper industry began its rapid development, more than doubling the value of its products in the following 20 years. This expansion is still continuing at a rate of about 4 per cent annually. An idea of the relative importance of this industry among other wood-using industries of the United States may also be gleaned from the fact that it consumes about 30 per cent of all wood used for industrial purposes in this country. This compares with only 2 per cent in 1900.

In 1960 the United States consumed 39 million tons of paper and paperboard products, of which nearly 80 per cent were produced from domestic pulpwood and old paper, and 20 per cent were imported in the form of pulpwood, various types of pulp, and newsprint, chiefly from Canada.

A study of trends in paper and paperboard consumption in the United States indicates that the 1960 per-capita consumption of 432 lb of wood-pulp products may increase to 550 lb by 1975,† or an increase from about 34 million tons in 1960 to 64 million tons in 1975.

Various United States Government agencies [22, 25] have estimated that the 1965 consumption of paper and paperboard products will require an equiv-

* The term *paper industry* is used here to designate the industries engaged in the production not only of paper but also of various types of board and other items made of wood pulp.

† The world-wide per-capita consumption of wood-pulp products in 1955 was estimated at 47.4 lb, ranging from 1.9 lb in India to 280 lb in Canada.

321

alent of 53 million cords of pulpwood, compared with the 1960 consumption of about 40 million cords. Further projection indicates that by 1975 an equivalent of 65 to 72 million cords of wood will be required to satisfy the country's needs for wood-pulp products. An estimate prepared by the Stanford Research Institute for the Weyerhaeuser Timber Company in 1954,[17] *and now considered too conservative*, presents the following picture of the future pulpwood requirements in the United States.

Table 17-1. Estimated Pulpwood Requirements for Domestic Pulp Production, 1952 to 1975, Millions of Cords, Roughwood Base [a]

	1952	1960	1965	1970	1975
Pulpwood equivalent of wood pulp consumed..	29.5	38.9	44.1	48.9	54.8
Pulpwood equivalent of imports of wood pulp..	3.0	4.1	4.7	5.3	6.0
Pulpwood consumed in U.S.................	26.5	34.8	39.4	43.6	48.8
Net imports of pulpwood...................	2.0	1.4	1.3	1.3	1.3
Requirements for domestic pulpwood........	24.5	33.4	38.1	42.3	47.5

[a] SOURCE: Stanford Research Institute. America's Demand for Wood, 1929–1975. Palo Alto, Calif. 1954.

It is anticipated that hardwoods will progressively become more plentiful than softwoods and that as a result a marked increase in the consumption of hardwoods for pulp will take place, rising from less than 7 per cent in 1940, and 21 per cent of the total in 1960, to more than 30 per cent by 1975.

I. PULP MANUFACTURE *

Pulp has been defined as the crude fiber material, produced either mechanically or chemically from fibrous cellulosic raw material, that after suitable treatment can be converted into paper, paperboard, pulp molded products, rayon, plastics, and other products.[4] Wood pulp is produced either by (1) mechanical attrition, in which case it is called *mechanical,* or *groundwood, pulp* (see pages 337–340); (2) by chemical fiberization of chipped wood, commercially accomplished by one of three methods, viz.: the *acid sulfite process, soda process,* and *sulfate process* (see pages 340–358); or (3) by a combination of chemical treatment and mechanical attrition, in which instance the resulting pulp is designated *semichemical,* or *chemi-groundwood, pulp* (see pages 358–362).

On the basis of quantity produced in 1960, sulfate pulp accounted for about 57 per cent, sulfite pulp for 15 per cent, groundwood pulp for 13 per cent, semichemical pulp for 8 per cent, and soda pulp and exploded fiber for the remaining 7 per cent.

* For the latest statistical information on pulpwood and pulp and paper production and consumption, see current numbers of "Wood Pulp Statistics," United States Pulp Producers Association; "Pulpwood Statistical Review," American Pulpwood Association; and "Current Industrial Reports," U.S. Bureau of the Census.

In 1959 there were 364 wood-pulp mills in the United States, with an annual production capacity of 28.3 million tons of pulp. In that year 54 per cent of pulp capacity was located in the South, 29 per cent in the North and Central states, and the remaining 17 per cent in the West. This compares with only 6 per cent of the pulp capacity in 1920 in the South and most of the remainder in the North and Central states.

A. Papermaking Materials

Pulp can be made from the fiber * of many different kinds of plants. The principal factors which determine the commercial utility of a given plant for pulp and papermaking are (1) suitability of fibers for conversion into pulp; (2) fiber yield per unit volume of raw material; (3) quality of the resulting pulp for papermaking purposes; (4) dependability of supply; (5) cost of collection, transportation, and conversion; and (6) degree of deterioration in storage.

The three principal materials used by pulp and paper industries in the United States are wood, old-paper stock, and cotton and linen rags. Other papermaking materials used in this country are jute, hemp, straw, cornstalks, and sugar cane, or bagasse. Of these, wood † is by far the most important source of pulp on the basis of the quantity produced, representing more than 90 per cent of all the fiber consumed. Both conifers and hardwoods are used; the former, however, are utilized to a much greater extent, because of the greater average length of their fibers and the larger percentage of long fibers per given volume of wood. The average length of fibers in a coniferous wood is about 3 mm, in hardwoods only about 1 mm (Fig. 17-1). The most important coniferous pulpwood species in this country are southern yellow pines, spruces, western and eastern hemlocks, jack pine, true firs, and Douglas-fir. The most important hardwood species are cottonwoods, aspen, beech, birch, maple, redgum, blackgum, yellow-poplar, and oaks (Table 17-2).

Spruces, which until 1939 were the leading pulpwood species on the basis of quantity used, have since been superseded by southern yellow pines, which now account for approximately 50 per cent of all the domestic pulpwood used. Spruces and true firs contribute less than 20 per cent, hemlock about 10 per cent, and hardwoods more than 20 per cent. It may be expected that the relative importance of spruce will continue to decline, while that of southern pines, hemlock, and hardwoods will continue to advance.

It is especially noteworthy that owing to the growing scarcity of the preferred softwoods and improvement in high-yield, semichemical processes of pulping, utilization of hardwoods has been steadily increasing. This is particularly true in the Lake states, where in 1959 more than 58 per cent of pulpwood was produced from hardwoods, principally aspen, and in the Northeast, where 38 per cent of pulpwood was made from mixed hardwoods. In the South, also, utilization of hardwoods is increasing, accounting for almost 18 per cent of the total pulpwood used in 1959. In the West hardwoods compose only about 4 per cent of the pulpwood consumed.

---

* The term fiber as used in this chapter refers to any vegetable cell found in paper, regardless of its exact technical designation.

† Including old-paper stock.

Wastepaper, commonly called old-paper stock, accounts for about one-third of the total pulp consumed in the United States. In certain grades of paper and paperboard it surpasses new pulp in quantity used. The largest tonnage of wastepaper is remade into boards, liners, and newsprint, although the best grades of old paper, properly freed from ink and color, are used extensively in book paper and even in the manufacture of the finest writing papers.

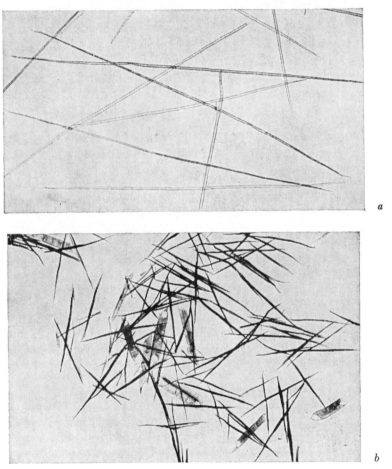

Fig. 17-1. Wood fibers: *a*, spruce fibers, $\times$ 20; *b*, poplar fibers, $\times$ 20. (*From The Manufacture of Pulp and Paper, McGraw-Hill Book Company, Inc.*)

Of other papermaking materials, cotton and linen reach the pulp mills in the form of rags. Jute, manila, and sisal hemp come in the form of old rope, sacking, and burlap. Hemp goes largely into special strong papers used in flour, cement, and other heavy-duty paper sacks. Jute is used in the manufacture of wrapping, buff drawing, and similar strong papers. Numerous attempts made to utilize cornstalks and sugar cane for papermaking have met with scant success in this country, owing primarily to the high cost of baling

Table 17-2. Pulpwood Consumption in the United States by Major Species, Selected Years, 1920 to 1960, Millions of Cords [a]

| Year | Total | Roundwood | | | | | | | | | Residues |
| | | Softwoods | | | | | | Hardwoods | | | |
		Total	Southern pine	Spruce and true fir	Hemlock	Jack pine	Other softwoods	Total	Aspen	Other hardwoods	
1920	6.1	5.2	0.3	3.9	0.9	b	0.1	0.7	0.3	0.4	0.2
1925	6.1	5.2	0.5	3.4	1.1	0.1	0.1	0.8	0.4	0.4	0.1
1930	7.2	5.7	1.0	3.2	1.2	0.2	0.1	0.9	0.5	0.4	0.6
1935	7.6	6.4	1.8	2.9	1.5	0.2	b	0.9	0.4	0.5	0.3
1940	13.7	12.0	5.0	3.7	2.8	0.5	b	1.5	0.6	0.9	0.2
1947	19.3	16.0	8.1	4.0	2.9	0.8	0.2	2.5	0.9	1.6	0.8
1950	29.7	23.1	14.0	4.2	2.9	0.7	1.3	4.7	1.2	3.5	1.9
1953	35.2	23.6	15.3	4.0	2.7	0.6	1.0	6.2	1.2	5.0	5.4
1960 [c]	40.3										6.9

[a] SOURCE: U.S. Department of Commerce, Bureau of the Census; U.S. Department of Agriculture, Forest Service; and the American Pulpwood Association.

[b] Less than 50 thousand cords, included with related species.

[c] Preliminary figures.

and delivery. Considerable quantities of bagasse and a limited quantity of cornstalks are, however, converted into certain types of building board and heat-insulating materials.

Another important source of paper fiber deserving special mention, although not used in the United States, is esparto, a grass common in Spain and North Africa and used in considerable quantities in Europe for high-grade book and printing papers. Various grain straws are important as a source of papermaking fibers in some countries, especially Argentina and Hungary, where the scarcity of other materials makes them an important source of pulp. In the United States the use of cereal straws is limited largely to the manufacture of corrugated board, boxboard, and insulating board, while flax straw is used in the manufacture of cigarette papers.

Wood for pulp comes from two sources: pulpwood, also called roundwood, and wood chips, obtained chiefly from the sawmill residues.

1. Pulpwood. Pulpwood mills in the United States procure pulpwood (1) from loggers, chiefly concerned with cutting sawtimber; (2) from the lands which they own or control; (3) from pulpwood dealers; (4) from pulpwood producers and farmers; and (5) from importers.

Of the total pulpwood cut, about one-fourth comes from timberlands owned or controlled by pulp-producing concerns, and the remainder from privately and publicly owned lands. The large acreage of uncut forest lands owned or controlled by active pulp concerns in the Northeastern, Lake, and Central regions is, to a considerable extent, held as a reserve for future use when more urgently needed. Such reserves, commonly called "security wood" are likely to remain untouched as long as pulp and paper companies can acquire their supply of pulpwood from outside sources at a figure lower or not appreciably higher than the cost of getting out wood from their own holdings.

The steadily rising pulpwood-extraction costs and pulpwood prices have focused industry's attention on cost-reducing techniques of harvesting and transporting pulpwood. A notable example is the substitition of chain saws for hand saws in cutting pulpwood. This has led to a considerable increase in production per man. For instance, a study conducted in a Southern pine region [12] revealed that introduction of chain saws has resulted in reduction of labor requirements for cutting pulpwood from 4 man-hours to 1 man-hour per cord.* The same study has shown that palletized handling of pulpwood also promises considerable cost reduction. On the West Coast installation of chippers capable of handling 24-ft logs, up to 42 in. in diameter, has in many instances led to elimination of breakdown mills and special equipment for removing knots and other defects. Chemical debarking (see page 332) is another example of an attempt at cost reduction in pulpwood production.

A recent development in pulpwood procurement in the South is the introduction of tree-length logging. In operations conducted by the St. Regis Paper Company, timber is felled, skidded, loaded, and trucked to the chip

---

* The *total* production time required to produce a cord of pulpwood in the South is generally calculated on the basis of 8-hour "man-days." In 1950 it required 1 man-day to produce 1 cord of pulpwood. In 1960 production was raised to the average of 1.7 cords, and it is expected that further improvements in harvesting and handling of pulpwood will lead to still further increase in production.

mill in tree lengths. At the mill site the stems are debarked and passed through a cutoff saw, where the material is sorted into sawlogs, poles, and pulpwood. The pulpwood is chipped, and the chips are conveyed into open freight cars for delivery to the pulp mill. It is claimed that this system of logging provides opportunity for integrated utilization of southern pine, because each part of the tree can be assigned its highest economic value. Tree-length logging also results in a 5 per cent, or greater, increase in pulpwood volume recovery, due to reduction of the minimum top diameter from 4 to 2 in., elimination of saw kerf, and utilization of tops and low-grade portions of sawlogs and poles for pulp. Finally, experience has shown that a typical logging unit, consisting of nine men, attains a production rate of about 6 cords per man-day under the tree-length system, compared with the average production on similar sites of $1\frac{3}{4}$ cords per man-day on conventional pulpwood-bolt operations.*

An important innovation in pulpwood production is the Busch Combine, a pulpwood harvester designed by T. N. Busch of International Paper Company in collaboration with the Timberline Equipment Company, and built by Garwood Industries, Inc.

This combine consists of a heavy-duty tractor with four-wheel drive and a two-piece chassis, which allows each section to pivot independently of the other. Mounted on the tractor are two hydraulically operated, scissor-like cutting heads, one for shearing the tree at the base and the other for bucking the stem into pulpwood lengths. The cutting heads are capable of shearing off stems up to 19 in. in diameter (Fig. 17-2a).

The tree is felled parallel with the tractor, over a steel pickup arm which lifts it onto the tractor carriage (Fig. 17-2b). A log-feed device installed on the tractor carriage pulls the tree, butt first, through a delimber, a 1-ft-wide, sharp, flexible steel belt, which strips the branches as the tree is pulled forward (Fig. 17-3a). A bucking scissor cutting head cuts the stem into bolts of proper length, which fall into a cradle attached to the front of the tractor (Fig. 17-3b). When the cradle, holding about $1\frac{3}{10}$ cords of pulpwood, is full, a binder chain is tightened around it, and the load is dumped on the ground in such a manner that pulp sticks stand on end. A vertical bundle thus formed is then bound with wire or a reusable strap, and the binder chain is released. It is stated that an operator and a helper can, on the average, cut and bundle 16 cords of pulpwood in an 8-hour day with this machine.

The bundles are lifted by a specially designed loader and placed on trailer frames (Fig. 17-4); these, in turn, are picked up by the truck trailers for delivery to wood yards or railroad sidings.

A combine designed to cut pulpwood into 8-ft lengths is now under development. It will deposit 8-ft pulpwood into $\frac{1}{2}$-cunit packages, which will be picked up in conventional manner by fork lifts for loading on trucks.

a. Delivery of Wood to the Mill. The method of delivery of pulpwood depends largely on the location of the mill. The most commonly used methods of delivery are (1) water transportation, including driving, towing in booms,

---

* Baker, T. N. Integrated Utilization via Tree-length Pulpwood Logging. *Forest Prod. Jour.,* **10**(8):389–391. 1960.

Fig. 17-2. *a*. Busch Combine Pulpwood Harvester. Two hydraulic cutting heads, resembling giant scissors, are mounted on tractor. One shears tree at base of trunk; the other bucks it into pulpwood lengths. Both heads are capable of slicing through trees up to 19 in. in diameter. *b*. Tree is sheared off at ground level and felled on steel arm which lifts tree onto carriage. (*Courtesy International Paper Company, Southern Kraft Division, Mobile, Alabama.*)

Fig. 17-3. *a.* The combine's delimber (*upper right*) is a sharp, 1-ft-wide, flexible, hydraulically operated steel belt. This wraps around trunk and strips off limbs up to 5½ in. in diameter. Belt is spring loaded, and progressively contracts to fit decreasing diameter of trunk. *b.* Carriage draws stem through delimbing device and into bucking blade. Pulpwood sticks fall into an attached cradle. (*Courtesy International Paper Company, Southern Kraft Division, Mobile, Alabama.*)

rafting, and delivery in barges and vessels; and (2) land transportation, including rail, truck, and sleigh.

Whenever feasible, water transportation is preferred, since it is usually the cheapest method of delivery. Rail transportation is the most expensive and is restricted to hauls not in excess of 600 miles and usually much shorter. In recent years trucking has become an increasingly important method of pulpwood delivery either directly to the pulp mills or to concentration yards and railroad loading points; it is usually confined to a radius of about 65

Fig. 17-4. Busch pulpwood loader delivers wood from combine to specially designed trailer frames. (*Courtesy International Paper Company, Southern Kraft Division, Mobile, Alabama.*)

miles or less and is frequently limited to dry seasons of the year. However, in some localities truck hauling to mills 200 miles or more distant is more economical than the combined truck-rail transportation. Pulpwood, depending on the method of transportation, is delivered either in cord-wood lengths of 4 to 8 ft, in log lengths of 12 to 24 ft, or in tree lengths.

b. Condition of Delivered Pulpwood. Pulpwood which is floated is usually delivered to the mill unbarked. Except in the South, pulpwood delivered by rail, truck, or barge, as well as imported pulpwood, is frequently barked by the pulpwood dealers. The advantages of barking pulpwood in the woods are as follows: (1) barked wood dries more rapidly and is less subject to decay and insect attack; (2) when pulpwood is to be shipped long distances, the shipping charges can be reduced; and (3) if pulpwood is cut during the growing seasons, i.e., in the spring or early summer, the bark can be pried loose more easily by hand.

c. Storage of Pulpwood. Whether pulpwood is used immediately upon its arrival at the mill or held in storage, and the extent of storage in the latter case, depends largely on the method of delivery, climatic conditions, and capacity of the mill.

When the wood is transported by rail, truck, or ship, it must be unloaded immediately to avoid tying up equipment. If pulpwood is delivered faster than it can be used by the mill, the excess is stored in log ponds or, if none is available, in storage piles. The mills depending on water transportation usually have holding areas or storage ponds for impounding pulpwood. In a cold climate or where water driving is confined to certain seasons, the mills accumulate their winter supplies of pulpwood during the spring and summer months. Pulpwood in that case is ordinarily placed in land storage piles. It is recommended that wood be debarked before it is piled, to reduce deterioration. Such piles may attain 80 ft or more in height, their size limited chiefly by the fire risk and the type of equipment available for handling pulpwood.

The two major hazards of pulpwood storage are fire and decay. A serious fire in the storage area may result in a prolonged shutdown of the mill due to lack of raw material. The fire-prevention methods, relating to the size of piles and the condition of storage areas, suggested by the Associated Factory Mutual Fire Insurance Companies, should be followed.

Decay in pulpwood storage is a serious problem because pulpwood generally contains more than 20 per cent of moisture. Decay results in appreciable decrease in the yield and quality of pulp. Underwater storage and adequate ventilation and sanitation in land storage help to reduce decay damage. Spraying the ends of the pulpwood bolts and dipping into an insecticide-fungicide combination have also been recommended. To be effective, the chemical should be applied as soon as possible after cutting, not more than 24 hours in the case of southern pines.

d. Measurements of Pulpwood. Pulpwood is measured either by volume or by weight. The units of volume measurements are: (1) The cord, containing 128 cu ft of space, or 75 to 100 cu ft of solid pulpwood, depending on diameter, crook and knottiness of the sticks, bark thickness, and closeness of piling. (2) The cunit, or C-unit,* consisting of 100 cu ft of solid, unbarked wood. The cunit approximates the actual cubic content of solid wood in the cord and can therefore be taken as the equivalent of a well-stacked cord. (3) The cubic foot. When the latter measure is used, pulpwood is scaled in cubic feet and then converted into cunits or cords. (4) The board foot. It is customary to consider 500 bd ft as a cord of wood.

A number of mills, particularly in the South, are buying pulpwood on the weight basis.[20] Only freshly cut unpeeled wood, however, gives consistent enough weights to make this method practical. Peeled wood does not lend itself to weight purchases, since it is delivered at varying moisture contents. Weight basis per cord is established at the mill by actually weighing a number of scaled loads.

The water-displacement method for determining wood volume and dry

---

* Not to be confused with "the unit," a measure commonly employed in the South to designate a stack of pulpwood 8 ft long and 4 ft high but consisting of bolts greater than 4 ft in length. The units may contain from 128 to 160 cu ft of solid wood. The term unit is also applied to 200 cu ft of sawdust and hogged fuel.

weight of bolts has also been suggested. The green volume, easily and accurately determined by the water-displacement technique, can be converted to corresponding dry weight by the use of conversion factors which are the ratios between green volume and dry weight.*

2. Preparation and Treatment of Pulpwood. When pulpwood is delivered to the mill in log length, it is first reduced to shorter lengths in the breakdown mill (also called cutoff mill), which generally consists of a log haul-up or a jack ladder, a slasher or swing-saw system, and a series of conveyers to carry the blocks and refuse from the mill.

A slasher is designed to cut logs into bolts of the desired length. It is made up of endless chains, equipped approximately every 20 in. with castings or spurs similar to those used on haul-up chains; a table with an inclined plane; and a series of circular saws. The chains carry the logs up the inclined plane against the saws, which cut them into the desired lengths. The pieces, called bolts or blocks, usually 2 to 4 ft long, are then conveyed to the barking drums or to storage piles. Where logs of varying lengths and diameters have to be reduced to short blocks of uniform length, a swing-saw system is generally used in place of a slasher. On the West Coast, the large western hemlock logs are reduced into cants in a band sawmill. In a number of western mills this method is giving way to hydraulic barking, followed by log-size chipping.

3. Barking or Rossing the Pulpwood. Before the blocks are reduced to chips all the bark must be removed, if this has not already been done in the woods. The machines for removing bark are called *barkers*, and the process of removing bark is referred to as *rossing*. The reasons for removing bark before pulping are :(1) the fiber content of bark is very low; (2) if left on the wood the bark would consume the chemicals and steam and occupy space without bringing any return; (3) bark produces dirty pulp and paper; (4) bark of some species contains too much resin.

a. In the Woods. During the active growing season, i.e., from April through June, depending on the part of the country, bark can be easily separated by the use of a long-handled chisel, called spud, or with a draw-shave knife. A number of portable barking machines have also been developed and are in commercial use.[5, 10]

Killing tress with chemicals prior to harvesting has also come into prominence in the Northeast, Lake states, and Canada. This method, called *chemical debarking*, consists of removal of a band of bark at the base of the tree and application of a chemical to the exposed sapwood. The treated trees die within a few weeks, and the bark becomes loose enough to drop off in handling the bolts or can be easily removed by mechanical means. A number of chemicals are effective, but only sodium arsenite, at about 40 per cent concentration, was found to be economical for large-scale operations. The average cost of treatment is $1 to $1.75 per cord of pulpwood. Chemical debarking was found especially effective with hemlock, spruce, aspen, and beech. It is not recommended for use in the South because of excessive decay and insect damage to the treated trees before they can be harvested.

A study of the economic aspects of chemical debarking was carried out

* Haygreen, J. Volumetric Method of Determining Dry Weight of Green Aspen Bolts. *Forest Prod. Jour.,* **9**(1):38–42. 1959.

by the New York State College of Forestry. It was concluded that chemical debarking is generally more costly than drum barking. It could be a valuable adjunct to the latter in situations where wood requirements are slightly higher than the capacity of a drum installation but not high enough to justify the addition of another drum.*

b. At the Mill. The barkers used at the pulp mills can be divided into *mechanical* and *hydraulic* types.

Mechanical barkers are of many designs. They can be classified according to the principle of bark removal as (1) those that are designed to separate

Fig. 17-5. Barking drums. (*Courtesy The Newsprint Service Bureau, New York, New York.*)

bark by cleavage of the cambial layer, and (2) those removing the bark by the cutting action of knives.

There are several methods for removing bark by splitting it at the cambium. These include crushing between rollers, hammering, abrasion with chains, and rolling and tumbling in drums.† Of these, drums of various designs are most commonly used for peeling pulpwood. Most of them work on the same general principle (Fig. 17-5). The drums consist of a large, slightly inclined shell up to 12 ft in diameter and 45 ft in length, with an inner surface made of boiler plate, channel iron, or specially designed corrugated bars. The blocks of wood are fed automatically into one end of the drum within which

* For detailed discussion of chemical debarking see reference 27 at the end of this chapter and mimeographed report by Christiansen, N. D., and J. Fedkiw, Chemical Debarking—Is It Economically Advantageous? State University College of Forestry at Syracuse University, Syracuse, New York. 1959.

† For detailed description of bark-peeling methods and machines, see references 5, 6, and 10.

they are tumbled against one another by a rotational motion of the barrel. This action, assisted by the streams of water played constantly on the blocks, removes the bark; clean blocks gradually work their way toward the opposite end of the drum, where they are discharged on the conveyers.

A variation of the barking drum is the *stationary barker*. One design consists of several trough-shaped pockets equipped along the bottom with several revolving double-ended cams. The cams give a rolling and tilting motion to the logs, resulting in removal of the bark. The bark removal is further facili-

Fig. 17-6. Hydraulic barker. Applies water at 1,350 psi and can handle a 50-ton log, 44 ft long and 7 ft in diameter. (*Courtesy Weyerhaeuser Company, Tacoma, Washington.*)

tated by a water spray which keeps the bark wet. The logs travel gradually from the first pocket into the last, from which they are discharged on a conveyer.

The *knife barkers* are the oldest type of barking equipment. As used in pulp mills, they usually consist of a heavy cast-iron disk mounted on a shaft and equipped with four to seven knives. Usually only short blocks are barked by means of knife barkers. The blocks are held against a rapidly revolving disk and are rotated either by hand or by means of an attachment until all bark is removed. The knife barkers are very wasteful, since they remove a considerable amount of wood with the bark; as much as 25 per cent of the wood is sometimes lost. For this reason, and also because knife barkers are very dangerous to operate, they have been largely replaced by drum and stationary barkers, except for removal of pieces of bark still adhering to the barked blocks and in mills where exceptionally clean wood is desired.

Hydraulic barkers employ water jets under pressure of up to 1,500 psi as a means of bark removal. Hydraulic barkers require a large quantity of water and a considerable amount of power to operate the high-pressure pumps (Fig. 17-6). This is compensated by their large capacity, resulting in reductions in labor cost, in cleaner pulp, and, most important, in saving wood lost in the mechanical barkers.* Hydraulic barkers, when employed at the sawmills, also permit utilization of slabs and edgings for pulp chips (see page 336).

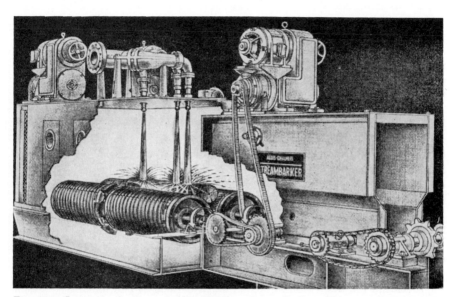

Fig. 17-7. Isometric drawing of Allis-Chalmers Streambarker. (*Courtesy Allis-Chalmers Manufacturing Company, Milwaukee, Wisconsin.*)

There are several designs of hydraulic barkers capable of handling logs of large sizes now in operation on the West Coast and in the South. One of these is the Bellingham type, in which the log is revolved by toothed wheels forming a trough. A stream of water is directed against the log through a nozzle mounted on an arm. The arm is attached to a carriage which travels from one end to the other end of the barker. In other designs water issues from a revolving ring equipped with a number of jets, or from disks, each with several jets, above and below the log.

One model of hydraulic barker, designed for standard-size pulpwood, is the Allis-Chalmers Streambarker. It consists of a chamber equipped with two bed rollers, one of which is fluted and the other spirally threaded (Fig. 17-7). Both rollers revolve in the same direction but at different speeds, so that the log is revolved and moved forward at the same time. The barking is accomplished by water streams under a pressure of 650 psi, issuing from

---

* A 5 per cent reduction in wood loss due to barking, in a mill operating 310 days and using 500 cords per day, at $20 a cord, will result in savings of $150,000 a year.

three nozzles located parallel with, and directly over, the axis of the logs moving through the machine. The first two nozzles, placed close together at the entering end of the barker, loosen the bark, which is then removed by a steel-wire brush placed in the center of the spiral roll. The third jet washes the log clean. Barking is facilitated by the sharp threads of the rolls which score the bark, thus providing a number of points where the water stream can start bark erosion. The logs are brought to and carried away from the barker by a conveyer system. This barker averages about 10 cords of coniferous wood and 6 cords of hardwood an hour.

4. Utilization of Mill Residues.[5] One of the most significant developments in the pulpwood industry is the growing use of mill residues for pulp chips. In 1960 an equivalent of more than 6.9 million cords of pulpwood was obtained in the form of chips. It is expected that by 1965 the chips may provide almost 8 million cords of wood. This is an amount greater than the total United States consumption of pulpwood in 1935.

The utilization of sawmill residues for pulp chips was initiated on the West Coast, where large logs of suitable species and large concentration of residues resulting from daily manufacture of lumber led to adoption of debarking of logs prior to milling and to large-scale production of pulp chips. Since 1958 more than 40 per cent of the total wood requirements for pulping on the West Coast were met by chips. This situation is expected to continue as long as the supply of large timber lasts. In the South chip production in 1960 reached 11 per cent of that region's total pulpwood requirement for that year.

With the development of less-expensive small-sized barking equipment, the conversion of mill residues to chips has extended to other timber regions with pulping facilities with the exception of the Lake states region. Further improvements and reduction in cost of small barkers should make installation of such equipment economically feasible at mills with production as low as 7,000 to 10,000 bd ft of lumber a day, provided they operate on species suitable for pulping and are located at reasonable shipping distances from the pulp mills. This development is warranted because smaller mills use a proportionately larger volume of logs with small diameter than larger installations. In turn, smaller logs produce a proportionately higher volume of chippable residue. For instance, twice as much chippable material results from the manufacture of 1,000 bd ft of pine lumber in the South from 10-in. logs as would from 20-in. logs. In a study conducted in Texas, yield of green chips was found to vary from 1,500 to 4,000 lb per 1,000 bd ft of lumber produced, depending largely on the size and quality of logs. Other studies indicate a yield of 0.4 to 0.7 cords of usable slabs and edgings per 1,000 bd ft of lumber. In the period from 1953 through 1958 more than 300 southern mills installed log barkers and chippers. In 1958 they produced the equivalent of 1.6 million cords of wood, worth 24 million dollars.

Sawmills producing chips remove the bark by passing the logs through a barking unit before sawing. The slabs and edgings developed in the manufacture of lumber, unless recovered for some other use, go to the chipper together with miscut lumber, end trims, and low-grade logs. The chips are screened and may be either stored or loaded directly onto a conveyance.

Chips may be delivered by railroad car, truck-trailer combination, or by barge.

At the pulp mill, chips may be stored in outdoor piles, a method not infrequently used on the West Coast; in specially constructed bins or silos; or dumped directly into hoppers with sufficient capacity for the entire load of chips from a trailer.

The chips are purchased by volume, on the basis of a "unit" (160 cu ft), or more commonly on the weight basis. Some mills convert weight to cords, using 4,200 to 5,200 lb as equivalent to a standard cord of pulpwood.

Some mills not equipped with log-barking equipment find it profitable to install slab barkers. These perform quite efficiently, but the yield of chips is only about 80 per cent that of whole-log barkers. Slab barking presents an opportunity for establishing a slab concentration yard, equipped with a barker and a chipper. Such an organization could handle the slab output of a number of mills within a 30 to 50 mile radius.

Another promising development is a new process for separation of bark from wood chips, developed by Battelle Memorial Institute, and put into operation at the Savannah, Georgia, plant of Union Bag–Camp Paper Corporation. In this process, known as the *Vac-Sink* process, a mixture of wood and bark chips is subjected to a partial vacuum and then submerged in water. When the vacuum is released, the wood chips absorb water and sink, while the bark fragments remain largely unaffected. As a result, the waterlogged wood chips sink while the bark floats. This means possible separation of the two materials.

The *Vac-Sink* process permits production of clean wood chips from unbarked sawmill slabs without the installation of expensive and rather inefficient mechanical barkers. The Union-Camp plant is said to be capable of processing the equivalent of 100,000 cords of wood annually.

B. Mechanical Pulp

Mechanical pulp, otherwise called groundwood pulp, is produced by forcing rossed wood laterally against a revolving grindstone. By this means the wood is reduced to a fibrous mass which, after screening and subsequent thickening, is converted into pulp products. The process was perfected in Germany by Keller, who built the first successful wood grinder in 1840.

1. Woods Used. Softwoods are preferred for the production of groundwood. The principal species used are spruces and balsam fir, but hemlocks, true firs, jack pine, and southern pines are also utilized. Hardwoods, unless specially treated,* are not as well suited for mechanical pulping as softwoods, since they have short fibers which produce a weak pulp. However, groundwood of fair quality can be made from low-density hardwoods, and appreciable quantities of aspen are used for this purpose in the United States.

One of the most important factors influencing grinding is the moisture content of the wood. Better grinding conditions, with less power consumption, prevail when moisture content is more than 30 per cent, preferably 45 to 50 per cent.

* See Cold-soda and Chemigroundwood processes, page 362.

2. Grinders. Machines designed for reducing wood to mechanical pulp are called pulp grinders. Depending on the method of loading, pulp grinders may be divided into two major types: *intermittent* and *continuous*.

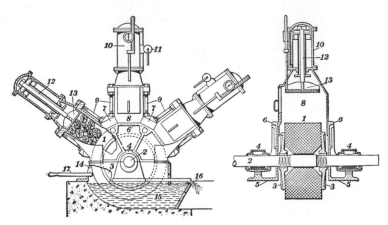

Fig. 17-8. Three-pocket, hand-fed pulpwood grinder. Grindstone *1*, is mounted on a shaft *2*, clamped between flanges *3* and supported by bearings *4*, fixed to bar plates *5*. Stone is covered by steel plates *6*, connected by cross braces *7*. Wood is placed in pocket *8*, supported over stone by steel bolts *9*. Piston, in a hydraulic cylinder *10*, mounted over each pocket is actuated by flow of water under pressure from valve *11*. Piston rod *12* transmits pressure through pressure foot *13* on sticks of wood. Stone is cooled by white water flowing through pipe *14*. Pulp collects in pit *15*, and falls over dam *16* into stock sewers. Stone is sharpened with a burr mounted on a lathe *17*. (*From The Manufacture of Pulp and Paper, McGraw-Hill Book Company, Inc.*)

Intermittent grinders can be further classified as hand-fed, or pocket, grinders and magazine grinders. The *hand-fed* grinder consists of a *grindstone* mounted on a steel shaft (Fig. 17-8). The wood is placed into pockets, of which there may be three or four, and pressed against a revolving grindstone by means of hydraulic cylinders. These grinders use 24- to 48-in. sticks of wood and are capable of producing 5 to 7 tons of air-dry pulp in 24 hours. The resulting ground mass is washed off the stone by a continuous stream of water. The wet pulp, termed *slush*, collects in the pit below the grindstone whence it is conducted to screens. The *pocket grinders* are found mainly in the small and medium-sized mills.

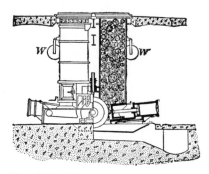

Fig. 17-9. Two-compartment magazine grinder. (*From The Manufacture of Pulp and Paper, McGraw-Hill Book Company, Inc.*)

The main feature of the *magazine grinder* (Fig. 17-9), of which there are many makes, is a grindstone mounted on a shaft, with a magazine holding a quantity of wood placed over it. The sticks of wood are forced against the stone by means of hydraulic cylinders.

The ground pulp is washed into the pit by water, as with the hand-fed grinders. The main advantage of the magazine grinder over the hand-fed type is the lower cost of grinding, achieved by (1) increased holding capacity of the magazines, requiring less attention, and (2) increased production per foot of face of stone. The size of sticks used in the magazine grinders is usually 48 in., and the capacity is 16 to 25 tons of air-dry pulp in 24 hours.

Continuous grinders were developed to increase production by cutting down on the time required for loading wood in the intermittent types. An

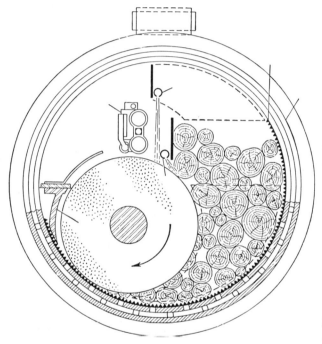

Fig. 17-10. Roberts (ring-type) grinder. (*From The Manufacture of Pulp and Paper, McGraw-Hill Book Company, Inc.*)

example of a continuous grinder of the pocket type is the *Warren and Wright chain grinder*, in which the wood is loaded in a central magazine placed over the stone. The wood is pulled down against the stone by heavy chains, equipped with spiked points, running in grooves along the sides of the magazine.

Another type of continuous grinder is called the *ring type*. One design known as the *Roberts grinder* (Fig. 17-10) consists of a box-like, cast-iron chamber, in which a massive ring revolves in a nonconcentric circle around a standard grindstone. The pulpwood is drawn into the space between the ring and the stone and is forced against the face of the stone at a predetermined pressure.

Grindstones used in the grinders are either (1) natural sandstones quarried in Virginia or Ohio, or imported from England, Canada, or Japan; or (2) artificial grindstones made of artificial abrasives, generally silicon carbide or aluminum oxide, bound together with cement to a reinforced concrete core, or made up into vitrified segments which are connected by heavy screws or bolts to the reinforced concrete core. Today artificial stones are by far the more important. They are more uniform in quality, are available in larger sizes, require less frequent reburring, and are good for 10,000 to 60,000 cords, contrasted with an average life of 1,200 cords for a good natural stone.

Before a new stone can be put into operation its surface must be scarred by a burr. The burr is rotated against the surface of the stone, forming grooves of a definite design on the face of the stone. During the grinding of pulpwood the ground fibers collect in these grooves and are protected there from regrinding until washed from the surface of the stone into the pit. The quality of pulp produced by a grindstone is greatly affected by the type of burr used. When the stone becomes dulled it must be *reburred*, or *sharpened*. The frequency of sharpening depends on the hardness of the stone and the quality of pulp desired.

3. *Properties and Uses of Mechanical Pulp.* Mechanical pulp differs from semichemical and chemical pulps in that it contains practically all the wood substance, including lignin and hemicelluloses. Furthermore, the pulp consists of bundles of fibers as well as of fiber fragments, rather than the individual cell units. Mechanical pulp is characterized by low initial strength. It deteriorates rapidly in strength and turns yellow with age. It is also low in brightness, but this property can be considerably improved by special bleaching techniques. On the plus side, mechanical pulp is characterized by outstanding printing qualities.

The yield of mechanical pulp is about 90 per cent of the dry weight of wood, as compared with 50 per cent or less in chemical processes. Wood density has a considerable bearing on the yield of pulp and on its strength. Conventionally made groundwood from the high-density hardwoods produces pulp of lower strength and poorer color. On the other hand, the high-density hardwoods yield 25 to 30 per cent more pulp per cord than either spruce or such low-density hardwoods as aspen.

Groundwood in most cases is mixed with varying amounts of semichemical or chemical pulp. It is used principally in products which do not require high strength or are intended to be used only a short time. The largest single use for mechanical pulp is newsprint, in which it constitutes 75 to 85 per cent of the total composition. Other important uses of groundwood are in papers for cheap books, catalogues, magazines, toilet paper and toweling, low-grade wrapping paper, and wallpaper. Large quantities of this pulp are also used in some grades of boxboard, wallboard, and paperboard, and in molded wood-pulp products such as plates and egg and fruit containers. Mechanical pulp is also used as an absorbent for explosives in the manufacture of dynamite.

Mechanical pulp possessing a greater strength than groundwood pulp produced by the ordinary method may be obtained by steaming or boiling the logs in retorts at a moderate pressure, prior to grinding. This treatment causes discoloration of the pulp, with the result that pulp produced from the steamed wood is brown in color, the product being known as *brown pulp*. Steamed groundwood pulp is somewhat more expensive to produce, and its dark color limits its use to container boards and wrapping papers simulating kraft paper. The compensating factor, as has been noted before, lies in its greater strength.

Equipment is also available for production of mechanical pulp from chips. An example of this is the *Asplund defibrator*, in which wood chips $1\frac{1}{2}$ by $\frac{3}{4}$ in. in size are used. The chips, fed from a hopper, are first compressed into cylindrical plugs and then subjected to the action of high-pressure steam (140 to 170 lb) for 45 to 60 seconds. The action of the steam renders the chips soft for the grinding operation that follows. From the preheater the chips fall through to a screw-type conveyer which delivers them to metal grinding disks, one stationary and the other rotating. The chips are delivered by a screw-type conveyer to the center of the rotating disk and are then thrown into the grinding zone by centrifugal force. The fiber thus produced is blown into a discharge chamber from which it is discharged into a cyclone separator. In the separator a water shower forms a fiber-water suspension. Asplund defibrators are rated to produce 12 to 14 tons of ovendry fiber per 24 hours of operation.

A number of pulping methods combine mechanical grinding of wood, either in bolt or chip form, with chemical pretreatment. These are described under the heading Semichemical Pulp (see page 359).

C. Chemical Pulp

The object of chemical pulping is to separate wood fibers from each other with minimum mechanical damage; this is accomplished by suitable chemical action which removes the more soluble cementing materials, largely lignin and hemicelluloses, leaving behind a fibrous mass (pulp), consisting of more or less pure cellulose. Chemical pulp is made commercially either by the *acid sulfite process* or by one of the two alkaline methods, known as the *soda process* and the *sulfate process*, respectively.

1. Chipping. In the manufacture of chemical pulp, bolts are reduced to small chips about $\frac{1}{2}$ to 1 in. in length and $\frac{1}{8}$ to $\frac{1}{16}$ in. in thickness. Chipping is accomplished by machines called *chippers* (Fig. 17-11). They consist of a revolving disk equipped with several heavy knives. The blocks of wood are fed into a chipper through an inclined chute making a 45° angle with the horizontal. As a result the chips are sliced on an angle presenting a maximum of open fiber ends to the action of chemicals. Chippers of commercial size will reduce 5 to 30 cords of wood per hour. The largest chippers used on the West Coast have 175-in. disks, driven by 1,500 horsepower motors. They are capable of taking logs up to 42 in. in diameter and 24 ft in length.

The chips as they come out of a chipper are not uniform in size and contain 2 to 5 per cent of wood dust and up to 3 per cent of coarse material. In order to separate the fine material and obtain chips of a fairly uniform size, the chips are put through a screen of either rotary or shaker type (Fig. 17-12). The dust is blown to the boilerhouse, and the large slivers remaining on the screen are put through chip crushers or rechippers, where they are either crushed or cut into smaller chips.

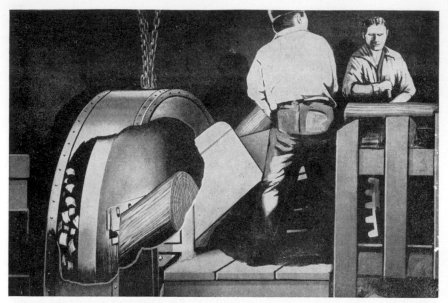

Fig. 17-11. Pulpwood chipper. Debarked pulpwood is reduced to chips as it is fed against a 7-ft disk fitted with four knives and revolving at a speed of 300 rpm. (*Courtesy Hammermill Paper Company, Erie, Pennsylvania.*)

Fig. 17-12. Oscillating screens. These screens allow acceptable chips to drop to a conveyer below that delivers them to storage bins over the digesters. Large pieces are returned to a crusher. (*Courtesy Weyerhaeuser Company, Tacoma, Washington.*)

2. Sulfite Pulp. The first man to suggest the use of sulfur dioxide with the addition of a base for cooking wood was an American chemist, Benjamin Chew Tilghmann, who in 1866 and 1867 was granted English patents on a process of cooking wood with a solution of sulfur dioxide with or without the addition of the bisulfite of alkali, such as calcium or magnesium bisulfite. Extensive experiments were carried on by him and his brother at the mills of W. W. Harding & Sons, in Manayunk, Pennsylvania, but his process, because of inferior equipment, was never a commercial success, and the experiments were finally discontinued. It was the Swedish chemist C. D. Ekman who apparently quite independently developed a commercially successful sulfite process and established the first sulfite mill at Bergvik, Sweden, in 1874. In 1880 a German, A. Mitscherlich, patented a sulfite process which, though it differed little from the original Tilghmann process, was commercially successful. At about the same time Eugene Ritter and Carl Kellner of Austria also patented a sulfite process. It was not until about 1887 that the sulfite process was reintroduced in the United States.

a. Woods Used. Since wood resins are only partially eliminated by the sulfite cook, highly resinous woods are not suitable for this process. The principal species used in the Eastern and Lake states regions are black, red, and white spruces; balsam fir; and to some extent eastern hemlock. In the West the preferred species are western hemlock, white fir, and Sitka spruce. Very little sulfite pulp is made in the South, although at least one Southern mill has produced sulfite pulp from southern pines on a commercial scale. Small amounts of jack pine have also been used in the Lake states. Such pine pulp is said to produce a high percentage of screenings and requires special treatment for removal of pitch.

Substantial quantities of hardwoods are now also pulped by the sulfite process. The most commonly used species are aspen, poplars, redgum, blackgum, birch, and maple. Hardwood sulfite pulp has shorter fiber and lower strength than softwood pulp. It nevertheless finds increasing use in the manufacture of a variety of papers.

b. Liquor Preparation. The sulfite process is an acid process in which the active reagents which reduce wood to pulp are an aqueous solution of sulfurous acid (H_2SO_3) and a bisulfite of calcium, magnesium, a mixture of these, sodium, or ammonium. Of these calcium bisulfite [$Ca(HSO_3)_2$] is the oldest and still the most widely used base; it is derived from limestone ($CaCO_3$). The cooking liquor, as it is called, is therefore a water solution of a bisulfite, containing an excess of sulfurous acid.

The first step in preparation of the acid cooking liquor is the production of sulfur dioxide (SO_2) gas. This is accomplished either by burning sulfur (S) in the rotary or spray type burners or by roasting pyrites (FeS_2) in the roasters. Sulfur dioxide gas is also produced by the fluidization technique, in which finely ground pyrite particles are completely surrounded by upward moving gas heated to from 1600 to 1700°F.

The hot sulfur dioxide gas is cooled by passing it through lead pipes cooled with water or by spraying water directly into the gas stream.

To produce cooking liquor, some of the sulfur dioxide gas must be allowed to react with a base, such as limestone, to form a bisulfite (calcium bisulfite in this case), while the remainder is dissolved in water, forming sulfurous acid. The most widely used method to accomplish this objective is the *two-tower Jenssen* system. The interconnected towers are made of reinforced concrete, lined with acid-resistant tile. The towers are filled with limestone, and the sulfur dioxide gas is forced by a fan at the base of the tower, called a "strong tower." It is so called because the concentration of the gas and acid is greater in it than in the adjacent "weak" tower. The limestone in the weak tower is sprayed at the top with water, which percolates slowly downward while a weak stream of sulfur dioxide from the strong tower rises countercurrently. The resulting weak acid is pumped to the top of the strong tower, where it comes into contact with a rising stream of sulfur dioxide gas. The liquor, known as the *raw acid*, consists of calcium bisulfite with an excess of sulfurous acid. It is conducted to a sulfur dioxide recovery system, where it combines with the unabsorbed portion of SO_2 gas liberated from the digesters during the cooking operation. The resulting hot, *fortified*, or *cooking, acid* is then stored in a pressure, vessel until needed for use in the digesters.

In an efficient system as much as half of the sulfur in the cooking liquor is recycled; i.e., it comes from a recovery system, and only enough fresh sulfur is added to compensate for that lost in venting gases and in spent cooking liquor. The most commonly used recovery system is the hot-acid or *chemipulp* system, in which the heat in the digester relief gases is retained by introducing the hot gases directly into the raw acid under pressure.

The chemical reaction which accompanies acid formation, when limestone is used as a base, may be presented in a simplified form as follows:

$$S + O_2 \rightarrow SO_2$$

$$H_2O + SO_2 \rightarrow H_2SO_3$$

$$2H_2SO_3 + CaCO_3 \rightarrow Ca(HSO_3)_2 + H_2O + CO_2$$

The carbon dioxide gas (CO_2) that forms during this reaction escapes through the top of the tower. If, instead of limestone, milk of lime, $Ca(OH_2)$, is used, the reaction may be represented as follows:

$$Ca(OH)_2 + 2H_2SO_3 \rightarrow Ca(HSO_3)_2 + 2H_2O$$

In many mills, *milk of lime* [$Ca(OH_2)$], made by slaking burned lime with water, is used instead of limestone. In the Baker system the slaked lime slurry is run to the top of a tower, from which it flows by gravity through a series of compartments, each containing several perforated plates. The milk of lime reacts with the sulfur dioxide gas which rises from the bottom of the tower through the plate perforations. The resulting weak acid passes through an overflow pipe to the compartment below, where it is fortified by absorbing more gas. From there raw acid is pumped to the top of the recovery system.

The chief disadvantage of this system is the tendency to plugging of the plate perforations. This has been overcome to some extent by using two towers, so that shutdown could be avoided when plates in one become plugged.

When milk of lime is used, the reaction accompanying acid formation may be represented as follows:

$$Ca(OH)_2 + 2H_2SO_3 \rightarrow Ca(HSO_3) + 2H_2O$$

Although calcium is the most abundant and the cheapest base available for sulfite pulping, other, more soluble bases present certain advantages that merit their use in spite of the higher cost of chemicals. The bases that are now used on a commercial scale are magnesium, ammonium, and sodium. The principal advantages of these more soluble bases are reduced cooking time, with a consequent increase in digester production, and easier recovery of heat and chemicals, with a resulting reduction in stream pollution. Their applicability to pulping species which are difficult to pulp with the calcium-base liquor is another important advantage.

c. *Cooking.* Cooking subjects the wood chips to the action of acid liquor at elevated temperature and pressure long enough to produce pulp, i.e., to free the cellulose fiber from the cementing and encrusting substances. During the cooking process, lignin combines with SO_2 or—HSO_3, forming a calcium derivative (or other salt derivative if other bisulfites are used) of ligno-sulfonic acid, which splits off from the cellulose-lignin complex and passes in that form into the waste liquor. Since the sulfite process presents conditions favorable for acid hydrolysis, the hemicelluloses, which are also removed from the wood during cooking, are hydrolyzed to simple sugars such as mannose, xylose, and glucose. Other by-products of sulfite cooking are acetic and formic acids and methanol. Acetic acid apparently results from a partial hydrolysis of the noncellulosic cell-wall constituents, while the origin of formic acid is unknown. Methanol arises mostly from lignin and to some extent also from the hemicelluloses. Since cellulose is relatively stable under conditions prevailing during sulfite cooking, the major portion of it remains behind in a more or less unaltered state.

Sulfite cooking is carried on in large, vertical steel vessels,* called *digesters* (Fig. 17-13). A digester consists of a cylindrical shell 10 to 17 ft in diameter and 40 to 70 ft in height, with a dome-shaped top and cone-shaped bottom. The shell is lined on the inside with acid-resistant brick, laid in an acid-resistant mortar. The upper end of the shell is equipped with a detachable head, which is removed every time the digester is charged. The head is secured to the rim of the digester shell proper with clamps or bolts that hold it firmly against the internal pressures which develop in the digester during cooking. At the bottom of the digester is a large pipe, opened by means of a blow valve, through which the contents are forced or blown into a pit at the completion of a cook.†

The digesters are fitted with connections to a relief line (usually in a head), couplings for steam and acid lines, and temperature- and pressure-recording and controlling instruments.

* Bronze, lead-lined iron, ceramic-lined steel, and stainless steel vessels are also used.
† Some older digesters used in the *Mitscherlich process* were placed horizontally and were equipped with a large manhole, through which the pulp was washed out or removed with rakes.

The usual capacity of the digester is from 12 to 18 tons of fiber, equivalent to 20 to 35 cords of wood, although digesters with a capacity as high as 40 tons are in use. Chips are loaded into the digester by gravity from a chip bin placed above, or, in some of the newer mills, by belt conveyers, which carry the chips to the top of the digester from chip bins or silos located on the ground floor. Cooking liquor, approximately 2,000 gal for each ton of pulp produced, is added. Then the digester is closed and the temperature gradually raised until the pressure builds up to 70 to 110 lb.

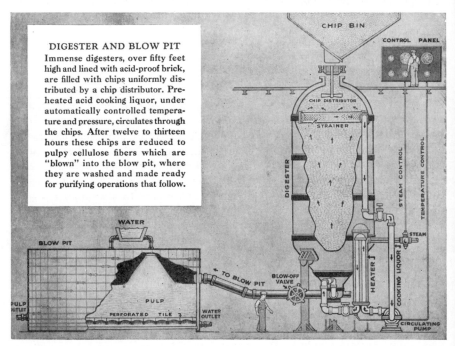

DIGESTER AND BLOW PIT
Immense digesters, over fifty feet high and lined with acid-proof brick, are filled with chips uniformly distributed by a chip distributor. Preheated acid cooking liquor, under automatically controlled temperature and pressure, circulates through the chips. After twelve to thirteen hours these chips are reduced to pulpy cellulose fibers which are "blown" into the blow pit, where they are washed and made ready for purifying operations that follow.

Fig. 17-13. Digester and blow pit. (*Courtesy Hammermill Paper Company, Erie, Pennsylvania.*)

The methods used in sulfite cooking may be classified as (1) the *indirect*, or *Mitscherlich*, process; (2) the *Morterud indirect* system; (3) the *direct-cooking* process, also known as the *Ritter-Kellner* method; (4) the *Chemipulp* hot-acid method; and (5) the *Va-Press* process.

(1) *Mitscherlich Process.* In the Mitscherlich process, now obsolete in the United States, the chips are steamed in the digester for several hours at temperatures not exceeding 100°C. The object of this initial steaming is to obtain better penetration of the cooking liquor into the chips. At the end of the steaming period the steam valve is turned off and the digester filled with cold cooking liquor. Heat for cooking is supplied by copper, lead, or stainless-steel coils placed in the digester. Since steam does not come into direct contact with the liquor, less acid and in more dilute concentration is used than in the direct-cooking system.

The actual cooking time in the Mitscherlich process is 16 to 24 hours, and the complete cooking cycle may take 30 to 40 hours. Because of the length of time required to complete cooking and the high cost of maintaining the steam coils in repair, this process has been replaced in the United States and Canada by faster methods of cooking.

(2) *Morterud Process.* In the Morterud process, the liquor is continuously circulated through the digester and an outside heater. Since heating in this process, as in the Mitscherlich method, is accomplished indirectly, the liquor is not diluted. Similar savings in acid consumption are therefore achieved, but in addition, more uniform temperature conditions prevail in the digester, and a considerable reduction in steam consumption is effected. The cooking time in this process is 6 to 10 hours.

This method is used extensively in Europe. In the United States it is employed in production of certain special grades of sulfite pulp and is used extensively with the alkaline processes of pulping.

(3) *Direct-cooking Process.* In the direct-cooking process steam is introduced directly into the digester; this allows rapid building up of temperature with resulting reduction in time of cooking. The digesters are filled by allowing chips to drop loosely into the digester; to ensure closer packing, some mills employ mechanical distributors consisting of rotating paddles or blow steam or compressed air. The acid is pumped into the bottom of the digester or run into the top before the head is secured. Since the steam condenses in the digester and dilutes the cooking liquor, more concentrated acid must be used in this process than in the indirect-cooking systems.

The actual cooking procedures vary considerably from mill to mill. In the older systems it takes 8 to 12 hours to complete a cook; in a more recent modification, known as the "quick-cook" process, cooking takes 6 to 8 hours. The temperatures are gradually raised to a maximum of 300°F (149°C), the rate of temperature rise being controlled largely by the time necessary to obtain full penetration of the chips by the liquor. If temperatures are allowed to rise too rapidly before the chips are completely saturated with liquor, the chip centers may become hard and turn reddish or brownish in color. The pressure which accompanies the cooking temperatures reaches a maximum of 75 lb. Since pressure in direct-cooking procedures may be due to a large extent to the pressure of the gas liberated during cooking, it cannot be used as a reliable indicator of temperature conditions in the digester. Direct temperature readings, therefore, are indispensable in this process for the maintenance of proper cooking conditions in the digester. Temperatures and pressures are generally controlled automatically by operating the relief or the steam valves. In either case, gases formed during cooking are blown off (*relieved*) intermittently or continuously. This is done to ensure a safe working pressure in the digester, to aid circulation of liquor, and to increase the efficiency of the SO_2 recovery.

(4) *Chemipulp Process.* More than 75 per cent of the sulfite mills in the United States and Canada now employ a modification of the direct-cooking process, generally known as the Chemipulp hot-acid cooking, or recovery, system. This system requires the use of an accumulator, a spherical shell provided with standard digester lining. The accumulator has a capacity of from

$1\frac{1}{3}$ to 3 times the volume of acid required to fill the digester. When the digester is filled with chips and the head is firmly secured, the acid from the accumulator is pumped into the digester. The air from the digester is allowed to escape into the atmosphere. The air vent is closed and the relief lines leading to the accumulator are opened just as soon as the digester is filled with the acid. The hot relief gases from the digester are mixed with the acid in the accumulator. The excess of gases not absorbed in the accumulator is conducted to the absorption tank. After the first few hours of cooking, the connecting lines between the digester and the accumulator are closed, and relief gases are conducted directly to the absorption tanks. The Chemipulp system results in a material saving in steam consumption, reduction in cooking time, higher acid strength, and more efficient recovery of sulfur dioxide.

(5) *Va-Press Process.* A method based on replacing air in the chips with steam or a soluble gas prior to pulping, and thereby obtaining improved penetration of the cooking liquor, has been developed by the Pulp and Paper Research Institute of Canada. Known as the Va-Press process, it is applicable to all chemical pulping processes but is especially successful with the soluble-base sulfite liquor. Higher yields and improved quality of pulp are the chief advantages of this development. Other important advantages are lower consumption of steam and chemicals, reduction in cooking time, and the possibility of cooking mixtures of wood species with varying moisture contents.

d. Blowing the Digester. Cooking in all systems using a calcium base is continued until the SO_2 content in the digester, as determined by test samplings of the acid, reaches a certain point or until the desired quality of pulp, determined from samples blown from the digester, is obtained. When cooking is completed, the digester is emptied under pressure of about 50 lb through the blow valve at the bottom of the cylinder. The contents are forced into a large acid-resistant tank called the *blow pit*. This operation is referred to as *blowing the digester*. The blow pit is equipped with a perforated wood or chrome-nickel-stainless-steel bottom, through which the cooking liquor, containing dissolved lignin and other noncellulosic wood components, is drained into a sewer or is conducted to a plant for recovery of by-products. The pulp is then washed until the water shows no traces of the cooking liquor, leaving a clean, light-colored, fibrous pulp, which is transferred to a storage tank. This operation generally requires several hours.

In the magnesium process, as practiced by the Weyerhaeuser Company, the digesters are not blown but instead are back-filled with waste liquor and dumped into large receptacles. The diluted pulp is separated from the liquor by washing in steel boxes with acidproof lining. The waste liquor, containing 12 to 14 per cent solids, is carried to the recovery plant.

e. Recovery of Gas. The efficiency of recovery of sulfur dioxide in the form of gas is very important from the standpoint of reduction in the consumption of sulfur per ton of pulp. In commercial practice it takes 150 to 300 lb of sulfur to produce a ton of pulp, depending on the efficiency of the recovery system, the quality of pulp, and the species of wood used.

In the sulfite process the first part of the cooking cycle consists of penetration of chips and combination with lignin by the active chemical reagent

(SO$_2$), without appreciable solution of any cell-wall constituents of the wood. This stage is followed by acid hydrolysis and dissolution of sulfonated lignin. To obtain a reasonable rate of chemical reaction, two to four times the amount of SO$_2$ actually consumed is added to the cook. This excess of SO$_2$ is gradually drawn off as a gas from the top or the side of the digester and returned to the acid-making plant to be reused in preparation of fresh sulfite liquor.

The relief gases frequently contain considerable quantities of liquor. These are separated by passing the relief through separators and coolers. The gas is piped to the acid-storage tanks for absorption, while either the separated liquor is discarded or that portion of it which was obtained at digester temperatures below 135°C is mixed with the acid before it is strengthened with relief gases. The relief liquor obtained above 135°C is not used because it contains considerable amounts of decomposed disulfite. In the plants employing the Chemipulp system, appreciable quantities of relief are recovered in the pressure accumulator. The so-called "sulfite turpentine" is sometimes recovered from the condensed relief gases in amounts of from 0.36 to 1 gal per top of pulp.

f. Sulfite Spent Liquor. Calcium-base sulfite waste liquor is one of today's greatest industrial wastes. Most of the mills dispose of it by discharging it into lakes and streams. For each ton of sulfite pulp made, 2,500 gal of the liquor is produced. This liquor has a solids content of 9 to 12 per cent, representing about one-half of the total wood substance used and containing appreciable quantities of chemicals. About half of the solids are lignin, about a quarter are various sugars, and the remainder various chemicals, principally calcium and sulfur.

When discharged into a body of water, besides being a complete economic loss, sulfite liquor creates serious pollution problems. The chief trouble comes from the wood sugars, which, by increasing the supply of food for bacteria and molds, create a deficiency in the dissolved oxygen content of the water and an appreciable increase in the carbon dioxide conditions highly unfavorable to other forms of aquatic organisms.* Federal and state regulations prohibiting pollution of streams make economic disposal of the waste sulfite liquor one of the most pressing problems facing sulfite mills.

A number of approaches to the economical utilization of sulfite liquor have been tried, and some are now in commercial use.

(1) *Precipitation Method.* Of a number of processes proposed, the Marathon-Howard method is successfully used in this country. In this process most of the liquor and the carbohydrates are precipitated with lime. The carbohydrates are filtered out and discarded. The ligneous residue can be dried, at a rate of about 1,250 lb per ton of pulp, and used as fuel or employed in the manufacture of vanillin and other lignin derivatives. In addition, 50 to 75 lb of sulfur and 80 to 130 lb of calcium oxide per ton of pulp are also recovered.

---

* The 5-day biochemical oxygen demand test (B.O.D.) is generally used to determine the consumption of oxygen in the streams by a waste effluent. The comparable B.O.D. values expressed in pounds of oxygen consumed per ton of pulp produced are: groundwood pulp, 3 to 8 lb; kraft pulp, 55 to 70 lb; semichemical pulp (without recovery), 125 to 275 lb; sulfite pulp (without recovery), 400 to 600 lb.

(2) *Evaporation and Burning Methods.* Sulfite waste liquor contains solids capable of yielding 8000 to 8500 Btu per lb, or two-thirds as many heat units as a pound of coal. The cost of evaporating large quantities of water, corrosion problems, and excessive scale formation present the major obstacles to this method of using sulfite liquor. Nevertheless, several methods of evaporation are now in use, particularly in Sweden.

(3) *Other Uses of Sulfite Waste Liquor.* More than 70 million gallons of unconcentrated sulfite liquor are used annually in the United States as a dust binder on unpaved roads. Sulfite liquor can be used for ethyl alcohol and yeast production for stock feed (see Chap. 20). Other uses include extracts for tanning leather, manufacture of dyestuffs, as a source of printing inks, and production of adhesives. The ligneous fraction is suitable for molded plastics, and the entire organic residue has been found to give beneficial results as a fertilizer and soil conditioner. All these uses, however, account for only a very small fraction of the available waste liquor, estimated at more than 8 billion gallons a year.

(4) *"Soluble-base" Sulfite Liquor.* The difficulties encountered in the economical recovery of useful products from the calcium-base sulfite liquor led to the commercial development of the soluble-base methods, employing magnesium and ammonia in place of calcium. In the magnesium-base process the problem of waste liquor is minimized by the recovery of magnesium oxide dust and sulfur dioxide gas. The magnesium oxide is recovered by evaporating the waste liquor and then burning the concentrate. The burned chemical is collected in cyclones and redissolved in water. The sulfur dioxide gas is released in the burners and is carried off into the absorption towers. A nearly 100 per cent reuse of chemicals and very low pollution effect for the effluent is claimed.

In the ammonia-base process, as practiced by the Crown-Zellerbach Corporation, the waste liquor is converted by dehydration to a product (Orzan-A) stated to have beneficial fertilization and soil-conditioning properties.

g. Yield, Quality, and Uses of Sulfite Pulp. The yield of pulp in the sulfite process depends on the species of wood used and on the severity of treatment. The average yields range between 45 and 50 per cent of the dry weight of wood. The sulfite process requires 1.7 to 2.3 cords of rough wood for each ton of pulp produced.

The chemical composition of sulfite pulp is influenced by the extent of delignification, the species of wood, and to a lesser degree by variations in the quality of wood of a given species. Cooking procedures leading to a more complete removal of lignin result in a reduction in the total yield of pulp and an increase in its alpha-cellulose content. Consumption of bleach is also reduced with the reduction in lignin content of the pulp.

If strong, unbleached sulfite pulp is required, it is cooked for a shorter time than the grades of pulp intended for further purification and bleaching. Unbleached sulfite pulp is fairly light in color and can be used for many grades of paper without bleaching; it is extensively used with groundwood pulp in newsprint and other inexpensive papers. The bleached sulfite pulp is used in the higher grades of writing and printing paper and, when further purified with an alkali, in the manufacture of rayon and other chemically derived cellulose products.

3. Alkaline Pulps. The two principal alkaline processes for the manufacture of pulp are (1) the *soda process*, developed by Watt and Burgess in England in 1853, and (2) the *sulfate process*, proposed by C. F. Dahl, of Germany, in 1879.

In both processes caustic soda (NaOH) is the active pulping agent responsible for dissolving the noncellulosic components of the wood, and in both soda is reclaimed, to be used again. However, in the sulfate process sodium sulfide (Na_2S) is also present. Since only about 85 per cent of the soda can be thus recovered, the loss is made up by addition of soda ash (Na_2CO_3) in the soda process and salt cake (Na_2SO_4) in the sulfate process.

The chemical reactions involving dissolution of the noncellulosic components of the wood by an alkali pulping agent are still poorly understood. Alkaline delignification apparently involves three consecutive stages: (1) absorption of alkali by the wood, probably by acidic-phenolic groups in the lignin complex; (2) chemical combination between the lignin and the absorbed alkali; and (3) chemical hydrolysis which causes the alkali-lignin complex to disperse in the liquid.

The alkaline liquors remove the intercellular lignin rapidly and before any action on the lignin in the secondary wall takes place. This is in direct contrast to the sulfite liquor, which removes both types of lignin at about the same rate but more slowly than the alkaline pulping agents. This difference in selective action may be responsible for the fact that less degradation and hence stronger fiber result from the alkaline cooking, especially by the sulfate process. Alkaline cooking also results in removal of a much greater percentage of the noncellulosic carbohydrates, which are severely degraded and appear in the "black liquor" in the form of sodium salts.

a. The Sulfate Process. The sulfate process differs from the soda process in that, in addition to caustic soda (NaOH), the cooking (*white*) liquor also contains some sodium sulfide (Na_2S) as well as small amounts of other, relatively inactive, compounds such as sodium carbonate, sodium sulfate, and others. The process is called the sulfate process because sodium sulfide is obtained from sodium sulfate [*salt cake* (Na_2SO_4)].

(1) *Woods Used.* Any kind of wood can be pulped by this process. The method is, however, especially adapted to pulping resinous woods. In this country, southern yellow pines provide the bulk of the wood pulped by the sulfate process; tamarack, spruce, and hemlock contribute 12 to 15 per cent each. Of late, jack pine and Douglas-fir have also been used in increasingly large quantities.

(2) *Cooking.* Cooking in the sulfate process is carried on in welded-steel digesters, usually unlined. However, the more severe cooking methods now employed, with the correspondingly greater corrosion problems, have led to the increased use of stainless-steel and brick-lined digesters. The digesters are either rotary or stationary. The rotary digesters are usually of the tumbling type. The end of the digester is capped with a heavy steel cover. The digester is filled by turning it to a vertical position; the cover is removed, and the

chips and cooking liquor are run in together. The cover is then bolted, the steam turned on, and the digester set into rotation. The cooked pulp is discharged through a blow valve at the end of the digester.

Vertically mounted digesters are in more common use. Compared with those used in the sulfite mills, they are taller and narrower, the common size being 16 ft in width and 65 ft in height. Larger digesters are also in use.

Either direct or indirect methods of cooking are employed. In both methods the liquor is generally recirculated to obtain more uniform cooking conditions. Directly heated digesters are less expensive to install but consume more steam. The most commonly used system for circulating liquor in the indirect-cooking method is the already described Morterud system or one of its numerous modifications.

The total cooking cycle in the sulfate process, as practiced in the United States, varies from 4 to 6 hours depending upon the type of pulp wanted. For less exacting uses, such as container stock, shorter cooks of $1\frac{1}{2}$ to 2 hours are used, while for bleachable grades 4 to 6 hours of cooking is required.

The cooking cycle in the sulfate process may be divided into three stages: the initial, or liquor-penetration phase; the full-pressure period; and the gassing-down period. In the first stage steam pressure is built up to about 60 or 70 lb. This stage lasts 30 minutes to 2 hours, and during this time the air and all noncondensable gases must be removed from the digester as fast as they form; failure to do so results in false pressure that interferes with steam penetration and causes incomplete delignification (undercooking) of the pulp. During this period turpentine (when highly resinous woods are used) begins to distill with the steam. Toward the close of this period, when the pressure reaches 40 to 60 lb, the rate of delignification is noticeably increased.

Following the initial stage, the pressure is brought up quickly to 100 lb or more (340°F or above).* No relieving is necessary during this stage of cooking because generation of gases is small, and therefore the temperature and pressure will correspond quite closely. The full-pressure period is associated with the removal of more resistant secondary-wall lignin.

When the cooking is completed, which in some mills takes up to $2\frac{1}{2}$ hours, the pressure is gradually reduced to about 60 lb; this is called gassing down. The digester is then blown by opening a blow valve at the bottom.

In the more modern mills, to conserve as much steam as possible, the digesters are blown into a blow tank, a large steel shell elevated above the digester. The blow tank is connected to the blowpipe of the digester. The sudden release of pressure when stock enters the blow tank through a blow-pipe causes violent evolution of steam and helps in disintegration of the cooked chips into a fibrous state. The steam is passed through a separator to remove the turpentine and then is used for heating the water employed in washing the pulp. The stock, which at this stage contains the fiber, the

---

* Experiments at the U.S. Forest Products Laboratory, Madison, Wisconsin, have indicated that in the range of 320 to 366°F there is no critical temperature which is more beneficial than others from the standpoint of yield of pulp. Higher temperatures speed the reaction and hence decrease the time of cooking; at the same time they decrease somewhat the strength of the resulting pulp.

organic matter dissolved during the cooking, all the alkali added to the cook, and considerable quantities of water, is allowed to drop from the blow tank to the washing equipment. In mills which are not equipped with blow tanks the stock is discharged into a blow pit.

In the most recent installations the trend is toward the continuous digester systems. The principal advantage of the continuous process over the batch process lies in more accurate and continuous control of the quality and quantity of the end product. One such system is the *Kamyr continuous digester*

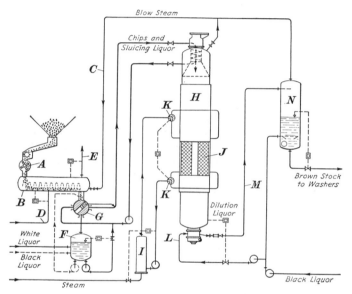

Fig. 17-14. Kamyr continuous cooking system: *A*, rotary feeder; *B*, steaming vessel; *C*, blow steamline; *D*, low-pressure steamline; *E*, air and turpentine relief; *F*, liquor mix tank; *G*, pressure chip feeder; *H*, digester; *I*, heater; *J*, strainer; *K*, flow reverse valve; *L*, black liquor line to digester; *M*, blow line; *N*, blow tank. (*From The Manufacture of Pulp and Paper, McGraw-Hill Book Company, Inc.*)

(Fig. 17-14). In this system the chips are metered volumetrically from the chip hoppers and conveyed from these into a low-pressure steaming vessel, where they are held for about 5 minutes. From there the steamed chips are conducted into a high-pressure digester feeder, consisting of a rotary, tapering plug, and capable of feeding 3,400 cu ft of chips per hour. In the feeder, as the plug turns, the chips are sluiced with the cooking liquor into the vertical part of the feeder, where they are soaked with circulating hot liquor and from where they are forced into the top of a digester 90 ft long and 13 ft in diameter. In the digester the chips move downward through circulating hot liquor and finally are discharged through a strainer into the blow tank. Travel through the digester takes about 2¾ hours.

(3) *Washing.* A thorough washing of pulp is very essential in the alkaline processes because the efficiency of the washing system determines not only the purity of the resulting pulp but also the completeness of chemical re-

covery. Since all the waste liquor goes to a recovery system, the minimum dilution of the waste liquor commensurate with good recovery of chemicals is aimed at in an efficient washing process. The system is considered to be efficient when the total washing loss of salt cake per ton of fiber is less than 50 lb and the amount of water used in the removal of chemicals does not exceed 2 lb of water per pound of pulp.

In the earlier practice, the washing of pulp was carried on in *wash pans* and in *diffusers*. The *wash pans* are open steel tanks with false bottoms in which washing takes place by gravity displacement of water flooded on the top of the pulp. The *diffusers* are large steel vessels, equal or larger in volume than the digesters, provided with perforated false bottoms through which the liquor drains. The washing is carried on with hot water under pressure. The washed stock is transferred into a storage chest, where the stock is kept agitated and further amounts of soda are removed by soaking. The most commonly used method of washing sulfate pulp in modern mills is by means of *vacuum washers*. In this method pulp stock from the blow tank passes over the vibrating screens to remove knots and undercooked chips, is then diluted with black liquor and delivered to the vacuum washers. These consist of a wire-wound cylinder rotating in a vat of pulp. The suction draws a layer of pulp to the wire and filters out most of the black liquor. The mat of fibers formed on the wire is drenched with hot-water showers, and the wash water is likewise drawn through the wire. The pulp is then washed off the cylinder with a shower of *white water* into a stock chest, where it is diluted for further washing in the secondary vacuum washer. The filtrate from the first cylinder is conducted to the black-liquor storage to be treated for the recovery of chemicals; the filtrate from the second washer is usually discarded.

(4) *Recovery System.* The success of operations in a sulfate mill depends to a large extent on the efficiency of the recovery system. The black liquor separated from the fiber during washing contains most of the chemicals originally used in cooking, as well as more than one-half of the total dry weight of the original wood.

The recovery of chemicals in the sulfate process involves the following series of steps: (1) concentration of liquor to a consistency at which the organic material it contains is in a combustible condition; (2) burning of the concentrate, with maximum recovery of heat and sodium salts, and reduction of sodium sulfate (Na_2SO_4) into sodium sulfide (Na_2S); (3) conversion of sodium sulfide and sodium carbonate (Na_2CO_3) into caustic soda (NaOH); and (4) separation of impurities (mud) from the sodium salts to form the clear white liquor used in cooking.

In modern plants black liquor is passed into a battery of multiple-effect, vertical-tube evaporators. As the liquor progresses from one cell (called "effect") to another, its boiling point is lowered by reducing the pressure (with a vacuum pump). With this system the only heat needed from outside sources is that required to raise the liquor to the boiling point in the first effect. Steam evaporating from that cell is sufficient to boil the liquor in the second, since this is kept under reduced pressure. Additional effects are similarly operated. After the system is placed in operation, it is customary to have the most concentrated liquor in the first effect and the fresh liquor in the last; this is because more concentrated liquor is more viscous and requires more heat to continue further evaporation of water. The concentrated liquor as it is discharged from the evaporators contains 50 to 60 per cent of solids.

In some mills further concentration of liquor is accomplished by passing it through a series of tubes in which the liquor is brought into direct contact with the flue gas from the recovery furnace at a temperature of about 700°F. This brings the solids content of the black liquor to about 60 to 70 per cent.

The concentrated liquor is pumped into a recovery furnace in which the remaining water is evaporated and most of the sodium in the liquor is converted into sodium carbonate, while other solids and gases are ignited and burned.

One of the commonly used furnaces is the *Tomlinson reduction furnace.* In this furnace enough sodium sulfate is added to the concentrated black liquor to make up for the loss (about 150 to 350 lb per ton of pulp). The liquor is then sprayed from an oscillating nozzle on the walls of the furnace heated to about 1700°F. The dehydrated chemicals fall to the lower part of the furnace.

Regardless of the method of smelting used, the smelt from the furnace is next dissolved in water. The resulting solution is called *green liquor* because it is usually tinted green by iron compounds. Green liquor consists mostly of sodium carbonate and sodium sulfide.

Green liquor is treated with lime [$Ca(OH_2)$], which converts sodium carbonate in the green liquor into caustic soda.

$$Na_2CO_3 + Ca(OH)_2 \rightarrow 2NaOH + CaCO_3$$

This step is known as *causticizing of soda ash.* In the older mills it is carried on as a batch process by agitating green liquor and quicklime in a tank. When the reaction is completed, the agitation is stopped, and the sludge is allowed to settle out. The clear solution on the top of the tank is decanted, forming the white, or cooking, liquor. The precipitate (mud) is then washed out and filtered in order to recover the residual chemicals contained in it.

In the more modern mills the batch method of causticizing has been replaced with one of the several continuous systems. In these systems all the steps of causticizing and clarification of white liquor and washing of mud are carried on as a continuous process. The most widely used continuous system is the Dorr process.

The clear white liquor, consisting principally of caustic soda and sodium sulfide, is pumped into a storage tank until needed. In some mills the lime left behind in the mud is reburned and used again with fresh lime in causticizing further quantities of green liquor.

The sulfate process is accompanied by a very disagreeable odor, largely due to formation of *methyl mercaptan* (CH_3SH). This odor is a sufficient nuisance to prohibit use of this process in heavily populated areas. No extensive odor control is practiced at most mills. It is, however, possible to reduce odors either (1) by passing the waste gases and black liquor through an oxidation tower, where the sulfur compounds are oxidized and removed from the liquor and gases, or (2) by burning the gases from digester relief and blow tanks in a furnace and then oxidizing the combusted gases. A less efficient method is to dispose of the combusted gases by means of a very high smokestack.

(5) *By-products of Sulfate Pulping.* There are several by-products of sulfate pulping that have considerable commercial value. These are *tall oil, sulfate turpentine,* and *alkali lignin.*

Tall oil is the name used to designate the fatty and resinous acids which result from the resinous materials contained in pines, spruces, and other resinous softwoods. The composition of tall oil varies with the species. The crude tall oil is present in the black liquor, and since it is lighter than water, it tends to rise to the surface of the tank, where it can be skimmed off. The general procedure is to remove it by withdrawing the black liquor, at about 30 per cent solids, from the second effect of the evaporator system to a settling tank. After the soap-like material is skimmed off, the black liquor is returned to the evaporator. The amount of skimmings varies from 100 to 300 lb per top of pulp, depending on species, the type of wood, and the duration of the pulpwood storage.

Some mills use the skimmings as fuel. In most instances, however, the skimmings are acidified with sulfuric acid to form a dark, oily liquid, called *crude tall oil.* Southern pines yield 50 to 90 lb of this oil, while spruce yields only 10 to 20 lb, per ton of pulp. Crude tall oil can be purified by distillation. The resinous products can be crystallized and used as a sizing agent for paper and in lacquers. Some tall oils are rich in phytosterols. This presents a possibility of utilizing tall oil as a cheap source of vitamin D and certain hormones.

Sulfate turpentine is recovered from the digester relief gases by passing them through a condenser. The condensate, which contains water and turpentine, is run into a settling tank, from which the turpentine is removed by draining from the surface, while water is removed from the bottom.

Crude sulfate turpentine has a disagreeable odor owing to the presence of sulfur compounds. In some mills crude turpentine is used as fuel; in others it is refined by distillation. This distillation also yields some pine oil. The resulting products are used mainly by the paint industry. The yield of turpentine varies from 1.5 to more than 4 gal per ton of air-dry pulp, depending on species, percentage of heartwood, and age of the wood.

Alkali lignin is a ligneous product that can be recovered from black liquor by mild acid treatment. This results in precipitation of about 50 per cent of the total lignin content of the liquor. A purer product is obtained by fractional precipitation with carbon dioxide. This process is used mainly in soda pulping of hardwoods.

The purified alkali lignin has a number of present and potential uses. It has been successfully used in the negative plates in lead-acid storage batteries, as an emulsifier for asphalts, and in plastics. Alkali-lignin derivatives have possibilities as flavoring, medicinal, and preservative products. A number of the commercial alkali lignins from the soda and sulfate processes are sold under such trade names as Meadol, Tomlinite, and Indulin.

(6) *Properties and Uses of Sulfate Pulp.* The sulfate process has replaced the sulfite method as the most widely used and also the fastest growing pulping process. Although it can be used for pulping either hardwoods or softwoods, its chief use is in pulping southern pines. In the Northern states it is used to some extent in pulping jack pine and spruce, and on the West Coast for cooking Douglas-fir, spruce, western hemlock, true firs, and lodgepole pine. Aspen is the principal hardwood pulped by the sulfate method, but other species such as gums, oak, birch, and maple can be readily used. The average yield of softwoods is from 45 to 48 per cent of the dry weight of wood, and that of hardwoods 45 to 55 per cent.

Sulfate pulp, when made from softwoods, has outstanding toughness, strength, and durability. This is due in part to the length of softwood fibers and in part to the milder effect on the strength of the fibers of chemicals used in this process compared with that of chemicals used in the acid sulfite and soda processes.

Unbleached sulfate pulp is sometimes called "kraft" pulp, from the German word meaning "strength." The use of this term, however, should be limited to that subdivision of the sulfate process in which the pulp is intentionally undercooked to produce very strong fiber. The principal use of kraft sulfate pulp is in the manufacture of wrapping paper, high-strength paper bags, and heavy paper used for corrugated and solid-fiber paperboards, fiber drums, and cans.

Sulfate pulp is brownish in color, but it can be bleached to the same brightness as sulfite pulp for production of high-grade paper. Bleached sulfate is stronger than bleached sulfite pulp. It is used principally in the manufacture of white paper where strength is important. Some of the uses of bleached sulfate are in the manufacture of envelopes; writing, printing, and onionskin papers; strong white wrapping papers; food containers; and folding boxes.

Hardwood sulfate pulps are somewhat lower in strength than unbleached spruce sulfite pulp but are stronger than the corresponding hardwood pulps made by the soda process and in some cases stronger than bleached sulfite. These pulps find their principal use as a filler in the manufacture of printing and writing papers. Special dissolving grades (see page 366) of hardwood sulfate pulp, suitable for rayon manufacture, have also been made.

b. Soda Process. The soda process is much older than the sulfate process and is also simpler. As in the latter, the success of operations depends on the efficiency with which the chemicals are recovered. The principle of this process is based on the ability of caustic soda to dissolve lignin and to hydrolyze the noncellulosic carbohydrate components (hemicelluloses) of the wood.

The active reagent in the soda process is caustic soda (NaOH). The chemical reactions that take place in the formation of cooking liquor may be indicated thus:

$$CaO + H_2O \rightarrow Ca(OH)_2$$

$$Ca(OH)_2 + Na_2CO_3 \rightarrow CaCO_3 + 2NaOH$$

The white liquor in the soda process contains also small amounts of sodium carbonate (Na_2CO_3). The presence of this chemical is considered undesirable, but its occurrence in white liquor is believed to be unavoidable because of the incomplete reaction between lime and soda during causticizing. Since the introduction of the sulfate process, most soda mills have begun to add small amounts (8 to 12 per cent) of sodium sulfide.

(1) *Woods Used.* The soda process is practically restricted to the pulping of hardwoods, principally aspen, cottonwood, beech, birch, maple, redgum, and yellow-poplar, although almost any other available hardwood may be used. Sometimes, to produce long-fibered stock, conifers such as spruce, hemlock, pine, and balsam fir are also used. The conifers require more alkali and longer cooking and yield less pulp than do the hardwoods.

(2) *Cooking.* The stationary digesters used in the soda process are similar in design to those used in the sulfite process, except that they are not lined with brick. In the older soda mills, rotary digesters, either horizontal or spherical, are extensively used.

In the older mills cooking is generally carried on by the direct-heating method. Maximum pressures of from 125 to 130 lb and temperatures of from 165 to 174°C are reached in the digesters as quickly as possible and are maintained until cooking is completed. To do away with the effects of false pressure due to accumulation of air and gases, the digester is relieved several times in the course of cooking. Cooking is continued for 2 to 6 hours. In more modern installations direct heating of the liquor is combined with some method of continuous circulation of the liquor and generally indirect heating (the Morterud system). Continuous systems of cooking are also in use.

When cooking is completed, the pulp is blown into a blow pit or a blow tank, where the expended cooking liquor is allowed to drain off. Washing of soda pulp is carried on in rotary vacuum washers, as in the case of other chemical processes.

(3) *Recovery System.* The black liquor obtained from the blow pits and the first wash of the pulp contains all the organic matter dissolved in pulping wood, as well as most of the alkali used in cooking. This liquor is conducted into a recovery plant, where it is put through multiple-effect evaporators similar in design and operation to those employed in the sulfate mills.

The concentrated black liquor, containing about 60 per cent of solids, is put through a furnace, where all organic material is burned. The resulting black ash, largely a crude form of sodium carbonate (Na_2CO_3), is combined with fresh sodium carbonate to make up the loss. Unlike the sulfate process, no smelting is required in the soda recovery system. The ash is dissolved in water and treated with quicklime (causticized) by a batch or a continuous process similar to that employed in the sulfate mills. The solution, clarified by settling, forms the white, or cooking, liquor; it contains about 10 per cent caustic soda. The precipitate (mud) is washed to recover residual chemicals.

The loss of chemicals in a modern plant equals about 60 lb of soda, and in the older plants two to three times that amount, per ton of fiber.

(4) *Yields, Characteristics, and Uses of Soda Pulp.* The yield of soda pulp is about 45 per cent or less of the dry weight of wood. The resulting pulp is quite dark in color but is readily bleached to good white.

Since hardwood fibers are short, the soda pulp produced from hardwoods is usually mixed in varying proportions with sulfite pulp made of coniferous woods. The short hardwood fibers act as a filler, with the result that paper containing soda pulp is more opaque and bulky. Soda pulp is used largely in the manufacture of book and magazine stock and in writing and absorbent papers.

D. Semichemical Pulp

The standard commercial processes of pulping wood generally result in yields of less than 50 per cent of the original wood by weight. In order to increase yields, a number of processes have been devised in which the wood chips are reduced to pulp by a mild chemical treatment, followed by mechanical refining of the softened chips to a fibrous mass. Pulps resulting from such two-stage treatment are called *semichemical.*

Semichemical pulping yields 65 to 85 per cent of pulp, based on dry weight of wood, or 10 to 40 per cent more than the conventional chemical methods. This is because the mild cooking employed in the semichemical pulping removes less than 50 per cent of the lignin and only 30 to 40 per cent of the hemicelluloses.

All woods can be reduced by some form of semichemical pulping. The hardwoods, however, are more suitable than softwoods, because the former produce pulp with higher strength characteristics than would be expected from their fiber lengths, while the reverse is true in the case of softwoods. Coniferous species also consume more liquor in the semichemical pulping than when pulped by the sulfate process.

Any of the standard pulping agents can be used in the semichemical processes. Neutral sodium sulfite (sodium sulfite + sodium carbonate), however, is used most frequently. Others are acid sulfite, modified soda and sulfate liquors, nitric acid, as well as a number of other chemicals and combinations. The general term *semichemical pulp* must therefore be prefaced with the name of the particular agent used, e.g., *neutral sodium sulfite semichemical pulp.*

The cooking of wood chips can be done in any conventional type of digester. Rotary digesters are in common use in mills with daily tonnage requirements of 50 to 75, or where different types of raw material, i.e., hardwoods, softwoods, and agricultural fibers, are used, each type requiring separate cooking. Vertical, stationary digesters are used in the mills with larger tonnage requirements for the same type of pulp.

In more recent installations the trend is toward a continuous cooking process, embodying a combination of continuous digestion and refining under pressure. A typical example of such equipment is the Asplund defibrator (see page 341) combined with a reaction chamber for cooking the pulp. The resulting equipment is called the Chemipulper (Fig. 17-15); it provides for a combination of continuous cooking under controlled conditions of time, pressure, and temperature, and controlled refining under pressure. The chemical treatment can be either alkaline, acid, or neutral. The reaction chamber, which is interposed between the feed end and the defibrator, is made up in sections; the number of sections provided is varied according to the degree of treatment required.

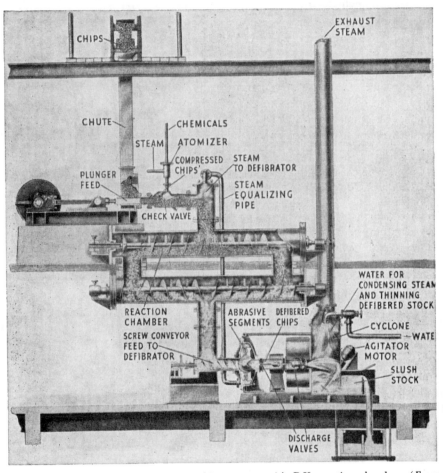

Fig. 17-15. Chemipulper continuous cooking system with B-K reaction chamber. (*From The Manufacture of Pulp and Paper, McGraw-Hill Book Company, Inc.*)

Another continuous system combines the advantages of the Va-Press presteaming operation (see page 348) with controlled impregnation of chips by continuously circulating liquor of predetermined temperature. The chemically impregnated chips are then conveyed into a digester for continuous cooking. By controlling the presteaming, impregnation, and cooking operations, pulps with a wide range of characteristics can be produced with the same equipment.

When digesters are used the cooked chips are blown into a blow tank, from which they are conveyed to the refining equipment. Since it has been found that hot chips fiberize more easily than cold, in the newer installations the transfer of chips to the refiner is accomplished in a minimum of time. In mills using continuous pulping equipment the cooking chips are removed through a blow valve and conveyed to the refiners in a continuous flow.

In some mills, chips are ground in the presence of hot cooking liquor, and the resulting pulp is then washed in vacuum-type washers. In other mills, chips are washed prior to refining.

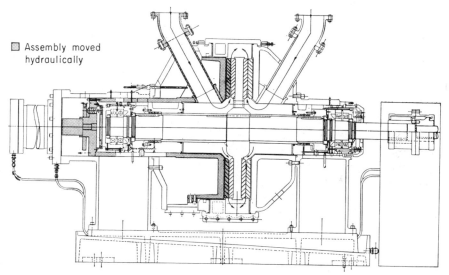

☐ Assembly moved hydraulically

Fig. 17-16. Working arrangement of the Sprout Waldron single-disk refiner. (*Courtesy Sprout Waldron Company.*)

Refining of chips can be done by a number of mechanical means, but the most commonly used equipment is the *disk refiner*. One type, the *Sprout Waldron single-disk refiner*, consists of a stationary disk and a rotating disk between which the chips are fiberized. The desired minimum clearance between the two disks, which determines the degree of fiberization, is manually adjusted by the spring-set screws at either end of the drive shaft. When this machine is in operation, the desired refining clearance between the disks is automatically maintained by hydraulic pressure since the rotary disk is free to float between the stationary and hydraulically positioned heads (Fig. 17-16). Another type, *the Bauer Bros. refiner*, consists of two disks rotating in opposite directions. The chips may be passed through one refiner, or two, or more, depending on the degree of softness and the quality of pulp desired.

1. Characteristics and Uses of Semichemical Pulps. Perhaps the greatest potential value of semichemical pulping lies in the opportunities it offers in the expanded utilization of low-grade hardwoods. The high yield compared with the standard chemical pulping processes is another important factor favoring semichemical pulping.

Pulps of different physical characteristics can be produced depending on the raw material used and the cooking liquor and cooking procedures employed. Semichemical hardwood pulps are characterized by low wet, and rather high dry, strength. Strength properties are related to the yield of pulp, strength being low at yields of more than 80 per cent. However, even such pulps have adequate strength for such uses as corrugating medium, liner boards, insulating boards, and roofing belts. Pulps of sufficient strength for finer papers are produced in lower yields. Bleachable grades are cooked to a lignin content of about 10 per cent.

E. Special Semichemical Pulping Methods

1. Cold-soda Method. The *continuous cold-soda* method, as originally developed at the U.S. Forest Products Laboratory,[6] consists of soaking the hardwood chips in a solution of sodium hydroxide at room temperature for about 1 to 2 hours and fiberizing the treated chips in a disk mill. The resulting pulp has properties between the conventional groundwood and neutral sulfite semichemical pulp. The yield of pulp ranges from 85 to 92 per cent of the original wood. This pulp is suitable as a filler in printing papers and qualifies for use in corrugated board.

In a commercially successful continuous pulping system, the hardwood chips are sprayed with soda solution. The chips then pass through a screw press, and the compressed chips are released under the surface of a caustic soda solution, where they rapidly absorb the liquid as the chip mass expands. The impregnated chips are conveyed through a drainage section into another screw press, where the excess liquor is expressed and the chips are coarsely macerated. The chips are then fiberized in a disk-type refiner.

2. Chemigroundwood Process.[15] The term *Chemigroundwood pulping* is given to a groundwood process in which the wood bolts, generally hardwoods, are treated with a neutral sodium sulfite solution prior to grinding. In this process bark-free wood blocks, cut to length for grinding, are placed in a treating cylinder. After the door has been secured, 28 in. of vacuum is drawn for about 30 minutes, after which the neutral sodium sulfite liquor is admitted under hydrostatic pressure of 100 to 150 lb, and the temperature raised to 135 to 150°C. These conditions are maintained for 5 to 6 hours, after which the pressure is released and the liquor blown out.

The pulpwood bolts are then ground on any conventional type of grinder, and the pulp treated in the manner usual for mechanical pulp. A yield of pulp of up to 87 per cent can be obtained by this method. The resulting product is considerably stronger than the conventionally ground spruce pulp. Most important, this method makes grinding of hardwood possible at a cost lower than that of softwoods. This is due to the reduction of about 60 per cent in power requirements, as compared with conventional grinding, and to almost 25 per cent greater yield of pulp per cord, due to the greater density of hardwoods.

F. Treatment of Pulp

Before either mechanical or chemical pulp can be converted into paper or shipped as such, it must undergo a series of treatments, first of which is *screening*.

1. Screening. Screening is necessary to remove dirt, foreign material, knots, and other uncooked or unbroken pieces of wood from the pulp and to separate fibers into different grades, based upon their size. This operation is accomplished by means of a series of screens and can be divided into two stages: *coarse screening* and *fine screening*.

The object of coarse screening in the case of mechanical pulp is to remove unground slabs and slivers of wood. This is accomplished by flowing the pulp over a grating made of iron bars set on edge, about 1 in. apart. The coarse material that fails to pass through the grate is removed and usually returned to the grinders. The pulp is then passed over sliver screens consisting of perforated steel plates with holes from $\frac{3}{4}$ to $\frac{5}{32}$ in. in diameter. The screens are either stationary or oscillating. The efficiency of operation depends largely on the size of the holes and thickness of the pulp stock; the smaller the holes and the thinner the stock, the freer of slivers will be the accepted stock. On the other hand, fine screening results in a considerable loss of fibers through clinging of fibers to the slivers.

The object of coarse screening in chemical pulp is to remove uncooked chips and knots without breaking them into a finer fibrous mass. The reason for this is that the uncooked portions of the pulp are of poor quality and color and if allowed to break into fibers will adversely affect the quality of the pulp. The chemical pulp from the storage tanks is first thinned with water until the content of the solid matter is about 0.75 per cent or less. This stock is then passed through a rotating screen having $\frac{3}{16}$- to $\frac{1}{2}$-in. perforations, thus removing unbroken pieces.

After coarse screening, the pulp, in some mills, is passed through a *riffler*, an apparatus where sand, pieces of mineral, particles from the lining of the digesters, and other impurities are allowed to settle. This is accomplished by passing the diluted pulp through a long trough, containing pockets of a rough felt fabric at the bottom, into which the impurities settle. The riffler is placed level, the current being produced by a pump that draws the stock from the riffler. The velocity of flow is regulated so that the fibers will not settle to the bottom with the heavier impurities.

Different grades of paper require different qualities of pulp. The purpose of fine screening, therefore, is to separate the fibers into several grades according to their length and diameter. This is brought about by controlling (1) the size of the perforations in the screen, (2) the force used in passing the fibers through the screen, and (3) the consistency of the stock.

Fine screens are of either the centrifugal or the diaphragm type. Centrifugal screens consist of a steel-plate shell containing an impeller (agitator), surrounded by a cylindrical screen plate. The rapidly rotating impeller throws the pulp stock, which enters at the top or the side of the machine, against the screen. The centrifugal force developed by the impeller causes the good stock to pass through the perforations in the screen. The accepted stock falls by gravity through the space between the plates and the casing and is carried out through a discharge spout. Stock too coarse to pass through the screen is thrown off the agitator and discharged at the bottom of the screen.

Diaphragm screens, also called *flat screens*, consist of flat metal plates, perforated with slits, which form the top of a shallow box, the bottom of which is made of rubber diaphragms. These diaphragms are vibrated by a cam mechanism which causes the diaphragms to rise and fall, thus creating a partial vacuum in the space under the screen plates. The operation of a diaphragm screen, therefore, depends on a combination of gravity and suction. Enough stock is admitted at one end of the screen to keep the upper surface of the screen plates completely flooded. The accepted stock, or the fiber that passes through the screen openings, is discharged from the flow box under the screen, while the coarse fiber is removed at the tail end of the screen or battery of screens.

2. Thickening the Stock. The consistency of the stock after fine screening is from 0.25 to 0.6 per cent solids. The process of thickening the stock to 3 to 6 per cent is called *slushing, deckering, dewatering*, or *concentrating*. Thickening can be accomplished by means of *a decker*. There are many designs of deckers, but a common type consists of a cylinder covered with a fine-mesh wire screen revolving in a vat into which the thin stock from fine screens is delivered. The pulp accumulates on the wire mesh, while the water runs through it inside the cylinder. The water thus removed is called *white water* and usually is reused in thinning the stock for screening. In many mills the deckers in which water is drained by gravity have been replaced by more efficient vacuum filters. In these the water is withdrawn by vacuum applied to the underside of the filter deck on which a sheet of pulp is formed.

3. Bleaching. Pulp as it appears after screening and slushing is off-color owing to the presence of small amounts of noncellulosic substances (lignin, hemicelluloses, and various extractives, such as tannins and resins) that remain in pulp after cooking. These impurities are chemically combined with the cellulose fiber and can be removed from the pulp only by chemical treatments that will first dissociate them from the cellulose. The operation that accomplishes this end is called *bleaching*. Bleaching not only whitens the pulp but also purifies it by removing some of the colorless noncellulosic impurities which, if left in the pulp, would interfere with the subsequent manufacturing operations.*

Although a number of bleaching agents can be employed, only a few are commercially important. In bleaching of chemical pulps the most important agents are chlorine, and sodium and calcium hypochlorites.

There are in general three systems of bleaching chemical pulp: single-stage, two-stage, and multiple-stage processes. The choice of the system depends to some extent on the process by which the pulp was produced, the degree of whiteness desired, and the size of the output of the mill.

* The visual whiteness of pulp is usually expressed in terms of "brightness," which is the numerical value of the reflectance of pulp to light in the blue and violet portions of the spectrum, as measured by specially calibrated reflection meters.

Single-stage bleaching consists of adding bleaching liquor to pulp stock in a vat where the stock can be agitated until the desired degree of whiteness is attained; the soluble waste products are then washed from the pulp. In the case of the sulfite and soda pulps the chief drawbacks of the single-stage bleaching are the excessive consumption of bleach required to bring the desired results and the deterioration of the fiber that may result if bleaching is prolonged. Sulfate pulp can be bleached only to a light-brown shade by the single-stage procedure.

The excessive consumption of bleach in the single-stage method can be considerably reduced by employing *two-stage bleaching*. When pulp is bleached, some of the oxidation products dissolve long before the desired whiteness is attained; these dissolved products, however, continue to consume bleach. In the two-stage method the oxidation products are removed by an intermediate washing of the pulp, and therefore less bleach is required. Some improvement in the strength of pulp may also result because of the lower concentration of the bleach used. Although in some plants calcium hypochlorite is employed in both stages, a more common procedure calls for chlorine-hypochlorite bleaching. In this method, in the first stage, chlorine gas is injected into the pulp suspension and is allowed to act until exhausted. The pulp is then neutralized with caustic soda and washed; this treatment removes nearly all residual lignin and most of the color. It is then subjected to the second stage of bleaching, consisting of treatment with a weak solution of calcium hypochlorite. It has been found by experience that up to a certain point more color can be removed per pound of chlorine used if chlorine gas is employed in place of bleaching powder. However, to obtain satisfactory whiteness the pulp must be subjected to a final treatment with the calcium hypochlorite. In the case of sulfate pulp, when whiteness comparable to the bleached sulfite pulp is required, caustic extraction is substituted for neutralization with lime, following the first-stage chlorination. This treatment serves to extract tannin-base dyes which are not affected by chlorination.

A more recent development in bleaching is the *multiple-stage method*. The actual procedures vary, depending on the type of pulp and the degree of brightness desired. The general purpose of multiple-stage bleaching is to achieve the highest degree of fiber purification by successive application of chemicals of different degrees of selectivity. A typical procedure might consist of (1) reaction of chlorine gas with lignin, followed by (2) application of sodium hydroxide to remove the alkali-soluble constituents of the pulp, (3) treatment with a hypochlorite solution to destroy the coloring matter, and (4) application of sulfur dioxide to remove iron and other mineral constituents of the wood fiber. Following each stage of treatment, the pulp is washed to remove all the soluble materials. Washing during these bleaching operations requires from 26,000 to 50,000 gal of water per ton of pulp.

The bleaching of groundwood presents considerable difficulty because groundwood contains nearly all the original cell-wall constituents. In the past a considerable tonnage of groundwood was brightened with calcium and sodium bisulfite, sulfur dioxide, and similar reducing agents. The color of mechanical pulp so bleached gradually darkens, due to atmospheric oxidation. More recently a more satisfactory bleaching of groundwood has been obtained with sodium peroxide and zinc hydrosulfite. Bleached groundwood is used chiefly in lightweight paper to increase opacity without appreciably decreasing the tensile strength of the paper.

4. Alpha-cellulose Pulps. All grades of chemical pulp contain appreciable amounts of hemicelluloses and small quantities of lignin. These materials have no detrimental effect in papermaking and frequently are considered desirable because of the increased yield of pulp and improved sheet formation. For certain uses of chemical pulp, however, these materials cannot be tolerated and must be removed by further alkali treatment, generally with sodium hydroxide. Two methods are in general use: (1) the hot process, in which dilute alkali (0.5 to 2.0 per cent) is employed at temperatures ranging from 100 to 160°C; and (2) the cold method, used when the highest degree of purification is desired. In the cold method alkali solutions in concentration of 3 to 25 per cent are employed, at temperatures of 20 to 50°C. The resulting pulps are called *alpha, alpha-cellulose,* or *dissolving pulps.* Sulfite pulp is preferred, though high-grade alkaline pulps can also be purified. Hot treatment results in 85 to 90 per cent alpha-cellulose content, while cold treatment produces pulps with 98 to 99 per cent alpha-cellulose content. The purification treatment reduces the final yield of pulp to 30 per cent, or less, of the original dry wood base.

Alpha pulps are used in making vulcanized parchment, resin-impregnated papers, and for special grades of paper where permanency is a major requirement. A highly purified alpha pulp, called *dissolving pulp,* is used in the manufacture of rayon, cellophane, nitrocellulose, cellulose acetate, and other cellulose derivatives (see Chap. 18). Pulp used in rayon and cellophane manufacture requires an alpha-cellulose content in the range of 89 to 91 per cent while that for cellulose acetate and nitrocellulose must have an alpha-cellulose content of 94 to 98 per cent.

5. Lapping. If the pulp is not used immediately at the mill but prepared for shipment or storage, it is necessary to extract most of the water from the screened stock and make the pulp into sheets which can be placed in bundles, or laps. This further extraction of water is called *lapping.* One type of machine for lapping pulp still in use in many mills is called the *wet press.* It consists of a cylinder and a vat similar to that used in a decker. The pulp formed on the cylinder is picked up by an endless felt belt and is passed through a series of squeeze rolls. A smooth roll atop the felt collects pulp from the felt. When sufficient pulp thickness is built up on the roll, the pulp is cut so that it comes off the roll as a rectangular sheet. This sheet is placed on a table in front of the press roll and folded to make a sheet of pulp about 18 by 24 in. The folded sheets of pulp are called *laps;* they contain 55 to 70 per cent water by weight, based on the air-dry weight of the pulp. The newer pulp driers consist of a cylinder or a Fourdrinier wet end (see page 374), coupled with either ordinary drying rolls or with a special drying chamber where the pulp is dried by radiation without extensive contact with metal surfaces.

G. Use of Old Paper in Papermaking

Paper stock, or wastepaper, is the second most important source of fiber in pulp and paper manufacture. The annual consumption of old paper ranges from 8 to 10 million tons, representing almost one-third of the total domestic wood-fiber requirement. The reasons for the extensive use of old-paper stock are (1) the immense supply available, (2) the low cost of material, (3) the low cost of converting the stock into pulp, and (4) the relatively high quality of the resulting product.

There are some 40 grades of wastepaper recognized by the trade. However, on the basis of consumption all wastepapers can be separated into three major groups: (1) mixed wastepaper, consisting mostly of old fiber containers, newspapers, and wrapping papers; (2) printed book, magazine, and ledger paper, composed largely of bleached chemical pulps; and (3) clean, fresh wastepaper, made up of paper shavings and cuttings without printing. Within each of these groupings there are a number of grades, with wide price variations. The first of the three categories accounts for more than two-thirds of the total wastepaper tonnage. It is used principally by the container industry in the manufacture of solid-fiber and corrugated boxes. It is also used in wallboards; in roofing, wrapping, and bag papers; and in newsprint. Wastepapers used by the container industry are generally not deinked. The better grades of old paper are used in the manufacture of book, printing, and writing paper. The percentage of old-paper stock in these papers may range from 10 to 80 of the total pulp content.

Conversion of old paper into pulp is largely a mechanical process, accomplished in a variety of specially designed breaker-beater machines, the description of which is beyond the scope of this text.* The paper may be shredded or now more commonly placed in the beaters in bales.

In production of white pulp from papers it is necessary to remove the ink. The process, called *deinking*, consists of treating wastepaper pulp with a suitable chemical solution, the most commonly used being sodium carbonate and sodium hydroxide, and then washing to remove the loosened ink pigments. The pulp is then screened to remove foreign material, and bleached, if necessary.

II. PAPER MANUFACTURE

Paper has been defined as matted or felted sheets of fibers, principally vegetable, formed on a wire screen from a water suspension. Paper derives its name from the word *papyrus*, a sheet originally made by pasting together thin sections of an Egyptian reed (*Cyperus papyrus*).

The two main subdivisions of paper products are paper and board. The distinction between them is not sharp, paper being lighter in weight, thinner, and more flexible than board.

* The one commonly used machine is the Hydrapulper, consisting of a bowl-shaped tank equipped with a multivane motor and special attachments to remove "junk" and rags. For other types, see references 6, 7, and 15, at the end of this chapter.

A. Beating

Paper manufacture begins with the operation called *beating*. In this operation the fibrous material (pulp) suspended in water is given a mechanical treatment which results in cutting, splitting, and crushing the fibers. The beaten stock acquires a characteristic slimy feel and is transformed into a compact and uniform mass of pulp, which can be spread into sheets of desired characteristics; its ability to retain water when allowed to drain is considerably increased; and the fibers develop a property of strong adhesion when allowed to dry in contact with each other. Some chemists hold the opinion that, in addition to purely physical changes, beating produces a chemical change which results in the formation of a certain amount of "hydrated" cellulose, a gelatinous mass which acts somewhat as a binding material and tends to increase the tensile and bursting strengths of the paper by cementing the crushed fibers together. This latter view has not been substantiated by any experimental evidence, and it is now believed that the principal effect of beating is physical.

Clarke,* in an attempt to reconcile all the known phenomena resulting from "beating" the stock, has postulated a composite hypothesis which recognizes (1) loosening of the primary wall and "fuzzing" of the underlying cell-wall structures, thus enabling the wood elements to interlock; (2) deformation of the fiber, allowing better fit among the neighboring fibers; and (3) the creation of smaller bodies which can fit in the spaces between the fibers, thereby further increasing contact between them. The further effects of beating become evident in increased fiber flexibility and increased external surface of the fiber. The latter improves specific adhesion between adjacent fibers in the paper sheet.

The results of beating are determined by the *freeness value*, which is a measure of how fast water drains through the pulp. The freeness value, i.e., the rate of water drainage, decreases with increased beating. Beating is done in a *beater* or by means of *disk mills*.

1. The Beater. There are many kinds of beaters, but the most prevalent is the Hollander or its numerous modifications (also called the *breaker beaters*, or *pulpers*, when used for breaking dried sheets of pulp). A typical beater of the Hollander type (Fig. 17-17) consists of a large open tub, 15 to 25 ft long or more, 6 to 11 ft wide, and 2½ to 5 ft deep, built of wood, concrete, or cast iron. A central partition, called the *mid-feather*, extends to within 2 or 3 ft of each end of the tub; this partition allows a continuous flow of the pulp around the tub in one direction. On one side of the tub, between the partition and the wall of the tub, is a roll mounted on a shaft. Beneath the roll is a metal bedplate; both are equipped with metal bars or knives. An integral part of the beater is the *backfall*, shaped on one side to conform to the curve of the beater roll and sloping steeply on the other side. As the roll revolves, the pulp passes between the roll and the plate and is thrown over the backfall, forming a head and causing the *stuff* to travel away from the roll, around the tub, and back to the roll again.

* Ott, Emil (ed.). High Polymers, Cellulose and Cellulose Derivatives. Interscience Publishers, Inc., New York. 1943.

The Hollander beater is criticized on a number of points, chief among which are large power consumption per ton of product, poor and uneven mixing of stuff, lack of ready control over the beating action, and necessity for large floor space. Attempts to correct these defects have resulted in numerous variations of the Hollander type, as well as in a number of special types of beaters, radically departing in design from that of the breaker beaters. At present, various modifications of the Hollander still constitute the greater number of beaters.

Fig. 17-17. Beater. To reduce size of fibers so they will form a close, firm sheet, pulp is mixed with water and beaten between metal bars for several hours. Sizing and coloring materials are also commonly added to the furnish at this point. (*Courtesy Hammermill Paper Company, Erie, Pennsylvania.*)

One special type of beater used for beating dried pulp is known as the *kneading pulper,* a machine adopted from the breadmaking industry. Another commonly used type is the *Hydrapulper* (see footnote, page 367). A pulper of radically different design is the rod-mill beater (Fig. 17-18), consisting of a hollow cylinder 6 to 12 ft in diameter, partially filled with steel rods approximately 3 in. in diameter. The rotation of the cylinder causes the rods to roll over one another, exerting a crushing action on the fiber.

2. Disk Mills. Disk refiners similar to those employed in the fiberization of pretreated wood chips in the semichemical mills are employed in some mills in place of beaters and the Jordan (see page 373). The pulp stock is fed either by gravity or under pressure, and the degree of refinement is controlled by an adjusting screw which controls the distance between the grinding plates. The *single-disk Bauer Bros. refiner* is an example of this type of equipment (Fig. 17-19).

3. Loading and Sizing. At this stage of paper manufacture various products are added to the stock in the beater in order to reduce the absorbent

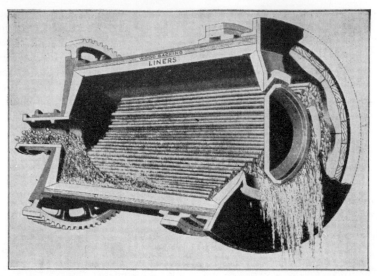

Fig. 17-18. Rod-mill beater. (*From The Manufacture of Pulp and Paper, McGraw-Hill Book Company, Inc.*)

qualities of the pulp and to produce a smooth-finished paper surface. Paper made of pure fiber would be so soft and absorbent that it could not be used for writing or printing. An example of such a paper is filter paper. A thin sheet of such paper would also be so transparent that writing or printing could be read from the other side of the sheet. The various materials which are added to pulp stock in the beater are called *fillers* and *sizing*.

Fillers are mineral substances, of which the most common are clay, talc, agalite, crown filler, and pearl filler, the last two being proprietary compounds whose main constituent is calcium sulfate. More expensive fillers include titanium dioxide, zinc sulfide, and similar highly opaque materials. Starch may also be added to increase bursting and tensile strength and improve the finishing characteristics of paper. The fine particles of the filler occupy

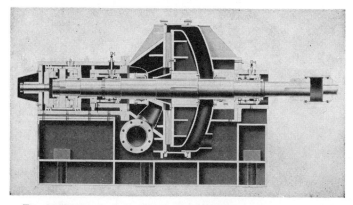

Fig. 17-19. Single-disk refiner. (*Courtesy The Bauer Bros. Co.*)

the spaces between the fibers, making the resulting paper smooth and opaque. The addition of a filler to pulp is called *loading*, or *filling;* the latter term is preferred because the former suggests adulteration. Although this was true to some extent in the past, in modern papermaking, fillers are added because they improve the opacity of the paper, prevent "show- and strike-through" and give more "body" to the sheet. Too much filler tends to decrease the strength and to increase the softness of the paper, while too little results in formation of paper with a poor surface and with excessive hardness. If fillers are not properly "anchored" in the paper, they will cause *dusting.*[19]

In addition to filling, most papers are *sized,* to make them less absorbent to ink and moisture. This is accomplished by adding some water-repellent material which will coat the fibers and fill in the spaces between with a varnish-like substance. An example of paper that has not been sized is blotting paper, which is entirely unsuited for writing. The most common materials used in sizing are rosin and wax. Rosin is first added to a hot soda ash (Na_2CO_3) solution and the mixture cooked for 4 to 6 hours with constant stirring. This causes the rosin to saponify, the resulting product being a thick fluid called *rosin soap.* The rosin soap is then diluted until an emulsion is formed. This emulsion is introduced into the beater, and a certain amount of alum (aluminum sulfate) is added. The chemical reaction which follows produces curdling of the rosin size, which is then deposited on the fibers in the form of very small particles. Of the wax sizes, paraffin is the most commonly used; it is introduced into the beater in the form of a water emulsion.

Some types of paper, in place of the *internal sizing* and filling just described, are given surface treatment. The term *surface sizing* includes any treatment given to the *web* after its formation on the wet end of the paper machine; the purpose is to improve the surface characteristics and to reduce the absorptive properties of the paper. A secondary effect of surface sizing is the increase in bursting strength of a paper. Formerly, the most widely used materials in surface sizing were gelatin glues and starch. During the last decade the paper industry has seen an increased trend away from the use of starch and gelatin for surface coating and toward thermosetting types of resins. These are generally used as emulsions and are frequently applied at the conversion plants, but may be applied on the machine or even added in the beater. Some of the commonly used emulsions are polyvinyl acetate, polyesters, vinyl chloride, and acrylic resins.

4. Wet Strength. One of the most characteristic properties of ordinary paper is loss of strength when a sheet of paper is wetted. Completely wet paper can be disintegrated into pulp by a little mechanical action. If wet paper is allowed to dry without being disintegrated, it will regain much of its original strength. The proportion of original strength that a paper retains when completely wet is termed its *wet strength.*

Papers retaining a high percentage of their dry strength when wet are called *wet-strength papers.* A clear distinction should be made between the wet strength and the water repellency of a paper. True wet strength is exemplified by highly absorbent papers which permanently retain much of their dry strength when completely saturated. In contrast, water-repellent sheets may resist initial penetration of water but will retain their strength only for as long a time as is required for moisture to permeate the entire fibrous structure of the sheet. Of course, paper with high wet strength and good water-repellent characteristics could be produced.

Vegetable parchment is the oldest type of wet-strength paper. Parchments are produced by treating sheets of paper with strong sulfuric acid for a brief period of time. The paper is then washed and the acid neutralized. The resulting sheet is resistant to boiling in water and hot acid and alkali solutions. Other wet-strength processes in use for many years are (1) tub sizing of paper with glue or gelatin, followed by a hardening action with alum or formaldehyde, and (2) the incorporation into a paper of regenerated cellulose made by the viscose method.

The advent of synthetic resins and their introduction into papermaking have resulted in the production of papers with vastly improved wet-strength characteristics. The new processes have to a large extent replaced the older methods and have resulted in the extension of the use of wet-strength papers into a variety of new products.

Of a large number of synthetic resins proposed for use in the manufacture of wet-strength papers, the urea and melamine resins are considered outstanding, but some of the previously mentioned resin emulsions are also used.

As was pointed out, synthetic resin may be added to pulp in the beater, or the sheets of paper may be impregnated during their passage through the paper machine. The former method is now favored because surface treatment, especially when urea resins are used, tends to impair the tearing strength and increase the brittleness of paper.

5. Coloring. Since the natural color of paper produced even from bleached pulp is yellowish, most papers are colored. White paper is produced by adding blue to neutralize the yellow, and other shades are obtained by adding an appropriate dye or pigment to the pulp in the beater or by applying it directly as a coating to paper surfaces.

Pigments include all color materials which are insoluble in water; they are deposited on the surface of the fiber but do not penetrate into the cell wall and are held in the paper by mechanical means with the aid of size and alum. Most of the pigments are mineral colors. Pigments are fast to light but produce a dull surface. The dyes penetrate into the cell-wall structure of the fiber. Most of the papermaking dyes belong to the aniline group of dyes and are classed as direct dyes, basic dyes, and acid dyes. The direct dyes may be used without the aid of mordants; the basic colors are precipitated from solution by the addition of an alkali; while acid colors require addition of an acid mordant. In general, the direct colors have less tinctorial power but are faster to light than either the basic or the acid colors.

Results obtained by coloring are greatly affected by a number of factors, including the type of pulp to be colored, the degree of beating, sizing and filling, and the characteristics of the water. Color matching is, therefore, one of the most exacting operations, requiring a thorough understanding of color reactions and considerable ingenuity on the part of the color technician.

6. Refining. When pulp has been beaten and all the ingredients that go into the making of paper added, the beater furnish is usually passed to a refining machine. There are several types of these but the most successful are the conical refiners, a prototype of which, still in use in many mills, is the *Jordan*. Several improved Jordan-type refiners, operating at much greater speed than the original machine, are now in use.

Fig. 17-20. Jordan. (*Courtesy Shartle Bros. Machine Co., Middletown, Ohio.*)

Fig. 17-21. Jordan plug. (*Courtesy Shartle Bros. Machine Co., Middletown, Ohio.*)

The Jordan consists of a cone-shaped shell bearing bars on the interior surface and a cone-shaped, revolving plug (Figs. 17-20 and 17-21), which fits into the shell and is also equipped with bars. The pulp is fed into the Jordan in the form of a very dilute suspension and, as it passes between the two sets of bars, is given a refining treatment. The stock is then ready for fabrication into paper. In some mills the Jordans have been replaced by disk refiners (see page 370), or the two pieces of equipment are used in com-bination.

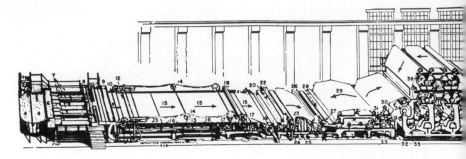

A FOURDRINIER PAPERMAKING MACHINE AT

1 REGULATING BOX		15 PULP CARRIED BY FOURDRINIE⯑
2 PULP		16 WIRE CARRYING ROLLS
3 WATER		17 FLAT SUCTION BOXES
4 SAND TRAP		18 DANDY ROLL
5 DRAIN PLUG		19 SUCTION ROLL
6 DISTRIBUTION TROUGH		20 SHEET CARRIED BY FIRST FE
7 SCREENS		21 TO VACUUM PUMP
8 COLLECTING TROUGH		22 DOCTOR
9 HEAD BOX		23 FIRST PRESS
10 SPRAY TO ELIMINATE FOAM		24 SUCTION ROLL
11 PULP CARRIED BY APRON		25 WATERMARKING
12 SLICE OR DAM		26 DOCTOR
13 SHAKING MECHANISM		27 SECOND PRESS
14 DECKLE STRAP REGULATES WIDTH OF SHEET		28 SHEET CARRIED BY SECOND

Fig. 17-22. Fourdrinier papermaking mac⯑

B. The Fourdrinier Paper Machine

The modern papermaking machine (Figs. 17-22 and 17-23) is commonly called the *Fourdrinier*. The details of design of the Fourdrinier vary considerably, but all of them can be divided into four basic parts for convenience of description: (1) the Fourdrinier proper, or the wet end; (2) the press section; (3) the drier section; and (4) the calender end.

The wet end of the machine is preceded by screens, generally of the revolving type. They are composed of a revolving cylinder made up of curved,

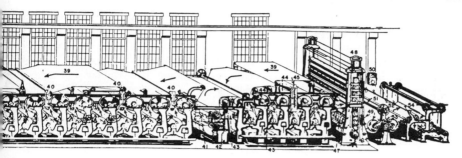

MERMILL PAPER COMPANY, ERIE, PENNSYLVANIA

IRD FELT RETURNING	43 BOTTOM DRYER FELT
CTOR	44 DRYERS
IRD PRESS	45 STEAM REGULATOR
EET CARRIED BY THIRD FELT	46 DRYER
COND FELT RETURNING	47 AIR COOLING PIPES FOR CALENDER ROLLS
OOTHING PRESS	48 CALENDERS
ST DRYER	49 PAPER WINDING ON REVOLVING REEL
T DRYER	50 MOISTURE CONTENT CONTROL BOX
COND DRYER	51 EMPTY REEL READY FOR WINDING
TTOM DRYER FELT RETURNING	52 SLITTER CUTTING PAPER FOR TWO SIZE ROLLS
DRYER FELT RETURNING	53 FULL REEL REWINDING
AM REGULATORS	
E LIQUID	54 ROLLS WOUND ON IRON CORES READY TO BE CUT INTO SHEETS
FACE SIZING PRESS	

rtesy Hammermill Paper Company, Erie, Pennsylvania.)

slatted screens. The purpose of these screens is to remove lumps, slivers, and as much foreign material as possible. The screened stock, consisting of about ½ to 1 per cent fiber and the rest water, passes to a rectangular tank, called *headbox*, from where the pulp slurry is conducted onto an endless fine-mesh wire screen of the Fourdrinier wet end. The moving screen is given a sidewise shaking motion, and as the stock travels along it some water drains through the mesh, leaving on the wire screen a felt of interlocked fibers called the web. In order to remove more water from the wet stock, the Fourdrinier end is equipped with suction boxes or rolls, in which a vacuum is created by means of suction pumps. This removes enough water to ensure the paper against being crushed by the press rolls.

When a watermark is desired on the paper, this is usually accomplished in the Fourdrinier part of the paper machine by means of a *dandy roll*. A dandy roll is a metal roll 7 to 24 in. in diameter; its primary function is to smooth down the surface of the sheet while it is still moldable. The roll is covered with fine wire cloth to which brass letters and designs may be attached when a watermark is desired. The dandy roll runs freely on the surface of the wet paper, exerting pressure equal to its own weight. The raised

Fig. 17-23. Fourdrinier machines with removable Fourdriniers for 100 ft by 190-in. wire. Buckeye Cellulose Corporation. (*Courtesy The Sandy Hill Iron and Brass Works.*)

design on the roll makes the wet paper thinner in places where the design comes in contact with it. The outline left on the wet sheet shows clearly when the finished paper is held against the light.

Next the wet sheet of paper is transferred by means of suction or the *couch roll* onto a fabric belt, called *felt*. The felt carries the wet sheet of paper through the press section of the machine. The *press section* is composed of a series of rolls between which the paper passes. Before entering the presses the sheet of paper contains about 90 per cent water. The object of the presses is to squeeze out as much water as possible without injury to the wet sheet. As the paper leaves the presses, it contains 60 to 70 per cent water. Frequently watermarking is done at this point rather than by a dandy roll.

The wet paper is then carried over the surface of steam-heated, cast-iron cylinders called *driers*. If surface sizing is used this is applied before the paper leaves the driers. It is applied to the bottom side of the sheet by a roll rotating in a bath of sizing solution, while at the same time sizing is applied to the top side by spraying. In the course of the passage of the paper over the driers the water content is reduced to 7 to 10 per cent based on dry weight. The size and number of driers and their arrangement depend on the speed of the machine and the product manufactured. It is customary to allow one 48-in. drier for every 20-ft-per-minute of paper speed for newsprint stock. Better grades of paper require more drying surface.

The remainder of the machine is composed of *calenders, slitters,* and *winders.* The calenders consist of a set of very smooth and heavy rolls, between which the paper is passed. The weight and friction of these rolls impart to the paper a smooth finish, which varies according to the pressure exerted. In some mills paper is colored at this point by flowing dyes from water boxes on the surface of the calender rolls, from which they are passed on to the paper.

As the paper leaves the calenders it is wound on a *reel.* When one reel is full the sheet is broken and led to a second empty reel, where winding is continued. The paper from a full reel is then rewound on winders. As it passes from the reel, the original wide sheet of paper is slit into strips of the desired width by means of revolving disks or slitters. The paper rolls from the winders are then wrapped for shipment.*

C. The Cylinder Paper Machine

The cylinder paper machine differs from the Fourdrinier type chiefly in the method by which a sheet of paper is formed. The essential feature of this machine is a cylindrical metal mold covered with a coarse bronze backing wire, over which is stretched a finer mesh wire on which the paper sheet is formed. There may be one or several cylinders, depending on the product manufactured. The cylinders are partially submerged in vats which contain diluted paper stock. As they revolve, a thin film of paper stock forms on the wire mesh. The water drains from the stock through the wire surface. The wet stock is picked off the mesh by the underside of an endless felt which passes over the unimmersed portion of the cylinders. The ability of the felt to pick up the wet pulp is due to the fact that wet stock, when in contact with two surfaces, always sticks to the smoother surface, in this case that of the felt.

If there is more than one cylinder, the felt continues to travel, and passing over the second cylinder picks up another layer of wet pulp, which sticks to the first layer. The number of laminations thus formed is limited only by the number of cylinders used. The first and the last layers formed are called the *liners* and are usually composed of a better grade of paper stock, which may contain coloring matter; the interior layers, or the *fillers,* are usually made of a poorer grade of stock. The paper or board is then passed through a number of pressure rolls and drying cylinders and through a calender, similar to those of the Fourdrinier paper machine.

* A modern newsprint paper machine produces a sheet of paper a little more than 300 in. wide at a rate of 2,000 ft per minute, or about 60,000 sq ft of paper a minute. A modern kraft-board machine produces lining board for corrugated cartons at a rate of about 250 sq ft a second, or 15,000 sq ft a minute, and 25 million square feet in 24 hours of operation, an equivalent of about 525 tons of product a day. At the other end of the scale are the tissue-paper machines, which form a sheet of paper 72 in. wide at a rate of 100 ft a minute, or about 2 tons of paper a day.

The one-cylinder machine is exceptionally well adapted to the making of tissue papers and the multicylinder machines to the manufacture of laminated products such as boards and bristols.

D. Special Paper Machines

There are two other papermaking machines used for special purposes. They are the *Harper Fourdrinier* and the *Yankee* machine. The Harper Fourdrinier is used for making lightweight papers, especially tissue, crepe, and cigarette papers. It differs from the standard Fourdrinier machine in that the Fourdrinier, or wet, end of the machine is reversed end for end, so that the flow box is between the Fourdrinier and the press rolls. This arrangement allows a longer run of felts carrying wet paper before the latter enters the presses.

The Yankee machine is designed for making machine-glazed papers, with a glossy finish on one side only. This machine has a wet end of either the cylinder or the Fourdrinier type. It differs from the standard machines in the method of drying. The Yankee machine employs only one large drier, 9 to 15 ft in diameter, with a highly polished surface. Pressing of the wet paper by means of rolls against the surface of the drier imparts a fine-glazed surface to one side of the paper.

E. Classification of Paper Industry Products

There are hundreds of grades of paper and paperboard, each sufficiently distinct in properties and uses to merit a name. A complete list of these with definitions may be found in reference 4, at the end of this chapter. Only a few major classifications are discussed in this text.

1. Papers. *Absorbent papers* is a term applied to papers with a high absorbent quality, such as blotting and filter papers and paper toweling. These papers are spongy, loosely fitted, and unsized. The higher grades of absorbent papers are made of soft rags; the lower grades contain mostly wood fiber.

Book paper is any kind of coated or uncoated printing paper for books, magazines, catalogues, etc. Many grades of book paper are available, ranging from low grades, similar in composition to newsprint, to high grades composed of rags, alpha, and bleached sulfite pulps. In between are grades made of bleached sulfite or sulfate pulp, with various percentages of soda pulp and wastepaper or small amounts of groundwood.

Building papers is a general term applied to thick paper made of old stock consisting of rags, wool, old papers, etc., sometimes mixed with asbestos. Building papers include sheathing papers used in house construction and felt papers used either as such for heat insulation or saturated with tar or asphalt as a roofing material.

Hanging or *wallpaper* is used for papering walls and ceilings. It is similar to newsprint but in heavier sheets. Most grades consist of 70 to 90 per cent groundwood and the remainder unbleached sulfite pulp. More expensive grades contain bleached sulfite or sulfate pulp and smaller amounts of groundwood pulp.

Newsprint is a class of paper developed for newspapers and other printing purposes where permanency is not required. It is made of 80 to 90 per cent groundwood and 10 to 20 per cent unbleached sulfite, or partially bleached sulfate fiber.

The United States consumed about 7 million tons of newsprint in 1957. This represented nearly 60 per cent of the world newsprint consumption. Canada supplies about 78 per cent of the United States requirements. Plans for additional newsprint capacity are expected to increase the domestic production from 1.7 million to about 2 million tons by 1960. This will account for about 25 per cent of the total domestic requirements for newsprint. The additional capacity will result from greater use of southern pines and from increased utilization of hardwoods in the East and the Lake states for newsprint production. Any further increase in domestic production is not expected to do more than keep up with the growing demand. Only the much greater use of hardwoods for newsprint could bring about any change in this picture. It is therefore anticipated that the United States will remain dependent on Canada for at least three-quarters of its newsprint requirements.

Writing paper is a class of paper with a good writing surface; it includes bond, ledger, envelope, drawing, and similar types of paper. Depending on the grade, its content varies from all rag, or combination of rag, bleached sulfate, and kraft, in the best grades, to sulfite and soda or old paper in the poorer.

a. Papers for Packaging. One of the chief reasons for the rapid and continued growth of the pulp and paper industry is the use of its products in packaging. Paper is an extremely versatile packaging medium when used either alone or in combination with other materials. The basic packaging papers are wrapping (mostly kraft), glassine and greaseproof papers, surface-coated papers, and tissues.

Glassine and *greaseproof papers* are semitransparent or transparent papers (glassine has greater transparency), resistant to the penetration of greases and oils. They are made of chemical pulps which were subjected to longer beating than usual, thereby reducing the "freeness" (see page 368) of pulp. Paper is made from such pulp in the usual manner. It is dampened while being rewound and passed through steam-heated supercalenders which impart gloss and increase the transparency of the paper. The high water content of these papers is a factor contributing to their resistance to oil and grease; they are not, however, resistant to moisture-vapor passage.

Surface-coated papers. Many packaging papers are designed especially for their protective characteristics against corrosion, molds, gases, water, chemicals, and grease and oil. Protective coatings used on papers may be classified as (1) hot melts, employing waxes such as paraffin; (2) solvent and emulsion coatings, consisting of nitrocellulose and polyvinyls, polyethylene, acrylic, and many other resins; and (3) extruded or plastic-coated papers, in which the paper to be coated is combined with a thin sheet of such materials as polyethylene.

Tissue is a general term used to indicate a class of papers made in weights lighter than 15 lb per ream, containing 480 sheets measuring 24 by 36 in.; it includes such papers as carbon (rag and bleached chemical pulp), cigarette

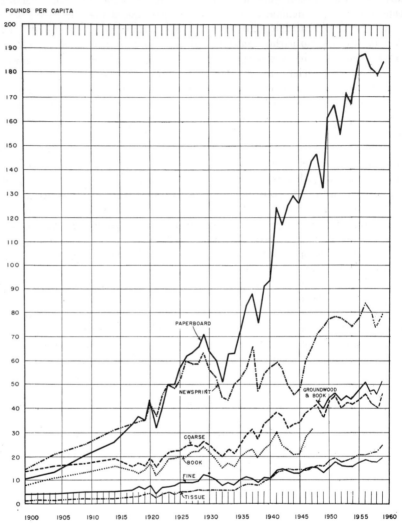

Fig. 17-24. United States new supply (production plus imports, minus exports) of paper. (*Courtesy United States Pulp Producers Association.*)

(flax), fruit wraps, facial tissues, and toilet papers (groundwood and chemical pulp).

Wrapping papers is a term applied to strong papers used for wrapping, paper bags, liners, etc. Depending on the grade, they are made of groundwood, semichemical, sulfite, or sulfate pulp; jute, old rags, or wastepaper. *Kraft paper* is a wrapping paper made wholly or principally of sulfate, or kraft, pulp.

2. Paperboards. Paperboards (fiberboards) and heavy papers are made either on a Fourdrinier or on one or more cylinder machines. Certain types of paperboard consist of several layers of different pulp stock laid down to form a thicker sheet. Such a board is made on a multicylinder machine. A paperboard of homogeneous composition throughout may be made on either a cylinder or a Fourdrinier machine. A Fourdrinier for insulating-board production differs from the conventional type in that the sheet is not wound but is held flat. The wet sheets are cut into pieces of desired sizes and dried in a continuous drier similar to that used in drying veneers.

The two principal divisions of paperboard on the basis of utilization are (1) boards for containers, and (2) boards for building and insulating uses.

a. Boards for Containers.[26] Up to 95 per cent of all the packaged freight in the United States moves in solid or corrugated fiber boxes.

Paperboards used by the container industry are generally of a combined type; i.e., the outer and the inner layers consist of different pulp stock. The board could be either solid or corrugated. Solid fiberboards differ from corrugated in that in the solid boards space between the two outer faces is filled with one or more layers of *filler chip,** as contrasted to one or more layers of corrugated material in the corrugated board. The outer and the inner faces of container board are called *liners* or *line boards. The corrugating medium* used in the corrugated boards is a sheet of paper, generally 9 points in thickness (1 point = 0.001 in.) fluted to form reversed arches.†

Boards for folding boxes. This grade includes a variety of boards used for making boxes which are shipped flat to be set up by the user to receive merchandise. These boards must be capable of bending (creasing) 90 to 180°, without rupture, during conversion into containers and during filling operations. They must also provide a surface suitable for printing and for reception of adhesives, if needed. The boards must be strong enough to ensure safe delivery of merchandise and to meet such special requirements as water and grease resistance and resistance to fading or discoloration in sunlight. To withstand creasing, the liners are generally made of long-fibered pulps, such as sulfite or sulfate.

Boards for setup boxes. These grades of board are made for boxes which are set up during fabrication and delivered in that form to the user. They differ from boards used in folding boxes in that bending characteristics are not as important; but the boards must be able to withstand rough handling.

* *Filler chip,* also called *chipboard,* is a grade of board made on a cylinder machine from wastepaper stock.

† *Corrugating medium* is generally made of a mixture of wastepaper, straw pulp, or semichemical or kraft pulp, by passing the steam-preconditioned sheet of paper through a set of corrugating rolls. These convert a flat sheet into one with uniformly spaced corrugations or flutes, in the form of arches. Three types of flutes are recognized, on the basis of the number of flutes per linear foot and the height.

	Height	Number of flutes per ft
A	. . .	36
B	$\frac{1}{8}$	51
C	$\frac{5}{32}$	41

Boards for setup boxes are made of short-fibered stock such as groundwood, wastepaper, or straw. Bleached pulp may be used for liners to improve their appearance.

Container board. This is a nonbending board made either from solid unbleached kraft on a Fourdrinier machine, or of kraft liners with a wastepaper filler produced on a cylinder machine. The board is usually sized with rosin for water resistance and commonly has a surface application of wax or, if intended for overseas military shipment, of some wet-strength resin.

Food-container board. There are a variety of boards intended for containers of such foods as milk, cottage cheese, ice cream, butter, etc. These boards are usually made of bleached chemical pulp, though groundwood may also be used. They are heavily sized with rosin and waxed. Good appearance, moisture resistance, and freedom from contamination are the principal requirements for this type of board.

b. Building and Insulating Wood-fiber Products.[21] There is a great variety of wood-fiber products available for building purposes. On the basis of their physical characteristics they may be classified as (1) *rigid*, (2) *flexible*, and (3) *fill* types. Only those made of wood fiber by a pulping process are discussed here. Those made from "exploded" fiber or by one of the several methods employed in wood particle-board manufacture are discussed in Chap. 12.

(1) Rigid Fiberboards. These are further classified as either *structural* or *nonstructural*. Structural boards are either homogeneous or laminated; they combine strength with heat- and sound-insulating properties and range in thickness from $\frac{1}{2}$ in. to more than 1 in., and are up to 4 ft wide and up to 12 ft in length. Several types of structural boards are recognized. Among them are *building boards, sheathing, lath, tile board, plank,* and *roof-insulation boards.*

Structural building boards can also be classified as (1) *wallboards* and (2) *insulating* boards. The primary characteristics of *wallboards* are rigidity, nonwarping qualities, and good finishing surface. They may also have good insulating properties and have one or both surfaces waterproofed. Wallboards covered with plaster are called *plasterboards. Gypsum boards* are wallboards made by rolling plaster between the sheets of heavy wrapping paper. *Insulating building boards* are made primarily for heat and sound insulation. They are not as rigid as wallboards, but may have a surface suitable for interior finish.

Rigid, nonstructural fiberboards, often called *slab insulation,* are made by lamination to form rigid blocks used mostly for cold-storage insulation.

All types of rigid fiberboards may contain groundwood, wastepaper stock, and unbleached chemical and semichemical pulps, as well as ground lumber residue, straw, cornstalks, bagasse, and similar vegetable material.

(2) Flexible Fiber Insulation, or *Felt.* This is manufactured in the form of *blankets,* also called *quilts,* and *batts.* It may be sold in rolls or in flat 4-ft sections. Fiber insulation ranges in thickness from $\frac{1}{2}$ to 2 in. and is available in a number of widths suitable for insertion between studs and joists. Flexible fiber insulation is made of loosely felted wood or some other vegetable fiber, usually covered on one or both sides with paper, fabric, or

aluminum foil. There are now many synthetic "foamed" products that are in direct competition with insulation made of wood fiber.

(3) *Fill Insulation.* This includes all insulation materials that are supplied in bulk form to be poured or blown into spaces between wall studs and floor and wood joists. The forest products most commonly used as fill insulation are wood fiber, ground and macerated newsprint, granulated cork, shredded redwood bark, and sawdust and shavings.

III. PULP AND PAPER INDUSTRY

The location of a pulp and paper plant is determined mainly by four considerations: availability of raw material, power, water, and the proximity to markets for the finished products. The ideal location would be where all these factors are favorable. Since this is seldom possible, the trend is to build integrated plants in which pulp and paper mills are located on the same site, so that the pulp is transferred to the converting mill by a pipeline in slush form. The new plants are usually located close to the source of the raw material because it is more profitable to ship a smaller bulk of finished products than to bring in a much larger bulk of raw materials over long distances. This situation exists in Canada and in the South, and to a lesser degree in the Far West of the United States. In all these regions, though pulp and paper plants are located close to the vast wood resources, they may be far removed from the main consumers of their finished products. On the other hand, there are numerous paper mills in this country, especially in the Central states, that are located far from any source of raw material. The original locations of these plants may have been determined years ago by availability of labor and power, proximity of markets, or personal preferences of the owners. Because of high investment in fixed assets,* pulp and paper plants cannot be dismantled and moved to another location. This explains why in the older regions, particularly in New England, there are numerous plants originally built close to the local sources of raw material but now operating almost exclusively on pulpwood or pulp imported from other regions of the United States or from abroad.

The selection of a site for a pulp and paper mill is governed to a large extent by consideration of the accessibility to means of transportation and the availability of an adequate water supply of good quality; the latter consideration is particularly important because of the large volume of pure water required in pulp and paper manufacture. The quantity of water used varies with the kind of pulp and paper made, ranging from 20,000 gal per ton of groundwood to 150,000 gal per ton of bleached sulfite, and from 15,000 gal per ton of book paper, where white water is efficiently reused, to as high as 200,000 gal per ton of paper in mills using large quantities of rags. The average paper mill requires 30,000 to 50,000 gal of water per ton

---

* A self-containing pulp mill requires an initial investment of more than $50,000 per ton of daily output, exclusive of forest lands. A recently completed sulfate mill in the South, with an initial capacity of 130,000 tons of pulp a year and a projected capacity of 170,000 tons, has been built for 60 million dollars. It uses 900 cords of pine a day and operates 24 hours a day on a 7-days-a-week schedule.

of finished paper. In the case of the sulfite pulp mills, an added reason for locating plants near a body of water was the universal practice of water disposal of the vast quantities of sulfite waste liquor. With passage of anti-pollution regulations in many sections of the country, this practice is, however, being gradually discontinued.

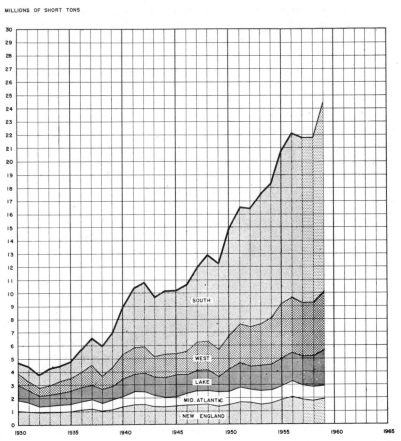

MILLIONS OF SHORT TONS

Fig. 17-25. Regional wood-pulp production. (*Courtesy United States Pulp Producers Association.*)

A. Regional Distribution of Pulp and Paper Industries in the United States

Pulp and paper plants in the United States are concentrated in four regions: (1) the Northeastern, including New England, New York, Pennsylvania, and New Jersey, the latter three states formerly grouped into a mid-Atlantic region; (2) the Lake states and Central region; (3) the Southern; and (4) the Pacific Coast (Fig. 17-25). These regions vary greatly as to the species of wood used, the supply of pulpwood available, the principal types of pulp manufactured, and the problems of conversion and distribution of pulp and pulp products.

1. Northeastern Region. Pulp and paper production in this region is concentrated mainly in New York, Maine, Pennsylvania, Vermont, New Hampshire, and Massachusetts. This region is characterized by large consumption of all grades of paper and other pulp products, owing to the concentration of population and large number of printing industries. The original concentration of pulp and paper mills in this area was logical, owing to the considerable areas of original spruce-fir forests, abundance of cheap power, low transportation costs, and nearness of markets. The present situation in regard to local supply of softwood pulpwood is, however, much less favorable than before. And even though the regional production of pulp has been increasing, due mainly to the expanding use of hardwoods, it is not adequate to meet the demand for pulp, which is growing at an even faster rate. The regional deficiency in pulp and paper is being met by importation of pulpwood, various types of pulp, newsprint, and other finished pulp products from Canada and to a lesser extent from the Scandinavian countries, as well as by substantial shipments of pulp and pulp and paper products from other regions of the United States.

The principal pulpwood species in the Northeastern region are eastern spruce and balsam fir; mixed hardwoods are also being used in increasing amounts. Groundwood pulp constitutes the largest tonnage of the output of this region, with sulfite pulp making up the bulk of the remainder. More recently semichemical hardwood pulp has also become of considerable importance. The greatest part of the pulp produced within this region is converted into paper and board in paper mills integrated with the pulp mills.

2. Lake States and Central Region. About 15 per cent of wood pulp is produced in the Central states, principally in Illinois and Ohio, and the remaining 85 per cent in Wisconsin, Michigan, and Minnesota. This region is characterized by a heavy concentration of paper- and board-converting plants, especially in Wisconsin, which leads the nation in production of paper and board products, and also in Michigan and Ohio. In the latter two states converting plants are far removed from the local supply of pulp, but they are strategically located in reference to the major industrial centers of consumption. The region as a whole consumes quantities of pulp far in excess of that produced locally. The deficiencies are supplied by importations from Canada and by shipments of pulp from the Pacific Coast and the South. Quantities of pulpwood are also being brought in from the Rocky Mountain states.

This region shows increasing reliance on hardwoods,* principally aspen, which at present accounts for approximately 50 per cent of the total pulpwood consumed in the local pulp mills. The principal pulpwood species, in addition to aspen, are spruce, balsam fir, hemlock, cottonwood, and jack pine. The most important types of pulp produced are sulfite, groundwood, and semichemical in about equal amounts, followed by sulfate. The semi-

---

* Feasibility of Using Lake States Hardwoods for Newsprint and Other Pulp and Paper Products. U.S. Department of Agriculture, Forest Service. 1959.

chemical pulp, though only recently introduced in this region, is gaining ground at a faster rate than the other types of pulp. The bulk of the pulp produced within the region is converted into paper and board products at mills integrated with the pulp mills.

3. Southern Region. This region comprises the area south of the Ohio and Potomac Rivers and includes also Louisiana, Arkansas, and Texas. The pulp and paper industry of the South is based largely on utilization of second-growth pine stands and to some extent worked-out naval-stores timber. Considerable quantities of hardwoods, principally redgum, blackgum, and oaks, are also available and are used in steadily increasing quantities.

Production of pulpwood in the South has expanded rapidly from 1 million cords in 1930 to more than 23 million cords in 1960. The 1960 pulpwood production represents nearly 58 per cent of the total pulpwood cut in the United States. Of this cut, southern pines account for 81 per cent and hardwoods for 19 per cent. About 11.5 per cent of the total pulpwood used, or 2.5 million cords, was in the form of chips derived from the pine mill residues. These were obtained from 900 sawmills, as compared with 642 mills in 1959. It is estimated that the South's share of the total pulpwood produced in the United States will continue to increase, reaching nearly 63 per cent of the total by 1975.

Correspondingly with pulpwood cut, pulp production in the South also represents nearly 58 per cent of the total domestic output, since practically all the pulpwood cut in the South is consumed by the regional mills. Owing to the resinous nature of southern pines, 80 per cent of all pulp produced in the South is sulfate, the remainder being about equally divided between the semichemical and mechanical pulps.

The bulk of pulp produced in the South is converted at the integrated paper and board mills. Substantial quantities of kraft pulp are shipped to the Northeast, Lake, and Central regions. The technological developments which made it possible to produce groundwood from the southern pines and hardwoods have led to the establishment of considerable newsprint capacity. Though not quite adequate to meet the total requirements of the South, this newsprint production accounts for more than 40 per cent of the total domestic newsprint capacity.

The rapid growth of the pulp and paper industry in the South has resulted in some areas of the region in competition between the pulp mills and the sawmills for the available supply of softwood timber. It is expected that because of the dynamic nature of this industry, and its more sound economic position, the southern pulp industry will continue to expand and to consume an increasingly larger share of the available timber. To some extent further expansion of pulping facilities may be based on the use of larger quantities of mill residues and of hardwoods. Considerable efforts are also directed by the pulp companies toward promotion of sustained-yield management and extensive planting on company-owned and other private forest lands. In co-operation with the Federal and state forestry agencies, the industry is also engaged in an extensive research program aimed at improving the quality of pulpwood species through genetic selection.

The spectacular growth of the pulp and paper industry in the South is due

to a combination of favorable factors. The most important of these is the
abundant supply of good pulpwood produced by rapidly growing species,
the future supply of which could be assured. This provides the basis for
stability, a factor of great importance in the pulp and paper industries, in
which capital investments in building and machinery make it prohibitive
for the industries to change locations. To this may be added lower construction and operation costs; an ample supply of labor; * availability of good

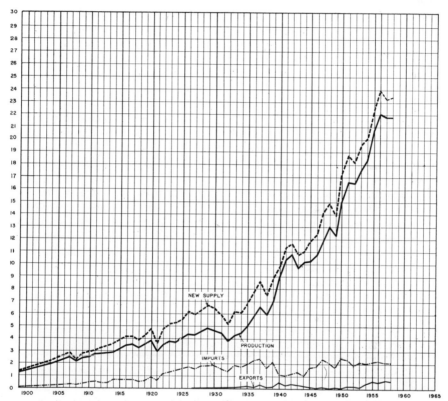

Fig. 17-26. United States production, imports, exports, and new supply of pulpwood.
(*Courtesy United States Pulp Producers Association.*)

water, chemicals, and power; good transportation facilities; and a favorable
tax structure. Finally, easy logging and more favorable climatic conditions,
permitting year-long woods operations, result in lower carrying charges for
pulpwood compared with the Northeastern and Lake states regions. The
multiple-use forestry idea, which looks to closer coordination between logging
for lumber and wood products, pulpwood production, and the naval-stores
industries, also promises important advantages for the pulp and paper industries of the South.

* Claims for advantages accruing from lower labor cost, formerly held by the South,
are no longer valid since the prevailing wage rates in the southern pulp and paper industries are as high or higher than in the East.

4. Pacific Region. Practically all pulpwood used in this region is of domestic origin. The principal pulpwood species used on the West Coast and in Alaska are western hemlock, Douglas-fir, western firs, and Sitka spruce. Of the total quantity of pulp produced in this region, approximately 70 per cent is manufactured in the state of Washington, and the remainder principally in Oregon. Recently, substantial pulping capacity has also been established in Alaska, and a start has been made in the Mountain states.

Fig. 17-27. Unloading wood chips at a pulp mill. (*Courtesy Agr. Ext. Service, North Carolina State College.*)

It is expected that further expansion of the pulp industry in Alaska will take place in the near future. More than 80 per cent of the total pulp produced in the Pacific Coast region is sulfite and sulfate, in about equal amounts; the remainder is mostly groundwood, with small quantities of semichemical.

Although most of the pulp produced on the West Coast is consumed by the paper and the board mills of Washington, Oregon, and California, the Pacific Coast is the only region at the present time which manufactures considerable quantities of pulp, principally sulfite, in excess of the needs of the local paper industries. Some sulfite pulp mills of this region produce pulp largely for sale in other sections of the country, as well as for export. Large quantities of bleached sulfite pulp are shipped from this region by water and rail to rayon-manufacturing regions of the eastern, central, and southern United States.

This region is characterized by extensive utilization of pulp chips derived from sawmill and plywood-mill residues. The 1958 utilization of chips represented about one-half of the total pulpwood consumed on the West Coast. Although the potential supply of chippable material in the West is estimated at about 8.8 million cords,[17] based on production of 20 billion ft of lumber, not more than two-thirds of this quantity could actually be made available. This is due to lack of barking and chipping facilities at the small mills and to the fact that the greatest quantity of residues originates from Douglas-fir, suitable mainly for sulfate pulping. The production of sulfate pulp, however, represents less than 40 per cent of the total pulp output of that region. The chips now account for three-fourths of the total wood needs for the sulfate process. Sulfite pulp, which also accounts for 40 per cent or more of the total pulp production on the West Coast, derives only 5 per cent of its total wood requirements from chips. This is because sulfite mills depend largely on western hemlock and to a lesser extent on spruce. Lumber production of both is relatively low, and the available chippable material is limited to about 0.5 million cords. Groundwood is not adapted to the use of mill residues and must therefore derive its raw material from roundwood.

On the basis of expected increase in population and the availability of softwoods, it is estimated that the paper and board industry on the West Coast will increase at least 50 per cent by 1965, as compared with the 1958 output.

Newsprint, coarse paper, fruit wrappings, and sanitary tissues constitute about 80 per cent of the western paper production. Production of the finer papers is lagging considerably behind current consumption. Western board mills account for almost 84 per cent of the total amount of container and bending boards made in the United States.

5. Canada.[9] Canada is second only to the United States in the quantity of wood pulp produced; it leads the world in the production of newsprint and exports more newsprint than all other countries combined. The Canadian wood-pulp industry is of special importance to the United States because Canada is the chief supplier of this country's imports of newsprint as well as the principal source of imported pulp.

The manufacture of pulp and paper is Canada's largest industry. The products of this industry also account for almost 25 per cent of the total value of the Canadian exports and represent one-third of all Canadian exports to the United States.

The principal product of the Canadian pulp and paper industry is newsprint, which makes up about 60 per cent of the total Canadian pulp production and almost one-half the world's output. Almost 80 per cent of the newsprint is exported to the United States. The second largest item, representing about 20 per cent of the pulp production, consists of different types of pulps for the export trade, about two-thirds to the United States. The remainder of the industry's capacity is devoted to production of a wide range of paper and paperboard products, chiefly for domestic markets.

The principal pulpwood species in Canada are spruce and balsam fir, which make up the bulk of the estimated 100 billion cu ft of potential pulpwood in that country. Canada exports about 2 million cords of pulpwood, 80 per cent of it to the United States. Because of the export embargo on pulpwood cut on the Crown Lands in all the Canadian provinces, the pulpwood exported to the United States comes from privately owned lands. Ontario, New Brunswick, and Quebec, in that order, provide most of the pulpwood for the export trade. In the future pulpwood exports from Canada to the United States are expected to remain at the current level or to decline slightly. This prediction is based on the anticipated increase in Canada's own needs for pulpwood and increased reliance of the United States pulp industry on domestic pulpwood resources, particularly the southern pines and hardwoods.

The total production of pulpwood in Canada reached 14 million cords in 1958. Since the Second World War there has also been marked expansion of the utilization of sawmill residues for pulp, especially in the coastal region of British Columbia where an equivalent of 1.3 million cords of pulpwood was produced in 1957 in the form of chips. This represented more than 45 per cent of the total pulpwood production in British Columbia, or an equivalent of the allowable cut of pulpwood from 1.6 million acres based on 61 cu ft of pulpwood per acre. Production of wood chips for pulp in eastern Canada, though on a much smaller scale than in British Columbia, is also quite substantial, reaching an equivalent of 200,000 cords of round pulpwood in 1957.

Canadian pulp-producing capacity had reached nearly 13 million tons in 1958, representing a gain of some 200 per cent since 1929. Of this capacity, 55 per cent is mechanical pulp, 20 per cent sulfite, 18 per cent sulfate, and the remaining 7 per cent special dissolving grades and soda and semichemical pulps.

The future trend of the pulp and paper industry in Canada has been projected in a publication by the Forestry Study Group of the Royal Commission on Canada's Economic Prospects to 1980.[9] This study group estimated that by 1980 manufacture of pulp should attain about 21 million tons of pulp. This would require an increase in consumption of pulpwood by the Canadian industry from the 1956 rate of more than 14 million cords to about 28 million cords by 1980. Newsprint will remain the main product of the industry, and the United States will remain the chief consumer of newsprint as well as of some grades of pulp. However, progressively larger quantities of newsprint and various pulps will also be exported to other countries.

SELECTED REFERENCES

1. Anon. Pulp, Paper, and Paperboard Industry. American Forest Products Industries, Inc., Washington, D.C. 1957.
2. Anon. Wood Pulp Statistics, 25th ed. United States Pulp Producers Association, Inc., New York. 1961.
3. Anon. Pulpwood Statistical Review. American Pulpwood Association, New York. 1960.
4. Anon. The Dictionary of Paper. American Paper and Pulp Association, New York. 1951.

5. Applefield, Milton. The Utilization of Wood Residues for Pulp Chips. *Texas Forest Serv., Bull.* 49. 1956.
6. Brown, K. J., and W. H. Monsson. Cold Soda Pulping of Aspen by Improved Methods. *Tappi,* **39**(8):592–599. 1956
7. Calkin, J. B., and G. S. Witham, Sr. Modern Pulp and Paper Making, 3d ed. Reinhold Publishing Corporation, New York. 1957.
8. Casey, J. P. Pulp and Paper Chemistry and Chemical Technology, rev. ed., vols. 1, 2. Interscience Publishers, Inc., New York. 1960.
9. Davis, John (Chairman). The Outlook for the Canadian Forest Industries. Royal Commission on Canada's Economic Prospects. Edmond Clouter Queen's Printer and Controller of Stationery, Hull, Canada. 1957.
10. Fobes, E. W. Bark Peeling Machines and Methods. U.S. Department of Agriculture, Forest Service, Forest Products Laboratory Report No. 1730. 1957.
11. Grant, J. Cellulose Pulp and Allied Products. Interscience Publishers, Inc., New York. 1959.
12. Guttenberg, S., and J. D. Perry. Pulpwood with Less Manpower. *South. Forest Expt. Sta., Occas. Paper* 154. 1957.
13. James, L. M. Marketing Pulpwood in Michigan. *Mich. Agr. Expt. Sta., Dept. Forestry, Spec. Bull.* 411. 1957.
14. Klemm, K. H. Modern Methods of Mechanical Pulp Manufacture (tr. from German by E. P. Chin). Lockwood Journal Company, New York. 1958.
15. Libby, C. E., and F. W. O'Neil. The Manufacture of Chemigroundwood Pulp from Hardwoods. New York State College of Forestry, Syracuse, New York. 1950.
16. Natwick, A. G., P. Luch, and J. T. Hallock (eds.). Making Paper, 3d ed. Crown Zellerbach Corporation, San Francisco, Calif. 1953.
17. Stanford Research Institute. America's Demand for Wood, 1929–1975. Palo Alto, Calif. 1954.
18. Stephenson, J. N. (ed.). Pulp and Paper Manufacture. McGraw-Hill Book Company, Inc., New York. Vol. 1, Preparation and Treatment of Wood Pulp, 1950; Vol. 2, Preparation of Stock for Paper Making, 1951; Vol. 3, Manufacture and Testing of Paper and Board, 1953.
19. Sudermeister, E. The Story of Papermaking. S. D. Warren Co., Boston, Mass. 1954.
20. Taras, M. A. Buying Pulpwood by Weight as Compared with Volume Measure. *Southeast. Forest Expt. Sta., Sta. Paper* 74. 1956.
21. Teesdale, L. V. Thermal Insulation Made of Wood Base Materials. U.S. Department of Agriculture, Forest Service, Forest Products Laboratory Report No. 1740. 1955.
22. U.S. Department of Agriculture, Forest Service. Timber Resources for America's Future. Forest Resource Report 14. 1958.
23. U.S. Department of Agriculture, Forest Service, Division of Forest Economics. Woodpulp Mills in the United States, by States and Type of Product. 1959.
24. U.S. Department of Commerce, Bureau of the Census, Industry Division. Current Industrial Reports: Pulp, Paper and Board. 1961.
25. U.S. Department of Commerce, Committee on Interstate and Foreign Commerce. Pulp, Paper and Board Supply-demand. House Report 573. 1957.
26. Werner, A. W. The Manufacture of Fiber Board. Board Products Manufacturing Co., Chicago, Ill. 1954.
27. Wilcox, H. Chemical Debarking of Some Pulpwood Species. The State University of New York, College of Forestry, *Tech. Bull.* 77. 1956.

CHAPTER 18

CELLULOSE-DERIVED PRODUCTS

Highly purified cellulose is used extensively as a basic material in the manufacture of a number of derived products such as rayon and acetate filaments and fabrics, transparent films, photographic films, artificial sponges, sausage casings, lacquers, plastics, and explosives. The two principal sources of cellulose suitable for the manufacture of these products are wood pulp and cotton linters. In the past the growth of industries depending on cellulose derivatives has been due in no small measure to the discovery that highly refined wood pulp is as suitable for these products as cotton linters. The principal type of wood pulp used is a bleached sulfite with 85 to 96 per cent alpha-cellulose * content, depending on the manufacturer's requirements. The importance of refined wood pulp, generally designated as dissolving pulp (see page 366), as a source of cellulose is evident from information presented in Table 18-1.

Table 18-1. Cellulose Consumption by the United States Rayon and Acetate Industries

Year	Total pulp, tons	Wood pulp		Cotton linters	
		Tons	Per cent of total	Tons	Per cent of total
1930	72,000	45,000	62	27,000	38
1940	238,000	178,000	75	60,000	25
1950	590,000	456,000	77	134,000	23
1955	634,500	546,900	86	87,000	14
1957	556,000	517,000	91	49,000	9
1958	501,000	479,000	96	22,000	4
1960	503,000	481,000	96	22,000	4

* Alpha-cellulose is generally defined as the more stable portion of cellulose that fails to dissolve in 17.5 per cent caustic soda, or 24 per cent potassium hydroxide, at room temperature (20°C).

It is significant, however, that as figures in Table 18-3 indicate, production of rayon and acetate fiber in 1960 declined to the 1958 level, while that of noncellulosic forms increased by almost 70 per cent. It can be assumed that in the future competition from the noncellulosic fibers will result in further deterioration of the relative position now occupied by the cellulosic fibers.

A parallel situation exists in the case of transparent films. Prior to 1950 cellulose-derived films enjoyed a virtual monopoly. In 1960, though the total production of cellulose and acetate films was still increasing, it constituted only 60 per cent of the total film production; the remaining 40 per cent were various noncellulosic kinds.

I. CELLULOSE FILAMENTS AND YARNS

Cellulose filaments and yarns can be produced by the cellulose nitrate, cuprammonium, viscose, and cellulose acetate processes. Of these cellulose nitrate is no longer used for this purpose in the United States, and only small quantities of cuprammonium fiber are still manufactured. Information on these two processes is therefore limited to data presented in Table 18-2.

Table 18-2. Flow Chart of Rayon Manufacture by Four Processes [a]

Viscose	Acetate	Cuprammonium	Nitrocellulose
Caustic steeping	Treated with acetic anhydride, acetic acid, and catalyst	Treated with basic copper sulfate and ammonium tartrate	Dissolved in nitric acid and sulfuric acid
1. Alkali cellulose: Pressing Grinding Aging Treating with CS_2 2. Cellulose xanthate: Dissolved in caustic soda Filtering Deaerating Ripening 3. Viscose spinning solution: Wet-spinning through spinnerets into acid bath Spinning Washing 4. Desulfuring Washing Bleaching Drying Twisting 5. Viscose yarn	1. Primary cellulose acetate: Hydrolysis Precipitation Washing Extracting Drying 2. Secondary cellulose acetate: Filtering Deaerating 3. Acetate spinning solution: Dry-spinning into warm air Twisting 4. Acetate yarn	1. Dissolved in cuprammonium solvent: Mixing Diluting Maturing Filtration Deaerating 2. Cuprammonium spinning solution: Wet-spinning into water Stretching 3. Copper removal: Acid wash Rinsing Drying Twisting 4. Cuprammonium yarn	1. Cellulose nitrate: Precipitation Washing Extracting Dissolving in alcohol-ether Filtration Deaerating 2. Nitrate spinning solution: Dry-spinning into warm air 3. Denitrating: Washing Bleaching Scouring Soaping Twisting 4. Nitrocellulose yarn

[a] Adapted from Rayon and Synthetic Yard Handbook, 2d ed. The Viscose Company, New York. 1936.

In the past the products of all four processes were collectively known in this country as *rayon* and sometimes as *artificial silk*. The term rayon was adopted by the trade in 1924 to designate a class of synthetic filaments and textiles made of them, the principal constituent of which is cellulose. A similar definition has also been formulated by the American Society for Testing Materials, which defines rayon as "a generic term for filaments made from various solutions of modified cellulose by pressing or drawing the cellulose solution through an orifice and solidifying it in the form of a filament."

Because of the marked difference in the chemical and physical properties of the filaments, the more recent trend is to confine the term *rayon* to the products of the viscose and cuprammonium processes. The cellulose acetate filaments and textiles are more commonly listed under that name, i.e., as *cellulose acetate* products.

A. The Viscose Method

The viscose method of producing rayon is based on the discovery in 1892 by Cross, Bevan, and Beadle that when cellulose is treated with a strong sodium hydroxide solution an *alkali cellulose* is formed. When this is treated with carbon disulfide, it forms an alkali-soluble *cellulose xanthate* (sodium cellulose salt of dithiocarbonic acid). Cellulose xanthate itself is an intermediate product and is of interest only because it is a practical and inexpensive means of bringing the original raw cellulose into a solution from which it can be reprecipitated (*regenerated*) with essentially the same chemical structure as the original cellulose but in a number of new physical forms, such as filaments suitable for spinning, films, or sponge-like masses. The process derives its name from the high viscosity of the cellulose xanthate solution in dilute sodium hydroxide ("viscose").

The steps in the manufacture of viscose rayon include the preparation of alkali cellulose by steeping, pressing, and shredding cellulose material. In steeping, raw cellulose is treated with 17.5 to 18 per cent caustic solution for about an hour at temperatures of 18 to 20°C. The object of this treatment is to remove all the hemicellulose and impurities from the cellulose and to cause it to swell and to absorb sodium hydroxide (i.e., to form alkali cellulose). When wood pulp is used, sheets of pulp approximately 12 by 18 in. are placed on edge in a rectangular tank. Each sheet of pulp is separated by perforated steel plates. The tank is filled with a caustic soda solution and the pulp is allowed to soak in this solution for about an hour at a definite temperature. When this treatment is completed, the caustic solution is drained, and the pulp is squeezed by means of a ram until the wet weight of the pulp is about three times that of its original dry weight. Pulp treated with caustic soda in this manner is called alkali cellulose. The wet sheets of alkali cellulose are then reduced to small particles, called *crumbs*, by means of a shredder. This machine consists of a cast-iron trough, equipped with two rotating horizontal steel arms fitted with steel-toothed shoes. The shredder not only breaks up the wet pulp sheets into small particles, but also blends and mixes the crumbs, equalizing the caustic soda content.

The next phase of manufacture consists of aging the crumbs. The crumbs are discharged from the shredder into large cans, which are placed in temperature-controlled rooms. There they remain for about 48 hours, while the alkali cellulose undergoes important physical and chemical changes that alter its molecular structure, decrease its viscosity, and increase its solubility. The aged alkali cellulose is then treated with carbon disulfide (CS_2), which transforms it into an alkali-soluble cellulose derivative known as cellulose xanthate. This operation is carried on in a rotating steel drum called a *churn*. During this reaction the cellulose changes its color from white to orange; at the same time the fluffy crumbs collect into tight balls ranging in size from marbles to baseballs. At the end of the xanthation reaction the excess of carbon disulfide is allowed to drain, and the cellulose xanthate is transferred into a mixer equipped with a stirring apparatus, where a known quantity of dilute caustic soda is added. The caustic soda gradually dissolves the cellulose xanthate, the resulting orange-colored viscous solution being the *viscose*, a complex mixture containing cellulose xanthate, caustic, carbon disulfide, and compounds formed by their interaction. The viscose is filtered through several thicknesses of cotton batting and the filtrate allowed to ripen for several days, until all the air bubbles escape, leaving a clear, air-free solution that is now ready for spinning (Fig. 18-1).

Ripening of viscose is accompanied by chemical and physical changes; the former consist of the spontaneous decomposition of the cellulose xanthate, by splitting off of the carbon disulfide and regeneration of the cellulose, and the latter results in improvement in the coagulation properties of the product. This spontaneous decomposition of the cellulose xanthate is too slow for industrial purposes; the regeneration of the cellulose can be made instantaneous, however, by treating the xanthate with mineral acids and acid salts. Such treatment, carried on in the coagulation baths, causes simultaneous regeneration and coagulation of cellulose in the desired physical forms.

More recently several continuous processes have come into use to replace the multiple-stage batch operations just described. In one continuous system, pulp sheets are passed through all the stages of operations on a conveyer, while in another, bulk pulp is treated in vats equipped with agitators and press rolls, or in a centrifuge, for removing the excess of caustic.

1. Rayon Yarn. Rayon filaments are produced by forcing the viscose through holes 0.002 to 0.005 in. in diameter in small cups or spinnerets made of precious metal, usually platinum. The spinnerets are immersed in an acid coagulating bath containing warm water, sulfuric acid, sodium sulfate, zinc sulfate, and usually some glucose or magnesium sulfate. This acid bath neutralizes the caustic soda, causing the breakdown of cellulose xanthate and regeneration of solid cellulose in the form of filaments, each hole in the spinneret forming a filament of pure cellulose.

These filaments are led to the Godet wheel rotating in a bath and from there to a spinning box, called a centrifugal pot, which rotates at 6,000 to 10,000 rpm and causes a group of filaments to be twisted into a thread of the desired twist. The diameter of the thread is controlled by the speed of the wheel, rotation of which causes the filaments to draw out prior to completion of coagulation.

In another type of spinning, called bobbin spinning, the thread as it leaves the hardening bath is wound without a twist, the twisting being done as a

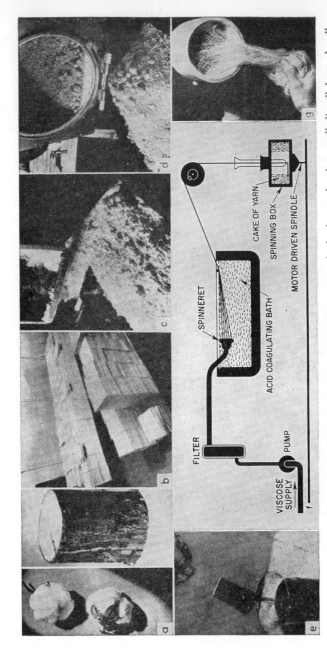

Fig. 18-1. Steps in manufacture of viscose rayon: *a*, cotton or spruce wood; *b*, sheets of pulp; *c*, alkali cellulose; *d*, cellulose xanthate; *e*, viscose spinning solution; *f*, spinning operation; *g*, yarn from spinning machine. (*Courtesy American Viscose Corporation.*)

separate operation. The yarn as it comes from the spinning machine still contains more than 70 per cent of bath liquids; these are washed out with water. It is then treated with an alkaline sulfide solution, which dissolves the sulfur deposited on the yarn, and is passed through heated driers. The yarn which is intended for weaving is bleached with a hypochlorite solution and then redried. This treatment restores the bright color to rayon which, when it comes from the spinning machine, is a dull yellow. Yarn intended for knitting is bleached after it has been woven into cloth. Rayon yarns have a high luster that is frequently objectionable in a finished product. Today, rayon yarns of normal, semidull, and very dull lusters are available. Delustering is accomplished by incorporating insoluble pigments into the rayon threads. Colored rayons are produced by spinning a solution of viscose to which a dye has been added.

A streamlined version of bobbin spinning is the continuous system, in which the filaments are extruded, processed, and finished on the same machine in the form of continuous knobless yarn. Continuous spinning produces a high-quality yarn with a considerable saving of time.

B. Cellulose Acetate

Cellulose acetate differs from the cuprammonium and viscose rayons in that it is not a regenerated cellulose but a cellulose derivative—an ester of cellulose—and hence a distinct chemical entity with physical and chemical properties different from the original cellulose.

Sheets of wood pulp used in the manufacture of cellulose acetate are first subjected to a mild bleaching and then treated with acetic acid. The pulp is then transferred into batch reactors, where it is mixed with acetic anhydride and glacial acetic acid in the presence of a suitable catalyst. Mixing is continued until the fibrous structure of the cellulose disappears. The charge, called the primary cellulose acetate, is dropped into a vat containing acetic acid to which 5 to 10 per cent of water is added. The mix is held in this vat for 10 to 20 hours to cause hydrolysis from the primary to the secondary acetate. Completion of this reaction is indicated by a drop from 44.8 per cent to about 35.0 to 41.0 per cent acetyl content.

When the desired degree of acetylation is achieved, the cellulose acetate is precipitated from the solution in flake form by running it into a tank containing cold water. The slurry is pumped into settling tanks, from which the clear liquor is siphoned off and conducted to the recovery plant, while the precipitated acetate is washed until neutral and then dried.

The dried acetate flake is redissolved in acetone and filtered. The fiber is formed by forcing the solution through spinnerets into a vertical cabinet or tubes, where the acetone is removed by circulating warm air, at temperatures above the boiling point of acetone (135°F). The coagulated filaments are wound on the spools, bobbins, or tubes (Fig. 18-2).*

* A cellulose acetate fiber of high tensile strength, known as Fortison, is produced by subjecting acetate yarn to mechanical stretching, which induces parallel molecular structure, and then regenerating the cellulose by saponification to remove the acetate groups. Fortison is akin to viscose rayon in that it is a regenerated cellulose. A number

ACETATE PROCESS

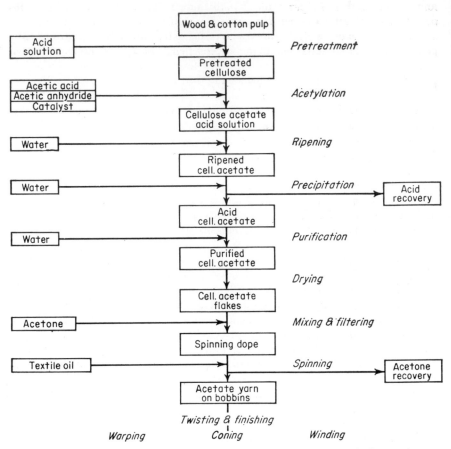

Fig. 18-2. Acetate process. (*Courtesy Celanese Corporation of America.*)

C. Staple Fiber

Staple fiber is made by cutting rayon and acetate yarns into short lengths and then spinning them into a thread. Originally introduced to make better use of the waste of rayon-fiber mills, it now accounts for about 40 per cent of the total production of cellulosic yarn. This development has resulted in a considerable increase in the amount of wood cellulose converted to textiles.

As now practiced the filaments are drawn from a number of spinnerets, without twisting, into a rope-like aggregation, called tow, which is then cut up into required lengths for spinning. The acetate tow requires no further treatment before cutting and spinning; that produced by the cuprammonium

of other "high-tenacity" cellulosic fibers are now made also from viscose solution. All these fibers have high tensile strength and are used in products where strength is the paramount consideration, principally in tire cords.

and viscose processes undergoes additional chemical purification, washing, and drying. If this is done before the tow is cut, the method is known as the dry process; if after cutting, the method is called the wet-cut system.

Staple fiber can be spun on any type of spinning machines, either as a pure fiber or in various blends with cotton, silk, wool, and synthetic fibers.

D. Properties of Cellulose Filaments and Yarns

With the advent of truly synthetic fibers such as nylon or dacron, the producers of rayon and acetate filaments and staple fiber were forced to develop products with specific technical properties to meet particular uses. An example is high-strength yarn, for use as tire cord and for conveyer and V-belts, which has 40 per cent more dry and 70 per cent more wet strength than the conventional rayon.

The viscose and cuprammonium rayons are quite stable in organic solvents but disintegrate completely in hot dilute and cold concentrated acids. Alkalies produce swelling and in concentrated forms affect fiber strength. Rayon loses strength at temperatures above 300°F and is weakened by lengthy exposure to strong light. It does not lend itself to insulating applications because of its relatively high moisture absorption. Under favorable conditions of temperature and humidity, rayon may be attacked by wood-destroying fungi and by molds, but it is highly resistant to moths. It is nontoxic and nonirritating to the skin.

Cellulose acetate yarns are unaffected by dilute solutions of acids and alkalies, but are decomposed by strong acids. Strong alkalies cause saponification; i.e., they will change the acetate into a regenerated cellulose. Cellulose acetate fabrics are attacked by strong oxidizing agents and a number of organic solvents and hence require special dry-cleaning methods. Superior to rayon and cotton in resistance to water absorption, cellulose acetate possesses excellent electrical insulating properties and is resistant to bacterial and fungal attacks. While it does not burn readily, it will melt under hot ironing conditions. It dries rapidly and has superior wrinkle resistance, properties which recommend cellulose acetate for draperies, ironed-in pleats, and similar uses where wrinkle resistance is important. Because of its low shrinkage and swelling, it is frequently blended with other fibers for stabilizing purposes. Easy removal of such water-soluble stains as fruit juices and ink by sponging with water is another outstanding characteristic of this fiber.

E. Uses

Rayon and acetate fibers are extensively used in textiles, alone or in a mixture with other fibers. This also includes woven and knitted fabrics, hosiery, and yarns for knitting and weaving.

Industrial uses of viscose rayon and acetate include tire cord; conveyer and V-belts; abrasive, polishing, and sanding wheels; safety belts; and reinforcement for paper and plastics.

Special forms of cellulose acetate are also used in lightweight raincoats, strong coated fabrics, and typewriter ribbons. Other forms are employed widely in the electrical field because of their low conductivity, dimensional stability, and flexibility.

F. Rayon Industry in the United States

The American rights to the viscose process were acquired by the Central Artificial Silk Company of Landsdowne, Pennsylvania, in 1905. These rights were later purchased by S. W. Pettit, under whose direction production reached some 300 lb of rayon a day. When Pettit died in 1909, his son sold the American rights to a company which, under the name Viscose Company of America, established a commercial plant at Marcus Hook, Pennsylvania, in 1910. The next year 365,000 lb of artificial silk were produced at that plant.

From that time on the American rayon industry began to expand rapidly, reaching a production of more than 10 million pounds in the year 1920 and the maximum production of 1,294 million pounds in 1955. The 1958 through 1960 production is presented in Table 18-3.

Table 18-3.   Man-made Fiber Production, in the United States,[a] Millions of Pounds

Fiber	1958	1959	1960
Cellulosic			
Rayon			
Filament yarn........	413.1	508.1	426.3
Staple and tow........	324.2	359.1	314.0
Acetate			
Filament yarn........	222.6	229.6	228 2
Staple and tow.......	75.0	70.0	60.0
Total cellulosic......	1,034.9	1,166.8	1,028.5
Noncellulosic			
Filament yarn.........	320.0	412.1	437.4
Staple and tow..........	170.5	233.2	239.8
Glass fiber..............	103.8	147.4	179.2
Total noncellulosic..	594.3	792.7	856.4
Total...................	1,629.2	1,959.5	1,884.9

[a] Adapted from *Textile Organon*, **30**(5):70. 1961.

The United States accounts for about 20 per cent of the world's output of cellulose filaments and staple fiber. It leads Japan narrowly on the basis of the total production of filaments and staple fiber. Japan, however, leads the world in production of staple fiber. Other important producers of rayon and acetate are West Germany, the British Commonwealth, the U.S.S.R. (including the satellites), Italy, and France.

II. TRANSPARENT CELLULOSE FILMS

Transparent cellulose films are produced from viscose, cellulose acetate, ethyl cellulose, and nitrocellulose. The most important of these is cellophane, which is made by the viscose method. It accounts for about 95 per cent of the total output.

A. Cellophane

The manufacture of cellophane is identical to yarn manufacture, except that cellulose-containing solutions are forced through long slits instead of spinnerets. The sheets are then carried through a regenerating bath, bleached, washed, and passed through a bath containing a softening agent such as glycol. Finally the sheets are dried on hot rolls and reeled. Cellophane is quite permeable to water vapor. To render it resistant to the passage of water vapor, some grades are treated with a thin film of cellulose nitrate lacquer. Cellophane films can be dyed any desired color or embossed by means of pressure rolls. The chief use of this film is in the packaging and overwrapping of all types of merchandise.

Sausage casings, bands, and similar items are made by extruding viscose through annular slots into a coagulating bath. The film is formed as a tube and then put through the various purification stages in a continuous operation.

B. Cellulose Acetate Films

The cellulose acetate films are cast through a narrow slot on the polished surface of a metal drum, which rotates in a drying chamber. This causes evaporation of the solvent and coagulation of the film. The film is peeled off the drum while still partially wet, carried between hot rolls, and wound on a reel. The cellulose acetate films are considerably more water resistant than cellophane. They have good electrical resistance, are not embrittled when exposed to sunlight, are self-sealing under heat and pressure, but are permeable to gases. The last property makes them particularly suitable for overwrapping fruits and vegetables and other products which require gas permeability to permit ripening and to prevent premature rotting and molding. The largest outlet for acetate films outside of packaging is in the manufacture of photographic and X-ray films and sound and visual recording tapes.

C. Nitrocellulose Films

Nitrocellulose films are made from cellulose nitrate. This form of cellulose is produced by treating cotton linters and shredded wood pulp with nitric and sulfuric acids in the presence of water. Cellulose nitrate remains in the fibrous form throughout this treatment. In this respect its manufacture differs from that of other cellulose esters, which go into solution as they are formed. Because of this, the physical form of the wood pulp is of great importance in the manufacture of cellulose nitrate. It has been found that purified wood pulp nitrates most readily when it is shredded in about its own weight of water and then redried. Another method consists of converting wood pulp into a lightweight tissue paper, which is then nitrated. The dried cellulose nitrate is dissolved in alcohol and ether and forced through narrow slots into a water bath, causing coagulation.

Nitrocellulose is still the leading material for making commercial motion-picture films. Its chief drawback is high inflammability, and for this reason it has been replaced by cellulose acetate in amateur motion-picture films. The high rate of inflammability also prevents the use of this product in the packaging field, in spite of its many excellent characteristics, such as high water resistance, dimensional stability, and toughness.

D. Ethyl Cellulose Films

Ethyl cellulose films are made from an alkali cellulose, i.e., cellulose treated with concentrated alkali solutions, which subsequently is converted into a cellulose ether with ethyl chloride at an elevated temperature. The films are cast from a solution in the same manner as cellulose acetate. Ethyl cellulose films are flexible at low temperatures, have good water-resistance and excellent electrical-resistance properties. They are self-sealing under heat and pressure. These films find application in packaging and in electrical insulation.

III. CELLULOSE SPONGES

In the manufacture of cellulose sponges, a viscose solution is mixed with long-fibered vegetable hair and crystals of salt, insoluble in the cold viscose. The batch is poured into molding cans and heated. This causes the crystals to dissolve, leaving holes. The size of the holes is determined by the size of the crystals used. The sponges are washed, dried, and cut to the desired sizes.

IV. EXPLOSIVES

Nitrocellulose is used as a stabilizing agent in blasting gelatins and as a major explosive ingredient in smokeless powders. Either cotton linters or alpha-cellulose wood pulp, prepared as described on page 366, is used for this purpose.

The cellulose used in blasting gelatins is nitrated under low temperature conditions until it attains a nitrogen content of about 12.3 per cent. Cellulose of high molecular weight with a degree of polymerization, i.e., the number of hydroglucose units in a molecular chain, as high as 3,000 to 3,500 is used. Blasting gelatin is prepared by mixing one part of cellulose nitrate with nine parts of nitroglycerin. The relatively small quantity of cellulose nitrate eliminates the danger of explosion due to sudden shock; at the same time it contributes to the power of explosion. The lower grades of dynamite are made in a similar manner with the addition of charcoal, sodium nitrate, and sulfur.

Cellulose nitrate used in the manufacture of smokeless powder has a degree of polymerization of about 500 and a nitrogen content of 12.6 to 13.3 per cent. Inert ingredients are added to reduce the energy of the explosion by retarding the rate of burning. Smokeless powders, such as *Cordite,* are made by gelatinizing cellulose nitrate with nitroglycerin.

V. LACQUER

The first and still the most important cellulose derivative used in lacquers is cellulose nitrate. Lacquers are made by mixing cellulose nitrate with resins, plasticizers, and organic solvents. More recently lacquers have been made with mixed cellulose esters and with ethyl cellulose. The new formulations produce lacquer films with greater resistance to light and heat, higher water resistance, and lower inflammability. These products are more expensive than nitrocellulose lacquers. However, because of its extreme toughness, even at a very low temperature, ethyl cellulose is expected eventually to capture the major part of the lacquer field.

In addition to their use in lacquers, the cellulose derivatives find a wide application in a variety of other protective coatings, such as peel and hot melt coatings, fabric and paper coatings, and as important ingredients in water-base paints.

VI. CELLULOSE PLASTICS

Numerous molding and extrusion powders are made with cellulose derivatives, for the manufacture of a great variety of plastics. These are discussed in Chap. 24.

SELECTED REFERENCES

1. Howsman, J. A. Non-apparel Applications of Viscose Rayon. *Textile Res. Jour.,* **28**(9):805–810. 1958.
2. Leeming, J. Rayon: The First Man-made Fiber. Chemical Publishing Company, Inc., New York. 1956,
3. Mauer, L., and H. Wechsler. Man-made Fibers. Rayon Publishing Corporation, New York. 1953.
4. Moncrieff, R. W. Man-made Fibers, 3d ed. John Wiley & Sons, Inc., New York. 1957.
5. Ott, E. (ed.). Cellulose and Cellulose Derivatives, 2d ed. Interscience Publishers, Inc., New York. 1954.
6. Paist, W. D. Cellulosics. Reinhold Publishing Corporation, New York. 1958.
7. Stamm, A. J., and E. E. Harris. Chemical Processing of Wood. Chemical Publishing Company, Inc., New York. 1953.

CHAPTER 19

CARBONIZATION AND DESTRUC-
TIVE DISTILLATION OF WOOD

This chapter deals with the manufacture of charcoal in kilns and portable ovens and with the dry, or destructive, distillation of hardwoods and softwoods. The steam-and-solvent distillation of pine wood for production of naval stores is discussed in the chapter on Naval Stores (Chap. 21).

The term *carbonization* is used in this chapter to denote the production of charcoal irrespective of the method employed, while the term *destructive distillation* implies also the recovery and refining of distillates, even though charcoal production may be the principal objective.

Wood carbonization and destructive distillation of wood have passed through several stages of development in this country. At first charcoal was the principal product of wood carbonization; it was used in large quantities by the iron and chemical industries, and also for cooking and fuel in city slums. Little or no attempt was made to recover the by-products of carbonization. With the advent of modern methods of wood distillation and the decline in use of charcoal in the iron industry, the by-products of destructive distillation became the principal items, and charcoal the by-product, selling for 4 to 6 cents a bushel.* Today, charcoal again is the principal product of the wood-carbonization and destructive-distillation industries, the distillates having lost ground to the more efficiently produced synthetic chemicals. Charcoal, although it still has many industrial applications, has now emerged as a specialty item for outdoor, and to some extent indoor, cooking.

I. DESTRUCTIVE DISTILLATION OF WOOD

Destructive distillation of wood consists of heating wood in a retort or oven in the absence of air or in the presence of a limited amount of air. When so heated, the wood substance decomposes, yielding a number of

* A bushel of charcoal weighs about 20 lb.

products, some of which are of commercial value, with the charcoal forming the residue.

The distillation of wood passes through several stages. In the first stage, little gas is produced, and the distillate consists of water and very volatile oils. As the temperature of the oven rises to about 280°C, the interior of the wood rises only to about 100°C, the boiling point of water. Even at the end of this stage the amount of gas evolved is still small, but the amount of distillate decreases abruptly, signifying that the wood is becoming dry. In the second stage, the temperature in the interior of the charge rises rapidly from 100°C to about 275°C, and that in the oven to near 300°C. The yield of distillate, consisting of tar and pyroligneous acid, increases rapidly, and so does the yield of gas. In the third stage, when the interior of the charge has reached about 275 to 280°C, the heat of the reaction is sufficient to carry on the distillation without the application of external heat, and the temperature of the interior of the charge rises above that of the exterior. In other words, the reaction becomes exothermic at a temperature of about 275°C. It is in this stage that the bulk of the pyroligneous acid, gas, and tar is produced. At a temperature of about 350°C, the exothermic reaction is finished and the temperature decreases. In the fourth stage, to complete the distillation, additional external heat is applied to rid the charcoal, as much as possible, of tarry bodies. The heat is usually increased to about 400°C, but very little distillate is produced. It is erroneous to say that charcoal is pure carbon, for analysis shows that charcoal contains only about 82 per cent carbon.

The gases * evolved in wood carbonization contain a great deal of carbon dioxide, and the percentage of combustible gases (carbon monoxide, methane, hydrogen, etc.) exceeds that of carbon dioxide only when the inside temperature of the wood and the temperature of the oven are above 350°C.

In commercial operations, the conditions of firing may affect the yield and types of various products. While this behavior is not noted in small-scale laboratory experiments, where the oven is uniformly heated and precise control of temperature is possible, commercial experience indicates that the yields † of alcohol and acetic acid may be increased by careful firing over a longer period than is customarily used in the distillation cycle. This discrepancy is explained by the fact that the wood may be superheated, causing additional secondary reactions, such as the decomposition of methyl alcohol

---

* The average composition of the total gas evolved in hardwood carbonization, omitting nitrogen (N_2) and the first 5 per cent and last 5 per cent of the gas, is as follows:

Constituent	Per cent
Carbon dioxide, CO_2	50.74
Carbon monoxide, CO	27.88
Methane, CH_4	11.36
Hydrogen, H_2	4.21
Ethane, C_2H_6	3.09
Unsaturated hydrocarbons, such as ethylene, C_2H_4	2.72

† In commercial practice, the yield of acetic acid varies greatly, depending on the amount of control exercised over the conditions governing carbonization; alcohol yield is more constant with varying conditions.

and acetic acid. Increased speed of distillation can be obtained only by employing higher temperatures, and as a result of higher temperature, secondary reactions are increased.

A. Destructive Distillation of Hardwoods

1. Wood for Hardwood Distillation. Methods of handling wood prior to oven carbonization vary. Some large plants operate sawmills to recover salable lumber and utilize the low-grade material, slabs, and edgings for distillation purposes; other plants have mills especially for the purpose of cutting and splitting crooked logs and large limbs into the proper sizes for the ovens; and some plants buy wood in cordwood and slab forms.

It was once universal practice to use wood in cordwood lengths and to air-season this stock for a period of 1 to 2 years. This required a large investment in raw material, and, as competitive conditions became more severe, many plants installed predriers built of ceramic tile or similar material. These predriers are usually heated by the flue gases from the oven furnaces. The flue gases enter the predrier at the end nearer the oven, the end containing the driest wood. When predriers are used, green wood can be dried in 48 hours to a moisture content of about 25 to 30 per cent. Plants operating on an 18-hour carbonization schedule may have enough wood for three oven charges in the predrier at one time. In these instances, the wood may be dried to a moisture content of 20 per cent, or less, in 54 hours. Each oven has its own predrier. In order to facilitate predrying, some plants cut the wood into about 1-ft lengths. Even at plants with mills and predriers, it is not uncommon to see a storage yard containing cordwood sufficient to keep the plant in operation for 3 to 4 months.

The principal species used in hardwood distillation are sugar maple, beech, birch, oak, hickory, black cherry, ash, and elm. Other species, such as hophornbeam, sourwood, black walnut, and dogwood, are also used in some plants. In general, the lighter weight hardwoods are little used, because of low yields of products, but in some instances the heavier ones are little used because of inadequate supplies. Other species, such as black willow, are sometimes used to make charcoal for special purposes. Bark is not removed from the wood; pieces of greater diameter than 6 to 7 in. are usually split; unsound wood is rejected; and when purchased by the cord, the lengths vary from 48 to 63 in., depending on locality. Several plants are now purchasing wood on a green-weight basis.

It is generally believed that heartwood gives greater yields of products than sapwood and that wood from old-growth timber gives higher yields than wood from second-growth trees. Small-scale tests made by Hawley and Palmer [9] gave rather variable results for heartwood, sapwood, slabwood, limbs, and bark.

2. Carbonization Processes Using Wood of Large Size. In these processes, the wood used ranges from small blocks up to cordwood 63 in. long, in contrast to sawdust, chips, and other similar small forms of wood.

a. Cast-iron and Steel Retorts. While destructive distillation on a rather small scale was carried on in some beehive ovens, the first high yields of volatile products came with the adoption of cast-iron retorts. These retorts were horizontal cylinders about 42 in. in diameter and 8 ft long, having a capacity of about ⅝ cord of wood. Later, steel retorts having a capacity of 2 to 3 cords came into use. These retorts were charged by hand, and they were usually set in pairs, with a firebox beneath for heating. The gases and vapors evolved from the carbonizing wood were led through condensers. The condensed vapors yielded *pyroligneous acid* * and tar, and the pyroligneous acid was refined to produce methyl alcohol and acetate of lime. The residue, charcoal, was raked out of the retort after it had been allowed to cool for a sufficient length of time to prevent combustion in air. These retorts, especially in hardwood distillation, were superseded by the steel ovens currently in use.

b. Ovens. The modern steel oven is rectangular in cross section, approximately 6 ft wide and 8 ft high. The ovens vary in length, usually from 26 to 54 ft, and have a capacity of 5 to 10 cords of wood. The steel of the oven is about ⅜ to ½ in. thick. Such ovens are set in brick chambers and, to allow for expansion and contraction, are suspended from steel beams supported by the surrounding brickwork. One firebox is provided under each short oven, but on long ovens a firebox is used under each end to get more uniform heating. Hot gases from the furnace pass along the underside of the oven, around its sides, and over the top. The better ovens are equipped with doors at each end; others have doors only at one end. An oven with doors on both ends permits more rapid charging and discharging; therefore it is more efficient from the standpoint of heat consumption, because it does not cool so much in charging and discharging processes. Virtually gastight inner doors are wedged onto the oven, and, in order to prevent condensation on their inner surfaces, they are protected by outer storm doors.

The bottom of the oven proper is set at the ground level in order to facilitate charging and discharging, and rails are provided for carrying the steel-slat buggies that hold the wood and retain the charcoal after distillation is completed. Each of these buggies has a capacity of 2 to 2½ cords of wood. An oven 54 ft long holds four of these buggies, giving it a capacity of 8 to 10 cords.

Ovens are provided with one or two outlets to which condensers are attached. The condensers are usually of the upright tubular type, containing numerous copper tubes expanded into brass heads. The heads and tubes are enclosed in a steel shell. Water enters the condenser at the bottom and flows through an outlet at the top. The vapors and gases from the oven pass through the copper tubes, and the vapors are condensed. The condensate is collected in a basin at the bottom of the condenser, the basin being equipped with a trap that permits the condensate to escape but not the noncondensable gases. These gases are generally used for fuel and are led into mains that feed the oven furnaces, or they may be allowed to escape to the atmosphere when they cannot be used as fuel.

The time required to distill a charge of wood varies with the moisture content of the wood, the rate of heating, and other factors. The cycle may

---

* *Pyroligneous acid* is a condensate of vapors formed during carbonization of wood. The term is derived from the words *pyrolysis* and *lignin*.

be completed in as few as 14 hours, or it may require 30 or more hours.

c. *Treatment of Charcoal.* The distillation of hardwoods yields three immediate products: (1) noncondensable gases; (2) pyroligneous acid and tar, commonly called "green liquor," which is collected from the condensers; and (3) charcoal, the residue left in the retort or in the oven.

The noncondensable gases are generally used for fuel under the ovens. On the average, they have a calorific value of 300 to 325 Btu per cu ft of gas. Treatment of the raw pyroligneous acid is a complex procedure. It is described in detail on pages 410–412.

As soon as the distillation is completed, the charcoal is taken from the oven and placed in almost airtight, uninsulated steel coolers in order to prevent the charcoal from burning to ashes when it strikes the air. When taken from the oven, the charcoal is glowing, and immediately upon contact with air, it blazes with almost explosive violence, although the transfer time from oven to cooler is only a few seconds. The charcoal must remain in these closed coolers for a period of about 48 hours in order that it will be cooled sufficiently to prevent spontaneous combustion when finally exposed to air. Plants operating on a 16- to 18-hour schedule must, therefore, have three of these steel coolers; those operating on longer schedules need fewer.

When the charcoal is taken from these coolers, it must remain out in the open for at least another 48 hours before it can be safely placed in storage bins. In large plants, the several buggies from a complete charge are hauled up an inclined track above the storage bin and placed in a rotary dumper, or *tipple,* that has the appearance of a cylindrical hollow cage with a rectangular opening just slightly larger than the cross section of a buggy. The tipple, mounted on rollers, is then rotated 180°, and the buggies are completely emptied in this one operation. The charcoal remains in the bins for another 48 hours before screening, grading, and bagging in sacks for shipment. Freight cars loaded with charcoal are held on a side track for another 24 hours as a precautionary measure, for spontaneous combustion is still to be feared. In accordance with Interstate Commerce Commission regulations, all fine charcoal screenings must be held in storage for a minimum of 21 days before shipment.

Some plants pulverize all or part of the charcoal produced, mix it with a starch or corn-flour binder, and compress the mixtures into briquettes for domestic and campers' fuel. Others granulate a part of the charcoal produced and pass steam through it to form *activated carbon.*

d. *Retorts Heated by Inert Gas.* As early as 1921, a patent for carbonizing green wood by circulating heated inert gases through the wood in a retort had been granted in this country. Several modifications and improvements in this original process have since been made in Europe, but none is in commercial use in the United States.

The best known of these is the *Reichert batch process* [4, 16] developed in Germany. In this process, once combustion has been started, some of the inert, noncondensable gases resulting from wood carbonization are burned in a tubular furnace to heat the remaining gas to 500 to 600°C. These hot gases are then introduced at the top of the retort and flow downward through the charge of wood. They supply enough heat to cause exothermic reaction. It is claimed that when a battery of such retorts is used distillation proceeds without the use of additional fuel. Gases containing the by-products of distillation, and the charcoal, are removed at the bottom of the retort. Since the temperature of carbonization can be controlled by regulating the temperature of inert gases, it is stated that the yield of charcoal and distillation of by-products in this process is in excess of that obtainable in the conventional ovens.

In a related continuous process, known as the *Belgian Lambiotte* process, the wood is continuously charged into the top of the cylinder, and the inert heated gas is introduced at about half of its height. The movement of wood and gas is so regulated that the wood is dried in the upper half of the retort and brought to distillation temperature in the third quarter. The temperature in the lower quarter of the cylinder is reduced by cooling gases to about 100°C. A 15 to 30 per cent higher yield of acetic acid and methanol, as compared with the oven method, is claimed.

3. Carbonization Processes Using Wood in Small Sizes. The destructive distillation of wood of small sizes, particularly sawdust, is rather difficult. Solid wood is a poor conductor of heat, and a bulk of sawdust is even poorer in this respect because of the small air spaces between the particles in the mass. It is, therefore, extremely difficult to transfer heat to the center of a stationary mass of sawdust. In distillation, the sawdust must be agitated in one way or another, either by a stirrer or by a rotary retort. The stirrer or the rotary retort must have gastight connections with the stationary parts of the equipment, and if such a process is to be continuous, the sawdust must be introduced and the charcoal must be removed through virtually gastight arrangements. Although many methods of distilling sawdust have been proposed, none is in commercial operation. The distillation of wood chips is not as difficult as the distillation of sawdust. One commercial method for distilling chips will be described.

a. The Badger-Stafford Process.[12] One process, the Badger-Stafford, now in commercial operation * in the United States, was developed to utilize small pieces of wood, to take full advantage of the exothermic reaction in wood distillation, and to make the operation of carbonization a continuous one. Wood is reduced to chips that are predried in sloping, rotary steel driers 10 ft in diameter and 100 ft long. Flue gases from the sawmill powerhouse enter the driers at about 315°C and flow countercurrent to the wood.

The wood requires 3 hours to traverse the length of the driers, and the capacity of each drier is about 3 tons of wood delivered at a moisture content of about ½ per cent. The dried wood, while still hot, is then conveyed

---

* One Badger-Stafford plant was built and formerly operated by the Ford Motor Company at Kingsford, Michigan, to produce chemicals and charcoal. It now produces only methanol and charcoal briquettes. Another (the Cumberland Plant) was completed in 1957 at Burnside, Kentucky, at a cost of about 2.5 million dollars. It has no provision for chemical recovery. This plant went into bankruptcy before going into full production, but was refinanced in 1959.

to stationary, cylindrical-vertical-type retorts, 10 ft in diameter and 40 ft high, having insulated walls 18 in. thick composed of firebrick, diatomaceous earth, and insulating brick within the steel outside shell.

The dried chips are fed into the tops of these retorts continuously by means of barrel valves that are virtually airtight. The rate of feeding is controlled so that the heat liberated by the exothermic reaction will be sufficient to bring the incoming wood to the distillation point, thereby eliminating the application of external heat, once the process is started. The gases and vapors are led through a condenser on the side of each retort, and a constant pressure, equivalent to 0.1 in. of water, is maintained in each retort to control the rate of removal of noncondensable gases from the condensers.

Charcoal is discharged by mechanical shakers in the bottom of the retorts through a barrel valve. Its temperature is reduced in continuous coolers by internal, water-cooled pipes and by water sprayed on the outside of the coolers. It is then sent through conditioners, which are similar to the coolers, to remove the heat developed when the charcoal takes up oxygen from the air, in order to prevent combustion in storage. Temperatures in the retort vary from about 255°C at the walls to 515°C in the center.

4. Refining Hardwood Pyroligneous Acid and Tar. Formerly it was universal practice to separate tar and oil from the raw pyroligneous acid distillate and to neutralize the tar- and oil-free acetic acid with lime, forming calcium acetate. The combination was distilled in iron stills to free the acetate liquor from alcohol and oil. Acetic acid was then recovered by mixing the crude calcium acetate with sulfuric acid and distilling the mixture to get an 82 per cent acid. This product was redistilled to obtain glacial acetic acid. The weak alcohol from the still was put into a fractionating column to yield a 92 to 95 per cent methanol, or merely condensed to an 82 per cent crude methanol, necessitating subsequent distillation.

Competition from synthetic acetic acid and synthetic methanol caused the abandonment of the acetate-of-lime method of acetic acid production and led to the development of several methods of direct recovery of acetic acid from pyroligneous acid. The alcohol, oils, and tars are removed from the raw pyroligneous acid before the acetic acid is extracted from the remaining liquor.

In the United States acetic acid is recovered mostly with ethyl acetate. The acid is dehydrated and separated from the solvent, which is then reused, by azeotropic distillation.*

a. Steps in the Refining of Pyroligneous Acid and Tar. The flow sheet in Fig. 19-1 shows the direct recovery of acetic acid by the solvent method and gives a general view of the conventional hardwood-distillation process. The major products recovered, in addition to charcoal, are methanol and acetic acid. Minor products include allyl alcohol; methyl acetone (a mixture of methanol, methyl acetate, and acetone); wood tar and alcohol oils; hard-

---

* Azeotropic distillation methods rely on the fact that many compounds form constant-boiling mixtures with water and alcohol and that the boiling points of these mixtures are lower than the boiling point of water alone.

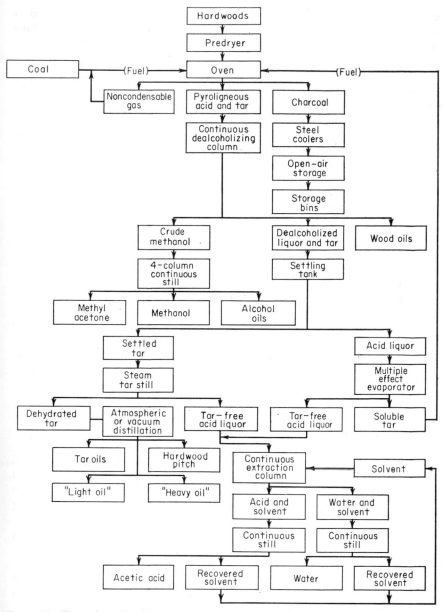

Fig. 19-1. Flow chart for direct recovery of acetic acid by solvent method from destructive distillation of hardwoods.

wood pitch; and formic and propionic acids. Not all distillation plants attempt to recover all these products, but, on the other hand, some recover products not named. The flow sheet, therefore, should not be regarded as directly applicable to every plant; procedures also vary. To cite only one example, many plants send the pyroligneous acid to settling tanks to remove heavy tar before extracting the alcohol. The resulting clear pyroligneous acid is then fed to the dealcoholizing column. One plant feeds some settled tar with the clear liquid to prevent fouling. The oils separated in the dealcoholizing column are burned as fuel or may be sold when markets are available. The crude alcohol is taken to a four-column continuous still, and, by treatment with caustic and acid, a water-white, pure methanol that can be sold directly is obtained. This column also yields alcohol oils, allyl alcohol, and methyl acetone. Methyl acetone and allyl alcohol undergo further distillation at some plants, although this is not indicated on the flow sheet. The allyl alcohol may be concentrated in a discontinuous still and the methyl acetone may be treated in the same type of still to get various combinations of acetone, methyl acetate, and methanol.

The dealcoholized liquor and tar are discharged from the dealcoholizing column into settling tanks, where heavy tar settles to the bottom. The settled tar is taken to a steam tar still, where acid liquor is removed. The resultant dehydrated tar may be distilled under vacuum or at atmospheric pressure to yield hardwood pitch and tar oils. These tar oils can be further separated into "light" and "heavy" oils. Dehydrated tar also may be subjected to a sodium bicarbonate extraction to separate acids from phenols and neutrals. Separation of phenols and neutrals can be accomplished by a sodium hydroxide extraction.

The acid liquor from the settling tanks is fed into a multiple-effect evaporator to free it of soluble tar. The soluble-tar residue from the evaporator is usually used as fuel under the oven. The tar-free acid liquor from the tar still is combined with the tar-free acid liquor of the evaporator, and the two are fed into the top of a continuous extraction column. The solvent used in this case is injected at the base of the extractor, and the two liquors are countercurrent.

From the extraction column, a water solution containing some solvent is taken to a continuous still, where the solvent is recovered and the water is discharged to the sewer.

From the top of the extraction column, the solvent containing the acetic acid in solution is fed into a continuous still to obtain crude acetic acid and solvent. This solvent and the solvent recovered from the water solution are sent to a tank for reuse in the extraction column.

5. Yields. Data regarding yields of hardwood-distillation products are highly variable; yields are usually expressed on the cord basis. It is reasonable to expect these variations because of differences in plant equipment and operation, products recovered, and the fact that the gross volume of the cord, as purchased, varies from 128 to 160 cu ft, depending on the individual operation. The ranges of yield * quoted on an average *cord* basis are:

---

* Dividing these figures by 1.375 will give average yields per ton of dry wood substance.

Product	Quantity			
Condensable gas	7,000	–11,500	cu ft
Charcoal	900	– 1,080	lb
Raw pyroligneous acid	200	– 276	gal
Acetate of lime	180	– 220	lb
Acetic acid equivalent for acetate of lime	103	– 125	lb
Crude methanol	9.5 –	11	gal
Refined methanol equivalent for crude methanol	7.25–	8.5	gal
Wood tar and oils	22	– 25	gal

In arriving at the acetic acid equivalent for acetate of lime, it is assumed that 100 lb of acetate of lime yields 57 lb of glacial acetic acid; to get the refined methanol equivalent for crude methanol, it is assumed that about 1.3 gal of crude methanol is needed to produce 1 gal of refined methanol.

Under present conditions, an average yield of the three principal products, for a standard cord of 128 cu ft, would be approximately 50 bu of charcoal, weighing 1,000 lb; 114 lb of glacial acetic acid; and 8½ gal of refined methanol.

The average cord of wood purchased by northern hardwood-distillation plants contains approximately 2,750 lb of dry wood substance. Therefore, to convert the above yields to an average cord basis, they should be multiplied by 1.375.

B. Destructive Distillation of Softwoods *

Destructive distillation of softwoods, at the present time, is confined to the resinous wood of southern pines, particularly longleaf and slash pine. Ordinary southern pine wood does not contain enough resin for distillation purposes; only extremely pitchy pieces are suitable. The principal source of material for this industry is the resinous heartwood from fallen trees, logs, and stumps that are so old that the sapwood has rotted away. Such material is called *lightwood* or *fatwood*. Other sources of material are the resinous, conical bases of limbs, known as *pine knots*, from which the stemwood of the tree has rotted entirely away, and slabs from logs cut from trees that have been tapped for naval stores.

The process of destructive distillation of softwoods is similar in principle to that outlined for hardwoods, but differences exist in the products obtained and in the method of refining distillates. The principal products recovered are charcoal, turpentine, pine oil, dipentene, pine tar, tar oils, and pitch. Softwood pyroligneous acid contains such small amounts of alcohol and acetic acid that no attempt is made to recover them; only the oily fractions of the distillate yield the products of value.

There seems to be little standardization in preparing final products from crude oily distillate. Many plants have developed special products for particular markets, and some of the products obtained depend on the temperatures to which the wood is subjected. The following description and the flow sheet shown in Fig. 19-2 are based on one method [13] only.

* See also Chap. 21, Naval Stores, for a description of the steam-and-solvent extraction process.

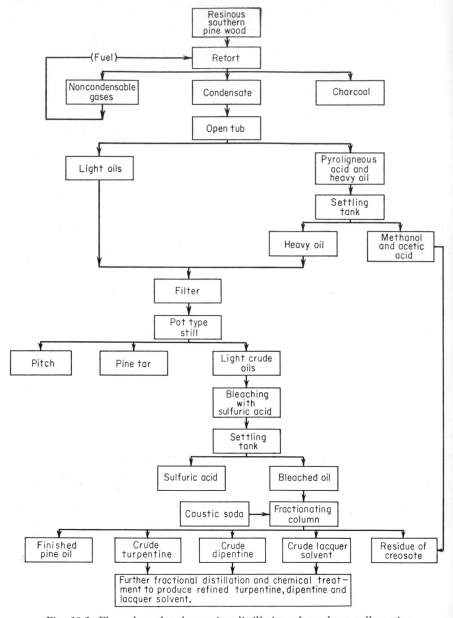

Fig. 19-2. Flow chart for destructive distillation of southern yellow pine.

Resinous pine wood is loaded into one-door, cylindrical steel retorts about 5 ft in diameter and 16 ft long. Two such retorts are heated by a brick furnace burning natural gas and the noncondensable gases given off during the carbonization period. The oven is heated for 16 hours and is allowed to cool for 8 hours before the charcoal is withdrawn.

The condensate is collected in an open tub, and that of the first half of the run is taken separately from the tub. It consists of the lighter oils. Pyroligneous acid produced in this period settles to the bottom of the tub and is drained off. Heavy oils are produced in the second half of the run, and they, with the pyroligneous acid, are put into settling tanks. Here the heavy oils settle to the bottom and the pyroligneous acid comes to the top.

After settling, the heavy and light oils are filtered and put into a pot still having a capacity of 110 bbl. Light crude oils and a residue of pine tar are obtained in this distillation. The pine tar is pumped into tank cars, to the barreling warehouse, or to storage. The light crude oils are agitated with sulfuric acid for bleaching, and the acid is permitted to settle out.

The oils are then put into a pot still having a fractionating column for another distillation in the presence of caustic soda. The fractionating column yields finished pine oil, crude turpentine, crude dipentene, and a fraction to be redistilled to produce a lacquer solvent. The residue in the still is a creosote, containing caustic soda. The creosote is neutralized by the addition of pyroligneous acid and may be sold as a wood preservative. The crude turpentine, crude dipentene, and crude lacquer solvent are pumped into separate storage tanks. These crudes are redistilled, one at a time, in a pot still with a fractionating column. The crude turpentine yields refined turpentine and crude dipentene; crude dipentene yields refined dipentene and pine oil; and the crude lacquer solvent, distilled in the presence of caustic soda, yields refined lacquer solvent and crude turpentine.

Yields from the destructive distillation of southern pine are extremely variable because of the lack of standardized methods and equipment in this branch of the wood-distillation industry. The variation in resin content of the wood also plays a major role in the amount of products recovered. One report gives the following yields for a cord of fatwood: 12 gal refined turpentine, 1½ gal pine oil, 50 gal tar, and 800 to 900 lb charcoal. Another report shows yields of 4 gal "wood spirits," 7 gal turpentine, 4 gal pine oil, 12 gal tar oil, 50 gal tar, and 960 lb charcoal.

C. Uses of Destructive-distillation Products

It is not feasible to list all products derived from the destructive distillation of wood and their uses. The following examples will serve to indicate the wide range and uses of some of these products.

1. Charcoal. The principal use of charcoal today is for domestic and campers' fuel for outdoor cooking. Other important uses of charcoal are in the manufacture of black-powder explosives, as metal casehardening compounds and for other metallurgical uses, in the manufacture of chemicals, such as carbon disulfide and sodium cyanide, and in its activated form as a purification agent and in water filtration.

a. Charcoal Briquetting. Considerable quantities of charcoal for domestic cooking are used in the form of briquettes. These are made of pulverized charcoal combined with starch solutions as a binder. The mixture is formed in a briquetting machine, and the product dried in a heated chamber. Most briquetting machines can form about 1 ton of briquettes per hour. The cost of such a machine is about $10,000, and the cost of the entire briquetting plant, including the building, is upward of $100,000. Larger machines can make 3 tons of charcoal briquettes per hour.

b. Activated Charcoal. Charcoal has a high capacity for absorbing gases, liquids, and even finely divided solids. This property may be enhanced by subjecting the charcoal to treatment with steam or carbon dioxide at temperatures ranging from 1500 to 1800°F. The resulting *activated charcoal* has much greater absorption power, brought about probably by the reaction of steam with the carbon, hydrogen, and oxygen in charcoal. This reaction produces more suitable internal-surface absorption conditions within the charcoal structure.

Activated charcoal is employed for the removal of objectionable odors, tastes, and colors from foods, medicinal products, and other substances. To some extent it is also used in water purification. However, most of the powdered water-grade carbon for treating municipal water now comes from the carbonized residue from paper mills, which is cheaper.

2. Acetic Acid. Acetic acid is used in making inorganic acetates such as lead, copper, iron, and sodium acetates; white lead pigments, methyl, ethyl, and amyl acetate solvents; and in manufacture of cellulose acetate fiber, motion-picture and photographic films, lacquers, plastics, and transparent sheets. It is also used for coagulating plantation-rubber latex, in perfume manufacture, and as a mordant in textile dyeing.

3. Methanol. Perhaps three-quarters of all natural methanol is used in denaturing ethyl alcohol. Other uses of methanol include the manufacture of shellac, dry-cleaning agents, formaldehyde, pyroxylin products, paints, varnishes, and textile-finishing agents. It is also extensively used as an antifreeze.

4. Hardwood Tar Oil and Pitch. Tar-oil products may be used for flotation oils, solvent oils, oils for paints and stains, disinfectants, rubber-softening compounds, wood creosote preservatives, and as inhibitor oils for stabilizing gasoline.

5. Softwood-distillation Products. The use of wood turpentine is detailed in Chapter 21, Naval Stores.

Pine oil is used for paints, flotation oils, disinfectants, and in the treatment of fabrics before dyeing. Dipentene is used as a solvent in reclaiming old rubber, and in the manufacture of paints and varnishes. Pine tar has a large number of uses, the most important of which are the manufacture of cordage, oakum, soaps, coating and binding materials, medicines, and disinfectants. Tar oils are used for paints, stains, disinfectants, soaps, and flotation oils.

D. The Wood-distillation Industry

The wood-distillation industry in the United States is faced with the twin problems of limitations placed by transportation costs on supplies of available raw material and shrinking markets for its products. The fact that the wood-distillation industry as a whole can and does utilize wood in the forms of tops, large limbs, low-grade logs, broken and crooked material, and stumps makes it an important factor furthering integrated utilization of our forest resources. Most of the larger sizes of wood used by this industry are unsuited for sawlog material. A few plants use wood from thrifty second-growth stands, and many purchase cordwood from nearby farm woodlands. The managements of large distillation companies are keenly aware of the need for future raw material, and the trend in the management of their forest properties is toward sustained yield.

1. The Hardwood-distillation Branch. In the past the plants for the destructive distillation of hardwoods were located in the Northeastern states, in Michigan and Wisconsin, and in the South Central states of Arkansas, Missouri, and Tennessee. Changing economic conditions, diminishing supplies of raw material at economical prices, increasing equipment and labor costs, and competition with synthetically produced methanol and acetic acid had led to a reduction in the number of operating plants from more than 50 in 1928 to 6 by 1952. A few plants formerly engaged in the recovery of chemicals have retained only the carbonizing equipment and now produce charcoal exclusively. It is estimated that the plants still recovering by-product chemicals use about 350,000 cords of wood annually.

Some observers feel that the hardwood-distillation industry today is limited more by charcoal market conditions than by the competition from synthetic methanol and synthetic and fermented acetic acids. The principal income of the industry is dependent on markets for the three principal products: charcoal, acetic acid, and methanol. The operating rate, however, seems to be determined by the volume of the charcoal sold. These three products are derived in a fairly constant proportion that cannot be altered. To expand production under such conditions, an enlarged year-round charcoal market is needed.

2. The Softwood-distillation Branch. The status of the softwood-distillation industry virtually parallels that of the hardwood-distillation branch with respect to available supply of wood, rising chemical-wood and labor costs, and competition in the marketing of its products. Supplies of old-growth, resinous stumps and other suitable pine wood are diminishing, and the second-growth trees lack the necessary resin content for successful distillation by current methods. The average annual consumption of wood, mostly in the form of stumps, for the past 15 or 20 years has been at the rate of about 2 million tons annually.

II. CHARCOAL PRODUCTION WITHOUT RECOVERY OF BY-PRODUCTS *

Most of the charcoal production in the United States comes from the destructive distillation of wood and from plants formerly operated for destructive distillation but now converted exclusively to production of char-

* For detailed discussion of charcoal production, marketing, and use, see U.S. Department of Agriculture, Forest Service, Forest Products Laboratory. Charcoal Production, Marketing and Use. Report No. 2213. 1961.

coal. It has been estimated that in 1958 about 70 per cent of the total charcoal production came from eight large plants. Much of the remaining 30 per cent came from more than 200 small producers operating some 1,500 kilns, most of which ranged in capacity from 2 to 11 cords of wood. More than half of them produced less than 100 tons of charcoal a year each.

Charcoal production in the United States is concentrated east of Mississippi, with the Lake states and the South accounting for two-thirds and the entire East for 98 per cent of the total production. California, with some 40 small plants, is the only Western state with appreciable charcoal production.

Charcoal production figures for the United States, as compiled by the U.S. Forest Service,[17] are presented in Table 19-1. The rapid decline of

Table 19-1. Charcoal Production in the United States [a]

Year	Production, tons	Year	Production, tons
1899	171,543	1944	306,192
1909	554,785	1952	251,784
1929	453,550	1955	237,770
1939	250,780	1956	264,990

[a] SOURCE: U.S. Department of Agriculture, Forest Service, Division of Forest Economic Research. Charcoal Production in the United States. 1958.

charcoal production from the peak of 554,785 tons in 1909 to 250,780 tons in 1929 was brought about principally by substitution of other materials for charcoal in the metallurgical and chemical industries.*

The oldest method of charcoal production without recovery of by-products was the pit- or dirt-kiln method. This was followed by the use of brick beehive kilns, and more recently by small beehive-type kilns made of metal and by rectangular masonry kilns. Of these, pit kilns are no longer used. A brief description of such an operation is included only for its historic interest.

* Quantities of lump wood charcoal are also imported, principally from Japan. The following data on the importation of charcoal from Japan were supplied by the Bureau of Foreign Commerce in San Francisco.

Calendar year	Total value, dollars	Total volume, lb	Value/ton, dollars
1958	7,001.00	275,617	50.80
1959	56,163.00	1,879,476	59.76

A. Charcoal Kilns

1. Pit Kilns. In this method of charcoaling, a central chimney 1 to 2 ft in diameter was formed by driving several poles into the ground. Wood in the form of pieces 3 to 4 ft long was stacked against the chimney; the sticks in the first layer were placed on end, leaning slightly toward the center. Then another layer of wood was laid flat on top of the first, and the stack built in the form of a beehive. As the building of the pile progressed, the central chimney was filled with kindling, twigs, and other combustible material.

The entire mound, with the exception of the central opening, was then covered with a layer of twigs, on top of which several inches of sod, turf, or clay were placed. The fire was started by dropping embers down the central chimney or by pouring kerosene on the kindling material and lighting with a torch.

Successful carbonization depended on the admission of air in quantities just sufficient to start combustion. Air vents at the bottom of the pile were then closed in order that carbonization might take place without allowing the charcoal formed to burn to ashes. When carbonization was completed, the pile was allowed to cool one to several days, depending on the size of the pile. The pit kilns held from 2 to 50 cords of wood. The carbonization required 2 to 5 days for 2- to 5-cord piles and as much as 2 weeks for 30- to 50-cord piles. The yield of charcoal from such woods as maple, beech, and oak was about 30 to 40 bu per cord of wood.

2. Beehive Kilns. Formerly, beehive kilns were used to a considerable extent, especially in Michigan, to produce charcoal for iron manufacture. Charcoal is no longer used for this purpose. A number of beehive kilns, however, are still in use, some built fairly recently.

A typical beehive kiln is a dome-shaped structure made of brick or stone, with air holes provided at the base and an iron trap door on the top. The charge is introduced through the top opening and the wood ignited at the bottom. The charging door is then closed and carbonization accomplished by the heat generated from the combustion of a part of the charge, air being admitted through ports near the base of the kiln (Fig. 19-3).

Such kilns have capacities of from 20 to 90 cords of wood. They produce a somewhat higher yield of charcoal, are easy to construct, and require less labor than the pit kilns.

3. Steel Drum Kilns. * A number of steel drum kilns have been patterned after the original French Magnein kiln. Some of them are characterized by easy portability, since they are constructed in several, relatively easy to move sections. The earliest successful units of this type built in the United States are known as *Black Rock Forest kilns*. The most recent version of it is the *New Hampshire kiln*.[2] This is a bell-shaped steel structure, made in two sections: the drum-like lower section and the conical cover (Figs. 19-4 and 19-5). Eight draft holes, equally spaced around the circumference near the bottom, are provided. Every other one of these is equipped with 6-ft

* For detailed description of these kilns, their operation, and cost analysis, see reference 2 at the end of this chapter.

Fig. 19-3. Brick beehive kiln. Large-scale production of charcoal in converted ceramic kiln. (*Reproduced by permission from Charcoal: Its Manufacture and Use, by Edward Beglinger and E. G. Locke, Economic Botany 11(2):160–173, 1597.*)

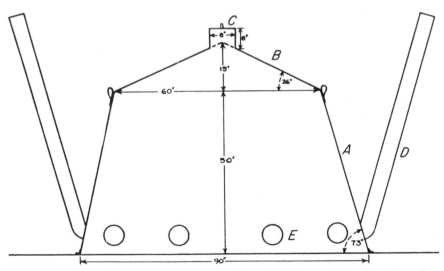

Fig. 19-4. Sketch of New Hampshire charcoal kiln. (*From Bull. 11, The New Hampshire Charcoal Kiln, New Hampshire Forestry and Recreation Commission.*)

Fig. 19-5. Swinging a kiln over stacked wood. (*From Bull. 11, The New Hampshire Charcoal Kiln, New Hampshire Forestry and Recreation Commission.*)

flue pipes when the kiln is in operation. The amount of air admitted through the holes and the draft created by the flues can be easily regulated.

In loading, 4-ft sticks of wood are stacked on end around a central stake and a stock of kindling around it. The kiln sections are lowered over the wood stack with a suitable hoist (Fig. 19-5). The capacity of these kilns is about ½ cord of 4-ft wood. The coaling time ranges from 8 to 30 hours, depending on the kind and dryness of the wood, the weather, and operational procedures.

4. Masonry Kilns.[10, 15, 18] Many types of masonry kilns have been designed and are now in use. The best known of these is the *Connecticut charcoal kiln,*[9] a 1- to 2-cord kiln, generally constructed of hollow cinder blocks. These kilns consist of a rectangular coaling chamber, a stove, which abuts one end of the coaling chamber, and the chimney, which connects to the stove and chamber. The stove is used to burn wood to induce a draft in the chimney. A part of the rear end of the chamber is left open; it is used for loading and unloading the kiln. During the coaling operation, the opening is closed with blocks laid without mortar. A number of air inlets are provided by placing a block on its side, so that a hole will run horizontally through the wall from the outside into the interior of the kiln. The roof may be formed of 4 by 8 by 16-in. blocks supported by a steel pipe threaded through the openings in the blocks and resting on the side wall of the kiln.

Larger kilns, of the same general design, with a capacity of 10 to 13 cords, have also been built and successfully operated.

A modified, composite-wall kiln made of cinder and concrete blocks has been designed and tested by Forest Service personnel[18] (Fig. 19-6). The most important feature of this design is the composite-wall construction,

Fig. 19-6. Model of a 7-cord, masonry block, composite-wall, charcoal kiln. (*Courtesy U.S. Forest Products Laboratory.*)

which minimizes air leakage, a common problem in single-wall construction. It is claimed that this design provides a kiln that can be easily modified to meet specific requirements of the producers. Reduced maintenance problems and better control of coaling conditions are the stated advantages of this design.

5. Special Methods of Charcoaling. Several methods of charcoal production from mill waste have been investigated. Of these only the continuous process for carbonization, originally developed by Svend Thomsen, appears to be of commercial significance.

a. Fluidizing-bed Technique. One method employs a fluidizing-bed technique, in which wood particles are fed continuously to a cell containing inert gas as a heating medium. The particles are kept agitated by upward movement of the gas stream, until they are completely carbonized.[5]

b. The Continuing Liquid-bath Method. In the continuing liquid-bath method, wood fines are completely submerged in and passed through molten metal. Extremely rapid heat transfer is possible in this method. Charcoal is withdrawn continuously from the discharge end of the apparatus.[5]

c. The Continuous Retort Process. The continuous retort process utilizes small retorts. Wood chunks are placed at the top of the retorts by conveyers and allowed to descend into the carbonizing zone by gravity. The carbonized wood falls into airtight cooling drums at the bottom. The gaseous products are burned to furnish the necessary heat. A single unit costing about $15,000 is said to be capable of producing 2 to 4 tons of charcoal in a 24-hour period.

d. The Continuous Process for Carbonization. A successful continuous process for carbonization of green or dry sawdust, shavings, and chips has been developed by Svend Thomsen and is now in operation by Conway Charcoal Industries, Inc., Conway, South Carolina.[7] The carbonizing unit consists of 13 metal tubes, equipped with screw feeds, which slowly convey sawdust through the tubes. The tubes themselves are enclosed in a heating chamber constructed of brick and concrete. The lower part of the heating chamber is a Dutch oven, in which a wood fire is started to get the temperature up to 800°F before the carbonization process gets under way. The heat from the Dutch oven passes into the upper part of the heating chamber, where the tubes are located. When the temperature in the heating chamber reaches 800 to 900°F, carbonization starts, and the gaseous products and tars are diverted from the tubes to the Dutch oven, where they are burned, supplying all the necessary heat for continuing carbonization.

The carbonized residue from all the tubes drops into a single exit tube that carries the charcoal to storage vats or to a briquetting machine. The rate of charcoal production ranges from 475 to 864 lb per hour, or 5.7 to 7 tons per 24 hours, depending on the type of material used and the species. It has been reported that a unit capable of producing 12 tons of charcoal per 24 hours is now in operation at another Conway mill.

6. Charcoal Briquetting. Much of the charcoal produced by the kilns is briquetted for domestic cooking. This requires pulverizing and briquette-forming equipment costing upwards of $100,000. Because of the relatively high investment involved in setting up a briquetting plant, the trend in the South and Southeast is for one plant to briquette the output of a number of associated and captive kilns.

B. Marketing of Charcoal

One of the most critical factors in small-scale production of charcoal is marketing. The most important outlet for charcoal today is for domestic, and principally outdoor, cooking. Because of this, nationally, the consumption of almost 60 per cent of charcoal manufactured is on a seasonal basis, with a sharp increase in consumption occurring during warm-weather months. This creates storage problems for the large producers. At the same time the yearly midseason summer shortage of charcoal, accompanied by higher prices, encourages small-kiln construction.

A marketing survey of charcoal consumption in Wisconsin for domestic purposes serves to illustrate the rising demand for packaged charcoal for domestic cooking. The following significant factors have been revealed by this survey.[19, 20]

More than 11½ million pounds of charcoal were sold in 1957 by 8,700 Wisconsin retailers. This was more than double the amount sold in 1955, while the number of stores retailing charcoal increased by 38 per cent during the same period.

In 1957 nearly two-thirds of the retailed charcoal was sold in 10-lb packages, and more than 90 per cent in the form of briquettes.

Grocery stores sold more charcoal than did any other single store type.

In 1955 they handled 47 per cent of the total volume of charcoal and 2 years later 60 per cent. In 1955 the largest wholesalers of charcoal were the fuel and lumber dealers. In 1957 nearly 60 per cent of all charcoal volume was handled by the grocery wholesaler.

SELECTED REFERENCES

1. Albin, T. C. Wood Chemical Plant Built to Meet Competition of Synthetic Products. *Chem. & Metall. Engin.*, **39**:382–387. 1932.
2. Baldwin, H. I. The New Hampshire Charcoal Kiln. *N.H. Forestry and Recreation Comm., Bull.* 11. 1958.
3. Beglinger, E. Hardwood-distillation Industry. U.S. Department of Agriculture, Forest Service, Forest Products Laboratory Report No. 738. 1956.
4. Beglinger, E. Distillation of Resinous Wood. U.S. Department of Agriculture, Forest Service, Forest Products Laboratory Report No. R496. 1951.
5. Beglinger, E., and E. G. Locke. Charcoal: Its Manufacture and Use. *Econ. Bot.*, **11**(2):160–173. 1957.
6. Benson, M. K. Chemical Utilization of Wood. National Committee on Wood Utilization, U.S. Department of Commerce. 1932.
7. Dargan, E. E., and W. R. Smith. Progress in Charcoal Production: Continuous Residue Carbonization. *Forest Prod. Jour.*, **9**(11):395–397. 1959.
8. Hawley, L. F. Wood Distillation. Reinhold Publishing Corporation, New York. 1923.
9. Hawley, L. F., and R. C. Palmer. Yields from the Destructive Distillation of Certain Hardwoods. *U.S. Dept. Agr., Bull.* 129. 1914.
10. Hicock, H. W., et al. The Connecticut Charcoal Kiln. *Conn. Agr. Expt. Sta., Bull.* 431. 1951.
11. Klason, P. Versuch einer Theorie der Trockendestillation von Holz. *Jour. Prakt. Chem.* (Leipzig), **90**:413–447. 1914.
12. Nelson, W. G. Wood-waste Utilization by the Badger-Stafford Process. *Indus. and Engin. Chem.*, **22**:313–315. 1930.
13. Nealey, J. B. Distilling Pine Product at New Orleans. *Chem. & Metall. Engin.*, **43**:20–21. 1936.
14. Ross, J. D. Carbonization of Douglas-fir Sawdust. Proceedings, Forest Products Research Society, Vol. 2, pp. 272–275. 1948.
15. Simmons, F. C. Guides to Manufacture and Marketing of Charcoal in the Northeastern States. *Northeast. Forest Expt. Sta., Sta. Paper* 95. 1957.
16. Stamm, A. J., and E. E. Harris. Chemical Processing of Wood. Chemical Publishing Company, Inc., New York. 1953.
17. U.S. Department of Agriculture, Forest Service, Division of Forest Economic Research. Charcoal Production in the United States. 1958.
18. U.S. Department of Agriculture, Forest Service, Forest Products Laboratory. Production of Charcoal in a Masonry Block Kiln-Structure and Operation. Report No. 2084. 1957.
19. Warner, J. R., and W. B. Lord. The Market for Domestic Charcoal in Wisconsin. *Lake States Expt. Sta., Sta. Paper* 46. 1957.
20. Warner, J. R., and B. L. Essex. Trends in the Wisconsin Charcoal Market. *Lake States Expt. Sta., Sta. Paper* 80. 1960.
21. Wise, L. E., and E. C. Jahn (eds.). Wood Chemistry, 2d ed., Vol. 2, Chaps. 19, 20. Reinhold Publishing Corporation, New York. 1952.

CHAPTER 20

WOOD HYDROLYSIS

Wood of most species of trees averages about 70 per cent carbohydrates, of which nearly 50 per cent is cellulose and the remaining 20 per cent hemicelluloses (see Chap. 3). Wood, therefore, is one of the most important potential sources of carbohydrates available to mankind on a world-wide basis.[3] Utilization of cellulose in its original or modified form is dealt with in the chapters on Pulp and Paper, Cellulose-derived Products, and Wood in the Plastics Industry. These industrial uses depend largely on the fibrous nature of the cellulose.

This chapter deals with hydrolysis of wood, also referred to as *wood saccharification,* a process of converting carbohydrates in wood to simple sugars. The hydrolysis is accomplished by treating wood in the form of sawdust, shavings, or chips with strong acids at room temperature, or with dilute acids at elevated temperatures. The resulting sugar solution, mainly glucose, with lesser amounts of mannose, galactose, xylose, and arabinose, can be evaporated to molasses, used to grow yeast, crystallized to glucose sugar, fermented to ethyl alcohol and glycerol, or utilized to produce furfural and hydroxymethyl furfural as well as a variety of alcohols and acids.

I. EFFECT OF WOOD CONSTITUENTS ON HYDROLYSIS

The amount and the kind of sugar obtainable from wood depend on the nature of the polysaccharides * found in the particular species of wood. The principal polysaccharide, regardless of the species, is cellulose, representing about 50 per cent of the weight of the ovendry and extracted wood substance. Cellulose, when hydrolyzed, yields a hexose sugar glucose. It is, however, extremely resistant to hydrolysis because of its crystalline structure, which offers low accessibility to the action of dilute acids.[6,7,16] This leads to treatment with either hot dilute acids at high pressure or cold concentrated acids.

* Polysaccharides are carbohydrates the formulas of which can be written $[(C_6H_{10}O_5)nH_2O]$ for hexosans or $[(C_5H_8O_4)nH_2O]$ for pentosans. When treated with a strong acid, polysaccharides are hydrolyzed to monosaccharides.

The other polysaccharides in wood, called hemicelluloses, lack the crystalline structure of cellulose and therefore hydrolyze into sugars more readily. There is considerable difference in the nature and amounts of hemicelluloses found in softwoods and hardwoods, and also in different species within each group. In general the predominant hemicelluloses in softwoods are a type that upon hydrolysis yield hexose sugars, which are easily fermentable into ethyl alcohol. Hardwoods, on the other hand, yield substantial quantities of pentose sugars, which are not fermentable to alcohol but can be converted into furfural or used to grow yeast.

Lignin is the main noncarbohydrate constituent of wood. It is resistant to the action of acid solutions used for hydrolysis and therefore remains as a residue, amounting to 20 to 30 per cent of the original dry weight of wood. The lignin residue has a high fuel value and many, largely unrealized, uses for production of chemicals and in the plastics industry.[15,16] It is the opinion of some chemists that in the future the utilization of lignin may prove to be the key to successful utilization of cellulosic residues by hydrolysis.[6]

II. SOURCES OF WOOD FOR HYDROLYSIS

The major products of the forest in the United States are relatively high-priced raw material.[18] On a pound-per-pound basis the better grades of pulp or lumber are comparable to or higher in cost than refined sugar. Therefore, economically, the processing of wood to carbohydrates is limited largely to residues and possibly to the utilization of small and low-grade timber that might be removed in the interest of better management.

The potential raw materials available at reasonable cost for wood hydrolysis can be grouped as (1) logging waste, (2) primary and secondary manufacturing residues, and (3) materials resulting from natural mortality due to overcrowding, disease, and insects (Table 20-1).

Table 20-1. Estimate of Raw Material Available for Wood Hydrolysis [a]

Raw Material	Millions of cu ft	Thousands of cords	Thousands of tons, bond-dry weight
Logging residue..........	1,362	17,700	21,900
Plant residue............	3,414	44,300	54,900
Mortality...............	3,389	44,000	54,200
Total................	8,165	106,000	131,000

[a] SOURCE: U.S. Department of Agriculture, Forest Service. Timber Resources for America's Future. Forest Resource Report No. 14. 1958.

During the past decade considerable progress has been made in reducing logging waste. This has been brought about largely by higher stumpage prices and by increased use of manufacturing residue and low-grade logs for pulp chips. About two-thirds of the primary and secondary manufacturing residues, or an equivalent of some 29 million cords of wood, are put to some use, largely for fuel. Increasing quantities of such residues are also used for pulp chips and in the manufacture of particle board. Currently very little waste material resulting from mortality finds any commercial use. Therefore, after allowances for the current uses of wood residues are made, it is estimated that an equivalent of more than 77 million cords of wood are now wasted annually because of the absence of markets for such material.[14] This indicates that when the chemical utilization of wood becomes economically feasible there should be available an ample supply of inexpensive raw material.

III. WOOD SACCHARIFICATION PROCESSES

The discovery that carbohydrate constituents of plants can be converted to sugars was made early in the nineteenth century. However, practical success in converting wood carbohydrates to sugars has been limited by economic and technical problems arising from (1) the difficulties of hydrolyzing the cellulose, (2) the unfavorable ratio existing between the rate of the cellulose hydrolysis and the rate at which the resulting sugars decompose, (3) the fact that wood hydrolysis results in formation of pentose and hexose sugars, and (4) high costs of equipment and operation. As a result wood-hydrolysis plants have been characterized by low yields and impure products, the latter due mainly to the decomposition of sugars.[18]

Industrial wood-hydrolysis processes can be classified as dilute-acid methods using *batch processes* and *percolation processes;* and as concentrated-acid methods.

A. Dilute-acid Batch Process

The single-stage batch process was the first wood-hydrolysis method used on an industrial scale. Two such plants were in operation in South Carolina and in Louisiana during the First World War, using 2 per cent sulfuric acid as the hydrolyzing agent. The method, which came to be known as the "American process," was simple, required relatively inexpensive equipment, but resulted in a low yield of only about 22 per cent of sugar on the basis of dry weight of wood, or the equivalent of about 20 gal of alcohol per ton of wood waste. These plants were capable of producing 5,000 to 7,000 gal of alcohol per day. The shortage of cheap raw material due to closing of sawmills, and low prices for molasses from other sources, forced these plants to close soon after the First World War.

Several multiple-stage batch methods have been more recently suggested.[3,7,15] In one, as reported by Cederquist,[14] wood in the form of shavings, sawdust, or chips is treated with 0.5 per cent sulfuric acid in the presence of steam at 190°C for about 3 minutes. The resulting sugars are washed out, and the wood is treated a second time with 0.75 per cent of sulfuric acid at 215°C and the resulting sugars washed out again. A yield of more than 50 per cent of sugars, based on dry weight of wood, is claimed. No commercial installations using this process are known.

B. Dilute-acid Percolation Processes

1. The Scholler Process. To date commercially the most successful percolation method is the *Scholler process* [7, 15, 16] and several of its modifications. In this process wood, in the form of sawdust or chips, is placed in an acid-resistant digester equipped with a porous-tile filter at the bottom, and hydrolyzed with dilute sulfuric acid.* The wood is first steamed at about 134°C, and the acid solution is then introduced in the top of the digester at 125°C. The acid is allowed to react for an hour. The sugar-containing solution is then withdrawn from the bottom of the digester. The residual chips are reheated to about 140°C, and fresh acid (less than 1 per cent concentration) is introduced at about 130°C and allowed to react for about one-half hour before the sugar solution is drained off. As many as 20 cycles of extraction may be used, each at successively higher temperatures, until a temperature of 180°C is reached. This temperature is then maintained for the balance of the cycles.

Because of the relatively short time interval between the formation of sugars and their withdrawal, considerably less sugar degradation occurs in this process as compared with the American process.

2. The Madison Wood-sugar Process. During the Second World War, the U.S. Forest Products Laboratory at Madison, Wisconsin, developed a modified Scholler process, generally known as the *Madison wood-sugar process.* It was hoped that this process might meet the unprecedented demand for ethyl alcohol, generated by war activities, should a shortage of this chemical develop.†

C. Hydrolysis of Wood with Concentrated Acids

A number of wood hydrolyzing processes using strong acid have been proposed. The most promising of these is the Schoenemann process,[1, 3] which is a modification of an early strong-acid method known as the *Bergius process.*[8, 15]

In the Schoenemann process chipped wood is prehydrolyzed by batch boiling with 1 per cent hydrochloric acid at 265°F under pressure.‡ This treatment extracts most of the more easily hydrolyzed hemicelluloses, result-

---

* From 1.0 to 2.0 per cent for the first cycle, and about 0.5 per cent for subsequent cycles.

† During the Second World War a plant based on the modified Scholler process, generally known as the Madison wood-sugar process, was built in Springfield, Oregon. Later this plant was converted to experimental production of molasses for stock feed. Because of a number of engineering and production problems, its operation did not prove economically sound for production of either alcohol or molasses, and the plant was closed. For details of this process, see references 11 and 13 at the end of this chapter.

‡ Schoenemann more recently proposed prehydrolysis of wood chips with hydrochloric acid of about 35 per cent concentration. Schoenemann, K. Improved Rheinau Process. *Chim. & Indus.* (Paris), **80**(2):140–50. 1958.

ing in a yield of about 22 per cent of mixed wood sugars on the basis of dry wood. These sugars can be easily evaporated to a consistency of feed molasses, or in the case of softwoods fermented to alcohol.

The prehydrolyzed wood chips, now consisting mainly of cellulose and lignin, are dried and passed to the main hydrolysis tower. Here cellulose is converted to sugar by counter treatment with 41 per cent hydrochloric acid at about 70°F.

The principle of this process depends on Hägglund's discovery [6] that hydrochloric acid solution which shows no more power to dissolve wood sugar regains its extractive ability when brought into contact with fresh, untreated sawdust. For this reason fresh acid is used on the nearly extracted chips, while an acid solution with progressively greater sugar concentration is employed in the digesters containing less hydrolyzed chips.

The resulting sugar sirup is converted to dextrose by a further mild hydrolysis, followed by refinement and finally crystallization. The hydrochloric acid is separated from the sugar-acid concentrate by a three-stage vacuum evaporation and is brought to the required strength of 41 per cent by addition of hydrogen chloride gas. Lignin, to the extent of about 30 lb per 100 lb of dry wood, is removed from the top of the tower and is dried and compressed. Unlike the lignin produced in high-temperature hydrolysis, it can be used in the production of phenol-formaldehyde resins.

Estimates made for conditions in the United States indicate that, where large quantities of suitable raw material are available, this process can be competitive with the production of dextrose from corn, assuming that the price of corn is not less than $1.35 a bushel. It is also stated that Schoenemann's process might be economically even more attractive if sawdust could be substituted for chips and further outlets could be found for lignin.

D. Wood-molasses Production

Molasses can be readily produced from wood sugars by suitable concentration. It has been suggested that in the United States the best opportunities for utilization of wood sugar lie in their use as a livestock feed supplement. Neither the production of alcohol by fermentation nor the use of purified sugar for human consumption appears to be economically or politically feasible.

In conjunction with this work a number of feeding experiments have been carried on at nine agricultural experiment stations.[9] The results indicate that wood molasses is roughly equivalent in food value to blackstrap molasses as a carbohydrate feed supplement.

IV. PRODUCTION OF ALCOHOL AND YEAST FROM PULPING RESIDUES

In the alkaline and acid pulping processes substantially all the hemicelluloses in the wood are hydrolyzed to sugars or are decomposed. The potential volume of such dissolved carbohydrates in the pulping liquor has been estimated (Table 20-2).[18]

Table 20-2. Potential Volume of Dissolved Carbohydrates in Pulping Liquor [a]

Process	1953 U.S. pulp production, short tons	Carbohydrate potential at 75% recovery, tons	
		Standard process	With prehydrolysis
Acid sulfite..............	2,300,000	345,000	
Alkaline kraft..............	9,435,000	. . . . . .	1,415,000
Alkaline soda............... ..	428,000	. . . . . .	64,000
Special alpha:			
(sulfite and kraft).....	677,000	100,000	
Total....................	12,840,000	1,924,000	

[a] SOURCE: Wiley, A. J., J. F. Harris, J. F. Saeman, and E. G. Locke. Wood Industries as a Source of Carbohydrates. *Indus. and Engin. Chem.*, 47: 1397–1404. 1955.

Of the potential total of 1,924,000 tons of carbohydrates based on the 1953 pulp output, about 500,000 tons of carbohydrates were actually produced. Several commercially successful methods for recovery of carbohydrates from the sulfite liquor have been designed and are now in use in the United States, Canada, and Europe. In the United States such utilization is based on fermentation methods for production of yeast or alcohol. Improved evaporation methods point to the possibility of recovering hemicellulosic products as such. In the case of hardwoods, an acid-catalyzed dehydration of pentose sugars presents the possibility of converting them to furfural. In the sulfite process the content of sugars derived from the hemicelluloses found in pulping liquor is about 400 lb per ton of pulp. Of this amount about 300 lb can be recovered.[18]

In the alkaline processes the hemicelluloses are broken down to saccharinic acid derivatives, which are burned during the recovery of pulping chemicals. In this case the sugars are not only a total loss, but they also consume valuable chemicals. The hemicellulosic products can, however, be recovered in the alkaline processes by prehydrolysis of wood chips. This procedure involves pressure steaming of chips. Since this results in formation of substantial amounts of acetic and formic acids, the treatment is actually a mild hydrolysis, bringing about the conversion of hemicelluloses to simple sugars. These sugars can be easily recovered for fermentation or the production of yeast.

V. PRODUCTION OF YEAST

Yeast proteins from wood sugars as a source of food for humans and livestock were of considerable importance to Germany during the Second World War. Considerable quantities of fodder yeast and satisfactory baker's yeast have also been produced in the United States and Canada using sulfite liquor as the source of sugar.[16]

The organism used for fermentation of wood sugars is a strain of *Torula utilis,* though other strains of yeast are also used. Food yeast propagated on wood contains about 50 per cent protein, 4 per cent fat, large amounts of vitamin B_2, and small quantities of vitamin B_1. Under good operational conditions 1 gal of 5 per cent wood-sugar solution will produce about 0.2 lb of yeast.

VI. PRODUCTION OF CARBOHYDRATE DERIVATIVES FROM WOOD RESIDUES

Wiley and others [18] suggest that the chemical utilization of wood in the United States may have a more promising future in the production of carbohydrate derivatives, without the intermediate production of sugar. They point out that the high-temperature hydrolysis of wood, using dilute mineral acids as a catalyst, results in conversion of pentosan hemicelluloses to furfural and of hexosans to hydroxymethyl furfural; the latter is then in turn converted to levulinic and formic acids. The anticipated yields from hardwoods, if complete conversion of carbohydrates to the above products is carried out, are approximately as follows:

Carbohydrate derivatives	Lb per 100 lb of dry wood
Acetic acid	5
Formic acid	4
Levulinic acid	14
Furfural	7
Lignin	30
Total	60

All these acids and furfural have important present and potential uses. The economic success of such chemical conversion of wood depends on the ability to produce them at a low enough cost and on the complete utilization of all the resulting products, including lignin. The feasibility of chemical utilization of wood will be considerably enhanced if higher uses of lignin than as a fuel can become a reality. Some of these possible uses for lignin are in glues and plastics and as an intermediate material in production of phenols and cyclohexanols.[18]

SELECTED REFERENCES

1. Anon. Wood Challenges Corn as Dextrose. *Chem. Engin.,* **61**(2):138–142. 1954.
2. Anon. Twenty Years of the Hydrolysis and Sulfite Alcohol Industries. *Gidrolyznaya i Lesokhimicheskaya Promishlenost,* **8**(8):1–3. 1958. Available in English translation through *SLA Translation Monthly,* **4**(3):127. 1958.
3. Food and Agricultural Organization. Sixth Meeting of the FAO Technical Panel on Wood Chemistry. FAO 108. Rome. 1954.
4. *Gidrolyznaya i Lesokhimicheskaya Promishlenost* (Hydrolysis and Wood Chemical Industry). Russian monthly journal, numerous articles on wood hydrolysis.
5. Gilbert, N. I., A. Hobbs, and W. P. Sandberg. Utilization of Wood by Hydrolysis with Dilute Sulfuric Acid. *Jour. Forest Prod. Res. Soc.,* **2**(5):43–47. 1952.
6. Hägglund, E. Chemistry of Wood. Academic Press, Inc., New York. 1951.

7. Hall, J. A., J. F. Saeman, and J. F. Harris. Wood Saccharification. *Unasylva,* **10**(1) : 7–16. 1956.
8. Harris, E. E. "Wood Hydrolysis," Chap. 21 in L. E. Wise and E. C. Jahn (eds.), Wood Chemistry, 2d ed. Reinhold Publishing Corporation, New York. 1952.
9. Harris, E. E. Hydrolysis of Wood for Stock Feed. U.S. Department of Agriculture, Forest Service, Forest Products Laboratory Report No. R1731. 1950.
10. Harris, E. E. Wood Saccharification, in Advances in Carbohydrate Chemistry, Vol. 4. Academic Press, Inc., New York. 1949.
11. Harris, E. E., and E. Beglinger. The Madison Wood-sugar Process. U.S. Department of Agriculture, Forest Service, Forest Products Laboratory Report No. 1617. 1946.
12. Kobayishi, T., and T. Ito. Report on the Wood Saccharification. Discussion Report 3. Forest Products Research Institute. Hokkaido, Japan. 1954.
13. Lloyd, R. A., and J. F. Harris. Wood Hydrolysis for Sugar Production. U.S. Department of Agriculture, Forest Service, Forest Products Laboratory Report No. 2029. 1955.
14. Matson, E. E., and E. G. Locke. Availability of Wood for Chemical Utilization. Presented at American Chemical Society Meeting, San Francisco, April 14, 1958. Mimeo.
15. Saeman, J. F., and A. A. Andreason. Industrial Fermentation, Vol. 1, Chap. 5. Chemical Publishing Company, Inc., New York. 1954.
16. Stamm, A. J., and E. E. Harris. Chemical Processing of Wood. Chemical Publishing Company, Inc., New York. 1953.
17. Technical Assistance Administration and Economic Commission for Asia and Far East. The Production and Use of Power Alcohol in Asia and the Far East. United Nations, New York. 1952.
18. Wiley, A. J., J. F. Harris, J. F. Saeman, and E. G. Locke. Wood Industries as a Source of Carbohydrates. *Indus. and Engin. Chem.,* **47**:1397–1404. 1955.

PART IV

DERIVED AND MISCELLANEOUS FOREST PRODUCTS

CHAPTER 21

NAVAL STORES

Naval stores, products such as turpentine, rosin, and pine oils, are derivatives of oleoresins obtained from several yellow pines. Initially the term naval stores was used to designate pitch and tar procured by destructively distilling wood of the Scotch pine (*Pinus sylvestris* L.) and appears to have had its origin with European shipbuilders of the late sixteenth century. These two commodities were indispensable to the maritime industry in the era of wooden ships and were used in prodigious quantities for preserving ropes and rigging, and for sealing the hulls and decking of newly constructed vessels as well as for maintaining the seaworthiness of those in service. But with the eventual evolution of men-of-war and merchant marines of steel, the demand for these particular products diminished rapidly, and processors directed their efforts toward the production and refinement of turpentine and rosin, for which many new and expanding markets were developing. While neither turpentine nor rosin is directly concerned with modern maritime construction, tradi-' tionally both are still marketed as naval stores.

There currently exist three distinct segments of the naval-stores industry, and while the products of each are essentially the same, they are so designated in the trade that their origin is clearly defined. Thus, turpentine and rosin derived from the gum (oleoresin) bled from living trees are marketed as *gum turpentine* and *gum rosin*, respectively. Similarly, when these chemicals are obtained by the action of steam and suitable solvents on macerated or chipped wood, they are known as *wood turpentine* and *wood rosin*. Finally, *sulfate turpentine* and *sulfate rosin* are important by-products of pulp mills employing the sulfate process in cooking chips of the southern yellow pines.

A. Historical

The practice of using sealers such as gum, tar, and pitch derived from trees may be traced to very ancient times. Frequent reference is made by naval-stores historians to Gen. 6:14, wherein Noah was directed, in the construction of the ark, "to pitch it within and without with pitch." According to at least one biblical journalist,* however, the pitch used in the ark was from some bituminous source and not from trees, as has been so commonly asserted. Notwithstanding, it is a proved fact that natural oleoresins were used by ancient Egyptians in the preparation of water-repellent varnishes and mummy lacquers which have endured through centuries.

In the Middle Ages production of wood tar and pitch centered in the region of the Baltic Sea, and Sweden, endowed with vast forests of pine, developed a thriving export trade in these commodities. Toward the end of the seventeenth century, however, Swedish merchants had gained complete control of the movement of these materials and had prohibited their export except in ships of Swedish registry. To a maritime nation such as Great Britain this action was intolerable. Fear of shortages and monopolies, and their potentially devastating effects upon her naval might and merchant shipping, led her to search for new sources of such supplies and to encourage naval-stores production in the American colonies.

The earliest authentic account of naval stores in the Western Hemisphere was chronicled by Cabeza de Vaca, historian of the Narváez Expedition of 1528 to the New World. His companions, he related, constructed crude boats on the shores of Florida's Apalachicola Bay and coated them with a pitch obtained from certain pines.

Commercial production of naval stores on the North American continent had its beginning in 1606, when French colonists in Nova Scotia began distilling a crude turpentine from the oleoresin of white pine (*Pinus strobus* L.). In 1608, settlers of coastal Virginia developed an export market in wood pitch and tar with England, and by the year 1700, chipping of longleaf pine for gum was a well-established enterprise in eastern North Carolina. A dearth of roads in those early colonial days restricted operations to the proximity of navigable streams, where the crude gum was gathered, barreled, and consigned to ports along the Atlantic seaboard, and then dispatched to England for processing. In the ensuing years demands for Carolina gum pyramided, and numerous distilling centers were put into operation to take care of both domestic and export trade in turpentine and rosin. By the middle of the eighteenth century North Carolina's exports alone exceeded 60,000 bbl of rosin, 12,000 bbl of tar, and 10,000 bbl of turpentine annually. Peak production was attained in 1880, when more than 1,500 stills were processing gum within her borders. With the approach of the twentieth century, however, lumbering became the major forest enterprise within her pine belt, and before long both worked-out turpentine timber and virgin stands of longleaf pine were virtually depleted.

Meanwhile chipping operations were being initiated in the vast longleaf (*Pinus palustris* Mill.) and slash pine (*Pinus elliottii* Engelm.) colonial forests to the south. For a short time South Carolina assumed production leadership but currently is contributing only a fraction of 1 per cent of the nation's total annual output of gum naval stores. The year 1875 marked the beginning of gum-naval-stores operations in Georgia. Within the short period of 5 years, distilling gum had become the state's leading industry, and cur-

---

* Keller, Werner. The Bible as History. William Morrow & Company, Inc., New York. 1936.

rently Georgia is producing nearly 80 per cent of the South's annual yield of gum turpentine and gum rosin.

The year of peak production of gum naval stores in the United States was 1909, when 750,000 bbl of turpentine and more than 2,500,000 drums of rosin were placed on the market.

Recovery of turpentine, rosin, and pine oils * from the wood of old pine stumps and roots by steam-and-solvent extraction was successfully accomplished on a commercial basis in 1909 and signaled the beginning of the wood-naval-stores industry † in the United States. More than a dozen such plants are now in operation in the longleaf pine belt and are currently producing appreciably greater volumes of both turpentine and rosin than are the processors of pine gum. Significant, too, has been the rise to a position of economic prominence of sulfate turpentine since 1935. Cheaper to extract than either gum or wood turpentine, sulfate turpentine is currently marketed in a volume more than equal to the combined volumes of the gum and wood derivatives. Figs. 21-1 and 21-2 delineate production trends in the three segments of the industry and serve the reader in placing them in their proper perspective. Each of these is fully discussed in subsequent sections of this chapter.

I. GUM NAVAL STORES

The gum-naval-stores industry in the United States is confined to the slash and longleaf pine belts along the Atlantic and Gulf coastal plains, a heavily forested area of approximately 30 million acres, virtually all of which is second-growth timber. Of this vast timber resource, however, only 1.1 million acres is currently being worked for naval stores. Ranked for many years with the South's leading forest enterprises, it is now of only minor importance in comparison with other flourishing forest industries of the region, and the gum harvest has come to be regarded as an intermediate crop or as a by-product of timber production. The low levels at which gum turpentine and rosin are currently being produced (Figs. 21-1 and 21-2) are largely the direct result of competition from other sources of naval stores. The substitution in numerous products and processes of other solvents and synthetic resins for turpentine and rosin, respectively, has also contributed to market losses. But even at current low production levels, the industry contributes to the support of nearly 27,000 wage earners and their families.

Unlike many forest enterprises which are seasonal, the production of gum is essentially a year-round undertaking. A normal season begins in March and ends in November, and ordinarily provides a cash income every 2 to 3 weeks during this period. Daily markets are available throughout the year. Unique, too, is the fact that when conditions are unfavorable and harvest

---

* Pine oils are not a component of gum from living trees. They are invariably present in the distillate of wood-naval-stores operations, however, and are believed to arise through the gradual oxidation of pinenes and other terpenes entrapped in heartwood of old stumps and roots.

† Not to be confused with the softwood-destructive-distillation industry, which currently produces about 2,500 bbl of turpentine spirits annually (see Chap. 19).

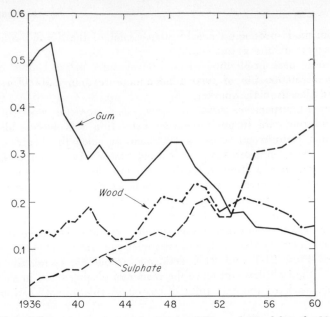

Fig. 21-1. United States turpentine production in millions of 50-gal barrels, 1936 to 1960

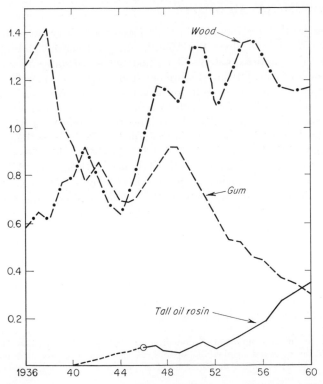

Fig. 21-2. United States rosin production in millions of 517-lb drums, 1936 to 1960.

and production costs exceed selling prices, operations may be temporarily suspended and then renewed when economic conditions again become favorable, without damage to the trees.

From the standpoint of gum production, slash pine is regarded as the more desirable species. Longleaf pine must attain an age of from 25 to 45 years before tapping is profitable, whereas good yields of gum are obtainable from slash pine trees of only 15 to 25 years of age. Moreover, slash pine is becoming manifestly more abundant, and in addition to the millions of trees that have been planted in the past several years, this species, under favorable conditions, preempts cutover lands formerly occupied by longleaf pine and now occurs naturally north of its original range in the pine belt. Longleaf pine, on the other hand, is the hardier of the two species and is much more capable of survival following fire.

None of the western pines is worked for naval stores although some of them are potential sources of turpentine and rosin. It has been determined that weekly yields of gum from ponderosa pine (*Pinus ponderosa* Laws.) commonly exceed those from slash and longleaf pines of similar size and vigor, although because of a shorter growing season, annual yields are appreciably smaller. High labor costs prevailing in the West, and limited markets, are other important factors which preclude working ponderosa pine, at least in the foreseeable future. On the other hand, the supply of old stumps used as raw material by the southern wood-naval-stores producers is nearing depletion, and the creation of a similar industry in the ponderosa pine belt may one day be feasible.

A. The Nature and Occurrence of Pine Gum

The oleoresins obtained from slash and longleaf pines are transparent to translucent, tacky, and viscous fluids. They are not finite chemical compounds but rather complex mixtures of substances, the principal constituents of which are terpenes and rosin-yielding acids * of high molecular weight. The complex is readily soluble in alcohol, benzene, ether, and numerous other organic solvents, but is insoluble in water. Analyses of gums from slash and longleaf pines made by Black and Thronson † over a 2-year period are presented in Table 21-1. It is readily apparent that the principal difference lies in the optical properties of the alpha-pinene fractions of the turpentines, which in slash pine is predominently levorotatory, and in longleaf pine, dextrorotatory.

Almost everyone has observed resin bleeding from wounds on the boles of many coniferous trees. Such exudations are traceable to an intercommunicating system of minute horizontal and vertical, tube-like, intercellular spaces in the woody stem, known as *resin canals* or *resin ducts*, which in the yellow pines occur in great profusion. Formation of these canals is occasioned by early postcambial separation of young, growing cells in such a manner that more or less cylindrical passages are formed. At maturity all these passages are enveloped in a layer of secretory tissue, *epithelium*, from

* Abietic acid ($C_{20}H_{30}O_2$) is the most important.
† See *Indus. Engin. Chem.*, **26**:66. 1934.

Table 21-1. Oleoresins from Longleaf and Slash Pines

Gum analysis	First year		Second year	
	Longleaf	Slash	Longleaf	Slash
Rosin (%)........................	77.7	77.6	79.1	78.0
Turpentine (%).....................	19.8	20.4	18.1	19.7
Rosin:turpentine ratio...............	3.92	3.81	4.36	3.97
Water and loss (%).................	2.5	2.0	2.8	2.3
Specific gravity of turpentine (15°/15°)	0.8700	0.8679	0.8689	0.8674
Refractive index of turpentine (20°)....	1.4681	1.4695	1.4705	0.4713
Optical rotation of turpentine (20°)....	+13.23°	−18.31°	+10.60°	−20.33°

which the resin is secreted. Vertical canals occur at random and frequent intervals in both springwood and summerwood portions of the ring, while the radial canals, which are of somewhat smaller diameter, are centrally disposed in the fusiform rays. Resin should not be confused with "sap," a watery solution containing sugars and other essential substances so necessary to the metabolism of a tree.

The biochemical and physiological processes concerned with the synthesis of resins are as yet not fully appreciated. It has been suggested that the precursors of resin-yielding materials undergo transformation in the ray parenchyma and epithelial cells, but how this occurs is still a matter of speculation. Numerous investigators have concerned themselves with factors influencing gum yield and have developed a number of pertinent correlations with tree inheritance, tree morphology, site, growth rate and vigor, and with certain physical and chemical relationships. Much has been learned, but there is yet much more to be determined before the whole story can be told.

B. Gum Producers and Processors

Woodlands in which trees are bled for gum are commonly referred to as *turpentine farms,* and the individuals who work them are known as *gum producers.* Ordinarily a producer disposes of his accrual of gum by selling it to a *gum processor.* Gum processors own and operate modern and efficient distillation plants but ordinarily are dependent upon producers for their crude gum. On occasion, however, a producer may not only arrange to have his gum custom-distilled but may also make his own negotiations for the sale of the refined products.

Most gum producers even now, as in the past, commonly require some financial assistance in order to sustain their operations. At one time this was provided by *factorage houses,* or *factors,* who were prepared to underwrite the costs of leases for timber and labor. They also served as a producer's agent and supervised the transportation and sale of his product. Some even acted as wholesalers. In the event of mismanagement of an operation, or upon default of payments due, a factor would often take over a turpentining operation or a distilling center. For services such as these he normally received a commission of 2½ per cent of the net return.

Factorage houses as originally conceived and operated are no longer extant. Several factors have erected and are operating modern distillation plants, while the newer gum processors, on the other hand, offer factorage services to their clientele. The producer who contracts for supplies and services with a processor agrees to sell his gum to that processor and permits him to deduct a certain percentage of the value of the delivered gum in payment of the obligation.

Operations of gum producers range from 500 to 1,000,000 trees. Most of them, however, are farmers who work 1,500 to 2,500 turpentine faces annually. In fact, about 65 per cent of the producers account for only 25 per cent of the total harvest. There are currently 20 modern gum-processing plants in the turpentine belt and nearly 4,400 gum producers working approximately 34 million trees on 1.1 million acres of forest land.

C. Leases

Gum producers having only limited timber holdings, or no timber at all, commonly resorted to the practice of leasing timber for the production of gum. Comparatively few such contracts are negotiated these days, however, because of the almost prohibitive rates of interest currently in demand. One of the more common types of contract, the *cash-lease* agreement, only a few years ago cost the lessee about 3 cents per face per year. A similar contract today calls for a charge of 15 cents per face. The once popular *percentage-work-basis* lease, an agreement wherein the producer paid a timber owner a specified percentage of his gross sales of gum for the use of timber, is now only infrequently negotiated because few producers are able to operate at a profit under the prevailing rates of 20 to 30 per cent.

D. Woodlands Operations

There are many factors to be considered in the establishment of a gum-naval-stores operation, not the least of which is the quality of the stand. Ideal trees for turpentining are those which yield copious amounts of gum throughout the life of the operation. These are young, vigorous, rapidly growing specimens, characterized by large volumes of sapwood, wide growth increments, and large crowns which clothe the boles in a dense canopy for one-third to one-half of their length. Experience has taught the gum producer that he can expect to obtain only nominal yields from trees with small crowns, narrow growth rings, and large amounts of heartwood. He has also learned that young, thrifty trees yield more gum than do older ones of similar height and diameter. This is but another way of stating that the better the site, the better the yield. Trees on good sites may be profitably worked after they have attained a diameter of 9 in. dbh; a 10-in. minimum diameter limit is recommended for those growing under less favorable conditions.

Large turpentine farms are divided into standard working units known as *crops*. The standard crop comprises 10,000 turpentine faces. In the early days of tapping, it was a common practice to work two and occasionally even three faces on large trees of virgin forest growth, and a worker could maintain about 10,000 a week during the season. In the second-growth stands of today, however, trees are smaller, somewhat more scattered, and usually unsuited for multiple facings. Under these conditions an operating crop seldom exceeds 8,000 faces, and is not infrequently smaller. For convenience of the woods crews, crops are often further divided into *drifts*. A drift, usually comprising some 2,000 to 5,000 turpentine faces, may be a predetermined acreage with clearly marked boundaries, or it may be a particular area delineated by such natural limits as streams, old fields, and burns, or by roads and firebreaks. Four to eleven crops, depending upon their size and location with respect to one another, constitute a *ride*. This term merely alludes to the total area which may be adequately supervised by an overseer.

When there were still adequate reserves of virgin pine timber to sustain large-scale lumbering operations, young, vigorous second-growth was preempting cutover lands, but these stands were regarded as suitable only for the production of gum. Little or nothing in the way of silvicultural treatments was prescribed for this growing stock, for its potential as timber crop had not been recognized. Early thinnings were made merely to promote crown development with its attendant increase in the flow of gum. Past experience had revealed that turpentining reduced height and diameter growth by as much as 25 to 50 per cent and that worked-out trees were frequently unworthy of salvage. Decay and insect damage were often prevalent, and turpentining itself resulted in a high rate of tree mortality. Moreover, butt logs were likely to be infiltrated with gum behind the working face and thus rendered unfit for manufacture into lumber. Nails and bits of metal left in the faces further militated against their use in sawmills and in pulpwood chipping operations.

Within the past few decades, however, the forest economy in the turpentine belt has taken on its own new look. Depletion of virgin timber forced the sawmills to turn to second-growth trees. This, coupled with a tremendous expansion of the pulp and paper industry with its insatiable thirst for pulpwood, rising stumpage prices, and a shrinking gum-naval-stores market, placed gum production in a noncompetitive position with other forest enterprises. Managers of pinelands found it expedient to reshape their forest-management programs and to regulate their growing stock in a manner that would provide a maximum of income from their operations. Out of such considerations has evolved a new concept in naval-stores production; namely, in the management of stands of second-growth trees, gum can be harvested and marketed as an intermediate crop while bringing the stands to maturity as timber crops. The success of this sort of integrated forest operation has been due in no small measure to well-conceived management plans, the development of bark-chipping and acid-treatment techniques, and new tapping tools and equipment.

1. **Marking Trees.** Preparatory to the establishment of a turpentining operation, it is customary to select and then mark the trees to be faced.

Two systems of selection are in current use. The first of these is known as "diameter limit cupping," in which all trees in the stand over a specified minimum diameter, usually 9 or 10 in., are marked for turpentining. The principal objection to this system, however, is early liquidation of the timber just as it is entering a period of maximum production of wood. The second system, "selective cupping," has as its objective the integration of a naval-stores operation in timber-production programs. Tapping is restricted to those trees which are destined to be removed in the management of a stand. Selective cupping does not suggest the facing of small trees. If stands become overcrowded before the trees are large enough to work at a profit, a pulp-wood thinning should be made. In dense stands, those trees which should be removed to provide more growing space for the final timber crop should be marked for cupping and harvested immediately following the turpentining operation. When markets are favorable, all merchantable trees in mature stands should be marked for cupping except those to be held in reserve as seed trees. The mark, a bark blaze or a splash or stripe of paint placed on the side of a bole, not only apprises woods crews which trees are to be worked, but also where the face is to be started.

2. Recommended Turpentining Procedures. As an aftermath of the First World War, the Germans, and later the Russians, as a part of their national programs toward self-sufficiency in the production of raw materials for essential industries, sought ways and means to increase production in their naval-stores industries. In 1933, both Germans and Russians began exploratory research in the use of chemicals to stimulate gum flow in Scotch pine. Many chipping techniques were devised, and a variety of chemicals were employed, including acids, bases, and salts, in an effort to stimulate and increase the flow of gum. From the wealth of resulting data it was found that certain chemicals actually retarded yield as much as 40 per cent, while others increased yields from 5 to more than 200 per cent.

The initial work on chemical stimulation in the United States was begun in 1936 at the Olustee Experimental Forest, Lake City, Florida, as a coopera-tive project between the Southern Forest Experiment Station and the Bureau of Agriculture and Industrial Chemistry. Unfortunately this study was under-taken at a time when large inventories of both turpentine and rosin were depressing prices, and the industry showed but little interest in the project. Studies that continued through 1941 indicated that chemical stimulation was practical, and in 1942 and 1943, recommended procedures for woods opera-tions were released to the industry. Today's recommendations include bark chipping with acid treatment, and it is of interest to note that bark-chipping and acid-treating techniques have rendered the older and less efficient wood-chipping * methods essentially obsolete. Recently released statistics reveal that 76 per cent of the gum producers are practicing bark chipping and acid treating on 82 per cent of the turpentine faces currently in work.

3. Preparing Trees for Turpentining. The initial operation in prepar-ing trees marked for a naval-stores operation is tin emplacement. Tins are

---

* The reader is referred to the first edition of this book, Chap. 20, Naval Stores, for a review of the wood-chipping operations formerly in general use throughout the tur-pentine region.

metal gutters and aprons used to direct the flow of gum into a collecting receptacle. Several types of tins have been used for this purpose over the years, but none has proved to be more satisfactory than the preformed spiral gutter specifically designed for bark chipping.

Tin installation is a winter operation, and the tins should be in place prior to the renewal of growth activity in the spring. When spiral gutters are used, a portion of outer bark is removed from the base of each tree to provide a

Fig. 21-3. Bark shaver and tins. (*Courtesy U.S. Forest Service.*)

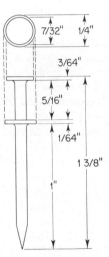

Fig. 21-4. Diagrammatic sketch of double-headed nail. (*Courtesy U.S. Forest Service.*)

smooth, contoured, fissure-free surface against which the tins can be snugly seated when properly attached. The removal of these patches of bark, or bark shaving as the operation is known, is easily and efficiently accomplished with a bark shaver (Fig. 21-3), a tool with which an experienced workman can shave some 700 to 800 trees daily. A two-man crew working behind the bark shaver can ordinarily seat a like number of tins in the same working day. The gutters are attached using the new double-headed nails (Fig. 21-4), which should be driven along the upper leading edge of the tin to ensure a good tight fit. Double-headed nails are easily withdrawn, and may be reused when it becomes necessary to raise the gutters and gum-collecting receptacles as the face is lengthened. Their extraction at the conclusion of turpentining operations also provides metal-free pulpwood bolts or sawlogs, which are wholly acceptable to wood processors.

It is recommended that 10-in. spiral gutters be used on trees 9 to 12 in. in diameter; 12-in. gutters are better suited for larger trees. The affixing of a curved or straight 7- to 8-in. apron (with double-headed nails) and a collecting receptacle for the gum concludes preliminary woods operations (Fig. 21-5).

Receptacles used for collecting the flow of gum are referred to as *cups* and are made of galvanized iron sheet, clay, or aluminum. The galvanized cup is most frequently used, although those in which the zinc has chipped

Fig. 21-5. Tins and collecting receptacle seated on a bark-shaved face. (*Courtesy U.S. Forest Service.*)

Fig. 21-6. Making a new streak with a bark hack. (*Courtesy U.S. Forest Service.*)

off or worn away impart objectionable discolorations to the gum. Ceramic cups are satisfactory but are often the most expensive to use because of excessive breakage occasioned by rough handling and by icing when water is permitted to stand in them during the winter months. Aluminum cups do not discolor gum but they are more expensive and much more subject to damage than their iron counterparts. Cups are available with capacities of 1 and 2 qt. The use of the larger ones reduces the number of dippings (see page 448) required during the season and thus effects a material saving in operational expenses.

An analysis [13] of four dippings of samples of crude gum taken from cups of 1- and 2-qt capacity, respectively, indicated no significant difference in product yield or grade. The average time that gum remained in the 1-qt cups was 28 days; for the 2-qt cups, 47 days. Upon distillation, turpentine yields were 21.51 per cent and 21.96 per cent, while rosin yields were 76.90 per cent and 76.28 per cent, respectively.

4. Bark Chipping. Periodic reference has been made to the turpentine face, which is, of course, the wound made by periodic chipping. New turpentine faces, as well as the renewal of older ones, are started in the middle of March unless adverse weather conditions prevail. In initiating a new face, a tool known as a *bark hack* (Fig. 21-6) is used to remove a ¾-in. horizontal strip of bark from the bole immediately above the tins and cup. It is essential for maximum gum flow that no residual bark be permitted to adhere to the exposed woody surface, or *streak*, and that the wood itself be unmolested. This operation is known as *bark chipping* to distinguish it from earlier facing practices when wood was removed to a depth of from 1 to 1½ in.

Wide faces yield more gum than do narrow ones. The length of the initial streak dictates the width of the eventual face; it should be equal to the diameter of the tree at breast height but no wider.

Subsequent streaks are superimposed at 2-week intervals. With an average of 16 per season, a virgin face will attain a height of from 12 to 14 in. at the end of the first year. Trees are usually worked for 3 to 5 years. During this working period the face continues to increase in height with each new streak and eventually becomes too high to be streaked with the conventional bark hack. When this situation develops,

Fig. 21-7. Spray puller hack. (*Courtesy U.S. Forest Service.*)

a workman uses a long-handled *spray puller hack* (Fig. 21-7), a dual-purpose tool designed to perform both bark chipping and spraying operations on high faces. When puller hacks are employed, a laborer chips out a converging, diagonal pair of bark strips, which results in a V-shaped streak.

5. Use and Application of Acid. Resin canals, as previously indicated, are normal to pine wood and occur in vertically and radially disposed systems. Since the canals of these two systems intersect at frequent intervals, it is only necessary to expose the opened ends of a number of the radial structures to drain off copious amounts of resin held in reserve. The discovery of this simple fact has been largely responsible for the general acceptance of bark chipping and acid treating, and the many economic advantages these techniques offer.

The removal of a ¾-in. strip of bark will expose the ends of more than 6,000 radial canals on a streak of average length. Merely removing bark, however, does not result in an adequate flow of resin, as the ends of many of the radial canals are at best only partially opened. Thus to ensure a steady flow of resin it is necessary to apply a small amount of a 50 per cent solution of sulfuric acid to the wound. Why? The answer is quite simple. When the acid comes in contact with the woody tissue, and especially those cells which terminate the resin passages, the constituents of their walls are partially digested by acid hydrolysis, thereby opening the passages and liberating the resin.

In woods operations, the workman chipping a streak is also responsible for applying the acid. The best and most widely used device for this purpose is a 16-oz polyethylene plastic squeeze bottle (Fig. 21-8) provided with mixing chamber and simple nozzle. The bottle, which at the outset is filled to two-thirds capacity with acid of the proper concentration, is also fitted with a hook so that it may be hung on a tree when not in use. When this type of bottle is squeezed, air and acid are forced into the mixing chamber and discharged from the nozzle in the form of a mist. Relaxation of pressure causes the bottle to assume its original configuration meanwhile drawing in air for the atomization

Fig. 21-8. Applying acid to a working face. (*Courtesy U.S. Forest Service.*)

of more acid when the spraying operation is repeated. Acid bottles fitted on spray puller hacks work on the same principle.

Strict adherence to a few time-proved procedures when spraying acid on a streak will assure resin of quality and maximum flow for the working life of a face. These are enumerated below:

1. An acidproof cover should be placed on the cup during chipping and spraying operations, not only to prevent debris from falling into the cup, but also to eliminate acid contamination of accumulated resin. Excessive acid discolors gum, and loss of grade results.
2. Acid should be applied to a streak immediately following its completion. The spray should be directed at the line of juncture between bark and wood. Special care should be taken to obtain even distribution across the streak. Large amounts of gum come from the upper corners of the streak; to miss them is to reduce yield.

3. A single pass of the spray bottle should be made across the streak holding the bottle at an angle of 45°. No more acid should be used than is necessary to wet the streak. Proper application of the acid will cause a reddish discoloration of wood and inner bark immediately above the last streak. If, upon making the next streak, it is found that the coloration extends upward and beyond the new cut, too much acid was applied. Inasmuch as excessive acid kills tissue and reduces yield, the new streak must be widened until fresh, green wood and bark appear above the discolored zone. On the other hand, if the upward penetration is less than ½ in., as disclosed by this change in color, yields are again usually something less than expected. In this case either too little acid was applied or the jet from the spray gun was misdirected. Ideal acid penetration should fall within the limits of ½ to ¾ in. upward across the streak.

4. Bark chipping followed by acid treatment should be repeated every 14 days during the tapping season. Little or nothing is gained by weekly chipping; in fact, the labor costs greatly exceed any income that may accrue from the small amount of additional gum obtained with this practice.

6. Dipping. The term *dipping* is a carry-over from the days when resin was collected by dipping or ladling it out of cavities hewed in the base of trees below their working faces. In its modern connotation dipping is concerned with emptying the cups as they become filled to near capacity. A laborer, usually referred to as the *dipper*, removes a cup from its supporting nail, inverts it over a bucket and pushes out the gum, or *dip*, with a dip-iron or wooden paddle. When an appreciable amount of resin is collected it is dumped into one of several wooden barrels or metal drums strategically located throughout the crop. When these are filled to capacity, they are sealed and dispatched to a processing plant for the production of turpentine and rosin. The dipper is also required to clean gutters and aprons when collecting the dip. Dipping is usually done at intervals of from 4 to 6 weeks depending upon weather conditions, productive capacity of the trees, and the size of the cups employed. Frequent dippings reduce losses from evaporation and oxidation, and from discoloration when rusty cups and tins are in use.

7. Scrape. Some of the flowing gum never reaches the cup, as it solidifies on the face in the form of white crystalline deposits. This is known as *scrape*. This hardening action of gum in all probability results from the evaporation of some of its more volatile constituents and from the oxidation of others. Some producers require their dippers to remove this scrape and punch it off into the cup with each dipping, for it is known that better yields of turpentine and rosin are obtained from scrape if it is collected in this manner. Others await the termination of the turpentine season before removing scrape and then take it all in one harvesting operation. Scrape collected at the end of the season is usually dirty and produces not only less turpentine but also rosin of a lower quality than that collected with each dipping. Dull-bladed tools that can be used to punch off scrape without chipping or gouging the woody bole are preferred.

In a normal season 5 to 10 per cent of the yield of slash pine gum is in the form of scrape, while for longleaf pine scrape makes up about 20 to 30 per cent of the total annual yield of gum.

8. Raising the Tins. As the height of a turpentine face increases over the working season, more scrape is formed with proportionate losses in liquid resin. But as the yield of turpentine from such material is relatively low, it is desirable to restrict its formation to a reasonable minimum. This is accomplished by periodically removing cups and gutters and relocating them immediately below the last-formed streak. In the industry this phase of the

Fig. 21-9. New hammer raising tool. (*Courtesy U.S. Forest Service.*)

woods operation is known as *raising the tins*. The frequency with which this is done depends upon several factors, not the least of which are the economic considerations involved. The additional labor costs for this operation are usually unwarranted when market prices are unfavorable or when there is but little price differential among various grades of refined rosin. Most producers raise their tins at the end of the first working season, and in alternate years thereafter. Tins should always be raised when the top of the working face exceeds a distance of 20 in. to the apron. The initial raise is often the one of greatest importance because cups on virgin faces are only a few inches from the ground where they are easily dislodged and spilled by grazing livestock, and where gum is constantly contaminated with wind-blown silt and sand, insects, and forest litter.

Tins should be raised at the close of a season's operation immediately following the harvesting of scrape. In this way the tins can be cleaned before they are reset and the hardened gum which has accumulated can be punched off and added to the scrape. The use of double-headed nails and the new hammer raising tool (Fig. 21-9) greatly facilitates this phase of the woods operation and requires the labor of only one person. When faces are worked out, removal of all metal poses no problem, and the metal-free butt logs are readily accepted for milling and/or pulping.

9. Yields. Seasonal yields from bark-chipped and acid-treated trees streaked at 14-day intervals are from 50 to 100 per cent greater than those from comparable trees wood-chipped weekly. The table below indicates what gum yield may be normally expected from trees of slash pine with faces equal to one-third of their circumference (dbh), when worked biweekly with trained labor using ¾ in. bark-chipped streaks sprayed with a 50 per cent solution of sulfuric acid. The data presented in this table are based on measurements of 1,306 trees, divided among 10 plots over a distance of about 200 miles between Dodge County, Georgia, and Levy County, Florida. Site index among these plots ranged from 72 to 99 ft at 50 years, while the trees ranged in age from 19 to 45 years.

Table 21-2. First-year Gum Yields from Single-faced Slash Pines for Six Diameter Classes and Five Crown Ratio Classes [a]

Dbh in in.	Gum yields in barrels for crown ratios of				
	0.20	0.30	0.40	0.50	0.60
9	172	190	208	226	244
10	209	227	245	263	281
11	246	264	282	300	318
12	283	301	319	337	355
13	320	338	356	374	392
14	357	375	393	411	429
15	394	412	430	448	466

[a] Yields expressed in 435-lb barrels of gum per crop of 10,000 faces.

E. Effect of Turpentining upon Tree Growth

Annual growth increments of trees worked for naval stores are appreciably less than those of round trees of similar size and morphological characteristics. The retardation of growth is due, at least in part, to a diversion of the products of photosynthesis to gum rather than to the formation of cell-wall substance; and to the turpentine face itself, which partially interrupts the flow and translocation of solutes.

Studies by Schopmeyer * reveal that volume increment in cubic feet of bark-chipped, acid-treated trees was 26 per cent less than that of unworked trees of similar size. In terms of stumpage values, however, he indicates that

---

* Schopmeyer, C. S. Gum Yield and Wood Volume on Single-faced Naval Stores Trees. *South. Lumberman,* pp. 123–124. December, 1955.

the return from a naval-stores operation can be from $1\frac{1}{4}$ to 9 times the value of the volume of wood lost during turpentining.

Sulfuric acid, when properly administered in accordance with recommended procedures, has no deleterious effect upon tree metabolism. As early as 1953 * Larson revealed that trees of slash pine in the Osceola National Forest in Florida were still healthy and yielding satisfactory amounts of gum after 9 consecutive years of bark chipping and acid treatment. The continuing good health of the many millions of trees which have been, and which are currently being, worked with acid, however, is inescapable evidence of its innocuous effects.

Mortality of naval-stores trees resulting from attacks of the black turpentine beetle seldom exceeds 1 per cent of the working stand. Such losses are not appreciably greater than those caused by this organism in unworked timber stands. Severe outbreaks of this beetle do occur at irregular intervals, but its activities can be satisfactorily suppressed by spraying at nominal cost.

Investigations pertaining to the strength properties of the wood from turpentined trees indicate that there is little or no loss in mechanical efficiency.

F. Fire Protection

Protection against fire is as essential to a turpentine operation as it is to the production of a timber crop. Its complete exclusion in pine forests, however, permits the accumulation of vast amounts of highly inflammable litter, which, if ignited, burns rapidly and fiercely with devastating results. On the other hand, the use of fire as a tool in forest management can have beneficial results when it is carefully controlled and properly handled.

The once-common practice of raking litter from the base of turpentine trees and subjecting the area to an annual winter burning is no longer generally extant because of prohibitive labor costs. Instead a prescribed burning is given each new area to be turpentined just prior to hanging the cups; a second similar treatment is recommended before *backfacing* † the trees, if such an operation is contemplated.

G. Naval Stores Conservation Program

In 1936 the Federal government instituted the Naval Stores Conservation Program for the purpose of encouraging and promoting sound conservation practices in the pine forests of the Southeast. Administered by the Forest Service for the Production and Marketing Administration, it provided all turpentine farmers with an opportunity to participate in cost-sharing payments when their woodlands are managed and the gum harvested in accordance with stipulated conservation practices. The regulatory measures and rate scale of this program have been revised periodically as new and better

* Larson, P. R. Turpentine Trees Remain Healthy after Nine Years of Acid Treatment. *South. Lumberman*, pp. 214–216. December, 1953.

† *Backfacing* is the initiation of a second face on the opposite side of a working or worked-out face.

woods practices have been developed, but its basic concepts remain unchanged. To be eligible for cost-sharing payments a producer must:

1. In harvesting worked-out timber, leave 6, 9, or 12 seed trees of 9, 8, or 7 in. in diameter per acre, respectively, unless there are at least 400 well-distributed stems per acre after the cut.
2. If clear-cutting is contemplated, plant at least 500 seedlings per acre within a period of 2 years following the logging operation.
3. Use double-headed nails with all tin raises on 1957 and 1958 virgin faces.
4. Use double-headed nails and/or spiral gutters or Varn aprons on all future raises, regardless of the age of the faces.
5. Remove all hardware from each tree when the turpentining operation is terminated.
6. Double-face no tree less than 14 in. in diameter.
7. When double-facing is practiced on larger trees, leave a pair of living bark bars between the faces, the aggregate widths of which must be not less than 7 in.

Payment is limited to a maximum of 2,500 faces per producer. Practices under which payments are made and the rates of each are delineated below:

Practice 1. Working trees 9 in. dbh or larger
  Rate: 2¢ per virgin face
      ½¢ second- to fifth-year faces

Practice 2. Working trees 10 in. dbh or larger
  Rate: 4¢ per virgin face
      3¢ second- to fifth-year faces

Practice 3. Working trees 11 in. dbh or larger
  Rate: 6¢ per virgin face
      3¢ second- to fifth-year faces

Practice 4. Working trees 12 in. dbh or larger
  Rate: 7¢ per virgin face
      3¢ per second- to fifth-year faces

Practice 5. Restricting turpentining to previously worked trees
  Rate: 7¢ per virgin face
      3¢ per second- to fifth-year faces

Practice 6. Working only selectively marked trees
  Rate: 8¢ per virgin face
      3¢ per second- to fifth-year faces

(Note: Trees must be marked 3 to 8 years in advance of cutting to qualify for this practice.)

Practice 7. Initial use of spiral gutters or Varn aprons and double-headed nails
  Rate: 2¢ per face for use in virgin installation or at any time of the first elevation in any drift
      ½¢ when double-headed nails are used without spiral gutters or Varn aprons

Practice 8. Removal of cups and tins from small trees
  Rate: 8¢ per face

(Note: This practice is limited to new producers to discourage working small trees with less than marginal productive capacity.)

Practice 9. Pilot plant tests of new methods and equipment

Rate: 8¢ to 11¢ per face depending on the nature of the practice

(Note: Such payments for tests are restricted to producers selected by the Forest Service to conduct controlled experimental work.)

H. Gum Processing

Gum processing, i.e., the recovery of turpentine and rosin as products of distillation from dip and scrape, is one of the oldest of the American forest-products industries.

Fig. 21-10. Model of old-type fire still. (*Courtesy U.S. Department of Agriculture.*)

Until the development and operation of central processing plants about 30 years ago, gum was processed for more than 100 years in small fire stills which so characterized the turpentine belt. These were kettles or retorts of copper (Fig. 21-10), of 10 to 30 bbl capacity, mounted over a stone or brick, wood- or oil-fired furnace, and connected by an apical elbow and arm to a copper coil, or worm, immersed in water. Their operation was inefficient and the products of distillation variable. Scorching of gum during distillation, a not too infrequent occurrence, imparted objectionable colorations to the rosin that resulted in degrade and lower prices, while losses of as much as 150 lb per charge were common in screening and filtering the hot liquid rosin.

Fire stills are now a thing of the past and have been almost completely replaced by some 20 or more modern central processing plants. For the most part these are owned and operated by gum processors and factors, each representing an initial capital investment of from $20,000 to $250,000.

Unlike the days when fire stills were in common use, raw gum is now thoroughly cleaned and washed to remove extraneous materials before it is distilled. In a typical operation at a modern central processing plant, gum is delivered to the unloading deck in sealed barrels. Each barrel is weighed and then dispatched to the receiving department. Here the heads are removed and the barrels inverted over steam-jacketed receiving vats (Fig. 21-11). After draining, a jet of live steam is directed into each barrel to remove the

Fig. 21-11. Delivery of crude gum to receiving vats. (*Courtesy U.S. Department of Agriculture.*)

last vestiges of gum. The barrels are then returned to the unloading deck and reweighed in order to determine the net weight of delivered gum.

In the cleaning and washing operations which precede distillation, raw gum is drawn off as needed from the bottom of the receiving vats and pumped into a blow case (Fig. 21-12), where it is melted with steam. The melt then flows into melter tanks, where trash, such as insect bodies, bits of bark, and pine straw, is removed by screening. Doors near the bottom of these tanks permit ready access to the screens for periodic removal of the accumulated debris. The screened gum passes from the bottom of the melter tanks to filter presses, where the residual solids, such as silt and sand, are removed. In completing the cleaning process the screened and filtered gum is pumped to a battery of wash tanks, where all water-soluble constituents are removed. The gum is then ready for distillation.

In distilling, a measured amount of gum is drawn into a still (Fig. 21-13), which is internally heated with steam. In the early stages of a run, turpentine and water vapors begin passing through the condenser in about equal

Fig. 21-12. Bottom of two receiving vats, *upper right;* blow case, *lower right;* melter tanks, *center and left.* The square doors on the melter tanks permit access to the refuse-retention screens. (*Courtesy U.S. Department of Agriculture.*)

Fig. 21-13. Measuring tank for clean, washed gum, *upper left;* insulated steam still and still head, *center;* horizontal, water-cooled condensing system, *upper right.* (*Courtesy U.S. Department of Agriculture.*)

volumes when the temperature rises above 212°F. Distillation continues until the ratio of the water-turpentine distillate reaches 9 to 1. The residual rosin is then pumped out and barreled or bagged while still in a hot, fluid state.

A continuous still of pilot-plant proportions was erected and operated by the Naval Stores Experiment Station at Olustee, Florida, several years ago and gave promise of further revolutionizing gum-processing procedures. Processors, however, have been loath to invest in this sort of equipment because of the drop in gum production over the past several years. One such plant was erected in New Jersey, but is currently inoperative.

I. Turpentine

When the turpentine-water distillate begins flowing from the condenser, it is directed to a settling tank, where the two nonmiscible liquids quickly separate by gravity. The watery layer is decanted off, but the turpentine is further dried by passing it through a layer of rock salt before pumping it to a storage tank.

As previously indicated, turpentine is not a single chemical compound, but rather is a mixture of several essential oils of the terpene series. Turpentine of American origin is composed principally of alpha-pinenes. Small amounts of beta-pinenes, champene, sylvestrene, and dipentene complete the complex.

Five standard barrels of dip produce one 50-gal drum of turpentine. Nine to ten barrels of scrape are required to produce a similar volume, since some of the turpentine constituents of scrape have been lost through evaporation.

The bulk of gum turpentine currently produced is marketed in tank-truck and tank-car lots of from 6,000 to 10,000 gals. Processors also offer the product in 50-gal drums. A considerable volume is also packaged in small bottles and tins ranging in volume from 4 oz to 5 gal and marketed in the retail trade under a variety of brand names.

Gum turpentine is offered to the trade in four grades, viz.: *water white, standard, one shade off,* and *two shades off*. Prices are usually quoted on the basis of the standard grade. Government specifications for the standard grade require that not less than 90 per cent of the complex shall distill over at 170°C, that the liquid shall be free of suspended matter and water and shall have a mild aromatic odor.

Consumption of turpentine in the United States for the year 1958 amounted to 565,100 bbl, of which 113,390 bbl constituted gum turpentine. Prime consumers were the paint, varnish, lacquer, pharmaceutical, and chemical manufacturers.

J. Rosin

Rosin, also known as *colophony*, is a brittle and friable solid with a faint aromatic odor. Its major chemical constituent is abietic acid, which reacts readily with metallic salts to form soaps, and with alcohols to form esters. It is insoluble in water but is readily dissolved in ether, benzene, and chloroform.

When the distillation of a batch of resin is completed, the hot molten rosin is immediately removed from the bottom of the still and pumped into light steel drums and/or multiwall paper bags seated on platform scales. The drum, which has a gross weight of 534 lb, contains 520 lb net weight of rosin. Bagged rosin weighs 100 lb net.

Before a batch of rosin cools and solidifies, however, small samples of the run are placed in molds and sent to the laboratory for grading. Rosin varies in color from a pale-yellow hue through shades of amber to dark red. Color is dependent upon many factors, including the source of the gum, the nature of the metallic equipment used in its collection, and methods of cleaning and distillation. Thirteen grades, all based on color, have been established by the Food and Drug Administration under the Federal Naval Stores Act. The lightest shades bring the best prices, but it is not always possible to produce them. These grades beginning with the lightest are: X, WW (water white), WG (glass white), N, M, K, I, H, G, F, E, D, and B. A large proportion of rosin offered for sale falls between the grades of N and H.

Type color standards, cubes made of tinted glass, have been deposited in the offices of the Department of Agriculture; a few other sets have been made available to inspectors and consumers. Most of the color-standard samples in current use, however, are matched sets made from rosin itself.

The equivalent of 399,910 barrels of gum rosin was produced in 1958; the bulk was consumed by manufacturers of soaps, paper, paints, varnishes and lacquers, plastics, adhesives, pharmaceuticals, synthetic resins, and chemicals.

K. Future of the Gum-naval-stores Industry

No one can predict with any degree of accuracy what the future holds for the gum-naval-stores industry, but the economic clouds which are gathering on the horizon have a foreboding look.

For the past several years gum-turpentine production has been in steady decline. Of the 626,820 bbl of turpentine produced in the United States during the 1957–1958 season, 49.8 per cent was sulfate turpentine, 29.6 per cent wood turpentine, and only 20.6 per cent gum turpentine. Sulfate turpentine is by far the least expensive to produce, and during the past several years has captured many markets formerly held by the gum and wood turpentine producers. In fact, it was suggested that by 1960, when production of sulfate pulps from southern pines was expected to exceed 10 million tons annually, enough sulfate turpentine could be recovered to meet the national needs for this commodity. Gum and wood turpentine must continue to be produced, however, as long as there exists a demand for rosin, but unless cheaper production methods can be devised, they will have to be disposed of at a loss to be competitive with the pulp-mill derivative.

The future outlook for gum rosin, on the other hand, is much brighter. Gum rosin's chief competitor is wood rosin. At the present time wood rosin makes up 64.1 per cent of the total annual production (gum rosin 21.4 per cent; sulfate rosin 14.4 per cent). Recovery of rosin from spent sulfate liquor is approximately 23.5 lb per ton of manufactured pulp. Thus, even in 1970, when it is anticipated that sulfate pulp production will reach 14 million tons, the mills could not produce more than 36 per cent of the predicted rosin requirements. This, coupled with the fact that the supply of stump wood will be exhausted prior to that date, suggests better times are ahead for the industry. There is, however, competition from another possible source. A pilot-plant operation is now under way in Oregon to determine the feasibility of developing a wood-naval-stores operation in the western ponderosa pine belt, where there is a stupendous supply of stump wood. Schopmeyer,[30] in commenting on this eventuality, states:

> If this move is made, the gum naval stores industry would have no market. On the other hand, if more gum producers improve the efficiency of their operations by adopting modern turpentining methods, and if the size of the turpentine trees can be increased by good forest management on more lands, the production costs of gum rosin might be reduced to a level below those of ponderosa pine stumpwood rosin. In that event, the present consumers of stumpwood rosin would gladly buy gum rosin, the stumpwood industry would cease operations, and the gum naval stores industry would reach its golden age.

Thus it is clearly evident that the gum-naval-stores industry has a two-fold problem. Not only must ways be found to reduce production costs, but gum production too must be materially increased. Does the answer lie, at least in part, in the development and management of naval-stores orchards? This is a new concept in gum production, but there are several in authority who believe so.

It is common knowledge that some trees are capable of yielding twice as much gum as others with similar silvical characteristics, and that the progeny of such trees are usually also superior gum-yielding specimens. Through careful breeding a seed source for superior gum-yielding trees is a distinct possibility. Vegetative propagation of such individuals is also possible. Such growing stock could then be set up in orchards, the management of which would more nearly approach the field of pomology than forestry. Planting stock would be limited to 100 trees to the acre to assure good crown development. The cultivation, irrigation, and fertilization of such sites might well pay good dividends. Under this kind of management slash pine trees would be large enough to face in their twelfth year, and by front- and back-facing for the next 12 years, the stand would be ready for a pulpwood harvest in its twenty-fifth year, and the land again planted with superior seedlings. No program of this sort is currently underway, but research at the Lake City Research Center over the past several years clearly points up the potential success of this revolutionary approach in future gum-naval-stores operations.

II. WOOD NAVAL STORES

Recovery of turpentine, rosin, and pine oils from the wood of old pine stumps was begun in 1909, when Homer T. Yaryan built and operated the first steam-and-solvent extraction plant at Gulfport, Mississippi. This plant was an immediate success and in its first year of operation produced 14,307 bbl of wood rosin, 1,700 bbl of turpentine, and 25,000 gal of pine oil. More than 10 such plants are currently operating in the South and in 1958 produced 1,195,990 bbl of rosin, 185,980 50-gal bbl of turpentine, 9,271,000 gal of pine oil, 1,247,000 gal of dipentene, and 4,795,000 gal of other monocyclic hydrocarbons.

Raw materials for the steam-and-solvent process are woods residues and consist of stumps and logs which have been in and on the ground for some 10 to 15 years after logging operations. While the essentially nonresinous sapwood has long since rotted away, the resin-rich heartwood is still in an excellent state of preservation. Heartwood of slash and longleaf pine stumps commonly yields as much as 25 per cent oleoresin; by comparison, that from logs seldom exceeds 10 per cent.

Stumps are removed with tractors or other specially designed mechanical devices. Blasting is seldom resorted to inasmuch as powder shatters the stumps and much of the resin-yielding wood is left in the ground. One specially devised stump puller consists of a boom crane fitted with a nutcracker jaw and mounted on a tractor of the crawler type. The jaw, when dropped over a stump, grasps it firmly at or near the ground and then pulls it from its anchorage much as a dentist would handle a troublesome bicuspid. The stumps and any suitable slash are transported to concentration centers, where they are split and trimmed, then loaded on either trucks or gondolas for shipment to an extraction plant. The larger companies work their own woods crews, who clear cutover lands on a lease arrangement with the land owners. At other plants wood is purchased from farmers upon its delivery at the yard.

The high quality of the products derived from steam-and-solvent extraction plants is the result of long years of chemical research and engineering skill, and the modern plant bears little resemblance to the original Yaryan installation. In the early days, steam-distilled wood turpentine was a strong, ill-scented distillate of uncertain physical and chemical properties. Many consumers of gum turpentine refused to use it, and others who did regarded it merely as a turpentine substitute. Wood rosin, too, was a consumer's nightmare. Although characteristically dark red, the color also varied. It was a product of questionable solubility and melting point and tended to crystallize and precipitate when blended with many petroleum and vegetable oils. When heated, it darkened appreciably and was generally considered to be inferior to gum rosin of similar grade. Pine oil in those early days was a straight-run distillate with limited use and limited markets.

Like the modern pulp mill, a steam-and-solvent extraction plant carries large wood inventories in the yard as a safeguard against the possible interruption of wood deliveries. Before the wood is suitable for treatment, it must be chipped or physically macerated in order to assure efficient extraction. Accordingly, the stumps are passed through a hog, or shredder, that reduces them to chips approximately 1½ in. long and ¼ in. in width. Better rosin recovery is possible if wood is chipped or shaved across the grain, since large numbers of resin ducts may be exposed in this way; but because of increased handling costs this is seldom done. Wood so milled is not screened, and thus slivers and wood dust usually accompany the chips to the extractor.

An extractor is a vertical steel tower 16 to 24 ft tall with a capacity of from 10 to 13 tons of shredded wood. The first phase of the extraction process is steaming. Live steam is admitted at the bottom of the vessel through sparging rings or perforated coils. The steaming phase requires from 3 to 4 hours, during which interval most of the volatile oils are removed. The condensate, which includes turpentine and other oils, is known as crude turpentine. Dry steam is then turned into the extractor in order to dry the chips before treatment with solvent. The solvent, a hot petroleum naphtha, is used to extract the residual rosin. Rosin extraction employs the counter-flow principle; hence several extractors are connected as a battery. Concentrated rosin extract removed from the last extractor is subjected to a preliminary refining operation during which oxidized resins and wood dust are removed. From this point the solvent-rosin extract is handled in diverse ways depending upon the degrees of refinement desired.

In some instances the extract is pumped under reduced pressure into evaporators, where the solvent and some pine oil are removed. The residual rosin, ruby-red in color, is then packaged for shipment and distributed as *FF* wood rosin. Some plants treat the hot liquid rosin with fuller's earth, an absorbent that removes much of the pigmentation present. In recent years, furfural, an oily liquid obtained from oat hulls, has been found very useful in the refinement of wood rosin. Solvent-rosin extract is mixed with furfural, and the two liquids are agitated. Since the two are immiscible, they separate into two layers when at rest. The furfural layer includes essentially all the pigment, and the solvent contains the clear yellowish-white rosin. Rosin obtained in this manner will yield grades from *I* to *X* and may be used in a manner similar to gum rosin of identical grades.

In a recent development, the steaming of chips has been eliminated, and turpentine, pine oils, and rosin are all extracted simultaneously with a suitable solvent. This complex is then fractionally distilled in order to separate the turpentine from various other terpenes of higher boiling point, such as dipentene, terpinene, and terpinalene, and from pine oil. The residual component in the still is crude wood rosin.

The resin-free chips in the extractor (irrespective of process used) are steamed in order to recover any solvent. This recovery, together with that from the evaporators, is used in subsequent extraction cycles. The resin-free chips are then discharged, and the extractors are recharged with fresh wood.

Attempts to use resin-free chips for kraft pulp have so far been unsuccessful. A plant at Pensacola, Florida, however, converts them into mechanical pulp, which is used to manufacture an insulation board offered to the trade under the name Temlock.

The yield of wood turpentine, pine oil, and rosin depends upon the resin content of the wood and the efficiency of the extraction process. Average recovery per ton of air-seasoned wood amounts to about 6 gal of turpentine, 1.5 gal of intermediate oil (high-boiling-point terpenes), 4.5 gal of pine oil, and 350 to 400 lb of grade *FF* wood rosin.

The refined turpentine has the same general properties as gum turpentine and is used for the same purposes. Refined rosin is now a competitor of gum rosin, grade for grade.

Pine oil is a mixture of terpene alcohols and some few terpenes of high boiling points. The production of pine oil now stands at about 5 million gallons annually. About 40 per cent of the output is used as a frothing agent in the separation of the constituents of certain heavy metal ores by flotation. Large quantities are consumed in laundries and in dyeing cotton, wool, silk, and rayon textiles. It is also employed in the manufacture of soaps, polishing compounds, paints, varnish remover, germicides, insecticides, and pharmaceutical preparations. The intermediate oils are used largely as solvents for special purposes.

III. SULFATE NAVAL STORES *

Sulfate naval stores, as previously indicated, are turpentine and rosin obtained as by-products in the manufacture of sulfate pulps. Cooks of loblolly, shortleaf, and Virginia pines yield 1.5 to 2.7 gal; those of slash and longleaf pines, 2.8 to 4.3 gal of turpentine per ton of air-dry pulp. Recoverable rosin approximates 25 lb per ton of pulp.

1. Sulfate Turpentine. In normal pulping operations entrapped air, together with steam and turpentine vapors, rises to the top of the digester above the level of the charge and develops excessive internal pressures, which stop the circulation of the cooking liquor. To prevent this situation, these vapors are periodically allowed to escape; that is, the digester is relieved

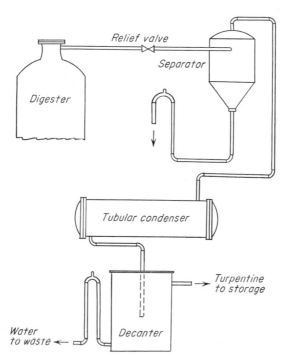

Fig. 21-14. Schematic diagram illustrating turpentine recovery in sulfate pulping process. (*Courtesy Technical Association of the Pulp and Paper Industry.*)

* See also Chap. 17, Pulp and Paper.

until desirable operating pressures are reestablished. Such vapors are often permitted to pass off into the atmosphere, or they may be captured and condensed to obtain valuable chemicals. In sulfate cooking the relief gases are directed into a cyclone separator (Fig. 21-14), where the hot vapors, small amounts of fiber, and black liquor are separated. The fiber and black liquor settle to the bottom of this tank, while the hot vapors are drawn off from the top and led into a condenser. The condensate, which flows into a decanting tank, is composed of water and an oily, vile-smelling liquid consisting of a mixture of turpentine, methyl mercaptan, methyl disulfide, pine oil, methyl alcohol, and acetone. This complex is fractionated by distillation between specific temperature limits, and the turpentine fraction thus obtained is treated with lead acetate, sodium hydroxide, sulfuric acid, ethyl diamine, or hypochlorites to remove the last vestiges of sulfur. The law requires that the refined product, which represents about 80 per cent of the original crude, must be marketed as "sulfate wood turpentine." This material currently sells for about 10 cents less per gallon than gum turpentine. The bulk of the sulfate wood turpentine recovered is used in the manufacture of paints.

Small amounts of pine oil are also recovered, but the oil is inferior to that produced by the steam-and-solvent process, and its use is largely restricted to flotation in the separation of ores and the manufacture of cleaning compounds for floors.

2. Sulfate Rosin. When chips of the southern yellow pines are cooked by the sulfate process, fatty and resin acids in the wood react with the alkaline cooking liquor to form soaps. When the spent cooking liquor is drawn off from the digester and permitted to cool, these soaps, dark brown viscous frothy masses, rise to the surface of the black liquor and are recovered by skimming. The skimmings are washed with hot water to remove any traces of suspended black liquor and then cooked with 40 per cent sulfuric acid to regenerate the fatty and resin acids. The complex thus obtained is known as *tall oil*. Refined tall oil is light yellow in color and free from obnoxious odors; it is composed of 45 to 55 per cent fatty acids, 35 to 45 per cent resin acids, 8 to 11 per cent unsaponifiables (largely phytosterols), and 0.5 to 2 per cent water. It is used in the manufacture of both hard and soft soaps, or may be processed into solid fatty acids that are suitable substitutes for the more expensive vegetable and animal fats. A product similar to linseed oil is produced by esterification with glycerin.

In some mills, however, the soap is recovered from the black liquor in multiple-effect evaporators at a point in the evaporation process when the solids concentration varies from 20 to 30 per cent.

Tall oil is also further refined by fractional vacuum distillation with superheated steam. A fatty acid fraction composed largely of oleic, linoleic, and some palmitic acids is recovered and manufactured into soft soaps and emulsifying agents. The resin acid fraction crystallizes, and the crystals, largely abietic acid, are removed by centrifuging. This is sulfate rosin. It is similar to gum rosin in most respects, but has a slightly higher melting point.

A total of 269,270 bbl (520 lb net) was produced in 1958. While sulfate rosin is a comparatively new naval-stores product, its potential is fully recognized, and it is expected that the rate of production will rise sharply in the years ahead.

IV. NAVAL STORES ABROAD

The most significant change in the world's production of naval-stores commodities has been the decline in the American output since 1907. In that

Table 21-3. World Production of Turpentine, Thousands of Gallons

Country	1946	1948	1950	1952	1954	1956	1957
U.S.A.	28,499	32,987	34,663	28,334	30,890	32,246	31,441
Portugal	3,100	3,033	3,062	2,373	2,822	2,932	3,717
France	3,504	4,376	4,632	5,084	4,500 *a*	3,199	3,766
Spain	2,900 *a*	3,062 *a*	2,755	2,671	2,572	2,082	2,694
Mexico	1,715	1,470	1,653	500	1,745	2,119	2,602
Greece	429	510	5,113	6,870	1,519	1,030	927
Japan	863	442	1,019	. . . . .	321	244	214
Honduras	. . . . .	. . . . .	. . . . .	. . . . .	. . . . .	35	5
Sweden	. . . . .	2,911	2,970	. . . . .	500 *a*	801	1,545
Norway	. . . . .	. . . . .	153	. . . . .	42	67	70
India	. . . . .	. . . . .	. . . . .	. . . . .	. . . . .	456	460
China	. . . . .	. . . . .	. . . . .	. . . . .	. . . . .	1,060 *a*	2,000 *a*
U.S.S.R.	. . . . .	. . . . .	. . . . .	. . . . .	. . . . .	9,242 *a*	9,000 *a*
Poland	. . . . .	. . . . .	. . . . .	. . . . .	. . . . .	6,108 *a*	6,000 *a*
Bulgaria	. . . . .	. . . . .	. . . . .	. . . . .	. . . . .	45 *a*	50 *a*

a Estimated.

Table 21-4. World Production of Rosin, Thousands of Pounds

Country	1946	1948	1950	1952	1954	1956	1957
U.S.A.	894,529	1,090,014	1,078,652	942,046	972,436	919,781	969,888
Portugal	88,186	87,375	94,798	69,022	51,281	113,625	116,543
France	94,086	115,581	123,457	128,965	125,000 *a*	84,656	98,766
Spain	70,000 *a*	75,000 *a*	72,752	63,616	64,815	59,524	67,901
Mexico	50,926	43,651	52,469	49,495	54,495	77,161	79,365
Greece	12,963	14,792	153,403	206,104	44,360	58,274	57,319
Japan	1,583	3,543	12,742	8,000	7,716	5,996	4,409
Cuba	100 *a*	100 *a*	100 *a*	100 *a*	100 *a*		
Honduras	600 *a*	600 *a*	600 *a*	600 *a*	974	1,480	1,381
Sweden	. . . . .	5,728	6,614	. . . . .	8,118	7,800	8,100
Norway	. . . . .	. . . . .	331	. . . . .	2,883	2,819	2,910
India	. . . . .	. . . . .	. . . . .	. . . . .	. . . . .	20,782	21,100
China	. . . . .	. . . . .	. . . . .	. . . . .	. . . . .	220,460 *a*	230,460 *a*
U.S.S.R.	. . . . .	. . . . .	. . . . .	. . . . .	. . . . .	35,982 *a*	33,900 *a*
Poland	. . . . .	. . . . .	. . . . .	. . . . .	. . . . .	28,007 *a*	27,500 *a*
Bulgaria	. . . . .	. . . . .	. . . . .	. . . . .	. . . . .	110 *a*	200 *a*

a Estimated.

year the United States supplied 79 per cent of all consumers' needs. Since then production elsewhere has risen steadily, and currently the United States is producing only about 50 per cent of the world's requirements for turpentine and rosin. Tables 21-3 and 21-4 delineate world-wide production trends for a 12-year period beginning in 1946.

SELECTED REFERENCES

1. Anon. Bark-chipped, Acid-treated Turpentine Butts Found Suitable for Pulping. *Jour. Forestry,* **48**(2) :99. 1950.
2. Anon. Gum Farming: A Cash Crop. *Fla. Forest Serv. Cir.* 8. 1952.
3. Anon. Marketing Gum Naval Stores. *U.S. Dept. Agr. Photog. Ser.* 34. 1959.
4. Anon. The 1958 Naval Stores Conservation Program Bulletin. *U.S. Dept. Agr.* NSCP-2201. 1957.
5. Anon. Turpentine Still Buildings and Equipment. *U.S. Dept. Agr., Misc. Pub.* 387. 1940.
6. Clement, R. W. Correct Use of the Spray Gun. *Naval Stores Rev.,* **62**(51) :16–17. 1953.
7. Clement, R. W. Double-headed Nails for Attaching Naval Stores Tins. *Naval Stores Rev.,* **65**(10) :15. 1956.
8. Clement, R. W. How to Install Spiral Gutters with Double-headed Nails. *AT-FA Jour.,* **16**(2) :4. 1953.
9. Clement, R. W. New Hammer Tools for Raising Tins. *Naval Stores Rev.,* **67**(8) :10–11. 1957.
10. Clement, R. W. New Spray Puller for Turpentining and How to Use It. *Southeast. Forest Expt. Sta., Sta. Paper* 77. 1957.
11. Clement, R. W. Southern Lumber Industry Profiting from Modern Gum Stores Extraction Methods. *South. Lumberman,* **195**(2441) :113–114. 1957.
12. Clement, R. W. The Bark Hack. *Naval Stores Rev.,* **62**(42) :12–13, 27–28. 1953.
13. Clement, R. W., and D. N. Collins. Larger Turpentine Cups Prove More Efficient without Effect on Product Yield or Grade. *Naval Stores Rev.,* **60**(13) :16–18. 1950.
14. Dorman, K. W. High Yielding Turpentine Orchards—A Future Possibility. *Chemurg. Digest,* **4**(18). 1945.
15. Dyer, C. D. Naval Stores Production. *Ga. Agr. Ext. Serv., Bull.* 593. 1955.
16. Everad, W. P. Modern Turpentine Practices. *U.S. Dept. Agr. Farmers' Bull.* 1984. 1947.
17. Gamble, T. Naval Stores History, Production and Consumption. Review Printing Company, Savannah, Ga. 1920.
18. Gruschow, G. F. Acid-treated Turpentine Butts Yield Quality Saw Timber. Reprint from July issue *South. Lumber Jour.,* **54**(7) :84–85. 1950.
19. Larson, R. L. Three-quarter-inch Chipping Proves Best. *AT-FA Jour.,* **16**(8) :13, 16. 1954.
20. McCulley, R. D. Management of Natural Slash Pine Stands in the Flatwoods of South Georgia and North Florida. *U.S. Dept. Agr. Cir.* 845. 1950.
21. McGregor, W. H. D. Flatwoods Farm Woodlot Improvement Pays. *Fla. Agr. Ext. Serv., Cir.* 125. 1954.
22. Mergen, F., and R. M. Echols. Number and Size of Radial Resin Ducts in Slash Pine. *Science,* **121**(3139) :306–307. 1955.
23. Mergen, F., P. E. Hoekstra, and R. M. Echols. Genetic Control of Oleoresin Yield and Viscosity in Slash Pine. *Forest Sci.,* **1**(1) :19–30. 1955.
24. Ostrom, C. E., and K. W. Dorman. Gum Naval Stores Industry. *Chemurg. Digest Reprint Ser.* 29. 1945.

25. Pomeroy, K. B., and R. W. Cooper. Slash Pine: A Tree of Many Uses. *U.S. Dept. Agr. Farmers' Bull.* 2103. 1956.

26. Ryberg, M. E., and H. W. Burney. Research in Equipment for the Production of Gum Naval Stores. *Fla. Engin. and Indus. Expt. Sta., Tech. Paper* 32. 1949.

27. Ryberg, M. E., and H. W. Burney. An Acid Spraying Puller for Naval Stores Production. *Fla. Engin. Soc. Jour.,* 5(3):15–27. 1951.

28. Schopmeyer, C. S. Acid-treatment: The Chippers Helper. *AT-FA Jour.,* 16(8):10. 1954.

29. Schopmeyer, C. S. Effects of Turpentining on Growth of Slash Pine: First Year Results. *Forest Sci.,* 1(2):83–87. 1955.

30. Schopmeyer, C. S. Gum Naval Stores Industry, Present and Future. *Naval Stores Rev.,* 64(5):8–9, 20–22. 1954.

31. Schopmeyer, C. S. Naval Stores Orchards for Future Gum Production. *AT-FA Jour.,* 19(2):12–13, 17. 1956.

32. Schopmeyer, C. S., and P. R. Larson. Effect of Diameter, Crown Ratio, and Growth Rate on Gum Yields of Slash and Longleaf Pine. *Jour. Forest.,* 53(11):822–826. 1955.

33. Shirley, A. R. Working Trees for Naval Stores. *Ga. Agr. Ext. Serv., Bull.* 532. 1946.

34. Snow, A. G., Jr. Effect of Sulphuric Acid on Gum Yields from Slash and Longleaf Pines. *Southeast. Forest Expt. Sta., Tech. Note* 68. 1948.

35. Snow, A. G., Jr. Turpentining and Poles. *South. Lumberman,* 177(2225):276, 278–279. 1948.

CHAPTER 22

MAPLE SIRUP AND SUGAR

The production of maple sirup and sugar is a typically North American industry, confined to the northeastern regions of the United States and to the provinces of Quebec and Ontario in Canada.

In the United States the importance of the industry rests in the fact that it is largely a farm activity. Sugar bushes are owned and operated mostly by farmers, who have come to rely on maple products as a source of small but regular income from their woodlots.

The six states leading in production of maple products are Vermont, New York, Pennsylvania, Michigan, Ohio, and New Hampshire. Vermont and New York account for more than 70 per cent of the total domestic output. In Canada more than 70 per cent of the production is in the province of Quebec.

In colonial days, because of the prohibitive price of cane sugar, maple sugar and sirup were classed as staple products. Most of the maple sugar was for home use, though considerable quantities of it were also used to adulterate cane sugar. This remained true until about 1875, when lower prices of cane sugar placed it within the reach of almost everyone. Since then the relative position of maple products and cane sugar has been reversed, and maple sugar and sirup have become articles of luxury, valued for their distinctive flavor. Prior to 1900 more maple sugar than sirup was made. Today, however, as the demand for sirup is increasing, more than 90 per cent of the total production in this country leaves the farms in the form of sirup.

The average number of trees tapped has been steadily declining from more than 17 million in 1918 to about 6 million since 1955. The production of maple sirup declined from more than 4 million gallons in 1918 to a 10-year average, 1948 through 1958 inclusive, of 1,675,000 gal. Maple-sugar production for the corresponding period decreased from 11.3 million pounds to less than 200,000 pounds.

Even though maple-sirup production is recognized as one of the oldest North American industries, relatively little scientific work had been done to improve it until a few years ago. Today a strong research program concerned with methods of maple-sirup production and with the basic questions of sap flow and the chemical composition of maple sap is being conducted by the

466

U.S. Department of Agriculture and the various northeastern agricultural experiment stations.

Surveys conducted in New York State, Ohio, Michigan, and Wisconsin have shown that production of maple sirup is one of the most profitable farm operations, returning as much as $3 an hour for all time spent in the operation, including cleaning the equipment. Moreover, since the maple season is short and comes in the early spring, it does not compete with other farm activities.

Modernization in the methods of collecting and processing sap has done much to increase profits, by reducing labor costs and contributing to the production of better grades of sirup. Lower income may be expected from groves with fewer than 500 buckets because fixed costs normally represent about 35 per cent of the total cost.

Since 1940 direct sale of the sirup to the consumer has been increasing, resulting in higher returns to the producer. Profits have been further augmented by conversion of sirup into maple confectionery.

I. WOODS OPERATIONS

A. Species of Maple Tapped

Although all native maples have sweet sap, practically the entire commercial production of maple products comes from the sugar maple and the black maple. The red, silver, and bigleaf maples are occasionally tapped locally, but it is generally felt that the amount of sap they yield and its low sugar content will not allow profitable commercial production. A comparison between the average sap flow of sugar maple and soft maple in Somerset County, Pennsylvania, was made by McIntyre (Table 22-1). He also re-

Table 22-1. Average Flow of Sap in Sugar and Soft Maple [a]

Diameter class, in.	Average number of buckets hung	Sugar maple		Soft maple	
		Number of trees	Aver. flow, qt	Number of trees	Aver. flow, qt
9–16	1.0	14	8.9	14	5.5
17–23	1.5	19	15.1	15	7.4
24–28	2.2	42	17.7	23	12.5
29–33	2.8	50	22.4	20	13.4
34–36	3.6	18	27.7	4	20.3
37–45	3.8	28	31.7	6	18.9

[a] McIntyre, A. C. The Maple Products Industry of Pennsylvania. *Pa. Agr. Expt. Sta. Bull.* 280. 1932.

ported that the sugar content of sap from soft maples was approximately 34 per cent less than that of the sugar maple, and the sirup and sugar were darker.

B. Sugar Bush

Woodlands containing a sufficient number of sugar maple trees for commercial tapping are commonly designated as a *sugar bush* or *sugar grove* (Fig. 22-1). At least three types of sugar groves can be differentiated: (1) those that are merely a part of a mature mixed-hardwood forest with a large number of sugar maples in the mixture, (2) more or less open and usually overgrazed groves consisting of old and frequently overmatured maple trees remaining from the original forest, and (3) second-growth mixed-hardwood

Fig. 22-1. Sugar bush. (*Courtesy Vermont Department of Agriculture, Montpelier, Vermont.*)

stands, from which a sugar bush can be easily developed by judicious thinning and liberal cutting, working toward the elimination of species other than hard maple.

Since the yield of sap from the maple tree is in direct relation to the size of its crown,* the ideal grove is considered to be one that contains the maximum number of trees per acre, consistent with full crown development. Other requirements of a good sugar bush include provision for ample reproduction to take the place of overmature or diseased trees and to provide adequate shade for the forest floor. When possible, grazing should not be

* In a study conducted at the Pennsylvania State Forest School, it was found that the average sugar content of the sap from the roadside trees with large, spreading crowns was 3.72 ± 0.047 per cent, as compared with that of the open-grown trees (crown length averaging 75 per cent of the total height) of 3.34 ± 0.034 per cent and the forest-grown trees (crown length averaging less than 50 per cent of the total height) of only 2.25 ± 0.21 per cent.

allowed in the grove, since cattle not only destroy the reproduction but also tend to harden the soil by compacting it. For maximum returns the grove should contain at least 500 trees of tappable size, i.e., 10 in. or more dbh.

Sap flow begins from the middle of February in the southern section of the commercial range and the middle of March toward its northern limits. It lasts an average of 4 to 6 weeks, although under extremely favorable conditions it may continue for 8 to 9 weeks. The length of the season depends entirely upon weather conditions. Ideal "sugar weather" consists of alternate freezing and thawing, as represented by warm days when the temperature may rise to 40 or 50°F and cold nights with temperatures below freezing. The run of sap may stop if weather conditions are unfavorable and begin again when suitable conditions prevail. During a season there are usually several such runs of a few days' duration each. Michigan and New York issue weather forecasts of suitable tapping weather. A similar service is contemplated by other states.

C. Maple Sap

Unfermented maple sap consists of 90 to 98 per cent water and 2 to 10 per cent solids, roughly 96 per cent of which is sucrose. When expressed in terms of the total sugar content, sucrose represents 99.95 per cent of the total, the remainder being raffinose and three unidentified oligosaccharides.* The sap also contains a large number of nonvolatile organic acids, of which malic acid exceeds all others by 10 times. It is thought that one or more of these acids may be responsible for the formation of "maple flavor."

The ash content of maple sap accounts for only 1 per cent of the dry solids. The minerals contribute to the astringency of the sirup, while calcium in the form of calcium malate forms the hard scale on the evaporating pans, known as *sugar sand*.

Maple sap at the time of flow is a sterile, crystal-clear liquid, with a sweet taste. Neither the brown color nor the typical "maple flavor" associated with the sirup is present in the sap. This can be demonstrated by freeze-drying aseptically collected sap. The resulting solids are white and sweet tasting but without any perceptible flavor.[8] The typical brown color and maple flavor are due to chemical reactions, brought about by heat, the exact nature of which are not known.

The finished sirup frequently contains small amounts of invert hexose sugars not found in the unfermented sap. The amount of these invert sugars is directly proportional to the extent of fermentation that has occurred. It has also been shown that up to the point of caramelization the depth of color formed is proportioned to the amount of flavor formed. This relationship no longer holds when the "caramel" flavor begins to be noticeable.

Toward the end of the season sap frequently acquires a greenish color and a peculiar taste. This colored sap, known as *buddy* sap, is attributed to physiological changes within the tree connected with the swelling of buds;

* Oligosaccharides are carbohydrates decomposable into a few (two to six) monosaccharide molecules.

the tendency for microorganisms to multiply rapidly at higher temperatures may also contribute to its formation. Other discolorations in the maple sap, known as *red sap* and *milky sap*, also called *sour sap*, are likewise caused by bacterial action. The souring of maple sap may be largely eliminated by washing the spouts and buckets clean frequently with hot water, collecting sap often and boiling it as soon as possible, and boring new tap holes if the old ones become infected.

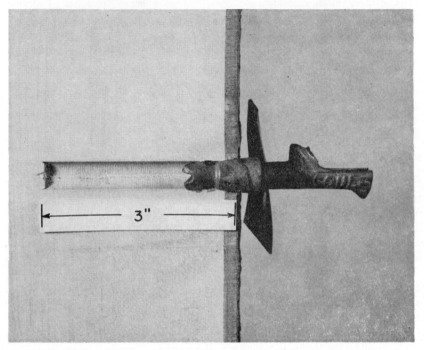

Fig. 22-2. The tap hole is bored to depth of 3 in. (*From Maple-Sirup Producers Manual, U.S. Department of Agriculture, Handbook No. 134.*)

D. Tapping

Tapping consists of drilling holes into the wood with a $\frac{3}{8}$ in. or $\frac{7}{16}$ in. wood bit. This can be done either with a carpenter's brace or with a breast drill. The former is slower but less fatiguing than the latter. For large operations, the time spent in tapping may be reduced considerably by using a power tapping machine, weighing approximately 30 lb. These machines are operated by a lightweight gasoline engine which drives a tapping bit inserted at the end of a flexible shaft. With this machine one man can tap as many as 3,500 holes in 12 hours.[5] The holes are drilled perpendicular to the tree or with a slight upward slant to facilitate drainage of the sap.

Studies at Michigan State University have shown that a tap hole 3 in. deep will produce up to 25 per cent more sap than one 2 in. deep (Fig. 22-2). Since, however, most of the sap flow is from the sapwood portion of the

stem, drilling holes deeper than the total width of the sapwood is of questionable value.

The first tap holes should be located 2 to 3 ft above the ground. The new
holes in successive years should be placed 6 to 8 in. from the previous year's
tap hole, working the tree in a spiral pattern until breast height is reached.
The compass location of the holes appears to be unimportant. It has been
shown that the average yield of sap and the weight of sugar contained in it
for the entire season are about the same regardless of the compass direction
in which a bucket is hung.[7]

In the past it was common practice after a tap hole "dried up," usually
after 3 or 4 weeks of operation, to "freshen" it by reaming it $\frac{1}{8}$ in. larger.
This was done on the assumption that the stoppage of sap flow was due to
the air-drying of wood tissue and hence could be remedied by exposing
fresh tissue. Research has shown, however, that drying up is caused by
growth of microorganisms in the tap holes. When microbial growth reaches
a count of 1 million per cubic centimeter, the sap will cease to flow, probably
due to the physical blocking of the vessels by the microorganisms as well
as to a bacterial invasion of the living tissue via pits. Reaming is not successful in restoring the flow because it fails to sterilize the hole by removing all
the microorganisms.[2, 4, 6] Control of microbial activities with chemicals, on
the other hand, appears to offer some promise. Of a number of chemicals
tried, paraformaldehyde inserted into the tap holes was the most effective,
resulting in two- to fourfold increase in flow of sap, compared with matched,
untreated tap holes inoculated with microorganisms. Further work on the
elimination of the formaldehyde from the sap is needed before this method
can be recommended for general use.*

Murphey † reports that "each tap hole bored into a maple dries out and
kills a strip of sapwood equal to the depth of the hole, a little wider than
its diameter, and from 2 to 5 ft up and down the tree, and usually much
more above than below the hole (Fig. 22-3). This wood remains permanently
dry and a tap hole bored into it later will not yield sap." This dead wood
will start decaying unless the tap hole is grown over within a few years.
Some operators plug the tap holes with sticks or corks. This practice is no
longer recommended, as the wedging of plugs tends to inflict additional
injury to the bark and wood tissues and to interfere with the natural heal-
ing process.

Trees under 8 in. in diameter usually cannot be tapped at a profit. Trees
more than 12 in. in diameter may have two tap holes and those more than

---

* Approval for the use of paraformaldehyde pellets, containing two parts of chemical
per one million parts, was granted by the Food and Drug Administration, Feb. 20, 1962.
The effectiveness of treatment on sap yield is dependent greatly on the average seasonal
temperature; the warmer the tapping season, the greater the yield. For more detailed
information on this subject, see R. N. Costilow, P. W. Robbins, R. J. Simmons, and
C. O. Willits. A Study of the Efficiency and Practicability of Different Types of Paraformaldehyde Pellets for Controlling Microbial Growth in Maple Tree Tapholes. *Mich.
Agr. Exp. Sta. Quart.* **44**(3). 1962.

† A. C. Murphey, The Maple Production Industry of Pennsylvania. *Pa. Agr. Exp. Sta.
Bull.* 280. 1932.

20 in. three. Heavier tapping is apparently not justifiable, since the slight increase in the yield of sap is more than offset by the rapidity with which a tree is completely encircled by tap holes, with a resulting decrease in the number of years a tree can be tapped. Moreover, the danger of decay is greatly increased by excessive tapping. The practice of placing two holes

Fig. 22-3. Cross section of maple log showing stained areas caused by fungus growth in old tap holes. (*From Maple-Sirup Producers Manual, U.S. Department of Agriculture, Handbook No. 134.*)

in such a way that they will drain in one bucket is considered unwise, since it definitely impairs the health of the tree.

E. Spouts

The *spouts*, or *spiles*, are tubes inserted into tap holes to conduct the flow of sap into a bucket. They also serve as supports on which to hang the sap containers. Today all spouts are made of metal. A good spout should be round in cross section, of even taper, and smooth, so that it can be inserted or removed easily without inflicting injury to the surrounding tissue. It should lend itself to easy cleaning and be strong enough to support a sap container when it is full.

F. Sap Containers

Formerly wooden buckets were universally used to catch sap. They were heavy, difficult to clean and tended to sour. Gradually they were replaced by metal buckets; of these aluminum pails are the best.

All buckets should be provided with a cover to keep out rain and falling debris. Covers are either attached to the spout or clamped to the bucket. In no case should the cover fit tightly over the bucket because the sap tends

Fig. 22-4. The plastic bag. Since spout is completely covered, it is free from contamination. (*From Maple-Sirup Handbook Producers Manual, U.S. Department of Agriculture, Handbook No. 134.*)

to sour quickly in a tightly closed container. Some operators paint the covers (or the sides of the buckets) with contrasting colors, turning the covers when the buckets are emptied. This procedure helps to keep an easy check on the containers which need emptying.

A recent development in sap containers is the use of plastic bags [8] (Fig. 22-4). Their use is too new to evaluate their performance. It is claimed that they present the following advantages: (1) because of small bulk they are easy to handle and to store; (2) they provide a self-cover; and (3) they can be easily emptied without detaching from the spout. It was also suggested that because of their transparency to sunlight radiation they may tend to keep the sap sterile. The possible disadvantages are their inadequate size for a day's run, their susceptibility to damage by rodents, and problems encountered in washing them.

G. Gathering the Sap

Collection of sap is the most expensive operation, frequently accounting for one-third or more of the total cost of sirup production.

When only a few trees are tapped, the sap may be carried away in galvanized iron gathering pails, with 15- to 20-qt capacity. In most sugar groves, however, some means of transporting the sap is required. This is usually done with metal tanks mounted on wagons or sleds and equipped with baffles to prevent splashing, a strainer, and a drainpipe.

Fig. 22-5. Maple-sap-gathering pipe line. (*Courtesy Vermont Department of Agriculture, Montpelier, Vermont.*)

To eliminate costly hand labor of sap collection, some operators have installed pipe-line systems. In some of these systems, the ends of the lines are provided with funnels into which gathering pails are emptied, allowing the sap to flow by gravity to the storage tank near the boiling house. In more recent years the sap has been piped directly from the tap holes to a storage tank. The earlier pipe lines involved the use of suspended tin tubing (Fig. 22-5). This not only involved an initial high cost but was subject to easy damage. The advent of plastic tubing made such pipe lines more practical,[8] since plastic tubing will not burst on freezing and hence can be placed directly on the ground. It is claimed that the cost of plastic tubing is comparable to the cost of new buckets and bags, that no sap is lost by spoilage or by freezing, and that the cost of collecting sap is materially reduced. Because of its newness no reliable information is available on the life expectancy of such tubing, sterilization problems, or possible effect on flavor of the sirup

Although best suited to locations with a favorable slope, such piping systems could be used on level ground or, with the aid of a pump, even at a level below the boiling house.

II. PROCESSING OF SIRUP

A. Boiling Houses

When only a few trees are worked and the yield of sap is small, the sap may be boiled on the kitchen stove, in a shed oven, or over an open fire. However, where 200 or more trees are tapped, it is necessary to have a separate building to house the evaporating equipment. In selecting a location for the sugar camp, particular attention should be paid to an adequate supply of good water. The building should be located, if possible, so that the sap can be handled from the gathering tanks by gravity.

The size of the house is determined by the size of the evaporating equipment, which in turn is determined by the number of trees to be tapped. The house should have sides high enough to ensure good ventilation, and a well-slanted, gabled wooden roof is considered more satisfactory than a metal one because it causes less condensation. A steam hood, made of galvanized iron or aluminum, or a steam vent pipe attached to the evaporator's cover should be provided to carry away excess steam.

The equipment of a boiling house includes *evaporators* and *sugaring-off pans*. The simplest type of evaporating equipment consists of a homemade sheet-iron or stonework structure known as an *arch*. As described in *U.S. Department of Agriculture Farmers' Bulletin* 1366,* the arch is: "3 or 4 ft. wide and from 8 to 15 ft. long inside measurements. The sides are usually 2½ to 3 ft. high. Between these two walls the bottom is bricked or cemented, and the walls are held together and in place by iron stays, which also support the pans. At the back the arch ends in a flue and a brick or metal chimney extending through the roof. In the front grate bars may be provided and an excavation made for an ashpit." Usually two pans are used; when partially concentrated, the sap is transferred into the second pan for finishing.

A number of well-constructed, patented evaporators are now available on the market (Fig. 22-6); these come with metal, brick-lined arches. The pans are commonly made of sheet iron coated with tin, galvanized iron, or copper. Copper pans, although they cost twice as much as galvanized iron, will last much longer and will generally produce a lighter colored sirup. The pans are partitioned to give a zigzag flow to the sap. In the modern evaporators the sap enters automatically at one end and progresses through the pan with a speed regulated by the heat under the pan. The fluid is drawn off continuously at the other end of the pan either in the form of finished sirup of a standard weight or as a partially condensed sap, which is brought to the standard concentration in the sugaring-off pans.

Where steam boilers are available, evaporation by means of steam is

* Bryan, A. H., W. F. Hubbard, and S. F. Sherwood. Production of Maple Sirup and Sugar, *U.S. Dept. Agr. Farmers' Bull.* 1366. Revised 1937.

recommended. It is claimed that if the condensed steam is returned to the boiler while hot, a considerable saving in fuel and labor will result. Other advantages claimed for steam evaporation are the uniformity of flavor and good color of the resulting products.

The rapid, atmospheric-tube-type evaporator designed to be operated with high-pressure steam is a new development.[7] This evaporator allows the sap, previously concentrated to a density of about 20° Brix,* to be converted to sirup of standard density in a few seconds.

Fig. 22-6. Wood-fired evaporator with covered top. Steam escapes through steam stack. (*Courtesy U.S. Department of Agriculture, Office of Information.*)

The size of evaporators is governed by the amount of sap handled. The patented evaporators come in sizes from 2 by 7 ft up to 6 by 24 ft. It is usually considered that the modern evaporator will evaporate about 2 gal of sap an hour for each square foot of bottom. Another rough estimate is that about 10 sq ft of bottom is necessary for every 100 buckets set out in the orchard. It takes 1 cord of wood to produce 8 to 10 gal of sirup.

B. Sugaring-off Equipment

The evaporating pans used for further concentration of sap, especially in the manufacture of maple sugar, are called sugaring-off pans. These are much shorter but deeper pans than the evaporators, set on separate arches

* See p. 477.

and provided with faucets for drawing off the sirup. The usual sizes of patented sugaring-off pans range from 2 to $2\frac{1}{2}$ ft in width, 3 to 6 ft in length, and 12 to 14 in. in depth.

C. The Boiling Process

The best grades of sirup and sugar, in terms of good flavor and light color, are obtained when the sap is boiled quickly but at as low a temperature as possible and when a batch of sap is reduced to the finished product without the addition of fresh sap. Long boiling and high temperatures tend to decompose the sugar and destroy the flavor of the sirup. Rapid concentration can be best accomplished when the depth of sap in the evaporator does not exceed $1\frac{1}{2}$ in.

When the sap is heated, scum forms on the surface; this should be continuously removed. As the concentration of the sap increases, its mineral contents begin to precipitate in the form of a fine sediment, variously known as *sugar sand, niter,* or *silica.* The presence of sugar sand in the sirup, although not harmful, detracts from the appearance of the product. The sugar sand can be removed by filtering the sirup through flannel before boiling is completed. After filtering, the sap is reboiled to the standard concentration. White of egg, whole milk, butter, or lard can also be used to remove sugar sand. These substances cause flocculation of suspended material, which can then be removed by skimming.

According to regulations of the Federal Food and Drug Act, all grades of table maple sirup, irrespective of color, must meet density specifications.[8] These require a density of 11 lb per gal of 231 cu in. of sirup at a temperature of 68°F, corresponding to 65.46° Brix * or 35.27° Baumé.* Determination of the density of sirup by weight is not recommended because it is subject to considerable variation unless exact temperature conditions are observed and precision instruments are used. The Brix scale is the easiest to use and is most accurate under field conditions.

It has been found that when a maple-sugar solution has been evaporated to the concentration of standard-density sirup (65.46 per cent of sugar or 65.46° Brix) its boiling point will be 6.85°F above the boiling point of water at a given barometric pressure. To use the information properly, it is imperative, therefore, to ascertain the temperature of boiling water on the day and the place of sirup making. It is not safe to assume the boiling point of water to be 212°F, since even relatively small changes in the barometric pressure may cause considerable depression or elevation of the boiling point and corresponding difference in the solids content of the sirup.

The color of maple sirup varies from light straw to a very dark reddish

* The Brix hydrometer scale relates the density of sirup to that of a sugar solution of known sugar percentage; i.e., 100 lb of sirup at 65.46° Brix will contain 65.46 lb of sugar.

The Baumé hydrometer values relate the weight of a unit volume of the solution being tested to some other liquid used as a standard. Since the Baumé scale does not express directly the solids content of a solution, its continued use for determining the density of maple sirup is not recommended.

brown. In most states sirup is graded on the basis of color as *fancy* or *grade AA, grade A, grade B,* and *grade C,* fancy being the lightest. The U.S. Department of Agriculture has developed a set of colored glass standards for the color grading of sirup. A darker colored sirup results from the growth of microorganisms, which cause increased alkalinity and inversion of sugar. Lighter colored sirup, therefore, results when equipment and sap are kept free of contamination. But, as has been stated, regardless of the color grade, all sirup must meet standard density requirements of 65.46° Brix at 68°F.

After the sirup has been properly strained, it is put into various containers: glass, tin cans, or, when produced on a large scale, in barrels. If the temperature of the sirup after straining is about 180°F, it can be safely packaged; otherwise it must be reheated. Since a given quantity of sirup expands and contracts with changes in temperature sirup should be packaged by weight rather than by volume. The most commonly used standard is 11 lb per gal, or 2 lb 12 oz per qt.

D. Maple Sugar

The returns from maple tapping may be increased by 20 to 160 per cent by converting sirup to sugar and maple-sugar confectionery.

Maple sugar is produced by further concentration of the sirup. The process is known as *sugaring off* and is conducted either in ordinary iron pots on the kitchen stove, or on a large scale, in sugaring-off pans. On the basis of hardness of the resulting product, maple sugar is differentiated as soft and hard sugar. Soft sugar is produced by boiling sirup until a temperature of 238 to 240°F is reached and then stirring continuously during cooling. The sugar is then put in tubs, tins, or glass containers. Hard sugar is made by boiling sirup to a higher concentration, until the thermometer reaches 242 to 245°F, and stirring it rapidly while it cools. Hard sugar is molded into cakes, the sizes ranging from 1 oz to 5 lb. The cake sugar is also called *maple concrete* and *brick sugar.* Formerly maple sugar was produced entirely on farms. The demand by commercial consumers for maple sugar of more uniform grade started a development in the industry which has largely removed the conversion of sirup to sugar, for commercial uses, from the farms to factories. Conversion factories in the United States are owned and operated almost exclusively by independent dealers.

E. Other Maple Products

Among other maple products is maple cream, made by boiling sirup to the concentration of soft maple sugar and then rapidly cooling it, stirring it vigorously all the time. A light-colored sirup boiled to the density of strained honey is sold under the name of *maple honey. Maple vinegar* is made of maple-sirup skimmings and slightly scorched or slightly soured sirup by adding small amounts of ammonium and sodium phosphate and inoculating the sirup with vinegar-making microbes.

F. Yields

A tree will produce 5 to 40 gal of sap a season, with an average of 5 to 20 gal. The sugar content of sap ranges from 0.5 to 10 per cent, with an average of 2 to 3 per cent; hence one tree will produce from 1 pt to 1 gal of sirup, or 1 to 7 lb of sugar, in a season. The average yield from year to year is about 3 pt of sirup and 3 lb of sugar. In a normal year it takes about 30 to 50 gal of sap to make 1 gal of sirup; 1 gal of sirup will produce $6\frac{1}{2}$ to 9 lb of sugar, with an average of $7\frac{1}{2}$ to 8 lb.*

SELECTED REFERENCES

1. Bell, R. D. Costs and Returns in Producing and Marketing Maple Products in New York State. Department of Agriculture, Economics, and Farm Management, Cornell University, Ithaca, N.Y. 1955.
2. Ching, T. M., and L. W. Mericle. Some Evidence of Premature Stoppage of Sugar Maple Production. *Forest Sci.*, **6**(3):270–275. 1960.
3. Marvin, J. W., and R. D. Erickson. A Statistical Evaluation of Some of the Factors Responsible for the Flow of Sap from the Sugar Maple. *Plant Physiol.*, **31**:57–61. 1956.
4. Naghski, J., and C. D. Willits. Maple Sirup. 9. Microorganisms as a Cause of Premature Stoppage of Sap Flow from Maple Tap Holes. *Appl. Microbiol.*, **3**:149–151. 1955.
5. Robbins, P. W. Production of Maple Sirup in Michigan. *Mich. Agr. Expt. Sta., Cir. Bull.* 213. 1949.
6. Sheneman, J. M., R. N. Costilow, P. W. Robbins, and J. E. Douglass. Correlation between Microbial Productions and Sap Yields from Maple Trees. *Food Res.*, **24**(1): 152–159. 1959.
7. Strolle, E. O., R. K. Eskew, and J. B. Claffey. A New Rapid Evaporator for Making High Grade Maple Sirup. *U.S. Agr. Res. Serv.* ARS-73-16. 1956.
8. Willits, C. O. Maple Sirup Producers Manual. U.S. Department of Agriculture, Handbook No. 134. 1958.

* Detailed information on the yield of sap *per tap hole* may be obtained from Robbins, P. W. The Yield of Maple Sap per Taphole, *Mich. State Univ. Quart. Bull.*, **43**(1):142–146. 1960.

CHAPTER 23

CHRISTMAS TREES

The custom of bringing green boughs into the house during religious festivals is ancient. The Egyptians, for instance, celebrated the shortest day of the year by bringing into their homes green date palms as a symbol of the triumph of life over death; in Norse mythology, green boughs symbolized the revival of Balder, the Sun God. It could, therefore, be said that the Christmas-tree custom is a heritage from the sun god religions, in which trees symbolized the revival of nature. The custom of decorating and lighting the trees apparently came from Germany. It is believed that it was first introduced in the United States by Hessian soldiers at the time of the American Revolution. Today the decorated Christmas tree has become an indispensable part of the Christmas celebration in most American homes.

It is estimated that the consumption of Christmas trees in the United States in 1957 and 1958 was in excess of 40 million.[11] Of these, about 28 million trees were cut in the United States and 12 million imported from Canada. Most of the domestic production comes from states bordering on Canada, with Montana, Washington, Minnesota, Maine, Vermont, and New Hampshire each supplying a million or more trees for out-of-state markets. Wisconsin, Michigan, New York, and Pennsylvania also cut more than a million trees each; most of them are sold locally. When divided by regions, the northeastern and Middle Atlantic states produce about 28 per cent of the total, with an equal amount coming from the Pacific Coast and Northwest states, about 25 per cent from the Lake states, and 15 per cent from the southern states; other regions contribute the balance, or about 4 per cent.

Conservatively estimated, the retail value of the harvest is about 100 million dollars, not counting boughs and wreaths.

The principal native-grown species used as Christmas trees in the East and Middle West are balsam fir, spruce, jack, Scotch, and red pines, and Douglas-fir. In the West, the principal species are Douglas-fir, true firs, and Colorado blue spruce; and in the South several species of southern pine, eastern redcedar, and plantation-grown Arizona cypress.

On a nationwide basis, A. M. Sowder [11] gives the following estimate for 1956:

480

Species	Trees	Per cent
Douglas-fir.........	7,162,342	28
Balsam............	6,006,224	24
Eastern redcedar....	3,085,754	12
Black spruce........	2,896,000	11
Scotch pine.........	1,560,000	6
White spruce.......	858,047	3
Red pine...........	802,657	3
Norway spruce......	448,302	2
White fir...........	441,251	2
Other species.......	2,598,596	13
Total............	25,369,223	100

Imports of Christmas trees from Canada have been steadily growing in number and value from about 5 million trees worth about $560,000 in 1955 to more than 12 million valued at about $6,350,000.[1] The United States accounts for 99 per cent of Canada's Christmas-tree export trade. The principal exporting provinces are Nova Scotia, Quebec, New Brunswick, British Columbia, and Ontario, in that order. In 1957 the trees were imported to 44 states.*

The principal species imported from Canada that year were balsam fir, 72 per cent; Douglas-fir, 18 per cent; spruces, 6 per cent; and Scotch pine, 4 per cent. The average wholesale value of Canadian Christmas trees has grown from 21 cents per tree in 1945 to 52 cents in 1957.

There is no accurate statistical information available on the domestic consumption of Christmas trees in Canada.[1] Estimated on the basis of one tree for each household, with the margin of error reduced by making no provisions for institutional use of Christmas trees, it is assumed that in 1957 about 4,100,000 trees were sold in Canada.

I. SOURCE OF TREES

A. Forest-grown Trees

Perhaps 90 per cent of the Christmas trees in the United States and Canada come from forest stock. In New England and New York, many trees are harvested from pasture lands. Such trees are very bushy and if allowed to grow to maturity will produce knotty lumber, but as Christmas trees they are highly desirable. In Minnesota the majority of trees come from the northern black spruce swamps. Because of unfavorable growing conditions, the trees are stunted and of considerable age; a tree 4 to 6 ft in height is frequently 60 to 100 or more years old. The wood from these exceedingly slow-growing trees is poorly suited for pulp and other wood uses, but because of their bushy appearance they make highly acceptable Christmas trees. A number

* For Christmas-tree trade practices in Canada, see Hawboldt, L. S., and G. R. Maybee. The Nova Scotia Christmas Tree Trade. *Nova Scotia Department of Lands and Forestry, Bull.* 14. 1955.

of companies in Minnesota specialize in trees 30 to 36 in. in height for the table trade. Some operators harvest such trees the year around, place them in cartons or crates, and keep them in cold storage until shipped. Trees may be painted prior to packaging. A recent survey of the black spruce swamps in Minnesota indicates that even with the present rate of harvesting of more than a million trees a year, the supply of stock suitable for Christmas trees is practically inexhaustible. In the Rocky Mountain and North Pacific states, Christmas trees, principally Douglas-fir, are either removed in improvement cuttings in national forests or produced in carload lots by private forest owners.

B. Christmas-tree Plantations

The relatively high price paid for well-shaped Christmas trees has resulted in a large increase in Christmas-tree plantations. For instance, in Michigan a rapid shift from wild to plantation trees has taken place since 1952.[7]

Stock	1952	1956
Wild.........	880,000	615,000
Plantation....	120,000	525,000
Total.......	1,000,000	1,140,000

Pennsylvania reported more than 40,000 acres established for Christmas-tree production, while Michigan nurseries reported sales of more than 40 million seedlings and transplants in 1956, and nearly 50 million in 1957, of major Christmas-tree species.* Other Northern states report similar growth in the acreage devoted to Christmas-tree planting. This suggests the possibility of sharply increased competition for available markets, with a resulting drop in prices for plantation trees. This may be particularly true in the case of Scotch and red pines, species most generally planted for Christmas-tree production. It is suggested that more emphasis should be placed on planting Douglas-fir, Colorado blue spruce, Austrian pine, and white fir, species which are now grown in relatively small numbers as Christmas trees in the North-eastern, Lake, and Central states.

1. Shearing and Shaping Christmas Trees.[3,4] With increased emphasis on tree quality resulting from growing competition in marketing Christmas trees, shearing and shaping of trees are becoming standard practices among progressive growers (Fig. 23-1).

The ideal Christmas tree should resemble an inverted cone, about two-thirds as wide at the base as it is high, tapering uniformly to the tip. This means a taper of 66⅔ per cent. However, tapers of 40 to 70 per cent in spruce and firs, and 40 to 90 per cent in pines, are considered acceptable. The purpose of shearing and shaping is to control the relationship between

* However, only some of the coniferous trees planted are later cut for Christmas trees.[7]

the height and the width by removing multiple leaders and branch deformities and cutting down on the length of leaders, and by stimulating the formation of side branches and density of foliage.

The work can be carried out with a pocketknife, sickle, machete, hand pruners, or hedge shears. The latter, with 8- to 10-in. blades, are considered the best all-around tools.

Shearing and shaping should be started when trees average 2 to 3 ft in height. The first operation includes removing all multiple stems and cor-

a   b

Fig. 23-1. Red pine Christmas-tree stock: *a*, before shearing in June, 1947; *b*, same tree in October, 1948, after two summer shearings. (*Courtesy School of Forestry, The Pennsylvania State University.*)

recting major deformities. Subsequent shearings are carried out to achieve the desired balance between the height and the width and to stimulate growth to fill in the gaps between the whorls.

Spruces and firs should be pruned when the trees are dormant, and pine species in the late spring or early summer, when the new growth is still succulent. The growth of branches can be retarded by removal of buds in the spring before new growth starts. This method is frequently employed by nurserymen but is considered too slow and too expensive for large-scale plantations.

II. HARVESTING

Harvesting of Christmas trees is done between October and mid-December, depending upon the distance to markets and the anticipated demand. Trees for out-of-state markets are frequently rough-graded by size and appearance. Several states have developed a standardized Christmas-tree classification as a basis for pricing. The recently established United States standards are discussed in the following section.

Some growers and producers attach tags to trees to indicate species, height class, and grade. In an attempt to eliminate theft of Christmas trees, several states require tags on all trees sold in the state, regardless of origin.

Trees transported long distances, especially if by rail, are usually tied in bundles of from two to a dozen, depending on the size and species. A wooden

Fig. 23-2. Bundling red pine Christmas trees. (*Courtesy U.S. Forest Service.*)

frame is generally used to hold trees while they are bundled (Fig. 23-2). The advantages of bundling are that more trees can be packed into a given space for transportation and they can be rehandled more conveniently and with less damage. Bundling also serves as a method of sorting by height and sometimes by grade.

Trees are shipped by truck, rail, and boat. Trucks are used for hauls averaging 200 miles or less. The variation in truckloads is considerable, depending on whether trees are bundled, tree size, size of the truck, and fullness of the load. This results in significant variation in freight costs. For instance, a study of Michigan and Canadian truckloads delivered to Michigan markets in 1956 showed the number of trees per truck varying from less than 100 to more than 800, with the Canadian trucks generally carrying more trees per load, with a resulting lower unit cost.[7]

Flatcars and boxcars used for longer hauls can carry 500 to 550 bundles, or from 1,000 to 5,000 trees, per car.

III. QUALITY AND GRADING STANDARDS

Although quality is extremely important in marketing Christmas trees, describing it for all species is difficult. There is no system of grading generally accepted by growers and sellers. The closest approach to it are "United States Standards for Christmas Trees," [12] which became effective November 1, 1957.

A summary of these standards is presented in Table 23-1.

Table 23-1. United States Standards for Christmas Trees [a]

Factor	U.S. premium	U.S. no. 1	U.S. no. 2
Density	Medium	Medium	Light
Taper	Normal	Normal (flaring or candlestick if tree is otherwise U.S. premium)	Normal (flaring or candlestick if tree is otherwise U.S. no. 1)
Balance	4 complete faces	3 complete faces	2 complete faces
Foliage	Fresh, clean, and healthy	Fresh, clean, and healthy	Fresh, fairly clean, and free from damage
Deformities	Not more serious than minor	Not more serious than minor (noticeable deformities permitted if tree is otherwise U.S. premium)	Not more serious than minor (noticeable deformities permitted if tree is otherwise U.S. no. 1)

[a] SOURCE: U.S. Department of Agriculture, Agriculture Marketing Service. U.S. Standards for Christmas Trees. 1957.

IV. MARKETING

The Christmas-tree industry is unique in that it is highly seasonal. The trees are harvested in a 2- or 3-month period preceding Christmas, and they are salable only once a year and generally only one to a family.

The market is dictated by consumer preference for certain kinds and sizes of trees. This has a decided bearing on local consumer sales, particularly in the years of oversupply, when a number of good trees may remain unsold because of preference for a particular kind of tree.

The sale price and profit realized vary greatly with locality, species, size, and appearance of trees, and the demand and supply. The greatest variable affecting the sale price is the stumpage price. Nationwide, this has been found to vary from a low of $0.10 for wild trees to $4 or more for well-grown plantation stock.

Distribution of the various factors affecting the final price was studied in Michigan by L. M. James.[7] The results of his study are presented in Table 23-2.

Frequently a greater profit can be realized in small towns located close to the source of trees, where the demand is supplied directly by the growers or by the retailers obtaining trees directly from the producers, who also deliver trees to the lot.

Table 23-2. Cost Items in the Average Retail Price of Christmas Trees Sold in Michigan, by Species, 1956. [a]

Cost item	Scotch pine	Balsam fir	Black spruce [b]	White spruce (wild)	White spruce (plan-tation)	Red pine	Nor-way spruce	Jack pine	Doug-las-fir
Stumpage............	$1.50	$0.70	$0.45	$0.60	$1.25	$0.60	$0.95	$0.45	$2.50
Cutting and loading [c]..	0.20	0.20	0.20	0.20	0.20	0.20	0.20	0.20	0.20
Freight [d]............	0.27	0.20	0.20	0.20	0.20	0.27	0.20	0.20	0.20
Concentrators' and wholesalers' services	0.18	0.15	0.15	0.15	0.17	0.15	0.17	0.15	0.25
Retailers' services.....	1.85	1.25	1.25	1.60	1.68	1.28	1.63	1.25	2.35
Total to consumer...	$4.00	$3.25	$2.25	$2.75	$3.50	$2.50	$3.15	$2.25	$5.50

[a] From James, L. M. Resurvey of Christmas Tree Marketing in Michigan. *Mich. Agr. Expt. Sta., Spec. Bull.* 419. 1958.

[b] Table-model trees not included.

[c] Average cost for Scotch pine and balsam fir is assumed to apply to all species.

[d] Average cost for Scotch pine is assumed to apply to red pine. Average cost for balsam fir and black and white spruce is assumed to apply to Norway spruce, jack pine, and Douglas-fir.

In the larger cities the marketing pattern is more complex and profits more uncertain. A large proportion of Christmas trees in the cities is sold to the retailers by commission agents representing one of the several large Christmas-tree companies. These companies operate both in the Eastern and the Western United States, cutting trees in Canada and the United States, and shipping them to all the larger consuming centers of the country. In general, the Illinois-Indiana line serves as a dividing line between the distribution of eastern and western trees, although occasionally trees from the Northwest are shipped as far east as Pittsburgh and even to New York City. The center of Christmas-tree consumption lies in the big cities of the East and the Middle West.

The demand for and supply of Christmas trees are difficult to regulate, particularly in the larger cities. For instance, in 1942 a total production of 800,000 trees in Michigan was barely sufficient to meet the demand. In consequence, few trees were unsold, and prices were up toward the end of the season. The following year the record production of 1,500,000 trees was too great, and at least 250,000 trees were unsold in Detroit alone. In the next year production fell back to the 1942 level, and the last carload in Detroit sold for double the price prevailing earlier in the season.[7]

Marketing of Christmas trees does not follow any regular pattern. Trees move from stump to consumer through a variety of channels. Trees may be sold by individuals who may also be producers, by wholesalers, by commission agents, and by retailers, who may be engaged in some other business such as running a gas station or operating a food or hardware store.

In a number of states, Christmas-tree growers have formed Christmas-tree growers' associations, some of which engage in cooperative merchandising of trees or perform the function of concentrating the output of a number of producers and growers for sale to the larger dealers. The National Christmas-tree Growers' Association, dedicated to furthering the over-all interest of the Christmas-tree industry, was organized at Butler, Pennsylvania, in

1955. The first convention of this association was held in August of 1950 at Purdue University.

In addition to Christmas trees, there is also an extensive market for evergreen boughs and wreaths. These may be obtained either from trees of poor form or from lapped branches. The marketing functions are frequently performed by the Christmas-tree cooperatives and result in considerable additional income.

SELECTED REFERENCES *

1. Babcock, N. M., and J. E. Nicolaiff. The Christmas Tree Industry in Canada. Department of Northern Affairs and National Resources, Forestry Branch, Misc. Pub. 10, 3d ed., Ottawa. 1958.
2. Barraclough, K. E., and H. K. Phipps. Christmas Trees: A Cash Crop. *N.H. Univ. Coop. Ext. Serv., Bull.* 123. 1954.
3. Bell, L. E. Shearing and Shaping Christmas Trees. *Mich. Univ. Coop. Ext. Serv., Bull.* 123. 1960.
4. Bramble, W. C., and W. R. Byrnes. Effect of Time of Shearing upon Adventitious Bud Formation and Shoot Growth of Red Pine Grown for Christmas Trees. *Pa. Agr. Expt. Sta., Prog. Rpt.* 91. 1953.
5. Cope, J. A., and F. E. Winch, Jr. Christmas Tree Farming. *Cornell Univ. Col. Agr. Bull.* 704. 1956.
6. Jackson, W. E. Growing Christmas Trees in Kentucky. *Ky. Univ. Coop. Ext. Serv., Cir.* 542. 1956.
7. James, L. M. Resurvey of Christmas Tree Marketing in Michigan. *Mich. Agr. Expt. Sta., Spec. Bull.* 419. 1958.
8. James, L. M. Production and Marketing of Plantation Grown Trees in Michigan. *Mich. Agr. Expt. Sta., Spec. Bull.* 423. 1959.
9. Metcalfe, W. Christmas Trees in California for 1955. Cooperative Extension Service. University of California, Berkeley, Calif. 1955.
10. Sopper, W. E., and J. H. Denniston. Production and Marketing of Christmas Trees in Central and Northeastern Pennsylvania. *Pa. Agr. Expt. Sta., Prog. Rpt.* 189. 1958.
11. Sowder, A. M. Christmas Trees: The Tradition and the Trade. *U.S. Dept. Agr., Forest Serv., Agr. Inform. Bull.* 90 (revised). 1957.
12. U.S. Department of Agriculture, Agriculture Marketing Service. U.S. Standards for Christmas Trees. 1957.

* The Christmas-tree literature is quite extensive. Information of local interest can frequently be obtained from the local state forestry agencies, the state cooperative extension services, and from the universities. A few examples are given here.

CHAPTER 24

WOOD IN THE PLASTICS INDUSTRY

Plastics are compounds or, more commonly, formulations of several ingredients which, when subjected to heat or pressure (commonly both), undergo chemical change while being cast, molded, extruded, or pressed into objects of finite form. Some finished plastic products are rigid, others are flexible, and a few exhibit elastic properties. Those in which the binding matrix of the formulation polymerizes under heat and pressure to form hard, infusible substances that can be neither remelted nor remolded at the completion of the polymerization reaction are classed as *thermosetting* plastics. However, not all thermosetting plastics are heat-resistant, and there are those which, although they do not soften, commonly char and even decompose when exposed to only moderately high levels of heat. A second class of plastics, *the thermoplastics*, are formulated with binders which, even after complete polymerization, soften when exposed to heat. Nevertheless, many thermoplastics exhibit excellent heat resistance and give satisfactory service when exposed to temperatures in excess of 200°F.

A. Compounding Ingredients

Chemically, commercial plastics are composed of several organic * compounds, each one of which plays an important role either in the forming processes or in imparting certain desirable physical or chemical properties to the finished product. The basic ingredients of plastics are classified as binders, fillers, plasticizers, solvents, lubricants, and coloring materials.

1. Binders. The binder is the principal constituent of a plastic; in fact, in some cases it may be the only substance used in the manufacture of a particular product. Most plastics are classified according to the binder employed. Thus, one made with a cellulose acetate binder is usually known as a cellulose acetate plastic. The principal binders used in the industry are

* In some plastics, inorganic materials such as asbestos, mica, or color pigments are incorporated.

thermosetting and thermoplastic synthetic resins, derivatives of cellulose, casein and soybean proteins, and to a lesser extent lignin. The important synthetic-resin binders include the phenol-, melamine-, and urea-formalde-hydes, acrylics, indenes, styrenes, and vinyl complexes. Among the notable cellulosic binders, mention should be made of cellulose nitrate, cellulose acetate, cellulose acetate-butyrate, and ethyl cellulose. Cellulose binders are prepared from either highly purified wood pulp or cotton linters.

2. Fillers. Fillers play a major role in plastic-products manufacture, particularly in the phenolic types, where they are used in quantity to increase the bulk of the molding matrix and to impart many desirable physical and chemical properties to finished articles. Fillers may be of either organic or inorganic origin, the choice of one being largely dependent upon the properties it will impart to the product.

The essential requirements of a good filler are cheapness and availability, good strength properties (particularly under impact and tension loads), low rate of moisture absorption, ease of wetting by resins and dyes, low specific gravity, and nonabrasiveness. When the specific use of the product to be manufactured is carefully considered, other desirable properties such as high dielectric strength, heat resistance, machinability, nonflammability, chemical inertness, particle size, color, and odor also become important.

Wood flour, walnut-shell flour, cotton linters, alpha-cellulose (see page 366), paper, asbestos, mica, graphite, diatomaceous silica, and glass fiber are among the filler materials most commonly used.

a. Wood-flour Filler. Wood flour, because of its availability, moderate cost, and many desirable physical properties, is the most widely used of the plastic fillers. The lighter colored woods of low density, such as white pine, basswood, cottonwood, yellow-poplar, balsam fir, and spruce, are preferred, but quantities of flour made from heavier timbers such as maple, oak, birch, and redgum are also used.*

Flour suitable for filler stock must be made in an attrition mill. Reduction by grinding is more a matter of fiberization, and the resulting particles retain much more of the strength characteristics of the original fibers than do those of flours made in ball or rod mills, where impact disintegration is employed. Particle size is, of course, variable, and the freshly ground stock must be screened to obtain a material of reasonably uniform texture. Nothing coarser than 60-mesh screenings is usually acceptable.

Moisture absorption by wood flour is higher than that by many other fillers but lower than in the case of straight cellulose fillers. Numerous substitutes have been tried, but most of them are either too heavy or fail to impart sufficient mechanical strength to molded articles.

b. Nut-shell Flours. Walnut-shell and pecan-shell flours are suitable fillers for many applications involving both cast and molded phenolic plastics. The shells are obtained from nut growers and processors and shipped to mills, where they are cleaned, deoiled, and reduced to flour. These flours contain appreciable amounts of both cutin and lignin. The cutin is some-what waxy and highly impervious to water, thus causing the fillers to resist

* See Chap. 13, Wood Flour.

absorption of the binder, although the flour itself exhibits good absorbent properties. The lignin serves as both binder and filter.

Plastics made with nut-shell filler are smooth and glossy, possess good bending and impact strength, but are only fair in tension and compression resistance. They are superior to wood-flour-filled plastics in both heat and water resistance and in dielectric properties.

3. Other Constituents of Plastics. Although not all derivatives of wood, other materials used in compounding plastics are mentioned briefly here in order that the reader may have a better appreciation of the composition of a plastic product.

a. Plasticizers. Plasticizers are multipurpose materials added to plastic formulations to increase their flow characteristics during molding processes and to render finished products less brittle or more flexible. Plasticizers are organic liquids or solids with low melting points; they must be compatible with other constituents in the formulation, yet chemically stable and essentially nonvolatile. Camphor is a nearly universal plasticizer for the cellulose nitrate type of plastics. Methyl and ethyl phthalates are most generally used with the cellulose acetate and cellulose acetate–butyrate compounds. These and many others are used with the ethyl cellulose type of plastics. A mixture of aniline and furfural is a suitable lignin plasticizer. A whole host of compounds is used to plasticize the various synthetic-resin plastics, but space does not permit their inclusion here.

b. Coloring Materials. A few finished plastic products are black or nearly so, but on the other hand, many of them are colorless or essentially so, and their general appearance and "eye appeal" may be materially improved by addition of dyes, lakes, or pigments to the molding matrix.

c. Lubricants. Materials are occasionally added to plastic mixes to increase flow during the molding operation and to prevent the cured plastic articles from sticking to the molds. Such materials are called lubricants; the most common are waxes, stearic acid, or metallic stearate salts.

d. Solvents. Several solvents, including numerous alcohols, acetone, and blends of ether alcohol, are used in the manufacture of plastics. Many of them are used as carriers for plastic coatings especially prepared for metal finishing. Cellulose nitrate dissolved in ether alcohol is the familiar pharmaceutical preparation known as "new skin." "Plasticwood" is cellulose nitrate dissolved in acetone with a wood-flour filler.

B. Lignin Plastics

Lignin, a product of metabolism of the tree, is Nature's own cementing matrix, which bonds together the cellular components in the woody stem. The chemist has been unsuccessful in his attempts to isolate this chemical complex without altering its physical and chemical properties, although he has clearly demonstrated that it possesses varying degrees of plasticity depending upon the treatment employed. This plasticity, however, becomes apparent only at very high pressures; hence attempts to use lignin as an adhesive have been unsuccessful, although it can and is used in molded plastics and in the production of densified hardboards.

A number of lignin-plastic materials have been developed beyond the pilot-plant stage. But even though suited for numerous industrial applications, they offer no special economic or physical advantages over well-entrenched competitive materials.

1. Hydroxylin. Hydroxylin is a generic term applied to lignin-plastic materials made from mill-run hardwood residues * by research chemists on the staff of the Forest Products Laboratory. Sawdust or finely divided wood is hydrolyzed with a dilute acid or the aromatic amine aniline to convert a portion of the carbohydrate constituents present to water-soluble sugars,

Fig. 24-1. A lignin plastic. (*Courtesy U.S. Forest Products Laboratory.*)

which in turn may be removed through repeated washings of the residual solids. In effect this merely increases the ratio of lignin to the cellulosic constituents in the hydrolyzed complex. Maples, redgum, oaks, hickories, and cottonwoods are species most commonly used.

a. Acid-hydrolysis Method. A typical formulation consists of a mixture composed of 100 parts by weight of finely divided dry wood residue, 250 parts of water, and 3 parts of concentrated sulfuric acid. These are placed in a rotary digester and cooked with direct steam for a period of 30 minutes at a pressure of 135 psi. The acid cooking liquor and its products of hydrolysis are then drained off, and the residual lignocellulose complex is water-washed until neutral to litmus, then dried, and finally milled to a particle size ranging from 40 to 100 mesh. The powder, which in effect is a lignin binder with cellulose filler, may be molded at 190°C with the mere addition of water. A product with better properties is obtainable, however, if a plasticizer such as a mixture of aniline and furfural is used.

* More recent work reveals that softwood residues may be similarly employed.

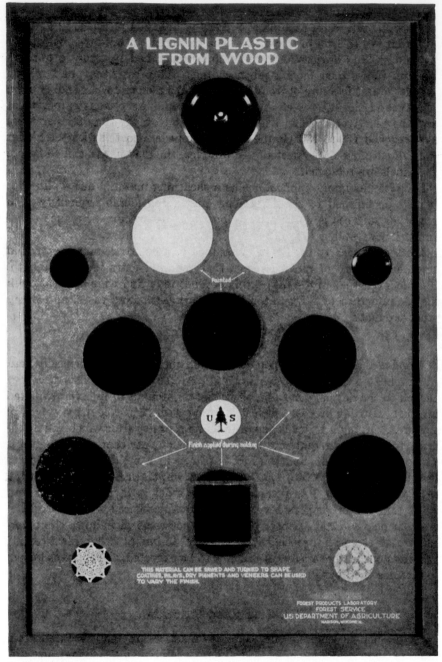

Fig. 24-2. General character of lignin-plastic material adaptable to many forms and products. (*Courtesy U.S. Forest Products Laboratory.*)

One such molding mixture consists of 100 parts of the powdered ligno-cellulose complex, 8 parts of aniline and an equal amount of furfural, 0.5 part of zinc stearate, and 2 parts of water. This mixture, when pressure-molded for 3 minutes at 150°C under 3,000 to 4,000 psi, produces a lustrous, jet-black plastic with a specific gravity of about 1.40.

b. Aniline-hydrolysis Method. As an alternate method of hydrolysis, aniline may be substituted for acid. Plastics made by this method exhibit greater mechanical superiority and better water resistance than those made by the acid-hydrolysis method. A typical furnish consists of 100 parts of dry wood residue, 100 parts of water, and 21 parts of aniline. This mixture is cooked for 3 hours with live steam at 160 psi. The spent cooking liquor is then removed, and the residual wood is washed until neutral, dried, and then milled to powder having a particle size ranging from 40 to 80 mesh.

A suitable molding formulation is composed of 100 parts of powder, 8 parts of furfural, 2 parts of water, and 0.5 part of zinc stearate. This mixture, when pressure-molded for 3 minutes at 155°C under 3,500 to 4,000 psi, gives a product similar in color and density to that made with acid-hydrolysis molding powder.

Although black in color, the hydroxylin plastics may be coated readily with enamels of nearly every hue, lacquers, paints, or finishes with incorporated metallic dusts. They are easily machined and practically indestructible under normal usage. Hydroxylin plastics have been found suitable for the manufacture of storage-battery boxes, distributor heads in the ignition systems of gasoline engines, many parts for electrical appliances, knobs, ash trays, and similar items.

While these plastics have created considerable interest, they are as yet not manufactured in any commercial quantity.

c. Hydroxylin Laminate Sheet. If wood chips similar in size to those used in the manufacture of wood pulp are used instead of finely divided wood, it is possible to produce a sheet of hydrolyzed wood suitable for the production of high-strength laminates.

Following acid hydrolysis of the chips, the residual wood is washed until neutral and then reduced to pulp in a disk mill. The pulp is then screened, and the accepted stock made into a sheet on a Fourdrinier paper machine. When dried, the sheets may be laid up and pressed at 2,000 psi and 175°C to form laminates of desired thickness.

To increase water resistance and dimensional stability and to minimize delamination, the sheets may be impregnated with a phenolic-type resin before laminating. While better physical properties are obtained in this way, the mechanical properties of the laminate are only slightly improved.

d. Hydroxylin Hardboard. In making hardboard, hydrolyzed chips are run through a disk mill or beating engine, then passed through coarse screens, and finally made into mats 1½ in. or more in thickness. The mats are placed between platens and pressed at from 300 to 500 psi in order to remove excess water. Following this, the pressure is reduced to 65 to 75 psi, and the platens heated to a temperature of 185°C. Polymerization of the lignin is complete in about 20 minutes, and a hardboard about ¼ in. thick and having a specific gravity of about 1.00 results. This hardboard is similar to other hardboards now on the market and may be used for the same purposes.

e. Other Processes. Several other methods for the production of hydroxylin were developed at the Forest Products Laboratory, but none was deemed promising enough to warrant commercial exploitation.

In one such process 100 parts of finely divided wood were treated with 3 to 10 parts by weight of chlorine. This mixture was heated until an exothermic reaction developed, at which time 10 parts of the aromatic amine aniline were added. The complex was then washed and dried. Synthetic resins, phenol, or furfural were added as a plasticizer. Upon molding, however, hydrochloric acid was liberated in sufficient concentration to etch severely the molding dies.

A second procedure, involving acetylation, consisted of heating wood with acetic acid at 100°C for 6 to 8 hours. At the end of the cycle the mixture was cooled (0 to 5°C), and acetic anhydride and an acid catalyst were added. The temperature of this complex was raised to 10 to 30°C for about 3 hours, and finally to 30 to 40°C for a similar period. The mass was then cooled, washed, and dried. While it was found that plasticizing agents were not required, the excessive production costs precluded industrial acceptance.

2. Benalite. This is a lignin-plastic hardboard manufactured by the Masonite Corporation and has the distinction of being the first American lignin-plastic material ever to be produced commercially. In the manufacture of Benalite, wood chips are placed in "guns" having a capacity of about 12 cu ft and subjected to live steam at pressures ranging up to 1,200 psi. During the short interval of from 30 to 45 seconds, acetic and formic acids developed from wood substance itself under such heat and pressure bring about a rapid hydrolysis of the hemicellulosic constituents present. The gun is then quickly opened, and upon the sudden return to normal atmospheric conditions, the hydrolyzed chips with their highly developed internal pressure explode into fiber masses and bundles. The products of hydrolysis are then removed, and the residual wood, consisting largely of reactivated lignin and hydrated but otherwise unchanged cellulose, is ready for rebonding into panel form.

In panel manufacture the fiber masses are formed into mats and placed in multiplaten presses capable of producing boards 4 ft wide, 12 ft long, and from $\frac{1}{10}$ to 1 in. in thickness. The press platens are fitted with heating and chilling devices and are carefully machined to assure a uniform pressure across their faces. When charged, the presses may be adjusted to exert a unit pressure of from 50 to 1,500 psi or more, depending upon density and thickness of the panel required. As platen temperatures rise, the reactivated lignin begins to flow, and by the time temperatures have reached 175 to 180°C, the fibers have been recoated with the binder and full cure has been effected. In practice, the platens are chilled before the panels are removed.

Benalite has many applications as a sheathing material and wallboard. It may be finished to simulate tiling, and a considerable quantity with enameled surfaces is used as a wallboard in bathrooms. Because of its dimensional stability it is also suitable for template stock.

Most of the early uses of the resinboard were for reinforcing around lugs and sections of molded pieces that require extra impact strength. It is believed that in the future this product may find wide application in the manufacture of refrigerator doors, sewing-machine cases, and many other products.

3. Marathon Lignin-enriched Filler. This product, commonly referred to as L.E.F., is a hydrolyzed-wood molding powder produced by the Marathon Paper Mills Company. In its manufacture hardwood chips are placed in a rotary digester with neutralized spent sulfite liquor from which the vanillin has been extracted and cooked for 30 minutes under 250 psi of steam pressure. During this period the hemicelluloses are hydrolyzed to soluble sugars, and the lignin in the liquor precipitates on the residual fibers as the cook approaches an acid state. The lignin content of the residual cellulosic constituents is thus enriched by as much as 20 to 35 per cent. After washing and drying, the molding powder may be mixed with 33 per cent of Vinsol resin, a plasticizing agent, and molded into products that exhibit both excellent strength properties and water resistance.

4. Hydrolyzed Redwood Plastics. One of the newer members of the lignin-plastic family is a redwood molding powder that requires no plasticizing additives in the production of finished goods. This material is prepared by cooking redwood chips in an enclosed vessel for 8 to 15 seconds with high-pressure steam ranging from 500 to 800 psi. Unlike in the production of hydroxylin or L.E.F., however, the water is removed by evaporation, and the products of hydrolysis and the extractives are retained. Among the latter are tannins and phlobaphenes, which under heat and pressure exhibit plastic and resin-forming properties, and which in the molding operation serve as binding agents. Additional binding resins may be incorporated if desired, but none is required. Formaldehyde has been found to improve the thermosetting properties of the powder, but its incorporation, too, is a matter of choice.

C. Urea-plasticized Wood

Wood soaked in an aqueous solution of urea and then dried may be easily bent when reheated to temperatures ranging from 212 to 220°F. This discovery has led to the development of a plastic material made of urea-impregnated sawdust. In this process sawdust is mixed with 25 per cent by weight of dry urea. Enough water is added to make a paste, after which the mix is dried to 2 or 3 per cent moisture content. From 1 to 5 per cent zinc stearate is added as a lubricant, and the mix is placed in a mold and subjected to pressures of from 1,000 to 1,500 psi and temperatures ranging from 180 to 185°C for from 3 to 5 minutes. The mold is then cooled to about 60°C, and the product removed. This urea-plasticized wood material is gray to nearly black, noncorrosive, and possesses machining characteristics and strength properties comparable to many other plastics. It exhibits poor water resistance, which is to be expected, since urea is known to increase the hydroscopicity of wood.

D. Cellulose Plastics

As previously pointed out, cellulose plastics are manufactured from esters, such as cellulose nitrate, cellulose acetate, and cellulose acetate–butyrate, and from the ether, ethyl cellulose. The bulk of these plastic materials is made from cotton linters, although highly purified wood cellulose has been used in ever-increasing quantities during the past few years. All these plastics are of the thermoplastic type.

1. Cellulose Nitrate Plastics. Celluloid, a cellulose nitrate derivative, was the first successful plastic material ever to be manufactured. Cellulose nitrate plastics are prepared by treating cotton linters with nitric acid in the presence of sulfuric acid, which acts as a catalyst. The Germans have developed wood-pulp nitrocellulose, but in the United States, linters are still preferred. Recently "dense cellulose," partially nitrated wood pulp in sheet form, has found favor with several manufacturers.

In processing, the nitrocellulose, of whatever origin, is dehydrated and plasticized with camphor in the presence of alcohol until a stiff, dough-like mass is formed. At this point the mixture is filtered and dyed or pigmented as required and finally rolled into sheets. The sheets are usually laid up and pressed into billets or blocks in order to facilitate their handling. These contain appreciable quantities of water and alcohol and must be seasoned before they are suitable for marketing. Seasoning rooms are usually equipped with temperature and humidity controls in order that the process may progress at a slow and uniform rate.

No cellulose nitrate compounds are currently available. Rather, the resins are sold to the consumer and fabricators in the form of sheets, tubes, rods, bars, films, and emulsions. They are available in an almost unlimited range of colors and are among the cheapest of all plastic materials produced.

Cellulose nitrate plastics are inflammable and are not suited for applications where open flames or high heat levels are unavoidable. They also have a tendency to become embrittled with age. On the other hand, they are tough, flexible, water-resistant, transparent, and resistant to the corrosive action of most chemicals. Vast quantities are used as liners in the manufacture of safety glass and in the production of photographic film. Among other large users are the manufacturers of fountain pens and mechanical pencils, combs, brush and mirror backs, hairpins, Ping-pong balls, toothbrushes, piano keys, decorative jewelry, cutlery handles, knobs, organ stops, slide fasteners, shoe heel covers, and dolls and toys of nearly every description.

Cellulose nitrate plastics are largely distributed under such trade names as Amerith, Celluloid, Nitren, Nixon CN, and Pyralin.

2. Cellulose Acetate Plastics. Cellulose acetate binders are made by treating cellulose with glacial acetic acid and acetic anhydride in the presence of a catalyst such as sulfuric acid. Such a mixture, when thoroughly aged, has the consistency of molasses and is composed largely of cellulose triacetate. At this point water is added to the mixture. This dilution effects a partial hydrolysis or deacetylation of the triacetate, and cellulose acetate settles out of solution as a white precipitate. This is permitted to mature for several days in a constant-temperature bath held at 86°F, after which the precipitate is filtered and dried.

Cellulose acetate plastics first appeared about 1910 in the form of photographic film, but it was not until 1927 that they became available in the form of thermoplastic molding compounds. Injection molding machines developed since 1934 have made this class of compounds a leader in the plastic-molding field. In the hands of a skilled chemist, cellulose acetate is indeed a versatile material. With selected plasticizers, it may be molded into many useful products with desirable physical properties. Because of the large number of suitable plasticizers and the various properties each of them imparts to the plastic matrix, many cellulose acetate formulations are available to industry.

Once a plasticizer or plasticizers have been selected, they are properly blended with the cellulose acetate in a mixing machine, usually in the presence of solvents such as acetone and alcohol. Dyes or color pigments are added if needed. When intimately mixed, the batch is run through warm rolls and sheeted. The sheets may be cut into molding-powder pellets, block-pressed, fed into extruders to form rods or tubes, or pressed through slots to form thinner sheets.

A complete list of all the cellulose acetate molding-powder compounds and the many and diverse articles made from them would occupy a space of several hundred pages, as cellulose acetate plastic materials are commonplace in our daily lives. The steering wheel of our automobile, cases of small table radios, housing for motors of electric shavers, our favorite bass and salmon plugs, knobs, handles, name plates, grommets, toilet articles, and flashlight cases are but a few of the many cellulose acetate products that we now use in going about our daily chores and pleasures.

Cellulose acetate plastic materials are distributed in the trade under the names of Fibertos, Hercules Cellulose Acetate Flake, Koppers Cellulose Acetate, Lumarith, Nixon CA, Plastacele, and Tennessee-Eastman Tenite I.

3. Cellulose Acetate–butyrate Plastics. This cellulose ester is made by treating cellulose with a mixture of butyric and acetic acids and anhydrides in much the same manner as cellulose acetate is prepared. The molding compounds are very similar to those of the cellulose acetates and are used for the fabrication of both injection-molded and extruded products. As a group, the acetate-butyrate plastics exhibit better dimensional stability, as they are more moisture-resistant and less affected by temperature than the straight acetates.

Typical uses of this plastic may be found in automotive steering wheels, shower nozzles, combs, electric-clock housings, cutlery handles, radio housings, sheets, and specialty films. It is sold only under the trade name Tenite II, a product of the Tennessee-Eastman Company.

4. Ethyl Cellulose Plastics. These are the most recent additions to the family of cellulose plastics. They have been on the market for a comparatively short time, but because of their many superior properties, they have become firmly established in industrial circles.

Ethyl cellulose is an ether of cellulose. In its preparation, cotton linters or purified wood pulp is treated with a strong base such as sodium hydroxide.

Under carefully controlled conditions, this alkali-cellulose complex is then treated with ethylating agents such as ethyl sulfate or ethyl chloride, which introduce ethyl groups into the cellulose molecule. The ether thus formed is washed; when dried it is a white, granular residue. With suitable plasticizers it may be molded, extruded, or cast. Like other cellulose derivatives, it is compatible with many brilliant dyes and color pigments.

Ethyl cellulose plastics have many outstanding properties. They are the lightest (specific gravity of 1.07 to 1.18) of the cellulose plastics and exhibit excellent resistance to heat, cold, water, alkalies, and weathering. Their dimensional stability leaves little to be desired, and from machinability and mechanical standpoints they compare favorably with any of the thermoplastic materials.

Ethyl cellulose is available in powder form for extrusion, injection, and compression processing; in sheets up to 0.020 in. thick; in thin films and foils; or uncompounded for use in lacquer coatings or protective films on metals. The various compositions of this material enjoy many of the same uses as other cellulose plastics.

E. Plastics Industry

Although no figures are available to indicate the total consumption of wood in the plastics industry, some idea of the rapid growth and present size of the industry as a whole may be gained from the fact that in 1933 there were only 101 establishments engaged in the manufacture of finished plastic articles from cellulose ethers and esters, vulcanized fiber products, and pressed pulps, while today there are more than 500 firms engaged in such production.

SELECTED REFERENCES

1. Anon. What Man Has Joined Together. *Fortune*, **13**:69–75, 143–150. 1936.
2. Anon. Plastics Catalog, The Encyclopedia of Plastics. Plastics Catalog Corporation, New York. Published annually.
3. Barron, H. Modern Plastics. John Wiley & Sons, Inc., New York. 1944.
4. Bois, P. Wood Residues in Compression-molded and Extruded Products. U.S. Department of Agriculture, Forest Service, Forest Products Laboratory Report No. 1666–7. 1955.
5. Ellis, C. Chemistry of Synthetic Resins. Reinhold Publishing Corporation, New York. 1935.
6. Fleck, H. R. Plastics: Scientific and Technical. Chemical Publishing Company, Inc., New York. 1945.
7. Guss, C. O. Acid Hydrolysis of Wood Waste for Use in Plastics. U.S. Department of Agriculture, Forest Service, Forest Products Laboratory Mimeo. Report No. R1481. 1945.
8. Harris, E. E. Utilization of Lignin Waste. U.S. Department of Agriculture, Forest Service, Forest Products Laboratory Mimeo. Report No. R1236. 1940.
9. Hunt, G. M. Wood Waste: A Challenge to Industry and Science. U.S. Department of Agriculture, Forest Service, Forest Products Laboratory Mimeo. Report No. R1631. 1946.

10. Lougee, E. F. Plastics from Farm and Forest. Plastic Industries Technical Institute, Chicago, Ill. 1943.
11. Paist, W. D. Cellulosics. Reinhold Publishing Corporation, New York. 1958.
12. Robinson, C. N. Meet the Plastics. The Macmillan Company, New York. 1949.
13. Sasso, J. Plastics for Industrial Use. McGraw-Hill Book Company, Inc., New York. 1942.
14. Simmonds, H. R. Industrial Plastics, 2d ed. Pitman Publishing Corporation, New York. 1941.
15. Simmonds, H. R., and C. Ellis. Handbook of Plastics. D. Van Nostrand Company, Inc., Princeton, N.J. 1943.
16. Stamm, A. J. Chemical Utilization of Wood. U.S. Department of Agriculture, Forest Service, Forest Products Laboratory Mimeo. Report No. R1601. 1945.
17. Stamm, A. J., and E. E. Harris. Chemical Processing of Wood. Chemical Publishing Company, Inc., New York. 1953.
18. Stamm, A. J., and R. M. Seborg. Resin-treated, Laminated, Compressed Wood. U.S. Department of Agriculture, Forest Service, Forest Products Laboratory Mimeo. Report No. R1381. 1955.
19. U.S. Department of Agriculture, Forest Service, Forest Products Laboratory. Forest Products Laboratory Urea-plasticized Wood. Mimeo. Report No. R1277. 1941.
20. U.S. Department of Agriculture, Forest Service, Forest Products Laboratory. Forest Products Laboratory Wood Plastics. Mimeo. Report No. R1382. 1941.
21. Weil, B. H., and V. J. Anhorn. Plastic Horizons. The Ronald Press Company, New York. 1944.

CHAPTER 25

MINOR FOREST PRODUCTS

In addition to the forest products described in the preceding chapters, there are numerous other commercially valuable products derived from forest vegetation. It is beyond the scope of this book to attempt to deal with all of them, but a few representative examples have been selected to illustrate the wide variety of other products that come from forests. In this connection it must be remembered that many important articles of commerce, such as rubber, cocoa, and quinine, although originally products of the forest, are now grown under intensive cultivation or have been replaced by synthetic products.

I. BARK

The bark of many trees contains a number of commercially usable products, including fiber, cork, and extractives such as tannins, waxes, and substances with medicinal properties.

Until the advent of large-scale barking of peeler and sawlogs, practically the only method of bark disposal was to use it for fuel. This presented no particular problem at the sawmills and plywood plants, where bark was burned together with the wood slabs and other wood residues. The yield of heat energy, however, was low because of the bark's high moisture content. At the pulp mills the disposal of bark by burning it for fuel was unprofitable because frequently supplemental fuel was required for its combustion.

Concentration of the millions of tons of Douglas-fir bark and of large quantities of other softwood and hardwood bark,* as a result of log barking, focused the attention of industry on the possibilities of profitable utilization of this by-product of wood conversion. On a nationwide basis most of the bark is still being burned. Nevertheless, a growing number of operations based on the utilization of bark for purposes other than as fuel are now in existence. In most instances, these uses are based on mechanical fractioning of the bark or on its value as a composting material. Increasing interest,

---

* Depending on the species, and the age and the size of the tree, 10 to 20 per cent of the log weight is bark.

however, is being shown in the opportunities offered by the utilization of bark extractives.

A. Cascara Bark [13,18]

Cascara bark is the trade name for the bark of the cascara tree, *Rhamnus purshiana* DC.,* a small tree 20 to 40 ft in height and under 15 in. in diameter. This tree is found in the Pacific Northwest states and in adjacent Canada.[6] Its commercial range, as reported by Starker and Wilcox,[18] is confined to northwestern California, western Oregon and Washington, and southern British Columbia.

Cascara bark is considered one of the most important natural drugs produced in North America and is widely used for laxative preparations and for tonics, both at home and abroad, under the name of *cascara sagrada* (from the Spanish *cascara* meaning "bark" and *sagrada* meaning "sacred"). The bark is collected during the growing season, which is usually from about the middle of April until the end of August. The collecting season varies greatly, however, with weather and locality.

The trees should be felled leaving an unpeeled stump about 6 in. high, with a sloping top surface to shed the rain, thus retarding decay. This practice is essential if the tree is to produce sprouts to provide a new crop 10 or 15 years hence. The bark is peeled from the trunk and from the branches down to about 1 in. in diameter. Ordinarily, before peeling, all moss and lichens are removed by scraping because clean bark commands a better price. Trees less than 5 or 6 in. in diameter should not be felled because they will not yield enough bark to make peeling profitable.

Bark should not be collected during rainy weather because of the danger of deterioration. If dried outdoors, it is placed on a canvas or a wooden platform, preferably in the shade, since direct sunlight causes staining, but it should be covered during the night against heavy dew or possible rain. The more suitable place for drying cascara bark is under a roof in a well-ventilated building. When dried, it is broken into small pieces and placed in burlap sacks, each containing 50 to 75 lb. The United States Pharmacopoeia states that cascara bark must be cured for at least a year before it can be used as a drug. This further aging is usually done by dealers in crude bark rather than by collectors.

Good peelers can produce an equivalent of 100 to 250 lb of dry bark per day. Bark may be sold either dry or green. However, dry bark weighs about 50 per cent less than green and sells for three times as much. The market price for dry bark fluctuates widely with the demand. The 1957 price was about 20 cents a pound and production about 2 million pounds, or only about 40 per cent of the 1947 production. This may indicate not only oversupply but a gradual downward trend in demand.

Earlier predictions that the available supply of forest grown cascara bark

* Other species of *Rhamnus* are found on other continents; of these *R. cathartica* L. and *R. frangula* L., native to Europe, northern Africa, and Asia, are the most important. Neither of these two species, however, has attained the popularity of *R. purshiana* since its acceptance by the medical profession in 1877.

is being rapidly exhausted led to recommendations for establishing cascara plantations. It now appears that the present annual requirements can be met from the wild trees, provided that adequate fire protection and more conservative methods of peeling are practiced. Properly located and well-cared-for plantations should yield a modest return on labor and investment. However, it must be borne in mind that a period of 12 to 15 years will elapse before any appreciable return can be expected. During this period the plantations must be given care and protection comparable to that required for a young fruit orchard. Young cascara bark is quite attractive to deer, cattle, and mountain beaver; grazing must therefore be excluded, and adequate fencing may have to be provided. Rodent control by trapping and poisoning may also be necessary. Trespassing and poaching also present a serious problem, and for that reason plantations should be located near habitation, where they can be watched.

A number of successful plantations of cascara have been established in Oregon, but no reliable information is yet available as to the amount of bark that can be expected from them. It appears that on favorable sites reasonably well-tended plantations may be expected to produce 12 to 16 lb of dry bark per tree, after 12 to 15 years. This would indicate a yield of 10,000 to 15,000 lb of dry bark per acre. Repeated crops can be expected at 10 to 15 year intervals if recommendations for cutting trees to favor sprouting are observed.

B. Cork

Commercial cork is obtained from the outer bark (phellem) of the cork oak (*Quercus suber* L.), a small to medium-sized forest tree widely distributed through countries bordering on the western Mediterranean.

It is estimated that the forests containing cork oak comprise an area of five million acres, two-thirds of which is in Portugal, a nation which also accounts for about 70 per cent of the world's trade in cork. Other important production centers include Spain, Algeria, Morocco, Tunisia, Spanish Morocco, Southern France, and Italy.

1. Cork Formation. In the cork oak, the cork cambium (the secondary phellogen) forms a continuous layer over the entire trunk. This layer produces each year a great number of cork cells toward the outside. If the cork trees were not stripped, this initial layer of cork cambium would persist for a long time, forming cork of considerable thickness. When, however, the first stripping is performed, the initial layer of cork, known as virgin cork, is removed, exposing a deeper cork-cambium layer. This layer dies, but a new one is formed in the bark behind it. This growth phenomenon is repeated at each successive stripping, with the result that new phellogen layers are developed in the inner bark (secondary phloem).

2. Collection of Cork. Portuguese laws require that cork trees should be at least 5 in. in diameter at breast height before the first stripping can be performed. At this size the trees are about 20 years old. The first, or virgin, cork is very coarse and rough and of very minor commercial value. Subsequent strippings are performed at intervals of about 9 years. The cork from the second stripping is finer in texture and hence of better quality, but the best grades of cork are obtained only beginning with the third stripping.

The average productive age of cork oak trees is about 100 years. The amount of cork produced varies with the size and age of the tree, site, local climatic conditions, rate of growth, and the intervening time between strippings. On the basis of a 9-year rotation, 40 to 200 lb of cork per tree is to be expected; however, individual trees have been known to yield as much

Fig. 25-1. Conventional method of stripping cork. (*Courtesy Armstrong Cork Company.*)

as 500 lb. The thickness of cork removed at each stripping ranges from $\frac{1}{2}$ to about 3 in.

The stripping operations are conducted from May to August, except for days when a hot wind, known as the *sirocco*, is prevalent. Stripping is avoided on such days to prevent the desiccation of newly exposed cork cambium. In Spain and Portugal cork is stripped with a long-handled hatchet (Fig. 25-1). In other places, especially in Algeria, crescent-shaped saws are sometimes employed. Two cuts are made around the tree, one near the ground and the other below the main branches. This is followed by two vertical cuts. Care is taken to cut only through the cork region without injuring the last-formed phellogen. The bark is pried loose with the wedge-shaped handles of the hatchet.

Frequently cork is also removed from the larger branches; this is of fine texture, although normally not very thick. In French government-controlled forests, however, stripping of branches is prohibited, and the minimum diameter of the trunk for stripping is limited to 6 in.

In general, trees grown at higher altitudes produce cork of a finer quality than those in the lowlands, but the quality of cork is never uniform even within the same locality. Usually only about 60 per cent of any stripping can be used for bottle stoppers and other *natural cork products*. The remaining 40 per cent, known as *refuge*, is good only for composition products, in the manufacture of which the cork is granulated or reduced to flour.

After the cork is removed from the trees, it is left in the woods to dry for a few days. It is then transferred to a central station, where it is boiled in large open vats. The object of boiling is to remove tannins, as well as to make the cork more pliable, facilitating the straightening of pieces. At the same time the rough outer layers are scraped off, thus reducing the weight of the cork by about 20 per cent. The scraped sheets are flattened, dried, and made up into rough bundles. These bundles are transported on the backs of burros to manufacturing or shipping centers. In the latter instance, bundles of cork intended for export are opened, graded according to quality in about 25 groups, trimmed, and baled for shipment.

When cork finally reaches the manufacturer, it is re-sorted into as many as 80 different grades. Aside from defects caused by such natural agents as insects, the basis for grading is porosity as expressed by the number and size of air passages (*lenticels*) and the presence of lignified tissues. The former decrease the impermeability of cork, and the latter account for a decrease in its elasticity. Specifications for some of the finer grades are so exacting that it takes a specialist with many years of experience to grade cork for industrial uses.

3. The Properties of Cork. Cork possesses a combination of many valuable properties which accounts for its industrial uses. The chief properties of cork are resistance to liquid penetration, compressibility, resilience, high frictional qualities, buoyancy, low thermal conductivity, chemical inertness, and stability. Most of these characteristics are traceable directly to its cellular structure.

Cork, as it appears under a microscope, consists of innumerable small cells which are more or less rectangular in the transverse and radial sections and polygonal in the tangential section (Fig. 25-2). These cells are so compactly grouped that an average cork cell is in contact with 14 other cells. The only breaks between the cells are due to the lenticels, small intercellular air passages surrounded by loosely arranged cells; these intercellular passages extend from the surface of the bark to the last-formed cork cambium. There is, however, no connection between these open spaces and the surrounding cells; therefore neither air nor moisture can penetrate into the adjacent cells, but can travel only along these open channels.

Cork cells possess very thin but strong walls of cellulose, heavily impregnated with a moisture-resistant substance called *suberin*. From 50 to 85 per cent of the total volume of cork is occupied by the cell cavities filled with air. The resistance of cork to liquid penetration is due to the compactness of its box-like cells with closed ends and the impenetrability of their suberin-coated walls. This property accounts for its use for bottle stoppers. The only leakage possible in a cork stopper is through the lenticels. This, however, is avoided by stamping out the corks in such a way that the lenticels run cross-wise through the tissue.

Since more than 50 per cent of the volume of cork is occupied by air, it is very light (specific gravity 0.15 to 0.25). The lightness of cork, coupled with its impenetrability to water, accounts for its buoyancy, a property which adapts it for use as floats. When under a load, the air in the cells is com-

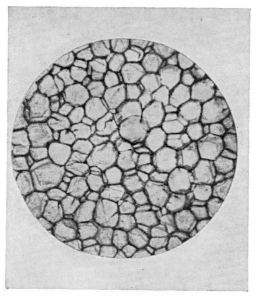

Fig. 25-2. Photomicrograph of cork, tangential view. (*Courtesy Armstrong Cork Company.*)

pressed, and since the cells are airtight, cork tends to resume its original size and shape as soon as pressure is released. Moreover, when compressed, cork does not spread laterally; the only dimension that changes is the thick-ness of the block. This ability to resist permanent deformation under a load, or its resiliency, accounts for many industrial uses of cork, e.g., gaskets, machinery isolation for absorption of shock and vibration, and seals and closures of many kinds.

Because of its myriads of air-filled cells, cork reacts very slowly to changes in temperature and to passage of sound; on this score it is used extensively for insulation. Moreover, because of its low thermal conductivity and chem-ical composition, cork will not burn or maintain combustion, except in direct contact with flame.

Cork is a highly frictional material, and this property is retained even when it is wet or coated with oil or grease. Because of this it is used in power transmission, in friction drives, and in feed rolls, especially in the textile and paper industries. Finally, cork is very resistant to the action of many chemicals, except alkalies and strong acid solutions. Under ordinary conditions cork is practically indestructible. It is not affected by atmospheric conditions and shows no signs of deterioration with the passage of time.

4. Cork Uses. * The industrial use of cork in the United States declined from a monthly rate of 11,000 tons a month in 1947 to less than 4,000 tons a month in 1958. The main reason for this decline in the use of cork is the wide fluctuation in price, while the supply remains fixed. This results in losses of markets to synthetic products with unlimited production every time cork prices go up, losses which are not regained on the downward swing of prices.

Cork is used either in its natural, or unmodified, state or in compositions achieved by grinding and mixing it with various binders.

Formerly the most important uses for natural cork were as bottle and jar stoppers and in insulating materials. The use of cork for stoppers has, however, been rapidly declining as a result of competition from bottle caps with plastic inserts and other types of container closures. A similar decline is being experienced in the use of cork for insulation, since it is being replaced by foamed synthetic products, and in such applications as tool handles, fish bobbins, floats, and washers.

Of cork compositions, important uses include linoleum, floor tiles, gaskets, and tack boards. All these encounter increasingly stiff competition from other products but are expected to maintain some of their markets because of the natural characteristics of cork and the improvements in their performance, brought about by new combinations of cork with other materials.

C. Douglas-fir Bark

The best known example of large-scale utilization of bark based on mechanical fractioning is the Weyerhaeuser process of separating Douglas-fir bark by grinding and screening into three basic fractions: cork, fibers, and amorphous "tissue powder." All these products are marketed under the trade names of Silvacon and Silvabark, with a number to designate particular grades and blends.[20]

The cork fraction of Douglas-fir bark consists of flake-like particles ranging up to $\frac{1}{4}$ in. in size and similar in their physical properties to commercial oak cork. This fraction represents approximately 20 per cent of the total Silvacon products. It is suitable as a soil conditioner and also makes an acceptable ingredient in phenolic molding compounds.

The "tissue powder" is a finely powdered, amorphous material obtained primarily from the highly friable parenchymatous tissues of the bark. It represents about 10 per cent of the Silvacon production. The material is thermoplastic in nature and can be used as a conditioning agent in insecticide dusts and in match-igniting compounds. The average size of particles in this fraction is 5.5 microns, with 90 per cent passing through 325-mesh screen.

The remaining 70 per cent of the output ranges from fiber-like material

* Information supplied by the Armstrong Cork Company, Lancaster, Pennsylvania, 1959.

to various combinations of fiber with cork and amorphous fractions. The fibrous material is suitable as a filler in phenolic and vinyl plastics, as an extender in phenolic adhesives, and as an additive in asphalt and magnesium chloride products.

Silvacon products also find use in oil-well drilling and as additives to improve cement nailability. Special blends are available for specific uses.

Intensive work at the Oregon Forest Products Laboratory indicates that Douglas-fir bark contains appreciable quantities of wax, with properties comparable to those of carnauba wax. A number of companies have indicated interest in its commercial extraction. Another promising extractive is di-hydroquercetin, a white crystalline substance with antioxidant and pharmaco-logical properties, characteristic of vitamin P. Its antioxidant properties are of interest in preventing rancidity in fats and oils, and as an inhibitor in plastics and rubber. Isolation of this chemical on a large scale is undergoing commercial development.

D. Agricultural Uses of Bark

Several companies in California report commercially successful composting of bark. One example of such utilization is described by the Ivory Pine Company. The shredded bark is placed in windrows some 18 ft wide, 6 ft high, and 300 ft long, and commercial nitrogen fertilizer is added. The piles heat up rapidly to about 135°F, but the temperature drops after 2 or 3 months, as the available supply of nitrogen is used up. Then more liquid nitrogen is added and heating is resumed. The decomposed bark is sold either in bulk form at about $7 a ton to commercial users, or is ground, screened, and sold in polyethylene bags through retail channels. An Oregon company sells hammer-milled but not composted bark by the truckload to nurserymen.

A number of companies have developed a profitable business for raw, fiberized bark as a potting medium for growing orchids, to replace an ex-pensive product made from Osmunda fern. Douglas-fir, white fir, redwood, and birch barks have been found acceptable for such use. The Ivory Pine Company reports the sale of 1,500 tons of white fir bark to orchid growers in the United States and abroad.

E. Redwood-bark Fiber [10]

The fibrous bark of redwood yields a number of commercially valuable products, the principal of which are (1) loose-fill insulation, (2) bark fiber used for mattress filling in combination with cotton, and (3) textile fiber for use with wool. Large residual quantities of bark dust, resulting from shredding of the crude bark, are used also as a soil conditioner to increase the moisture-holding capacity of the soil and to permit better aeration.

The bark of mature redwood trees ranges from 2 to 3 in. in thickness in the upper part of the stem to 6 to 12 in. in a basal log. The outer bark, which constitutes the greater proportion of the total bark present, is reddish in color and contains 5 to 10 per cent of long (up to 7 mm in length) and very strong fiber, while the thinner layer of the inner bark is whitish in color and contains 10 to 18 per cent of the fiber. Because of its thickness and its fibrous nature, the bark is peeled before the logs are placed on the sawmill carriage; otherwise it impedes the efficiency of the sawmill operation and affects adversely the quality of lumber by dulling saws.

In the procedure as developed at the Pacific Lumber Company mill, bark is pried loose with chisel-head crowbars. The long bark slabs are carried by a conveyer to a large circular swing saw, where they are cut into 5-ft lengths. These slabs are then delivered to a hammer-mill shredder by an overhead crane bucket. The shredded bark is put through a series of screening and textile-picker operations, in the course of which the finer fiber and bark dust are separated from the coarser bark. The coarse fiber is dried in a tunnel-like conveyer drier and baled in a hydraulic press into 300-lb packages; these are sawed into three sections for convenient handling. This product constitutes the "wool" insulation of the trade and finds considerable favor as loose-fill insulation in places where low temperatures are involved, such as cold-storage, locker, and ice plants, as well as a loose fill for home insulation.

The fiber and the dust collected in screening and refining operations are fed into a mechanical air separator, in which the fiber is separated from the granulated material (the dust). The highly refined fiber recovered in this operation comprises textile redwood fiber. Since this fiber is reddish brown in color, it can be used only in material having a dark shade. Blended with wool or cotton in amounts up to 50 per cent, redwood fiber produces a satisfactory material for inexpensive men's suits, overcoats, and mackinaws, low-priced blankets, mattress filling, and similar items.

Redwood bark dust, when cooked with an alkali, can be converted to a colloidal substance known under the trade name "sodium palconate." This substance and its derivatives have been found useful as a replacement for tannins in fixing mud in oil-well drilling operations.

F. Other Bark Uses

The bark of many other trees is used in medicine or for the extraction of tannin, dyes, and other materials. Among the native barks mention should be made of the stem bark of the northern white pine, which is used in the preparation of cold medicines, usually together with the bark of black cherry and the buds of Balm-of-Gilead. The barks of eastern and western hemlocks and Douglas-fir contain extractable tannins, while flavin, a dye, is obtained from the bark of black oak. A hemlock-bark extract, of a condensed tannin type, produced under the name of Rayflo by Rayonier Incorporated, is used as a mud dispersant in oil drilling.

II. TANNINS AND DYES

A. Tannins

Tannins are natural (vegetable), mineral, and synthetic substances capable of combining with the proteins in raw hides and skins to produce leather.

Natural tannins are widely distributed in the vegetable kingdom and are found in the wood, bark, leaves, and fruit of many plants in forest vegetation. They are complex, water-soluble, dark-colored compounds, with an astringent taste and an ability to produce colored solutions with ferric salts. *Mineral tanning* agents are inorganic salts such as chromium and zirconium. *Synthetic tannins,* known as "syntans," are mostly condensation products of phenol, cresol, and napthalene with an aldehyde, such as furfural.[6]

Most of the mineral tanning agents and synthetic tannins now in use lack certain properties necessary to produce heavy leathers and therefore cannot be regarded as true replacement materials for the vegetable tannins in all types of tanning. An exception is a recently developed "syntan," described as a condensed phenol–sulfonic acid type, derived from coal tar. It is a dark red, viscous material, resembling quebracho extract in appearance. It can be used alone or blended with vegetable tanning materials and is capable of tanning all grades of leather. Its chief disadvantage is cost, substantially higher than that of natural tannins. This development, however, again points to the prevailing trend of replacing products of natural origin with synthetics.

The bulk of natural tannins are used by the leather industry and by the petroleum industry as a dispersant to control the viscosity of mud in oil-well drilling.* Tannins are also employed in the manufacture of inks, in combination with iron. To a limited extent tannins are used as astringents in medicines and as mordants for fixing dyes.

The early American tanning industry was based on the use of oak and eastern hemlock barks. Tannin was extracted by the tanner himself using a system of leaching with water. Because of the resulting weak solution, tanning of heavier leather frequently required 12 to 14 months. Later, in the nineteenth century, concentration of tannin solution by boiling under vacuum was introduced. This made possible the use of vegetable materials containing lesser amounts of tannin and the use of more concentrated solutions, shortening the process of tanning. This development also made it no longer necessary for the tanner to extract his own liquor, leading to the establishment of an industry primarily engaged in manufacturing solid and liquid tanning extracts from vegetable materials.

Growing scarcity of oak and hemlock bark led to introduction of chestnut wood as the principal domestic source of vegetable tannin, supplemented with imports of quebracho and other materials of foreign origin. The destruction of chestnut by blight and subsequent exhaustion of the standing chestnut trees resulted in virtual elimination of domestic sources of natural tannin and complete dependence by the United States on foreign supplies. The number of producers of native tannins had declined to two in 1955; both of these have since ceased operations.

The most important source of vegetable tannin used today in the United States is quebracho extract from the wood of *Quebrachia* (*Schinopsis*) *lorentzii,* a tree widely distributed in South America.[11] Imports of this tannin

---

* For a detailed account of the use of bark extractives and wood-fiber products for oil-well drilling, see Miller, R. W., and W. G. Van Becum. Bark and Fiber Products for Oil Well Drilling, *Jour. Forest Prod.,* **10**(4):193–195. 1960.

come chiefly from Argentina and Paraguay, the former being the larger exporter. Other important sources of vegetable tannins are *wattle bark* (*Acacia* spp.), a tree successfully cultivated on a large scale in South Africa, British East Africa, and Southern Rhodesia, and *valonia*, the acorn cup and the hair-like appendages covering it, of the turkish oak (*Quercus aegilops* L.). Bark of mangrove trees (principally *Rhizophora mangle* L.), yielding up to 45 per cent tannin, is also imported in quantities. Because of its abundance and world-wide distribution, mangrove could become one of the most important sources of vegetable tannin. Other examples of imported tannins are *myrobalans*, the dried nuts of *Terminalia chebula* Retz., a tree indigenous to India, and *divi-divi*, the bean-like fruit of *Caesalpinia coriaria* Willd., a tree native to Central and South America and found also in India.

Appreciable quantities of bark extractives of western hemlock, redwood, and Douglas-fir are used as mud dispersants in oil-well drilling.

Since leather is considered to be one of the important strategic materials, particularly in time of war, many attempts have been made to find other acceptable sources of domestic tannin. These investigations have included experimental breeding of blight-resistant chestnut trees and exploration of the possibilities for utilization of the western hemlock and Douglas-fir barks and the wood of California and southern oaks.[2,4] Attempts have also been made to reintroduce eastern hemlock bark available in substantial quantities in the Upper Peninsula of Michigan. Specially processed waste liquor from the sulfite process has been used for the after-tannin treatment of leather. More recently research has led to its use in blends with other vegetable tannins. Methods of more efficient extraction and of more rapid tanning,* reducing the loss of natural tannins by fermentation and oxidation, are now being investigated. If commercially successful, these new approaches may lead not only to conservation of tanning materials but to production of better leather.[9]

B. Dyes

The first authentic records of dyeing are found in the tombs of ancient Egypt in the paintings depicting every phase of contemporary life, including the operation of the textile and dyeing industries of five to six thousand years ago. The origin of dyes used in the earlier days is unknown. Later, however, such familiar organic dyestuffs as indigo and madder made their appearance. From that time until the middle of the nineteenth century, natural dyestuffs remained the only source of dyes for coloring fabrics.

The whole art and practice of dyeing was completely revolutionized in 1856, when a young student of chemistry, William Henry Perkin, of London, accidentally oxidized a coal-tar derivative, aniline. The resultant product was the first artificial dye, mauve, or mauvine. This discovery was followed by others, leading to rapid development of a new aniline-dye industry. Today, the cheap and varied aniline dyes have almost completely supplanted natural dyes. Only a small group of the latter still retain a place in the world market.

---

* One of these is the solvent system, in which conventional tanning extracts are dissolved in a nonaqueous solvent, such as acetone.

Among them, several are products of the forest. The few examples described in the paragraphs that follow will serve to illustrate this type of forest product.

1. Logwood. Logwood, also known as *campechy wood,* is a product of *Haematoxylon campechianum* L., a member of the legume family. This tree is indigenous to the West Indies, Central America, and the warmer parts of South America. The best commercial stock is now obtained from Mexico, Honduras, Santo Domingo, Jamaica, and Haiti.

The coloring principle of logwood is a compound called *haematoxylin.* This substance is not, however, the dyestuff but only its progenitor. When oxidized, haematoxylin yields the actual dye, *haematine.* Since haematine is an adjective dye, it can be applied to material only through the medium of a second substance known as a *mordant.* Some of the more common mordants used with haematine are the salts of ammonium, iron, copper, and tin. Depending on the type of mordant applied, the color imparted to fabrics ranges from blue to black.

Logwood is used by dyers either in the form of chips or as an extract; the former method has largely been abandoned in favor of the latter. The principal products dyed with haematine are silk, wool, cotton, leather, and straw. It is also used for staining thin sections of biological materials, particularly wood, prepared for examination with a microscope.

2. Fustic. *Old fustic* is the product of *Chlorophora tinctoria* Gaudich., a tree belonging to the Moraceae family, native to Brazil, Mexico, and the West Indies. It produces fast olive-yellow and old-gold shades with chromium, iron, and copper mordants; when used with logwood extracts, it will form black and various compound shades of brown and olive.

Old fustic is used principally for dyeing leather and wool and to some extent for producing green silks, and for dyeing khaki cloth.

3. Quercitron. Quercitron is obtained from the inner bark of the native black oak. The bark is removed from the tree, dried, and ground between millstones. It often comes into the market as a mixture of wood fiber and fine powder of a yellow or buff color. As a rule, the more fine powder present, the greater the value of the sample, because the fibrous portion contains little coloring matter. A freshly made decoction of quercitron bark is transparent and of a dull orange-red color, but in a short time it becomes turbid and deposits a yellow crystalline mass. It should, consequently, be prepared only as required for immediate use.

A much stronger preparation of quercitron bark comes under the name of *flavin.* It yields brighter shades than the original bark. One method of preparation consists of boiling 100 parts of ground quercitron bark with 300 parts of water and 15 parts of sulfuric acid for several hours. After cooling, the resultant mixture is run through woolen filters and the paste is washed until free from acid. It is then dried and ground. The coloring properties of good flavin are 12 to 20 times as great as that of quercitron bark.

Quercitron is applied with the aid of chromium, aluminum, and tin mordants; the resulting shades are very similar to those produced by old fustic. Bark extracts of quercitron are used to a considerable extent in cotton and woolen printing. Flavin is used principally in wool and silk dyeing.

III. OILS, OLEORESINS, AND RELATED PRODUCTS

Oils, resins, gum, and related substances are widely distributed in both the animal and vegetable kingdoms; those of vegetable origin are obtained from various parts of plants, including seeds, fruits, leaves, trunks, and roots. The origin of these substances in the plant is not well understood. In general they are by-products of the metabolic processes of life; some arise as a result of injury and hence are the result of pathological processes. Among these plant products there are a number of commercially valuable substances derived from forest vegetation.

A. Oils

1. Tung Oil.[15] The tung oil of commerce, otherwise known as *China wood oil,* is expressed from the seeds of the tung-oil tree, *Aleurites fordii* Hemsl., a native of central and western China; another species, *A. montana* (Lour.) Wilson, native to southwestern China, also produces oil practically identical in composition and properties with tung oil. This species, however, is less hardy and less widely distributed than *A. fordii.*

Tung oil has been used in China for many centuries, but its important properties were not fully appreciated until the second half of the nineteenth century. The first large shipment of this commodity reached the United States in 1869, the year which marks the beginning of world trade in this valuable product. Ever since, the world's consumption of tung oil has increased annually; the United States consumes more than 70 per cent of the total production.

The increasing importance of tung oil in the paint and varnish industries and the uncertainty of supply and quality of oil from China led to extensive plantings of tung trees in the United States. The first successful planting of tung seeds was made in 1905 by the Division of Foreign Plant Introduction of the U.S. Department of Agriculture, at Chico, California. The seeds were supplied by the American Consul General, L. S. Wilcox, at Hankow, China. The seedlings raised at Chico were set out in experiment stations, parks, cemeteries, and private holdings throughout the southern and Pacific coastal regions.

The earlier commercial plantations of tung-oil trees were established without regard for the exacting climatic and soil requirements of this species. It was later determined that successful plantings in the United States are limited to a belt 75 to 100 miles wide, extending from eastern Texas, along the Gulf of Mexico, to the northern third of Florida and the southern third of Georgia (Fig. 25-3).

The tung tree requires at least 45 in. of evenly distributed rainfall per year. For best growth and production of nuts a uniformly warm temperature, day and night, is essential during the growing season. On the other hand, dormant trees require 350 to 400 hours of temperatures of 45°F or below. Tung trees can withstand temperatures down to 8°F and even lower for short periods. Deep, well-drained and aerated soils with a high moisture-holding capacity are necessary to maintain high productivity of the plantations at maturity.

Tung trees begin to bear fruit the third year after planting and reach commercial production by the fourth or fifth year. The maximum production of nuts is reached by the tenth year. The average life of the tree is at least 30 years. The U.S. Department of Agriculture has released five varieties of tung trees for commercial planting. These are grown as seedlings and are characterized by improved resistance to cold and by the higher and more uniform oil content of their fruit.

Fig. 25-3. Tung-oil plantation at the Florida Agricultural Experiment Station 4 years after planting. (*Courtesy University of Florida, Agricultural Experiment Station, Gainesville, Florida.*)

The average yield of tung-oil plantations, containing 100 to 140 trees per acre, is about ½ ton of fruit per acre. Most of the bearing orchards were planted with unselected seedlings and before much was known of the soil requirements of tung trees. It is expected that the more recent plantations should average 1½ to 2 tons of fruit per acre.

Most of the tung oil in the United States is produced on large, specialized plantations. Tung-tree cultivation, however, is well suited to small, diversified farming since the orchards need little labor during the time of year when other crops require attention.

Tung oil is expressed from the seeds, four to seven (usually five) of which are contained in a spheroid fruit with a pulpy exterior and a nut-like shell (Fig. 25-4). The fruit matures in September to early November and is allowed to fall to the ground, where it is left for 3 or 4 weeks to dry. It is then gathered by hand before the leaves fall, and sacked. Quite commonly the sacks are placed on trees for 2 to 3 weeks for further drying. Many of the mills, however, are equipped with driers, making prolonged field drying unnecessary.

In the United States, the oil is expressed in mills equipped with a decorticator and an expeller press. Analysis of the oil content of American-grown tung fruits indicates that about 20 per cent by weight of the whole, dry, ripened fruit is oil. It therefore takes about 5 lb of dry fruit to produce 1 lb of oil.

Tung oil is pale yellow to dark brown in color. It dries very rapidly to form a hard, glossy film. In addition this oil possesses preservative and water-

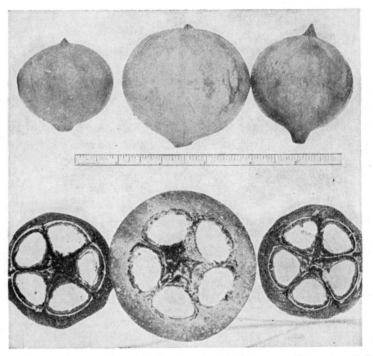

Fig. 25-4. Typical mature fruit of *Aleurites fordii*. (*Courtesy University of Florida, Agricultural Experiment Station, Gainesville, Florida.*)

repellent qualities which add to the durability of the finished product. Tung oil is used principally in the manufacture of spar varnishes, enamel and other paints, and in paint driers. However, development of synthetic varnishes and emulsion paints has resulted in considerable displacement of tung oil in the paint and varnish industry. Large quantities of tung oil are used in the manufacture of linoleum and oilcloth and in the waterproofing of cloth and other manufactured articles. It is also used in insulating compounds in the electrical industries, in brake linings, and as an undercoat for body finishing in the automobile industries. Other uses include printing and drawing inks, dressing for leather, and soap. In some sections of the South the tung meal remaining after the milling process has been utilized as a fertilizer.

Production of tung oil in the United States steadily increased from less than 5 million pounds in 1946 to more than 44 million pounds in the frost-free years of 1952 to 1953. In the 1955 to 1956 years, tung-oil crops were severely damaged by frost, resulting in only a 2-million-pound crop. Production was up to normal in the following 2 years. To encourage domestic production of tung oil, it was supported by a 20.5 cent a pound subsidy and protected through 1960 by an import quota, limiting imports to 11,800 tons. Most of the imports come from Argentina, which also has extensive tung-tree plantations.

2. Walnut Oil. This oil is obtained from the kernels of the English walnut, *Juglans regia* L. The production of walnut oil in the United States, as a by-product of shelling plants, has been estimated at about 50,000 lb a year. Crude walnut oil is used largely in the manufacture of soaps, in enamel paints, oil colors, and printing inks. The refined oil is used in foods. Pulverized walnut shells, called walnut-shell flour, are employed in plastics (see Chap. 24) and as a filler in synthetic-resin adhesives.

3. Turpentine. Oil of turpentine is by far the most important volatile oil produced in the United States. Its production and uses are discussed in the chapter on Naval Stores (Chap. 21).

4. Conifer-leaf Oil.[2,7] Conifer-leaf oil is extracted from the needles of different conifers, principally black spruce, white spruce, eastern hemlock, redcedar, and white-cedar. The oils of black and white spruce and those of eastern hemlock are very similar in composition and are usually sold in mixture; cedar oils are sold either separately or mixed. The conifer-leaf oils as a group have pleasant odors and chemically are composed mainly of terpenes and their derivatives.

The conifer-leaf oils are usually distilled in small quantities by farmers during the winter months. The average yield of oil is said to be up to 1 per cent of the weight of the green material. The yield is greatest from needles collected in January, February, and March. In northern New York the distillers of white-cedar are said to gross about $25 to $30 a week.

The distillation plants are usually small homemade affairs, consisting of a low-pressure boiler from which steam is piped to a large tub fitted with a tight cover. The cover is weighted down with pieces of iron or stone to press upon the boughs packed in the tub. From this tub a pipe leads to a barrel of water by means of which the steam and leaf oils are condensed. The oil floats to the top, from whence it is skimmed off.

Spruce- and hemlock-needle oils are used in perfuming greases and shoe blacking, also in liniments and other medicinal preparations. Cedar oil from *Juniperus virginiana* L. is used mainly in insecticides, while the oil from *Thuja occidentalis* L. is blended in liniments, insecticides, furniture polish, and floor oils.

5. Cedar-wood Oil. This oil is obtained from the heartwood of eastern redcedar by steam distillation of sawdust and other wood residues procured from mills making pencil slats, cedar-closet liners, chests, and other red-cedar articles. The yield of oil is from 1 to about $3\frac{1}{2}$ per cent by weight, depending on the relative percentage of heartwood, since sapwood contains essentially no oil.

Cedar-wood oil has a fragrant aroma and is used extensively in the manufacture of soap, deodorants, furniture polish, sweeping compounds, and perfumes. It is also used in the preparation of liniments and finds a wide application in the manufacture of insecticides because of its reputation as a moth repellent. Because of its high index of refraction, refined cedar-wood oil is used as a clearing agent in preparing objects for the microscope. Finally, cedar oil is used to extend sandalwood, geranium, and other more expensive oils. Some of the heaviest fractions are suitable for mineral flotation. The annual production of redcedar oil is estimated at 150,000 lb.

6. Oil of Sassafras. Sassafras oil is obtained from the roots and stumps of the sassafras tree by steam distillation. The oil is primarily obtained from stumps that are removed in clearing ground for agricultural purposes. The stumps are cut into small chips, which are placed in large retorts with a capacity of as much as 20,000 lb.

The root bark contains 6 to 9 per cent of oil and the root-wood only about 1 per cent. The average yield per ton of material is about 20 lb of oil. Sassafras oil is used in medicine, in the manufacture of soft drinks, and in soaps and perfumes. The industry is confined to the South and is best developed in Tennessee, Maryland, and Virginia.

7. Birch Oil. Birch oil is obtained from the bark of sweet birch. This oil is identical with wintergreen oil, and since it is more abundant in birch bark than in the wintergreen plant, the former has largely replaced the latter as a source of natural wintergreen.

Wintergreen oil is not found as such in birch bark but is formed by the chemical reaction between a glucoside present in the bark and an enzyme, when the bark is treated with water. Some operators utilize only bark that is peeled from the trunk and the larger branches in the late summer. Such bark is cut into small pieces and placed in a copper still, then saturated with water and kept warm overnight. The more common practice, however, is to utilize entire stems, limbs, and twigs. The stemwood is cut into 5- to 8-ft lengths, while branches are tied into bundles, and the material is then packed into the still, with the heavier wood on top. One cord of such material will yield approximately 4 lb of oil, while 1 ton of bark will produce about 5 lb of birch oil.

This industry is scattered through New England, Pennsylvania, and the southern Appalachian region, with the larger centers of distribution in Connecticut and Tennessee. The northern oil brings $3.60 to $3.75 per pound as compared with $1.65 to $2 per pound for the southern product.

Oil of wintergreen is used extensively in medicinal preparations, for flavoring confectionery and chewing gum, in soft drinks, and in the manufacture of disinfectants and insect powders. At the present time the natural wintergreen oil is being rapidly supplanted by a synthetic product derived from coal tar.

B. Oleoresins

The most important native oleoresins are *crude pine gum, Canada balsam,* and *Oregon balsam.* Crude pine gum, which is an exudation from the stems of several species of pines, is discussed under Naval Stores (see Chap. 21).

1. Canada Balsam.[12] Canada balsam is an oleoresin of the turpentine group, obtained from blisters in the bark of balsam fir, a coniferous tree indigenous to Canada and Eastern United States. The term *balsam* is incorrectly applied to this oleoresin, since the true balsams are characterized by the presence of benzoic and cinnamic acids, both of which are absent in Canada balsam. A more correct term would be *Canada turpentine,* the name under which this substance is listed in materia medica as well as in pharmacy.

This oleoresin is secreted normally in canals formed by the separations of the cells in the bark of the tree. It collects in small cavities under the epidermis, the cavities appearing as prominent blisters on the smooth bark of young trees and branches. Canada balsam is collected by puncturing and draining the blisters with a hollow metal tube or spout projecting from the rim of a can, into which the balsam flows. A better tool, producing cleaner balsam, consists of a steel-pointed glass syringe with a rubber bulb. The collection of balsam is made from June 15 until snowfall. It should be collected only during dry weather, since moisture getting into the balsam causes cloudiness, which decreases its market value.

Canada balsam is largely collected in the province of Quebec by the inhabitants of small villages. Whole families of balsam gatherers camp in the woods for 2 or 3 summer months while making collections. The average yield of oleoresin per tree is about 8 oz, and it takes an average of 150 blisters to yield a pint. A man working alone can draw about ½ gal of oleoresin a day, but with the assistance of two boys to climb the trees, the output may be increased to about 1 gal a day.

When enough resin is collected, the balsam is strained and put into 5-gal cans and transported to the nearest village; there it is exchanged for provisions or sold to a wholesale dealer.

Canada balsam is a transparent, viscid oleoresin, yellowish in color, with a distinct greenish fluorescence and pleasant turpentine aroma. When exposed to the air, it loses its volatile oil and thickens, ultimately drying into a transparent resinous substance. When used in thin films it is colorless. Canada balsam has an index of refraction of 1.518 to 1.521 at 20°C; it is soluble in ether, chloroform, xylol, benzol, and other substances but only partially soluble in alcohol.

Canada balsam is used to some extent in medicine as an antiseptic dressing for cuts and wounds; it is also used in the preparation of flexible colloidion, ointments, and plasters. When mixed with wax it is used to bind the component parts of pills and as a sealing wax to prevent deliquescent constituents from attracting moisture. The chief use for Canada balsam has been in the mounting of specimens for examination with a microscope and as a cement for glass in optical apparatus. Since the Second World War, however, Canada balsam has lost its preeminence as a glass-cementing material to synthetic preparations.

Canada balsam is sometimes adulterated by the addition of rosin and turpentine. Such adulteration can be detected by determination of the acid value of the dry resin, which is 120 to 124 for pure Canada balsam. If the acid value is more than 130, the presence of extraneous rosin or turpentine can usually be suspected.

2. Oregon Balsam. This is an oleoresin, similar to Canada balsam, obtained from Douglas-fir and found in bole cavities of wind-shaken trees. These cavities, in time, fill with oleoresin, and when an opening is cut into such a pocket, the oleoresin flows out. Trees containing such pockets can frequently be spotted by the one-sided crowns and bent trunks. As much as 1 to 3 gal is frequently obtained from a single pocket.

C. True Balsams

The term true balsam is applied to a group of resins and oleoresins whose fragrant aroma and pungent taste are due to the presence of benzoic or cinnamic acid, or both. The best known of these are storax, gum benjamin, balsam of Peru, and balsam of Tolu.

1. Storax.[16] Storax is a balsam obtained from *Liquidambar orientalis* of Asia Minor as an exudation from resin passages developed in the wood as a result of wounding. A similar substance may also be obtained from our native sweetgum. Storax is a semitransparent brownish-yellow liquid of honeylike consistency, with a strong aromatic odor and taste. It is used for perfuming soap and talcum powder and flavoring tobacco, and also in the preparation of adhesives and for medical purposes.

The possibilities of developing a native storax industry were first investigated by Dr. Eloise Gerry, who pointed out that it should be possible to obtain additional revenue through storax production. The production of storax from sweetgum was stimulated during the Second World War by the high prices for the oriental product. This industry is now centered in Clarke County, Alabama.

Production is started by removing a strip of bark, 8 to 10 in. wide, about 2 ft from the ground, leaving one or more longitudinal bars of bark 4 to 6 in. wide undisturbed. About every 2 weeks a narrow streak of bark, less than $\frac{1}{4}$ in. in width, is cut along the upper edge of the exposed face to stimulate the flow of resin. The exuded gum hardens on the barked face and is scraped off with a sharpened tool.

A tree will continue to produce for 3 to 5 years. If cuts are not made too deep, young, healthy trees will regenerate a new layer of bark over the face. The tapped trees may also be cut for pulpwood following cessation of gum flow. The yearly yield is $\frac{1}{2}$ to 1 lb of storax per tree. A good worker can handle about 2,500 trees and collect about 15 lb of gum per day.

The collected gum is heated until liquid and then strained to remove trash. The gum is highly inflammable when hot; it should not be allowed to either boil or scorch. The strained gum is poured into suitable metal containers equipped with tightly fitting lids.*

---

* Martin, I. R. (information collected by O. C. Helms). Bleeding Sweet Gum Trees for Sweet Gum, or Storax. County Agent's Office, Clarke County, Alabama. 1958.

IV. FRUIT

A. Pecans

Pecans are a product of a forest tree native to the South Central states, where pecan trees are found growing abundantly in the alluvial flood plains and delta lands of the lower Mississippi River and its tributaries and along the banks of all west Gulf rivers. In the last decade selected varieties of pecans have been planted in large numbers throughout the South, and today pecan nuts are an important source of farm income in many sections of the South.

The main productive belt of pecans from forest-grown trees includes Texas, Oklahoma, and Alabama, the three states which account for about 70 per cent of the total production. The improved plantation-grown varieties, however, come largely from the states east of Mississippi, with Georgia producing about 40 per cent of the plantation crop. The most popular improved varieties are the Stuart and the Schley.

Pecan trees begin to bear when about 10 years old. The improved varieties are largely marketed and distributed to consumers unshelled, while the unimproved varieties are mostly shelled commercially and used in confectioneries, baked goods, and ice cream. Pecan holdings range from a few trees to orchards containing 2,000 or more trees. Pecan orchards in general are developed in connection with some interplanted cash or feed crop for harvest; this practice is usually continued until the orchards reach bearing age.

B. Pinyon or Pine "Nuts"

Pinyon or pine "nuts" are the seeds of several species of pine native to the Southwestern United States. These include the Mexican pinyon, the singleleaf pinyon, the Parry pinyon, and the nut pine. Annual production ranges from less than a million pounds to bumper crops of more than 6 million pounds. The pinyon pines are found over an area of some 42 million acres in Arizona and New Mexico alone, with large areas also present in Colorado and Utah.

The pine "nuts" are collected by Indians, Mexicans, and white settlers, entire families often taking part in the harvesting. The harvest begins late in September and is continued until the first snow. The average market value of the "nuts" is estimated at about $500,000 a year.

V. VEGETABLE FIBER

A. Spanish Moss

Spanish moss, *Tillandsia usneoides* L., which grows in long, hair-like strands suspended from forest trees (Fig. 25-5), is a seed-bearing epiphyte of the pineapple family. Its natural range extends along the coastal plains from Virginia to Texas. Like all true epiphytes, Spanish moss derives its food from rain and dust in the air, using its host only for support. It is rarely propagated by seed, but rather by fragments of festoons blown by wind or spread by birds which utilize the strands in building their nests.

The commercial value of Spanish moss is based on the utilization of the central hair-like strand of fibrovascular tissue which is left when the outer cortical cells are removed.* The moss is gathered by farmers, who collect and sell it to supplement their farm income. The picking season begins in November and lasts until April. Several tons of green moss are piled in stacks or placed in trenches or pits and wet down until thoroughly soaked.

Fig. 25-5. Spanish moss on oak trees. (*Courtesy V. H. Sonderegger, Louisiana Division of Forestry.*)

This is done to rot off the outer layers of tissue which cover the elastic inner fiber that forms the core; at the same time the chemicals in the bark are said to tan and toughen the fibrous cord. The green moss is left piled for 4 to 6 months, during which time it is watered frequently to prevent heating.

When curing is completed, the top layer of the pile is discarded and the rest is spread or hung upon lines to dry. When thoroughly dry, the product is sold by the collectors to a gin operator. The rotted outer surface and all impurities are combed out by means of a toothed drum revolving against toothed concaves. The pure fiber that is left is pressed into bales weighing 60 to 150 lb or more and shipped to the factories under the name of vegetable fiber.

One short ton of swamp-cured moss, or 5 short tons of green moss, are required to produce 1,000 lb of ginned moss. The remaining vegetable matter

* Aside from its commercial and aesthetic value, Spanish moss is a wintering place for the cotton boll weevil. The U.S. Bureau of Entomology reports that on an average 365 weevils winter in a ton of moss.

may be offered for sale as a low-cost peat substitute. It has also been found that dry moss contains about 4.5 per cent of carnauba-like wax that can be extracted with a wax solvent prior to curing the moss.[5]

There are three main grades of Spanish moss produced: 2-X (light brown), 3-X (dark brown), and 4-X (black). The 4-X is more thoroughly cured and somewhat better cleaned than the other two grades and costs a few cents a pound more. The difference in quality and price between the three grades results from the selection of raw materials and the extent of curing, rather than from the processing of the moss subsequent to curing.

B. Wood Fiber

Whole-wood fibers are produced by several companies by fiberizing wood chips. As produced at the Weyerhaeuser Company,[20] Douglas-fir and western hemlock wood chips are preheated for about 1 minute and then fiberized in an Asplund defibrator (see page 341). The resulting fibrous mass ranges from a coarse product consisting of small "sticks" to light, fluffy material. This material may be further refined or it may be treated with asphalt and flameproofing chemicals and used as a fill insulation in commercial buildings and in houses.

Different grades of wood fiber (Silvacel) find a number of applications in specialty papers, paperboard, roofing felts, as a filtration medium in oil-well drilling, and as a cushioning material in packaging. Special grades of the Weyerhaeuser wood fiber, known as Silvalays, are compounded with various resins and used in a variety of molding composition products, including molded furniture.

VI. EVERGREEN HOLLY

Growing holly for Christmas greens offers farmers in some sections of the southeastern United States an opportunity to supplement their farm income. The American holly, valued for its rich green leaves and persistent round scarlet berries, formerly had a natural range from Massachusetts to Missouri, thence south to Florida and Texas.* Destructive harvesting has, however, practically eliminated this tree from Massachusetts to central New Jersey. The region most favorable for growing this species commercially extends in a strip 50 to 60 miles wide from Norfolk, Virginia, south to Florida, and westward in the Gulf region to Louisiana. Most of the holly used in the eastern United States comes from the Delmarva Peninsula, east of Chesapeake Bay.

Holly may be propagated by seed, cuttings, layering, and grafting. Propagation by layering or cuttings has a distinct advantage over reproduction by seed, because plants reproduced vegetatively run true to type. Pure plantations of holly on open lands are recommended in preference to under-

---

* The English holly, *Ilex aquifolia* L., and its many varieties, bears fruit more plentifully than the American species. The English holly is now successfully propagated in the Pacific Northwest. Holly farms have paid good dividends.

planting in woodlands.[15] In the East most holly, however, comes from forest-grown trees.

Harvesting of holly commences when the trees are 12 to 20 years old. Early in the fall, dealers scout the countryside and buy from the woodland owners their holly crops. Harvesting begins about December first and continues until Christmas time. The branches are cut off with a hatchet or a corn knife, 6 in. or more from the trunk. Not infrequently even the top of the tree is sawed off. The branches are collected either on a canvas or in large buckets. Some of the sprays are then made into wreaths, while loose sprays are packed into standard crates 2 by 2 by 4 ft. The destructive methods used in harvesting holly, appropriately known as "breaking of holly," lead

Fig. 25-6. Destructive harvesting of holly: *left*, tree being completely stripped; *center*, stripped tree; *right*, stripped tree starting to recover. (*Courtesy N. H. Fritz, Maryland Department of Forestry.*)

to serious damage to the trees (Fig. 25-6). Severely pruned trees will not produce another holly crop for the next 15 or more years and may die.

To reduce the destruction of naturally grown holly and to increase the cash returns to forest owners, the Maryland Department of Forestry is now sponsoring the production of "certified holly." The department holds harvesting demonstrations in the various parts of the state where holly grows naturally. At these demonstrations the proper method for pruning trees is demonstrated, and it is shown that large quantities of holly sprays can be obtained, still leaving the trees in a productive state (Fig. 25-7). Individuals interested in these methods of harvesting sign up as cooperators in the production of "certified holly." The department attempts to secure orders for holly from retailers and individuals, and these are distributed among the cooperators; the department also supplies the cardboard packing boxes. The holly is packed in boxes of three sizes: boxes 12 by 12 by 24 in. hold enough holly to decorate a small house; a 16 by 16 by 24-in. package, sufficient for a large house; and a 2 by 2 by 2-ft box is used for holly distribution to florists and other retailers. Each box contains only No. 1 grade holly, pruned from trees in accordance with instructions from the department. Boxes are inspected at designated stations, and if the holly meets all requirements, the inspector pays the cooperator cash for his materials. The boxes

are then sealed and a certification tag attached by the inspector. The middle-
man is thus eliminated and the savings passed to the forest owners, as the
department makes no charge for its services. The difference between the
selling price and that paid to the cooperator pays for boxes, inspection, and
clerical help. The stumbling block to further development of this conservation movement at present is the small number of orders that the department
is able to secure for the large number of cooperators. It is estimated that 99
per cent of the American holly sold each year is still produced by destructive
harvesting methods.

Fig. 25-7. Proper way of harvesting holly: *left*, clipping lower branches; *center*, clipping
higher up the tree; *right*, clipping branches from top of crown. (*Courtesy N. H. Fritz,
Maryland Department of Forestry.*)

SELECTED REFERENCES

1. Anon. Douglas Fir Bark Extractives. Oregon Forest Products Laboratory. Corvallis,
   Oregon. Undated.
2. Bailey, L. F. Leaf Oils from Tennesse Valley Conifers. *Jour. Forestry*, **46**(12) :882–
   889. 1948.
3. Bailey, L. F., and W. H. Cummings. Tannin and Secondary Products from Oak Slabs.
   Jour. Amer. Leather Chem. Assoc., **43**(5) :294–306. 1948.
4. Bandekow, R. J. Present and Potential Sources of Tannin in the United States.
   Jour. Forestry, **45**(10) :729–733. 1957.
5. Benett, R. B. Spanish Moss and Some Aspects of Its Commercial Possibilities. *Fla.
   Engin. and Indus. Expt. Sta., Tech. Rpt.* 1. 1954.
6. Chen, P. S. Syntans and Newer Methods of Tanning. The Chemical Elements, South
   Lancaster, Mass. 1950.
7. Cochrane, J. A. Cedar Leaf Oil Extractives. *Jour. Forest Prod. Res. Soc.*, **1**(1) :120–
   123. 1951.
8. Field, P. Residues from the Sawmill. *Forest Prod. Jour.*, **8**(11) :27A–30A. 1958.
9. Kremen, S. S. Sole Leather Tanning in a Solvent System. *Jour. Amer. Leather Chem.
   Assoc.*, **50**(4) :204–209. 1955.
10. Lewis, H. F. Utilization of Redwood Bark. Proceedings, Third Annual National
    Meeting, Forest Products Research Society. 1949.
11. Lobaugh, E. R. Tanning Materials. *U.S. Dept. Com. Business Serv., Bull.* 168. 1956.
12. Marriott, F. G. (Revised by C. Greaves). Canada Balsam: Its Preparation and Uses.
    Forest Products Laboratories, Mimeo. 123. Ottawa, Canada. 1947.

13. Office of State Forester, State of Oregon. Cascara Sagrada. Salem, Ore. Mimeo. 1958.

14. Perkins, C. Growing Christmas Holly on the Farm. *U.S. Dept. Agr., Farmers' Bull.* 1693. 1932.

15. Potter, G. F., and H. L. Crane. Tung Production. *U.S. Dept. Agr., Farmers' Bull.* 2031. 1957.

16. Ross, C. R. Storax from American Red Gum. *Amer. Forests,* **53**(9):404–405, 432. 1947.

17. Stamm, A. J., and E. E. Harris. Chemical Processing of Wood. Chemical Publishing Company, Inc., New York. 1953.

18. Starker, T. J., and A. R. Wilcox. Cascara. Reprinted from *Amer. Jour. Pharmacy,* **103**(2, 3). February and March, 1931.

19. U.S. Department of Agriculture, Forest Service, Forest Products Laboratory. Chemical Composition and Uses of Bark. 1957.

20. Weyerhaeuser Company. Silvacon, Silvacel, etc. Special Products Division Reports. Tacoma, Washington. Undated.

INDEX

Abietic acid, 456
Absorbent paper, 378
Absorbent sawdust, 275
Acetic acid, direct recovery by solvent
    method, 411
  extraction from pyroligenous acid, 410
  uses, 416
Acid stimulation in naval-stores practices,
    446–447
Acid sulfite semichemical pulp, 359
Acid-treated trees, vigor, 450–451
Activated carbon, 408
Activated charcoal, uses, 416
Adhesives in plywood manufacture, 142–
    146
Adit, definition, 78
Air-drying of veneer, 131
Alcohol from pulping residues, 429–430
Alcohol oils, 410
Aleurities fordii, 512
  montana, 512
Alkali cellulose, 394
Alkali lignin, 356
Alkaline pulps, 351–358
Alkaloids in wood, 51
Allyl alcohol, 410, 412
Alpha-cellulose, 49
Alpha-cellulose pulps, uses, 366
Angelique, 69
Aniline dyes, 372
Antichecking irons for crossties, 98–99
Apple barrel, 216
Arabinose, 425
Art of veneering, history, 120
Artificial silk, 393
Ash barrel, 213
Asplund defibrator, 341, 359
Associated Cooperage Industries of Amer-
    ica, Inc., 216
Attrition mills, 266–267
Ax handles, 316
Azeotropic distillation, 410

Back-cut veneer, 125
Badger-Stafford hardwood-distillation proc-
    ess, 409–410
Bag-molding in plywood industry, 151–152
Bagasse, 323

Balsam (*see* Canada balsam)
Balsam of Peru, 518
Balsam of Tolu, 518
Band headsaw, 112
Bark, agricultural uses, 507
  commercial products from, 500–510
  composting, species used, 507
  cork from (*see* Cork)
Barkers, drum, 71, 333
  hammer, 71
  hydraulic, 335–336
  knife, 334
  peeler, 71
  stationary, 334
Barking machines for posts, 71–72
Barrel saw, 204–205
Barrels, slack, 216–217
  apple, 216
  fish, 216
  glass-and-pottery, 216
  meat-and-poultry, 216
  nail-and-spike, 216–217
  No. 1, 216
  No. 2, 216
  potato, 216
  vegetable, 216
  tight, 212–213, 219–220
  ash, 213
  beer, 213, 219
  bourbon, 212
  Douglas-fir, 213
  gum, 212
  oil, 219–220
  pork, 220
  red oak, 212
  whisky, 219
    plywood, 213
  white oak, sap clear, 212
    sound sap, 212
    tight sap, 212
  wine, 219
Bartrev process for particle-board forma-
    tion, 257
Baseball bats, 317
Baskets, berry, 234
  climax, 235
  grape, 235
  market, 235

Baskets, round-stave, 234
  splint, 235
  till, 234
  veneer (see Veneer baskets)
Battery separators, 46
Beer barrel, 213, 219
Benalite, 494–495
Bending department in furniture plant, 174
Berry boxes, 234
Beta-cellulose, 49
Bevel siding, aluminum, 148
Bilge of barrel, 201
Bilge saw, 204–205
Birch oil, source, 516
Black liquor, soda process, 358
  sulfate process, 354
Bleached sulfate pulps, 357
Bleached sulfite pulp, consumption in
    rayon industry, 392
  use in production of films and filaments,
    392
Bleaching wood pulp, process for, multiple-
    stage, 365–366
  single-stage, 365
  two-stage, 365
Block-shingle machine, 188
Blood albumen glue, 143
Blow pit, 348
Blow tank, 352
Blowing of digester, 348
Board, insulating, 6, 248
Bobbins, manufacture, 314–316
Bolter mill, 312
Book paper, 378
Bourbon barrel, charring headings and
    staves, 209–210
Bowling pins, 317
Box construction, nails used, 231
Box-and-crate industry, status, 233
Boxes, 223–230
  classes, 226–229
  cleated plywood, 228
  corrugated fiber, 380
  nailed wood, 226–228
  place in agriculture, 228–229
  styles, 226–228
  wire-bound, 228–229
  woods used in assembly, 223–226
Breaker beaters, 368
Brick maple sugar, 478
Bridge ties, 94
Briquettes, 288, 423
Briquetting charcoal, 416, 423
Briquetting wood residues, 288
Bristol board, 378
Brown pulp, 340
Bucked staves, 204
Buddy sap, 469

Building codes, restrictive use of shingles,
    198
Building fiberboards, 382
Building industry, 34
Building papers, 378
Burr for facing pulpwood grindstones, 340
Busch combine pulpwood harvester, 327
Butt-treatment of western redcedar poles,
    62

Caesalpinia coriaria, 510
Calender machine, function, 377
Campechy wood (logwood), 511
Canada balsam, 516–518
  origin, 517
  procurement, 517
  properties, 517
  uses, 517
Canadian shingles, manufacture, 190–191
Carbohydrate derivatives from wood resi-
    dues, 431
Carbonization of wood, 404
Carbonization processes using wood resi-
    dues, in large sizes, 406–409
  in small sizes, 409–410
Cascara bark, 501–502
  collection, 501
  curing, 501
  uses, 501
Cascara sagrada, 501
Casein glues, 142–143
  properties, 142
Caul boards in plywood assembly, 148
Cedar-log market, 186
Cedar-wood oil, 515–516
Cell wall, morphology, 48
  woody, chemical nature, 48
Cellophane, 366
  manufacture, 401
  uses, 401
Cellulose, 49
  regenerated, 394
Cellulose acetate, 366, 393, 397–398
  fibers, 397
  films, 401
Cellulose acetate-butyrate plastics, 497
Cellulose fibers, 401–402
Cellulose films, 393, 399
Cellulose nitrate, 393
  plastics, 496–497
Cellulose plastics, 403, 496–498
Cellulose sponges, 402
Cellulose xanthate, 394
Cellulose yarns, 393, 399
Cellulosic lacquers, 403
Cereal straws for papermaking, 326
Chamfer in barrels, 211
Charcoal, 292, 404, 416–424
  briquetting, 416–423

Charcoal, manufacture and production
  from kilns, 417–422
  marketing, 423–424
  uses, 415–416
Charcoal kilns, 419–422
  beehive, 419
  masonry, 421–422
  pit, 419
  steel drum, 419–421
    Black Rock Forest type, 419
    New Hampshire type, 419
Charcoaling, special methods, 422–423
Charring bourbon whisky barrels, 209–210
Chemical debarking, 326
Chemical methods of wood pulp manufacture, 341–362
Chemical pulp, chipping wood for, 341
Chemical reactions, in production of sulfite cooking liquor, 344–345
  in recovery systems, soda-process, 358
    sulfate-process, 354–355
Chemigroundwood pulping, 362
Chemipulper, 359
Chemipulper continuous cooking system, 360
Chime in barrel, 211
Chippers, pulpwood, 341
Chipping in gum-naval-stores operations, frequency, 446
Chlorophora tinctora, 561
Christmas Tree Growers' Association, 486–487
Christmas trees, 480–489
  bundling for shipment, 484
  grading, 485
  harvesting, 483–484
  marketing, 485, 487
  principal species for, 481
  producing regions, 481
  shearing and shaping, 482–483
  sources, forests, 481
    plantations, 482
Cigarette papers, 378
Cinchona, 51
Circular headsaw, 111
Circular toothpicks, 305
Clamp carrier for assembly of plywood panel cores, 141
Class T ties, 102
Class U ties, 102
Classification of papers, 378–380
Cleated plywood boxes, 228
Climax baskets, 235
Clipping green veneers (see Veneers)
Colophony (see Gum rosin)
Commercial-veneer manufacturers, 135
Commercial veneers, 135

Commodity forest drain, 21
Comparisons of veneered and solid furniture, 181–182
Compartment kilns, 116–117
Components of plywood panel, 136
Compreg, 153–155
  formation, 154
  properties and uses, 154
Compregnated wood, 154
Cone-cut veneer, 126
Conifer-leaf oil, 515
Container-veneer manufacturers, 135
Container veneers, 135
Continuous liquid-bath method for charcoaling, 422
Continuous process for carbonization of small-wood residues, 423
Continuous retort process for charcoaling, 422
Cooperage, 201–222
  definition, 201
  historical account, 201–203
  slack (see Slack cooperage)
  tight (see Tight cooperage)
Cooperage industry, status, 218–220
Cord, 331–332
Cordite, 403
Cork, 502–506
  anatomy, 504–505
  collection, 502–504
  commercial sources, 502, 506
  formation, 502
  properties, 504–505
  uses, 506
Cork oak, 502
Cork-producing centers, 503
Cornstalks, use in papermaking, 323
Crate construction, nails used, 231
Crate industry, status, 233
Crates, diagonally braced, three-way corner, 229–230
  industrial, 232
  lumber-sheathed, 230
  place in agriculture, 231–232
Creosote (see Wood preservatives)
Crepe papers, 378
Cross bands in plywood panel assembly, 136
Crossties, definition, 9
  hewed, 104
  in maintained tracks, 104
  recommended piling procedures in seasoning, 90–100
  required per mile of track, 104
  species used, 102
  specifications, 101–104
Croze in barrel, 210
Cuprammonium fiber, 393

Cylinder paper machines, 377–378
  multicylinder type, 377
    laminate papers produced, 377
  one-cylinder type, 378
    tissues produced, 378
Cyperus papyrus, 367

Dandy roll, function on paper machine, 376
Debarking, chemical, 326
Debarking machines, for poles, 59
  for posts, 71–72
Decker use, 376
Deckering purpose, 376
Delignification of wood, 48
  (*See also* Chemical methods for wood
    pulp manufacture)
Demerara greenheart, 68
Destructive distillation of wood, 48, 404–
  417
  hardwood, 406–413
    Belgium Lambiotte continuous proc-
      ess, 409
    cast-iron and steel retorts used, 407
    processes for wood in large size, 406–
      409
    refinement of chemicals from, 408
    Reichert batch process in, 408–409
    species used, 406
    steel ovens, 407–408
    uses of products from, 415–416
    yield of chemical products, 412–413
  softwood, 413–415
    products from, 415
    yields of products per cord of wood,
      415
Diaphragm pulp screens, operation, 364
Dicorynia paraensis, 69
Differential density plywood, construction,
  155
Diffusers in sulfate process, 354
Digesters, blowing, 348
  operation, in soda process, 358
    in sulfate process, 352–354
    in sulfite process, 346–348
Dimension shingles, 193
Dimension stock, hardwood, 108
Dip, nature of, 448
Dissolving pulps, 366
Divi-divi, 510
Doors, flush, 6
  hardwood, production, 1950 to 1958, 6
  ponderosa pine, production, 1950 to
    1958, 6
Dorr process, 355
Double-butting shingle machines, 189
Double-dip strike-anywhere matches, 299
Double-end tenoner, 172
Double-headed planer, 117
Douglas-fir bark, 506–507
  cork from, 506

Douglas-fir bark, fibers from, **506**
  tissue powder from, 506
Douglas-fir barrel, 213
Douglas-fir pile timbers, 68
Douglas-fir plywood (*see* Plywood)
Douglas-fir poles, 57
Dowels, 313–314
Drawing pencils, 297
Drift, in gum-naval-stores practices, 442
  in mines, 78
Drum saw, 204–205
Dry distillation of wood, 404–406
Durability, of mine timbers, 86
  of poles, 61
  of railway crossties, 97
Dyes, 510–512

Equalizer, stave-bolt, 204
Esparto, 326
Ethyl cellulose, 402, 497
  films, 402
  plastics, 497–498
Evergreen holly, 521–522
  certification, 522
  destructive harvesting, 522
  inspection and grading, 522
  packing and shipment, 522
  principal producing areas, 522
Excelsior, definition, 306
  grades, 308
  manufacture, 308
    preparation of wood for, 306
  species used in production, 306
  uses, 308–309
  yield per cord of wood, 308
Excelsior machines, 306–308
Extenders for resin adhesives, 145
  use, 145–146
Exterior plywood, 157
Extractives in wood, 51
Extraneous components of wood, 51

Faber Pencil Company, 295
Face veneer, manufacturers, 134–135
  selling procedures in marketing, 134
Factorage houses, role in gum-naval-stores
    industry, 440–441
Factory lumber, 107–111
Fatwood, 413
Fiberboard industry, future developments,
  262
Fiberboard manufacture, dry process, 252–
  253
  wet process, 249–252
Fiberboards, 249–253
Fill insulation (*see* Sawdust, uses)
Fillers, in pulp stock, 369–370
  in synthetic resins, 146
Flakeboard, 254–255

Flat-cut veneers, 127
Flavin, 511
Flexwood, 156
Flooring, hardwood, 6, 111
  softwood, 111
Flour (see Wood flour)
Fluidizing-bed technique for charcoaling,
  422
Flush doors, 6
Forest products, estimated consumption in
  United States 1952 to 1975, 8–9
Forests as stabilizing influences, 13–14
Foundation piles, 67
Fourdrinier paper machine, 374–377
Freeness of pulp, 368
Furnace poles, 66
Furniture, 167–183
  advantages and disadvantages of wood as
    raw material for, 168
  period stylings, 120–121
Furniture construction, good, features of,
  180–181
  veneered versus solid, 181–182
Furniture manufacture, cabinet-room oper-
  ations, 174
  factor processes, 169–171
  finishing operations, 174–179
    sealing, 177–178
    staining and coloring, 175–176
    system used, 178–179
      mechanical, 179
    top coating, 178
  machine-room operations, 172–174
  wood lost, 179–180

Galactose, 425
Gamma-cellulose, 49
Ganghead saw, 113
Gas recovery in sulfite process, 344, 348–
  349
Gases evolved in wood carbonization, 405
Glass-and-pottery barrel, 216
Glassine, 379
Glazed papers, 378
Glesinger, E., 3
Glucose, 425
Glue reel for plywood panel assembly, 141
Glue spreader use in plywood panel, 146
  assembly, 147
Granulated cork from Douglas-fir bark, 506
Grape baskets, 235
Greaseproof papers, 379
Green-wood turnings, 316–317
Groundwood pulp, bleaching, 365
  concept, 337
  means of producing, 338–340
  species used, 337
  uses and properties, 340
Gum barrel, 212
Gum benjamin, 518

Gum naval stores, 437–458
  early American production, 436–437
  gum processing, 453–456
    fire stills formerly used, 453
    in modern central-processing plants,
      454–455
    intermediate crop of managed forests,
      442
  seasonal industry, 437–439
  seasonal yields from bark-chipped, acid
    treated trees, 450
Gum-naval-stores conservation program,
  451–453
Gum-naval-stores industry, 437–439
  future in United States, 457–458
Gum-naval-stores production, bark-chipping practices, 446
Gum-naval-stores woods operations, 441–
  450
  applying acid to working face, 446–448
    recommended frequency, 448
    recommended procedures, 447–448
  backfacing, 451
  bark chipping, 446
  characteristics of trees preferred, 441
  collecting dip, 448
  condition of worked-out trees, 442
  effect on tree growth, 451–452
  preparation of working face, 443–446
    bark shaving, 444
    hanging cups, 445
    installing tins, doubled-headed nails
      used, 444
  prior marking of trees for facing, 442–
    443
    diameter limit cupping, 443
    selective cupping, 443
  protection from fire, 451
  raising tins, 449–450
  raising working face, 446
  removing scrape, 448–449
Gum processor, 440
Gum producer, 440
Gum resin, chemical stimulation, 443
Gum rosin, 435
  chemical nature, 456
  grades, 457
  uses, 457
Gum turpentine, 435, 437
  chemical composition, 456
  grades, 456
  uses, 456

Haematine, 511
Haematoxylin, 511
Haematoxylon campechianum, 511
Half-round veneer cutting, 125
Hammer mills, 266
Hampers, 234

Hardboard, 6, 248
  hydroxylin, 493–494
  methods of manufacture, 259
  production 1950 to 1958, 6
  properties, 259–260
  treated, 253
Hardwood dimension lumber, 108
Hardwood doors, 6
Hardwood flooring, 6, 111
Hardwood-plywood floor tiles, 161
Hardwood-plywood industry, 164, 166
  regional production statistics, 166
Hardwood tar oil, 412
Harper Fourdrinier paper machine, 378
Harris, E. E., 30
Headings manufacture, for slack barrels, 215
  for tight barrels, 206–208
    dowel-jointing, 207
    species used, 206
    use of flagging, 207
Heat available from wood (see Wood fuel)
Heavy cooperage, 221
Hemicelluloses, 49–50, 357, 426
Hemp, 323
  manila, 324
  sisal, 324
Hewed ties, 103–104
High-density wood, 153–155
High-grade pencils, finishing, 297
Hogshead, tobacco, 217–218
Hollander machine, 369
Holly (see Evergreen holly)
Hoops, manufacture for slack barrels, 215
  types used in tight cooperage, 208
Horizontal veneer slicer, 127–128
Hot-plate platen driers, 138
Hot pressing in plywood assembly, 149–150
Household furniture (see Furniture)
Hudson, M. L., 58
Hydroxylin hardboard, 493–494
Hydroxylin laminate sheet, 493
Hydroxylin manufacture, 491–494
  acid-hydrolysis method, 491–493
  aniline-hydrolysis method, 493

Impreg, 154–155
  formation, 154
  properties and uses, 154–155
Indulin, 356
Industrial boxes and crates (see Boxes; Crates)
Inorganic constituents of wood, 51
Insulating board, 6, 248
  production 1950 to 1958, 6
Intercellular substance, 48
Interior plywood, 157
Irons, tie, antichecking, 98–99

Japanese plywood imports, 164
Jointing staves, 206
Jordan refiner, 373
Juglans regia, 515
Juniperus virginiana, 515
Jute, 323

Kamyr continuous digester, 353
Kegs, nail and spike, 216
Kiln-drying of veneer, 132
Kimpreg, 156
Knee-bolter saw, 187
Knot saw (shingle jointer), 189
Kraft paper, 380
Kraft pulp, 357

Laminated paper, 377
Lapping wood pulp, 366
Lead pencils (see Wood-cased lead pencils)
Lightwood, 413
Lignin, 50, 426
Lignin-enriched filler, 495
Lignin plastics, 490–491
Linen rags used in paper, 323
Liquidambar orientalis, 518
Lodgepole pine poles, 57
Loft-drying of veneer, 132
Logging, wood residues generated by, 21–22
Logwood, 511
Logwood dye, uses, 511
Longleaf pine, gum production, 439
Loose-cut veneer, 123
Lumber, air-drying, 116
  classification, 107–111
  factory, 108
  hardwood dimension, 108
  kiln-drying, 116–117
  manufacture, 111–118
    cross-cutting operations, 115
    log breakdown, 111–113
      headsaw use, band, 112
        circular, 111–112
        gang, 113
  measurement, 109–110
  production, 1776 to 1960, 4
    1904 to 1960, 5
  resawing, 113
  ripping, 113–115
  surfacing and shaping, 117–118
  yard, 108
Lumber dry-kilns, 116–117
Lumber grades, hardwood, 108–110
  softwoods, 108
Lumber industry, development in North America, 106–107
  status, 118

Madison wood-sugar process, 428
Mannose, 425
Maple cream, 478
Maple honey, 478
Maple sap, 469–470
  buddy, 469
  chemical constituents, 469
Maple sirup, density specification, 477
  grades, 478
Maple-sirup and maple-sugar production,
    466–479
  principal centers, 466
  processing sap for, 475–478
    boiling houses, 475–476
    boiling process, 477–478
    sugaring off, 476–477
  sugar species, 467
  woodland operations, 470–475
    gathering sap, 474
      containers, 473
    tapping, 470–471
  yields from raw sap, 479
Maple trees, effect of tapping on wood
    quality, 472
Maple vinegar, 478
Market baskets, 235
Masonite, 249
Match blocks, manufacture from pine, 299–
    300
Match boxes, manufacture, 304
Matches, 298–304
  early history, 298–299
  packaging, 303–304
  paper, 303
  round grooved-splint, 300–302
  square splint, 302–303
Meadol, 356
Measurement of lumber, 109–110
Meat-and-poultry barrel, 216
Mechanical pulps, manufacture, 337–341
  grinders, 338–340
    continuous-type, 339
    intermittent-type, 338
    pocket-type, 338–339
    Roberts ring-type, 339
  woods used, 337–338
  properties, 340–341
  uses, 340–341
Melamine-formaldehyde resins, 145
Metal-faced plywood, 162
Methanol, 410, 412
  uses, 416
Methyl acetone, 410, 412
Methyl mercaptan, 355
Middle lamella in woody tissue, 48
Mill residues for pulp production, 336–337
Millwork products, production 1950 to
    1958, 6
Mine guides, laminated, 92

Mine stulls, selection, 90
  specifications, 88
Mine-timber research, 92
Mine timbering, 77–84
  open rooms, 79
  square-set method for stopes, 80–82
  terminology, 77–78
  tunnels, drifts, and crosscuts, 79–80
  types used, 79–80
Mine timbers, advantages, 84–86
  annual requirements and trends, 91–92
  causes of deterioration, 87
  classes, 82–83
  costs, 91
  defined, 77
  peeling, 90
  preservation, 86–87
  production, 82–89
  purpose, 77–78
  qualities desired, 83–84
  seasoning, 90
  species used in production, 89–90
  storage, 90
Mining methods dictate types of timbering,
    79–82
Mining terminology, 78–79
Mitschlich pulping process, 346–347
Molded plywood, 151–152
Molder machine, types of products manu-
    factured on, 118
Mordant, use with logwood dye, 511
Morterund system for pulping wood, 358
Multiple-spindle carving machines, 174
Multiple-stage bleaching, 364–365
Multiple-use forest management, 14
Myrobalan nuts, 510

Nail-and-spike kegs, 216
Nailed wooden boxes, 226–228
Nails, cement-coated, 231
  etched, 231
  used in box and crate construction, 231
Narrow-gauge ties, 103
Narváez Expedition of 1528, 436
National Hardwood Lumber Association,
    108
National Wood Box Association, 233
National Wooden Pallet Manufacturers As-
    sociation, 236
Naval stores, in Biblical times, 435
  defined, 435
  gum (see Gum naval stores)
  history, 435–437
  production, abroad, 463–464
    domestic, 438
  products, 435
  sulfate, 461–463
  wood (see Wood naval stores)
Naval-stores industry, segments, 435

Nectandra rodioei, 68
Neutral sodium sulfite as pulping agent, 359
Newsprint, 379
N.G.R. stains in finishing furniture, 176
Nitrocellulose, 366
  explosive formulations, 402–403
  films, 401–402
Noncommodity forest drain, 21
Northern white-cedar poles, 58
Novoply, 254
Nut-shell flour, 489–490

Oil of sassafras, 516
Oil of turpentine (*see* Sulfate turpentine; Wood turpentine)
Oil barrels, 219–220
Old fustic, 511
Old paper stock, 323
  in paper making, 367
Olea europaea, 51
Oleoresins, 435, 516–518
Olive, European, 51
Open three-way corner, 229–230
Oregon balsam, 518
Orzan-A, 350
Osmose salts, 73
Ovens in hardwood distillation, 407

Packaging papers, 379
Painted poles, 65–66
Pallets, 236–246
  advantages in moving goods, 237–238
  classes, 238–239
  clinch-tite, 243
  construction details, 239–241
  crate and bin, 242
  crib-cargo, 242
  designs, 239
  double-faced, 239
  double-wing, 241
  Econ-O-Mee, 241
  eight-way block, 239
  expendable, 238, 241
  flush-stringer, 239
  four-way, 239
  four-way block, 239
  four-way notched-stringer, 239
  general purpose, 238, 241
  gluing of unitized loads on, 245
  grades, 244
  lifetime, 243
  maintenance, 246
  metal cap, 243
  no-block, 241
  nonreversible, 239
  patented, 243
  reversible, 239
  shipping, 239

Pallets, single-faced, 239
  special purpose, 238, 241–242
  strapping unitized loads on, 245–246
  structural details, 244–245
  styles, 239
  Take-It-Or-Leave-It, 242
  two-way, 239
  type designations, 241
  woods used in, 244
Panel assembly in plywood production, 137–150
Paper, 367–383
Paper manufacture, 367–378
  beating operations, 368
  calender machine used, 377
  coloring stock for, 372
    function, 377
  loading and sizing procedures, 369–370
  papermaking machines used, 374–378
Paper mill, factors governing location, 383–384
Paper reel, 377
Paper slitters, 377
Paper winders, 377
Paperboard, for building construction, 382
  for folding boxes, 381
  for food containers, 383
  for insulation, 382
  for setup boxes, 381
  for utility containers, 382
Paperboards, 381–383
  construction, 381
Papermaking materials, 323–326
Papreg, 156
Particle-board industry, future of, 262
Particle boards, 248, 253–259
  Bartrev continuous process for production, 257
  extrusion, 258–259
    Kreibaum process, 259
    Lane process, 259
  flat-pressing, 257–258
  formation, 256–259
  layered, manufacture, 254–255
  particle geometry, 254–255
  preparation of particles, 255–256
    drying, 255
    mixing, 256
      batch process, 256
      continuous process, 256
    screening, 255
  properties, extruded, 261
    flat-pressed, 260–261
Pecan orchards, 519
Pecans, commercial, 519
  production centers, 519
  varieties, 519
Pencil-slats, manufacture, 296–297

Pencils (*see* Wood-cased lead pencils)
Phenol-formaldehyde resins, properties, 144–145
Phenol-formaldehyde sheet resin, 144
Piles, 67–69
　Douglas-fir, 68
　factors in selection of timber, 68
　foundation, 67
　round-timber, 67–69
　southern yellow pine, 69
　species used in production, 68–69
　specifications, 69
Piling, definition, 68
Pine gum, biochemical processes in synthesis, 440
　nature and occurrence, 439–440
Pine nuts, 519
Pine oils, 415, 437, 460–461
Pine tar, 415
Pinus elliottii, 436
　palustris, 436
　ponderosa, 439
　strobus, 436
　sylvestris, 435
Pinyon nuts, 519
Pinyon pines, 519
Planer and matcher machine, 117
Plasterboards, 382
Plastics, binders, 488–489
　cellulose, 403, 496–498
　compounding ingredients, 488–490
　fillers, 489–490
　　nut-shell flour, 489–490
　　wood-flour, 489
　hydrolyzed redwood, 495
　lignin, 490–491
　nature and characteristics, 488
Plastics industry, 490
Plasticwood, 490
Pluswood, 154
Plywood, 136–153
　advantages over solid wood, 137
　construction, 136
　differential density, 155
　Douglas-fir, 6
　　production 1950 to 1958, 6
　　structural panels, 157
　　uses, 158–159
　　　concrete forms, 158
　　　roofing, 158
　　　sheathing, 158
　　　subflooring, 158
　　　wall boards, 158
　furniture and cabinet work, 161–162
　industrial uses, 162–164
　marine construction, 162
　metal-faced, 162
　mobile homes, 162

Plywood, molded, 151–152
　properties, 136–137
　redrying, 152–153
Plywood adhesives, 142–146
　mixing, 146
　spreading, 146–148
　types used in panel assembly, 143–145
Plywood assembly, use of caul boards, 148
Plywood industry, status, 164–166
Plywood panel backs, 141
Plywood panel cores, 141
Plywood panel faces, 136
Plywood panels, finishing, 153
　lumber core in, 141–142
　no-clap process in assembly, 149
　prefinished, 157
　pressing operations in assembly, 148–150
　　cold-, 148
　　hot-, 148
Plywood presses, 148–150
Plywood sheathing, 6
Plywood subflooring, 6
Pole industry, future of, 67
Pole-shaving machines, 59
Pole-type building construction, 64–65
　economics, 64
Poles, annual production, 56
　AWPA minimum acceptable retentions of preservatives, 61
　factors in selection, 55–56
　finishing, 59–61
　　framing, 59
　　incising, 59–61
　　　for butt treatments, 59–60
　　shaving, 59
　fish-net, 66
　furnace, 67
　history, in United States, 55
　painting, 65–66
　preservative treatments, 61–63
　seasoning, 58–59
　　air-drying, 58
　　vapor-drying, 59
　specifications, 63–64
Polewood species, 56–58
　Douglas-fir, 57
　lodgepole pine, 57
　northern white-cedar, 58
　ponderosa pine, 58
　southern yellow pine, 57
　western balsam firs, 58
　western hemlock, 58
　western larch, 58
　western redcedar, 57
Ponderosa pine, 58, 439
　doors, 6
Pork barrel, 220

Posts, 69–75
  annual consumption in United States, 69
  framing, 72
  hand setting, 74
  mechanically set, 74
  methods of treating, 72–73
  placement in service, 74
  preservative treatments, 72–74
  round-timber, bark removal, 70–72
  seasoning, 72
  sources, 70
  species used in production, 69–70
  steel, 75
Potato barrel, 216
Prefinished plywood panels, 157
Pres-to-logs, 290–291
Producer gas, 292–294
Profile lathe, 173
Progressive kilns, 116
Properties of wood common to all species, 42–45
Pulp, 337–366
  acid sulfite, 322, 341, 343–350
  alpha-cellulose, 366
  semichemical, 322
  soda, 322, 341, 357–359
  sulfate, 351–357
Pulp mill, factors governing location, 383–384
Pulp and paper industry, 383–390
  Canadian, 390
  economic position, 321–322
  regional distribution in United States, 384–389
Pulp refiners, Bauer Brothers, 361
  Sprout Waldron single-disk, 361
Pulp screens, 363–364
Pulp stocks, preparatory treatment to paper manufacture, 363–366
  bleaching, 364–366
  screening, 363–364
  thickening, 364
Pulping residues, derivatives, 429–430
Pulps, alkaline, 351–358
Pulpwood, chemical debarking, 332–333
  delivery to mill sites, 327–330
  demands for, 5
  measuring, methods, 331–332
    units, 331–332
      C-, 331
      standard cord, 331
  mechanical debarking, 332–336
  mechanical harvesting, 327–330
  preparation for pulping, 332
  procurement, 326
  production, 326–327
  storage, 330–331
Pulpwood barkers. hydraulic, 335
  knife, 334

Pulpwood chippers, 341
Pulpwood rechippers, 341
Pulpwood rossers (see Pulpwood barkers)
Pyroligenous acid, 407, 408, 415
  products from, 410
  refinement, 410–412

Quebracho, 509
Quercitron, 511
Quercus suber, 502

Radio-frequency bonding, 150
Railroad ties, 94–105
  Class T, 102
  Class U, 102
  classes, 94
  factors prolonging service life, 96–100
  general specifications, 101–104
  hewed, advantages, 104
  number required per mile of track, 104
  sawed, advantages, 104
  stacking for drying, 98–100
  (See also Crossties; Ties)
Rayflo, 508
Rayon, 366, 393–400
  general concept, 394
  industry in United States, 400
  manufacture, 393
  uses, 400
Rayon yarn, production, 395–397
Redcedar Shingle Bureau, 195, 197, 198
Redcedar shingle industry, character, 197
Reduction of wood residues through research, 32
Redwood-bark fiber, 507–508
Reeling green veneer, 131
Regenerated cellulose, 394
Rhamnus purshiana, 501
Rhizophora mangle, 510
Riffler, 363
Riffling pulp, 363
Rigid fiberboards, nonstructural, 382
  structural, 382
Rosin (see Gum rosin; Sulfate rosin; Wood rosin)
Rosin soap in papermaking, 371
Rotary-cut veneer, 121–125
Rotary pulpwood digesters, 359
Rough mill operations in furniture plants, 171–172
Round-timber piles, 67–69
Round-timber posts, 70–72
Roundwood products, production 1940–1957, 5

St. Regis Paper Co., 326
Sap flow in maple, 469–470
Sash production 1950 to 1958, 6
Sassafras oil, 516

Saw, barrel, 204–205
  bilge (drum), 204–205
  segment, 128–129
Sawdust, absorbent, 275
  graded for particle size, 273
  limitations in use, 273–274
  oxalic acid from, 279
  uses, 274–279
    abrasives, 277–278
    artificial stone, 277
    artificial wood, 277
    composition board products, 278
    composition flooring, 277
    concrete products, 277
    conditioning furs, 276
    curing cement, 276
    hardwood destructive distillation, 278
    gypsum compositions, 277
    leather manufacture, 275
    polishing metal, 276
    molded articles and plastics, 278
    stable and kennel bedding, 275
    stucco and plaster, 277
    sweeping compounds, 278
  volume of log reduced to, 272
Sawdust burner for domestic heating, 289
Sawdust fuels, 274, 287–288
Sawdust mulches, 274–275
Sawed veneers, 128–129
Scotch pine, 435
Scrape, harvesting in gum naval stores, 448–449
Sealers used in finishing panels, 177
Security wood at pulp mills, 326
Segment saw, 128–129
Semichemical pulping, chemiground-wood process, 362
  cold soda method, 362
Semichemical pulps, 361–362
Shakes, 199–200
Shavings, uses, 279–280
Sherrard, S. C., 30
Shingle-bolt splitter, 187
Shingle industry, future of, 199
Shingle jointer, 189
Shingle machines, 188–189
Shingle mills, 197
Shingle packers, 189–190
Shingle weavers (see Shingle packers)
Shingles, 184–200
  application, 194–195
  asphalt, 198
  Canadian, 190–191
  classification, 192–194
  commercial designations in American market, 194
  definition, 184
  dimension, 193
  dimensions of, 192–193

Shingles, early accounts, 184
  grades, 193
  life expectancy, 195–196
  manufacture, 186–191
  marketing, 197–199
  markets, 197–198
  plastics, 198
  preservative treatments, 192
  proper methods for laying, 194–195
  properties desired in woods for, 186
  seasoning and storage, 191
  species used in manufacture, 185–186
  staining, 191–192
  standard designations, 194
  undercourse grade, 193
  unit of measurement, 194
  Western redcedar, 185–186
  wood, advantages, 196
Silvacel, 521
Silvacon, 506–507
Silvalays, 521
Single-headed planer, jointer, 117
Slack cooperage, 201, 212–222
  assembly, 215–216
  barrels (see Barrels, slack)
  future of, 219
  head liners, 215
  manufacture, 214–216
    species used, 214–215
  standard classes, 216–217
  uses, 213–214
Slash pine, gum-naval-stores producer, 439
Sleepers, 94
Sliced veneers, 126–128
Soda process, 357–359
  cooking procedures, 358
  recovery system, 358
  woods used for conversion by, 358
Soda pulp, 351, 359
  uses, 359
Softwood distillation products, 416
  (See also Wood naval stores)
Softwood flooring, 111
Softwood lumber, classification, 109
  grades, 108
Softwood plywood, annual production, 1924 to 1959, 165
Softwood sawdust in steam extraction, 278
Softwood veneer manufacture, 135–136
Southern yellow pine pile-timbers, 69
Southern yellow pine poles, 57
Soybean glue, 143
Spalt, 188
Spanish moss, 519–521
  grades, 521
  uses, 521
Splinterboards, 254–255
Spools, 316

Spray puller hack, 446
Springboard (shingle jointer), 189
Square-set method of timbering stopes, 80–82
Stains, N.G.R., 176
  oil, 176
  water, 176
Standard hardwood lumber grades, 110
Staple fiber, 399–400
Staves, for slack barrels, 214
  tight cooperage, 203–206
    classification, 206
    jointing, 204–206
    manufacture, 203–206
    seasoning, 204
    species used, 203
Stay-log veneer cutting, 125–126
Steam-bending of wood, 174
Steel posts, 75
Storax, 518
  production in United States, 518
  uses, 518
Straw, cereal, 326
  flax, 326
Structural lumber grades, 108
Structural plywood, 157
Strychnos spp., 51
Subflooring, plywood, 6
Sugar bush, 468
Sugar sand (see Maple sap)
Sulfate naval stores, 461–463
Sulfate process, 351–357
  by-products, 356
  cooking procedures, 351–353
  recovery of chemicals, 354–355
  woods used, 351
Sulfate pulps, continuous digesters used to produce, 353
  properties, 357
  uses, 357
  washing, 353–354
Sulfate rosin, 435
  production, 462–463
Sulfate turpentine, 356, 435
  production, 461–462
Sulfite pulp, 343–351
  cooking processes, 345–348
  gas recovery in manufacture, 348–349
  uses, 350–351
  woods used, 343
  yields and quality, 350
Sulfite pulping, preparation of cooking liquors, 343–345
  milk-of-lime system, 344–345
  two-tower Jenssen system, 344
  spent liquor recovery systems, 349–350
    evaporation and burning, 350
    precipitation, 349–350

Sulfite waste liquor, "soluble base," 350
  uses, 350
Superpressed plywood, 154
Surface-coated papers, 379
Switch ties, 94
Syntans, 509
Synthetic resin adhesives, 143–145
  thermoplastic type, 144
  thermosetting type, 144

Tall oil, 356, 462
Tanks, wood, 221–222
Tannins, 508–510
  mineral, 509
  natural, 509
    extraction, 509
    sources, 509–510
    use in oil-drilling, 509
  synthetic, 509
Tape-splicing machine, 139
Tapeless splicer, 139–140
Tego film in molded plywood, 151
Tegowood, 154
Terminalia chebula, 510
Thuja occidentalis, 515
Tie irons, 98–99
Ties, 94–105
  dimensions, 103
    for narrow-gauge track, 103
    for standard-width track, 103
  factors prolonging life, 96–100
  hewed, 104
  mechanical wear, 97–98
  prestressed concrete, 95
  principal species used, 100
  replacement in tracks, 1948 to 1958, 95
  sawed, 104
  seasoning, 98–100
  switch, 94
  (See also Railroad ties)
Tight barrels (see Barrels)
Tight cooperage, 201, 203–221
  assembly, 208–212
    heading-up, 211
  charring bourbon staves for, 209–210
  future of, 220–221
  linings, 211
  manufacture, 203–212
  standard classes, 212–213
Tight-cut veneer, 123
Tilghmann, B. J., 343
Tillandsia usnoides, 519
Timber drain, 1776 to 1960, 3
Timber famine, 4
Timber ownership, permanency, 4
Timber piles (see Piles)
Timber products consumed in United States, 1944 to 1952, 7
Timber resources of United States, 3

Timber waste in mines, 82
Timbering in mines, 79–82
Timblend, 254
Tissue papers, 378, 380
Tobacco hogsheads, 217–218
Tomlinite, 356
Tomlinson reduction furnace, 355
Toothpicks, 304–306
  circular, 305
  flat, round-end, 304
  specialty, 305–306
Torula utilis, 430
Tray shakes, 200
Tung oil, 512–515
  production in United States, 513–514
  uses, 512, 514
  yields from plantations, 513
Turpentine (*see* Gum turpentine; Sulfate
    turpentine; Wood turpentine)
Turpentine "crop" defined, 442
Turpentine farms, 440
  leasing timber for, 441
  standard working unit, 442

Unbleached pulps, groundwood, 340–341
  soda, 351, 359
  sulfate, 357
  sulfite, 350
Urea-formaldehyde resins, 144–145
Urea-plasticized wood, 495
Utility veneers, 135

Va-Press continuous semichemical pulping
    system, 361
Values obtained from forests, 12–13
Vapor-drying process, 58–59
Vats, wood, 221–222
Vegetable parchment, 372
Veneer baskets, 233–236
  classes, 234–235
  species used in manufacture, 233–234
  standardization, 235–236
Veneer driers, 132–133
  operation, 133–134
Veneer flitch defined, 128
Veneer industry, 134–136
  hardwood, 134–135
  softwood, 135–136
Veneer-lathe checks, 123
Veneer logs, demand for, 5
  heating, 123–125
Veneer-plywood combinations, 156–157
  cloth-, Flexwood, 156
  foamated thermoplastic resins-, 157
  metal-, 157
  paper-, 156
Veneer reels, 131
Veneer slicers, 126–128
  horizontal, 127–128

Veneer slicers, vertical, 127–128
Veneers, air-drying, 131
  back-cutting on stay log, 125–126
  commercial, 135
  cone-cutting, 126
  container, 135
  cross bands, 136
  face, fancy, 134
  green, clipping methods, 129–131
    stack-and-bulk-chip, 130
    storage-deck, 129–130
  reeling, 131
  half-round cutting on stay log, 125–126
  kiln-drying, 132
  loft-drying, 132
  loose-cut, 123
  manufacture, 121–129
  rotary-cut, 121–125
  sawed, 128–129
  sliced, 126–128
  softwood, 135–136
  stay-log cutting, 125–126
  thicknesses, 129
  tight-cut, 123
  utility, 135
Viscose, 393
Viscose method for producing filaments,
    394–395
Viscose yarn, 395–397

Wallpaper, 378
Walnut oil, 515
Walnut-shell flour, 515
Water requirements of pulp and paper
    mills, 383–384
Water stains, 176
Watermark in paper, 376
Wattlebark, 510
Western larch poles, 58
Western redcedar poles, 57
Western redcedar shingles, 185–186
Wet-strength papers, 370–371
Whole-wood fiber, 521
Wood, inorganic components, 51–52
  major uses in United States, 4
  organic components, 49–51
  used in furniture manufacture, 168–169
Wood alcohol (*see* Methanol)
Wood ashes, chemical composition, 51
Wood briquettes, 288, 423
Wood-cased lead pencils, 295–298
  assembly, 297–298
  finishing, 297–298
  woods used, 296
Wood combustion, 283–293
Wood excelsior (*see* Excelsior)
Wood extractives, 51
Wood felts, 382–383

Wood-fill insulation, 383
Wood fillers, 176
Wood flour, 265–271
  classes, granularmetric, 269
    nontechnical, 269
    technical, 269
  major producing centers, 265
  mills used, 266–268
    attrition, 266
    beater, 267
    hammer, 267
    roller, 267–268
  prices paid for, 268–269
  species used, 265
  uses, 269–270
    in plastics, 489
Wood-flour industry, status, 270
Wood fuel, 281–294
  advantages, 282–283
  disadvantages, 282–283
  estimated consumption from 1940 to
    2000, 282
  heat available from, 283–285
  types, 286–293
    briquettes, 288–292
    charcoal, 292
    cordwood, 286–287
    mill residues, 287–288
    producer gas, 292–293
Wood household furniture (see Furniture)
Wood hydrolysis, 425
  sources of raw material, 426–427
  sugars from, 425
  using concentrated acids, 428–429
    Bergius process, 428
    Schoenemann process, 428
Wood industry, future requirements for
    timber, 6–10
Wood-molasses production, 429
Wood naval stores, 458–461
  industrial products from, 458–459
  sources of raw material, 459
  steam-and-solvent extraction, 437, 459
Wood pitch, 412
  uses, 416
Wood preservatives, 61–63, 72–73, 86–87,
    97
Wood properties, chemical, 47–52
  common to all species, 42–45
  differing among species, 45–46
  in relation to use, 41–52
  variation within species, 46–47

Wood residues, 16–39
  alcohol from, 39
  definition, 17
  principal sources, 21–26
  produced, by chemical conversion, 25
    by fiberboard manufacture, 26
    by furniture industry, 179–180
    in logging, 22, 26
    in primary manufacture, 23
    in secondary manufacture, 24–25
  research goals in utilization, 28–37
Wood rosin, 435
  grades, 460
Wood saccharification, 475
  processes, 426–428
    dilute-acid batch, 427–428
      American, 427
    dilute-acid percolation, 428
      Madison wood-sugar, 428
      Scholler, 428
Wood shavings, 279–280
Wood shingles, 196
Wood tar, refinement, 412
Wood-tar oils, 412
Wood-turning industry, importance, 310
  location, 309–310
Wood-turning machinery, 313
Wood turnings, 309–317
  green, 316–317
  manufacturing procedures and products,
    310–311
  species used, 309–310
Wood turpentine, 435, 460
Wood-using industries, importance in
    American economy, 11–13
  stabilizing influences, 14–15
Wood waste, definition, 17
  sources, 21–26
  (See also Wood residues)
Woody cell, walls of, 48
Woven-wood laminates, 156
Wrapping papers, 380
Writing papers, 379

Xylose, 425

Yankee paper machine, glazed paper pro-
    duced, 378
Yard lumber, 108
Yeast, from pulping residues, 429–430
  from wood sugars, 430–431